D1191335

Casebooks
in Earth Sciences
Series Editor: R.N. Ginsburg

*. . . . Its all underground, and it's all very expensive, and you are reconstructing history from hundreds of millions of years ago with eight-inch circles in the ground, essentially sending blind men into the ancient, lost country and trying to fill in hundreds of miles of buried forests and rivers and seas and dunes with these tiny pin sticks, like flag pins on a golf course. No, better, like trying to map the state of Colorado based only on what you can see for eight inches from, say, any fireplug in the state.**

R. Bass

*Excerpt from Oil Notes by R. Bass. Copyright © 1989 by Rick Bass. Reprinted by permission of Houghton Mifflin Co./Seymour Lawrence Publishing.

John H. Barwis, John G. McPherson,
and Joseph R.J. Studlick
Editors

Sandstone Petroleum Reservoirs

With 448 Illustrations in 683 Parts
Including 48 Color Plates

Springer-Verlag
New York Berlin Heidelberg
London Paris Tokyo Hong Kong

Editors

John H. Barwis, Shell Oil Company, Houston, Texas 77210, USA

John G. McPherson, Mobil Research & Development Corporation, Dallas, Texas 75244, USA

Joseph R.J. Studlick, Shell Oil Company, Houston, Texas 77001, USA

Series Editor

Robert N. Ginsburg, University of Miami, School of Marine and Atmospheric Science, Fisher Island Station, Miami, Florida 33139, USA

Cover art by David N. Helber and John G. McPherson.

Library of Congress Cataloging-in-Publication Data
Sandstone petroleum reservoirs / edited by John H. Barwis, John G.
  McPherson, and Joseph R.J. Studlick.
       p. cm. (Casebooks in earth sciences)
    Includes bibliographical references.
    ISBN 0-387-97217-X
    1. Oil reservoirs. 2. Gas reservoirs. 3. Sandstone
  4. Sedimentation and deposition. I. Barwis, John H.
  II. McPherson, John G. III. Studlick, Joseph R. J. IV. Title: Sand-
  stone petroleum reservoirs. V. Series
  TN870.5.S36   1990
  553.2′8 – dc20   90-9415

Typeset by Publishers Service, Bozeman, Montana, USA.
Printed in Hong Kong by South Sea International Press Ltd.

9 8 7 6 5 4 3 2 1

ISBN 0-387-97217-X Springer-Verlag New York Berlin Heidelberg
ISBN 3-540-97217-X Springer-Verlag Berlin Heidelberg New York

Series Preface

The case history approach has an impressive record of success in a variety of disciplines. Collections of case histories, casebooks, are now widely used in all sorts of specialties other than in their familiar application to law and medicine. The case method had its formal beginning at Harvard University in the mid-1800s when Christopher Langdell developed it as a means of teaching law. The technique was so successful that it was soon adopted in medical education, and the collection of cases provided the raw material for research on various diseases. Subsequently, the case history approach spread to such varied fields as business, psychology, management, and economics, and there are more than 100 books in print that use this approach.

The case method may not only have originated in law but may have spread directly to geology via an attorney. One of Langdell's law students was Oliver Wendell Holmes. After his legal training, Holmes spent six months in Great Britain recuperating from wounds incurred during his service in the American Civil War. One of his frequent companions there was Charles Lyell. We have no records of their conversations but can imagine stimulating discussions on the development of the paradigm as a learning tool in each of their disciplines.

The ideas for a series of Casebooks in Earth Sciences grew from my experience in organizing and editing a collection of examples of one variety of sedimentary deposits. The project began as an effort to bring some order to a large number of descriptions of these deposits that were so varied in presentation and terminology that even specialists found them difficult to compare and analyze. Thus, from the beginning, it was evident that something more than a simple collection of papers was needed. Accordingly, the nearly fifty contributors worked together with George de Vries Klein and me to establish a standard format for presenting the case histories. We clarified the terminology and some basic concepts, and when the drafts of the cases were completed, we met to discuss and review them. When the collection was ready to submit to the publisher and I was searching for an appropriate subtitle, a perceptive colleague, R. Michael Lloyd, pointed out that it was a collection of case histories comparable in principle to the familiar casebooks of law and medicine. After this casebook [Tidal Deposits (1975)] was published and accorded a warm reception, I realized that the same approach could well be applied to many other subjects in earth science. Consequently, Carbonate Petroleum Reservoirs was published in 1985.

It is the aim of this series, Casebooks in Earth Sciences, to apply the discipline of compiling and organizing truly representative case histories to accomplish various objectives: establish a collection of case histories for both reference and teaching; clarify terminology and basic concepts; stimulate and facilitate synthesis and classification; and encourage the identification of new questions and new approaches. There are no restrictions on the subject matter for the casebook series save that they concern earth science. However, it is clear that the most appropriate subjects are those that are largely descriptive. Just as there are no fixed boundaries on subject matter, so are the format and approach of individual volumes open to the discretion of the editors working with their contributors. Most casebooks will of necessity be communal efforts with one or more editors working with a group of contributors. However, it is also likely that a collection of case histories could be assembled by one person drawing on a combination of personal experience and the literature.

What can be learned from a compilation of representative sandstone reservoirs? Two pathways emerge from this kind of inventory: one well-traveled road leads to analogy; the other, less-followed route is toward groupings of classifications. Each well-documented case history can serve as a model and indeed much exploration and development rests on such similes—"It's just like the Frio (or Woodbine, or Rotliegendes)." At the same time that case histories serve as models for operations, they also can indicate what is not known but needed, the potentially fruitful directions for research. Organizing the chaos of examples is more demanding than using them as individual analogues, but it may also be more rewarding. This collection can, on the one hand, expand knowledge of the spectrum of variability within, for example, deltaic sandstones and it can also provide material for assessing the role of basin evolution. What kind of grouping is most appropriate for the engineering and geology of development? Is it possible to develop a grouping based on the nature and pattern of heterogeneity? One final dividend that could emerge from this collection of case histories is the stimulation of the known variety of reservoirs for imagining undiscovered new varieties.

Clearly the case history approach has been successful in a wide range of disciplines. The systematic application of this proven method to earth science subjects holds the promise of producing valuable new resources for teaching and research.

Miami, Florida   Robert N. Ginsburg
January, 1990   Series Editor

Preface

This book is about sandstone petroleum reservoirs and the influence of depositional environment on their architecture and performance. It is aimed at the wide range of earth scientists and engineers on whose expertise depends the discovery and production of oil and gas. The often disparate responsibilities of the many who are involved in these efforts mean that each discipline will usually consider a reservoir from a very different perspective. A geophysicist may view it in terms of acoustic properties or reflection characteristics. A sedimentary petrologist may see it solely as the mineral species that frame individual pores. A reservoir engineer might consider it primarily in terms of its drainage characteristics. Efficient management of the hydrocarbon resources a reservoir contains urgently requires close interaction among these disciplines. The most appropriate way to encourage this interaction is to examine in detail a shared focal point, the hydrocarbon accumulation itself. We hope the examples presented here will strengthen the too-often isolated approaches of exploration and production geologists, geophysicists, and petroleum engineers.

This book is therefore organized around the oil or gas field as a vehicle for understanding their reservoirs and is presented as a series of integrated case studies that demonstrate depositional influences at different scales. At regional and prospect levels, depositional origin is shown to control sand-body orientation, geometry and trap style. At intrareservoir scales, depositional events are documented as the mosaic of porosity and permeability zones and baffles that define the sand body's facies architecture. Within individual flow units, storage and delivery capacities and response to completion and stimulation techniques are strongly influenced by depositional as well as diagenetic texture and mineralogy.

The perception that depositional environment was an important influence on reservoir traits may be the primary reason that sedimentology and stratigraphy have become what they are today. Soon after World War II, oil industry geologists began to study recent depositional systems to help predict reservoir occurrence and behavior. By the late 1950s, many detailed lithofacies and depositional models based on modern analogues were already in use by exploration and production geologists. These models were three-dimensional and included braided and meandering streams, river- and wave-dominated deltas, and barrier island-tidal inlet systems. In the early 1960s, the vertical sequence approach to facies interpretation evolved from these models (not the other way around, as suggested by some geological historians). The role of depositional fabric in controlling porosity, permeability, and oil/water saturations was widely recognized. Interest in depositional systems viewed at all scales blossomed through the 1960s and 1970s, the most significant advances occurring in the use of modern environments for studying sedimentological process-response relationships and the advent of seismic stratigraphy. There nevertheless remains great variety in our understanding of depositional systems as reservoir builders. Turbidite depositional systems, for example, probably lag other settings by twenty years or so in terms of modern process-response analogs.

As subdisciplines in sedimentary geology became more specialized, facies-model building became fashionable and accepted as a legitimate undertaking in its own right. Yet the best "models" of all were being overlooked — true three-dimensional data from preserved depositional systems, complete with pressure and flow data ready to provide insight to sand body geometry and bed continuity. This information could never be gleaned from a two-dimensional outcrop or trench. Perhaps the most important clues to reservoir behavior were in the reservoirs themselves.

Yet despite the three-dimensional nature of reservoir data, the volume of reservoir we actually see is frustratingly small. For example, consider one square mile of a 100-foot thick sheet sandstone which is produced on 40-acre spacing. Even if the entire reservoir were cored in every well (a record for data availability), we would still see only about two ten-millionths of the rock in question. This highlights the need for petroleum geology to evolve far beyond reservoir description, using depositional models as a springboard for interwell interpolation.

We present these studies as examples of how reservoirs may be profitably characterized, rather than as a compendium of all possible reservoir types or reservoir-trap associations. Although the authors present reservoirs from depositional settings that range in basinal position from braided stream to submarine fan, their case studies are not primarily about the depositional environments themselves. This is not a book about facies analysis or stratigraphy, but assumes that readers have a good working knowledge of these topics and uses them as a basis for understanding reservoir behavior. We think that the efforts required for documenting such detailed case studies are warranted by exploration opportunities which carry increasingly higher risks and by higher exploration and production expenses associated with deeper and more remote drilling and/or deeper water operations. In addition, optimum planning of enhanced oil-recovery programs will require continuing improvements in our understanding of reservoirs at all scales. We hope that this collection addresses these concerns.

This casebook was suggested by Series Editor Bob Ginsburg as a companion volume to the earlier Carbonate Petroleum Reservoirs by Perry Roehl and Phil Choquette. In 1986, we contacted more than 300 companies and individuals worldwide who we thought would have access to integrated field studies and who might be willing to offer them as analogues. Of the many respondents, 33 agreed to meet informally and discuss the fields on which they had worked. With the generous support of 22 sponsoring companies, we met in New Orleans, Louisiana, in mid-1988 to examine these fields at a symposium on sandstone petroleum reservoirs. In this collegial atmosphere, we criticized each other's work and agreed on minimum data requirements for the final manuscripts. Only 22 of the fields met those final criteria, which required that individual reservoirs (not simply oil fields or formations) be related to basin history, depositional environment, and production performance. We have tried to provide analogues for other exploration and production efforts, not so much in terms of the sand bodies themselves, but as examples of lessons learned through the process of an integrated approach. We have also tried to include as many types of sand bodies as possible. With only 22 examples, we are very far from covering the range of facies variations observed in modern depositional systems. Previously unpublished information in the detail we required is simply not available.

Most of the examples presented are from North America, a geographic bias which largely reflects the availability of data; more than 2,200,000 wells have been drilled in the United States (Petroleum Information, pers. comm., 1989), compared to 310,000 for the rest of the noncommunist world (Petroconsultants, Geneva, pers. comm., 1989). Many of the fields outside North America which were originally considered were omitted for lack of data or withheld by their operators as too sensitive to publish. However, because the case-study approach deals primarily with facies analogues, and not with specific basin types or exploration plays, we think the geographic limitation offers no particular disadvantage.

Most of these case studies have not been published before, and many contain data that had not been released. Although some studies focused at particular scales, we have tried to broaden their approach, wherever data would permit, to embrace as many perspectives as possible, from reflection seismic data to thin sections.

We have attempted to standardize the order of presentation by chapter so that the reader will be able to compare and contrast each reservoir type as a function of depositional setting. Nevertheless, differences in style and organization are evident and reflect individual author's preferences. Copious illustration was encouraged and subsidized by our corporate sponsors. Each chapter is followed by a summary of the reservoir characteristics, again to provide field and reservoir data in a standard format. No explicit constraints were placed on environmental terminology or facies descriptions, even with terms such as "net" or "gross" sandstone or "net pay." Because the geological definitions of these words are nearly always based on economic considerations, they necessarily vary from place to place. Authors were therefore required only to define their terms as appropriate.

We thank Shell Oil Company and Mobil Research & Development Corporation for their logistical and financial support. We also wish to thank those who helped put this volume together: Jan S. Jackson and Nel Hohage for administrative assistance and advice, Pat G. Berwick and Barbara J. Mulkey for word processing and editorial consulting, John E. Bailey and David N. Helber for graphics work and for the cover design.

We greatly appreciate the advice provided by the following reviewers, plus the many others who remain nameless but who helped make this book possible:

Peter Andronaco	Timothy F. Lawton
Donald C. Beard	Rufus J. LeBlanc, Sr.
Edward A. Beaumont	Dale A. Leckie
Robert R. Berg	Bruce K. Levell
Ian D. Bryant	William T. Long
Jack G. Bryant	William D. Marshall
George F. Canjar	Edwin K. Maughan
Donna C. Caraway	Earle F. McBride
James M. Casey	Norman L. McIver
Alexander L. Chaky	Tor H. Nilsen

James H. Clement
Robert W. Dalrymple
William E. Galloway
Michael E. Guest
Walter A. Handy
James A. Hartman
Richard J. Hodgkinson
Joseph C. Hopkins
Clarence E. Hottman
John F. Houser
David P. James
John F. Karlo
Fred B. Keller
Ralph S. Kerr
Lee F. Krystinik
Mohan V. Kudchadker
William D. Lancaster
David T. Lawrence

Calvin A. Parker
Martin A. Perlmutter
Edward D. Pittman
Henry W. Posamentier
Riyadh A. Rahmani
Daniel E. Schwartz
Steve O. Sears
John L. Shepard
Robert M. Sneider
Anthony J. Tankard
Noel Tyler
Paul Weimer
Robert J. Weimer
Ernest G. Werren
John L. Whitworth
Michael D. Wilson
Charles D. Winker

Rijswijk, The Netherlands
Dallas, Texas
Houston, Texas
January, 1990

John H. Barwis
John G. McPherson
Joseph R.J. Studlick

Casebook Sponsors

We are pleased to acknowledge the financial support and encouragement of the following corporations:

Amoco Production Company, Houston, Texas

Amoco (U.K.) Exploration Company Ltd., London, England

Arco Oil & Gas Company, Plano, Texas

BP Exploration Inc., Houston, Texas

Chevron Oil Field Research Company, La Habre, California

Conoco Inc., Ponca City, Oklahoma

Dresser Foundation, Inc., Dallas, Texas

Elf Aquitaine Petroleum, Houston, Texas

Enserch Exploration Inc., Dallas, Texas

Exxon Production Research, Houston, Texas

Mobil New Exploration Ventures Company, Dallas, Texas

Mobil Exploration and Producing U.S. Inc., Denver, Colorado

Occidental International E&P Company, Bakersfield, California

Petroconsultants S.A., Geneva, Switzerland

Pogo Producing Company, Houston, Texas

Schlumberger Well Services, Corpus Christi, Texas

Shell Internationale Research Maatschappij B.V.

Shell Oil Company, Houston, Texas

Shell Western E&P Inc., Houston, Texas

Texaco Exploration and Production Technology Division, Houston, Texas

Union Oil of California, Brea, California

Veba Oel Aktiengesellschaft, W. Germany

Contents

Appendix

Introduction:
Reservoir Description Of Sandstones

Robert M. Sneider

Robert M. Sneider Exploration, Inc., Houston, Texas 77079

Introduction

The biggest challenge for hydrocarbon explorers and producers now and in the future is to significantly improve hydrocarbon recovery from newly and previously discovered reservoirs. A key to achieving this goal is to compile detailed reservoir descriptions of sandstone reservoirs. A reservoir description is a comprehensive picture of the three-dimensional distribution and continuity of the rocks, pores, and fluids of the reservoir and aquifer system, including barriers to fluid flow.

Table I-1. Key factors in reservoir analysis.

- EXPLORATION PHASE
  - Seismic Characteristics
  - Trap Configuration
  - Sand-Body Geometry
  - Hydrocarbon Migration Paths
  - Reservoir Volume
  - Nature of Seals
- APPRAISAL PHASE
  - Hydrocarbons in Place
  - Reserves
  - Production Rates
  - Aquifer Size and Strength
- PLANNING PHASE
  - Optimum Depletion Plan
  - Location and Number of Platforms
  - Location and Number of Wells
- DEVELOPMENT PHASE
  - Policies on Completions and Workovers
  - Optimum Distribution of Injection and Production
- RESERVOIR MANAGEMENT (SURVEILLANCE) PHASE
  - Infill Wells
  - Workover Wells
  - Surveillance Program
  - Redistribute Injection
  - Improved Recovery Methods

When is a reservoir description needed? The need starts during the exploration phase, when reservoir analogues are needed for play development and prospect evaluation. The need intensifies once a discovery is made and is being appraised as to the best estimates of hydrocarbons in place, recoverable reserves, drive mechanism, and rates of production. As a reservoir or field goes through the typical "life cycle" of discovery, appraisal, planning, development, and reservoir management or surveillance, a more complete reservoir description is both necessary and possible. Some of the critical factors to be evaluated (Table I-1) and questions to be answered (Table I-2) in the analysis of a reservoir during its life cycle give insight into the requirements of an effective reservoir description. Fulfilling these requirements means shortening the time of a reservoir's life cycle, the goal of which is to optimize hydrocarbon recovery.

Table I-2. Some typical questions in reservoir analysis.

1. What does the reservoir look like — geometry and continuity of pore space and fluids?
2. Will the reservoir have an effective natural water drive? Is there an aquifer and what is its size, geometry, continuity, and strength?
3. Where should wells and platforms be located?
4. Where and how should wells be completed? Where should perforations be shot?
5. Will recoveries be better by water or gas displacement?
6. Will water or gas injection be needed and when?
7. Will enhanced recovery processes be needed and when?

Casebooks in Earth Science
Sandstone Petroleum Reservoirs
Eds.: Barwis/McPherson/Studlick
© 1990 by Springer-Verlag New York, Inc.

Table I-3. Key reservoir description needs.

Reservoir description	Geophysical-geological-petrophysical contributions
External geometry of the reservoir	1. Structural attitudes, size, shape, orientation, continuity. 2. Structure maps: Combine log and seismic chronostratigraphic correlations. Requires careful well-seismic ties and time-depth conversion. 3. Isopach maps: Reveal burial history, preexisting oil/water contacts. 4. Fault orientation: Flow barriers, segmentation into fault-bounded units, high permeability fault-breccia zones in a fractured reservoir.
Internal geometry of the reservoir: both barrier units and pay intervals	1. Selection of core and test intervals to provide representative sampling of reservoir matrix types. 2. Chronostratigraphic correlation of wells (and of seismic data for thick reservoirs) to delineate internal stratal geometry. 3. Mapping of vertical flow barriers. Can particular reservoir zones be correlated from well to well? 4. Mapping of gross and net pay intervals (each discrete subzone or layer). 5. Does careful seismic correlation of thick reservoir zones show patterns of internal bedding that suggest facies and internal dips different from the top and base of the interval? 6. Does well and seismic correlation show unconformity truncation of beds within the reservoir? 7. What is the effect of faulting on reservoir continuity? Are there partial or complete fluid flow barriers? 8. Has fracturing in hard massive reservoir zones enhanced permeability? What is the fracture spacing? What are dimensions of fracture drainage blocks? Orientation of fractures?
Distribution of porosity, permeability, and capillary pressure-saturation properties	1. Make description of cores and cuttings to provide a petrophysical basis for log analysis and for selection of core analysis wells and intervals. 2. Do studies of cores reveal that thin shale or other impermeable laminae are common and may interfere with vertical permeability? 3. What are the distribution and continuity of porosity, permeability, ϕ_h, k_h? What are the k_v/k_h ratios? What is the directional permeability? Are the k_v measurements representative of the rock type? 4. If the reservoir is composed of several different formations having different rock types and therefore different capillary characteristics, these units need to be mapped individually, and separate relative permeability- and capillary pressure-saturation values need to be established for each unit. 5. Are reservoir zones likely to be discontinuous between wells, based on the depositional model, even though the gross pay interval extends from well to well?
Aquifer extent and permeability thickness	1. What is the thickness of effective porous-permeable aquifer rock in communication with the reservoir? 2. Do well and seismic data suggest that the aquifer is in communication with the reservoir? 3. How do you distribute permeability data for the aquifer over the area of predicted drainage? 4. Is there any reason to suspect that hydrodynamic flow may be present and may effect the fluid contact positions?
Distribution of clay minerals, minerals, and fines	1. What are the types, amounts, and distributions of clay minerals? 2. What minerals (grains and cements) will interact with the natural or injected pore fluids? What is their distribution? 3. What are the types and distributions of "fines"? Will they migrate and where?

Reservoir Description

The principal geophysical, geological, and petrophysical contributions to reservoir description are the external geometry of the reservoir; the internal geometry of the reservoir, including the distribution of fluid-flow barriers and pay types and intervals; the distribution of porosity, permeability, and capillary pressure-saturation properties; the aquifer extent, continuity, permeability, and thickness; and the distribution of clay and framework minerals, both grains and cements. Table I-3 provides some details of these reservoir description attributes. Prediction of these attributes requires that the physical processes which formed them be understood. Detailed studies of Holocene and ancient sand bodies in outcrops, cores, and the subsurface demonstrate clearly that most reservoir properties are the direct result of facies variations which are in turn controlled by depositional systems, their associated sedimentary processes, and the subsequent histories of diagenesis, burial, and structural deformation.

In the reservoir description process, a critical first step is the recognition of all correlative reservoir layers or subzones and the intervening dense, impermeable or low-permeability strata. Knowledge of the depositional and diagenetic processes controlling both the reservoir and the nonreservoir rock units is essential to the degree of confidence in correlating these units from well to well. Genetic sequence analyses of sand bodies and seismic facies, coupled with knowledge from well-documented outcrop studies, can add significantly to interwell correlations. Flow test and production data dovetailed with the three-dimensional picture of the reservoir-nonreservoir framework provide the best reservoir description of continuity-discontinuity of porosity, permeability, capillary properties, vertical and horizontal fluid-flow barriers, and fluids. All zones of unusual permeability contrast, especially zones of very high permeability, are critically important to all recovery processes.

The reservoir description for all displacement recovery processes should include the determination of net pay types, their continuity, and an estimate of the percent of pay that will be floodable based on the well spacing. This determines recovery estimates, the decision to infill, and the well spacing needed if infilling is desirable.

In sand bodies, shales are the dominant fluid-flow barriers and their distribution is the result of depositional environment and the associated sedimentary processes. Shales of all sizes must be described and mapped because of their influence on reservoir performance. Shale beds that are continuous between several wells both divide the hydrocarbon column into smaller and separate production intervals and act as vertical permeability barriers. These continuous shales can also help primary performance by preventing gas and water coning. Shales that are not continuous between wells also reduce vertical fluid flow. In reservoirs with primary gas-cap expansion drive, discontinuous shales lower effective vertical permeability and entrap oil on the shale bed tops, thus reducing oil recovery. In water or gas injection processes that are typical of pressure maintenance or enhanced recovery, discontinuous shales usually inhibit gravity segregation and thereby increase reservoir sweep efficiency.

A reservoir description, especially one made during appraisal, needs a correct assessment of the potential aquifer size (pore volume), continuity, and strength (pressure, thickness, and permeability). The aquifer's external and internal properties are derived from knowledge of the depositional environments, geophysical mapping, and flow tests. With knowledge of the aquifer properties, a reservoir engineer can estimate the potential importance of a natural water drive or the need for and the timing of pressure maintenance by water or gas injection. When the pore volume of an aquifer is at least 100 times the pore volume of a reservoir, a strong natural water drive results; however, if this pore-volume ratio is only 10, pressure maintenance by water or gas is required.

Summary

A reservoir description is a comprehensive picture of the three-dimensional distribution and continuity of the rocks-pores-fluids of the reservoir and aquifer system, including barriers to fluid flow. A good reservoir description designed to answer key reservoir performance questions is a fundamental tool for appraisal, planning, development, and reservoir management (surveillance) phases of production. In exploration, detailed reservoir description studies provide critical data needed by the explorationist to estimate reservoir, barrier, and seal quality and distribution from seismic data, well logs, cores, and samples.

Fluvial Environments

1

Braidplain and Deltaic Reservoir, Prudhoe Bay Field, Alaska

*Christopher D. Atkinson, Joseph H. McGowen, Salman Bloch, L. Lee Lundell, and Philip N. Trumbly**

ARCO Oil & Gas Co., Plano, Texas 75057; *ARCO Alaska Inc., Anchorage, Alaska 99510

Introduction

This chapter illustrates, by means of a case study of a super giant field, the excellent reservoir characteristics of braidplain and associated deltaic facies. It also discusses some of the difficulties encountered in most efficiently producing hydrocarbons from this type of sequence.

Braided stream systems produce deposits which can become excellent hydrocarbon reservoirs. Some of the world's largest sandstone reservoirs are composed of braided stream deposits. Typically, the sandstone and conglomerate sheets produced by bar migration and by braidbelt switching are laterally continuous and, relative to most other sand-body types, the deposits can be considered largely homogeneous. In addition, the coarse-grained and relatively clay-free character of most braided stream deposits favors high reservoir quality. This is basically the case for the reservoir discussed in this chapter. However, it must be noted that even reservoirs composed of braided stream deposits exhibit varying degrees of heterogeneity although often, as discussed in this chapter, this does not become apparent until the advent of secondary and tertiary recovery operations.

The Ivishak Sandstone, a largely braidplain facies, is the main producing reservoir of the Prudhoe Bay Field, a super giant that delivers 1.45 million barrels of oil per day (2.3×10^5 m³/D). This reservoir is currently supplying 17% of the total United States domestic production. Economic justification for the development of the field, which lies 250 miles (400 km) north of the Arctic Circle, was directly linked to the reservoir properties of the sandstones and conglomerates which comprise the reservoir. In particular, their overall high permeability (average 400 md) and laterally extensive nature combine to produce a reservoir capable of sustained high production rates. Such production characteristics more than compensated for the initial and continued financial investment required to develop the field in this remote location. Today, some 20 years after its discovery, the Prudhoe Bay Field represents the largest developed oil and gas accumulation in North America, with in-place reserves of 22 billion barrels of oil (3.5×10^9 m³) and 47 trillion cubic feet of gas (1.3×10^{12} m³).

Interest in the petroleum potential of the North Slope of Alaska (Fig. 1-1) began in the early 1900s with the discovery of surface oil seeps near Cape Simpson, east of Point Barrow (Leffingwell, 1919). In 1923, Naval Petroleum Reserve No. 4 (NPRA) was established to exploit potential resources in the area. Through 1963, most exploration was conducted by the U.S. Navy and the U.S. Geological Survey. Industry exploration accelerated in the early 1960s and, after a series of initial disappointments, shifted from the Brooks Range foothills to the coastal plain region east of NPRA and west of the Arctic National Wildlife Range (ANWR). By 1965, the presence of a large structure below Prudhoe Bay had been delineated using seismic data. Nearby exploration wells confirmed that the Permo-Triassic terrigenous clastic succession and the Lisburne Group carbonates contained prospective reservoir sections. The North Alaskan oil boom began on April 15, 1968, with the ARCO-Humble Prudhoe Bay State No. 1 well.

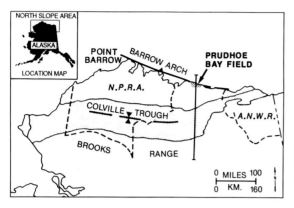

Fig. 1-1. Location of the Prudhoe Bay Field, North Slope, Alaska. Position of N-S cross section shown in Figure 1-2 is indicated.

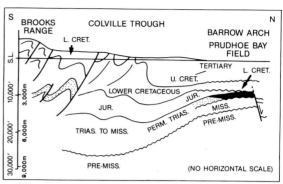

Fig. 1-2. North-south generalized cross section through the Alaskan North Slope from the Brooks Range to the Barrow arch (from Jamison et al., 1980, and reprinted by permission of American Association of Petroleum Geologists).

Initially targeted for the Lisburne Group limestones, the well encountered oil and gas at approximately 8,200 feet (2,500 m) subsea in the Triassic Ivishak Sandstone of the Sadlerochit Group. The confirmation well, Sag River State No. 1, proved conclusively that economic reserves existed in the North Slope and set the stage for the development of the largest oil and gas field in North America.

Regional Setting

The Prudhoe Bay Field lies approximately midway between NPRA to the west and the ANWR to the east

(Fig. 1-1). Structurally the field occurs on an anticlinal uplift which forms part of the east-west trending subsurface Barrow arch (Figs. 1-1, 1-2). This basement uplift, which generally parallels the trend of the coastline, represents a major subsurface tectonic feature across the North Slope. The arch had a history of recurrent movement from the Pennsylvanian through the Late Cretaceous (Jones and Speers, 1976), and several unconformities are associated with its sedimentary cover (Fig. 1-2). Of these, the Lower Cretaceous Unconformity (LCU) is the most significant and truncates the reservoir on the eastern flank of the field. The overlying Cretaceous shales form the main seal for the Ivishak Sandstone reservoir.

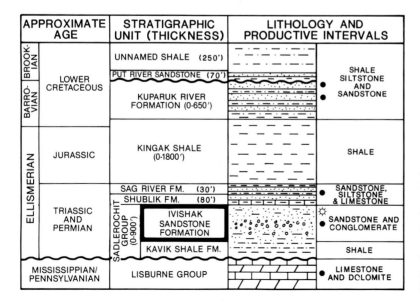

Fig. 1-3. Stratigraphy of the Ellesmerian-Brookian succession (Mississippian-Lower Cretaceous) in the Prudhoe Bay area (modified from Jamison et al., 1980, and Carman and Hardwick, 1983; reprinted by permission of American Association of Petroleum Geologists).

The Ivishak Sandstone is part of the Sadlerochit Group (Fig. 1-3), which includes the underlying Kavik Shale (Jones and Speers, 1976). The Sadlerochit Group rests unconformably upon carbonates of the Mississippian/Pennsylvanian Lisburne Group and is overlain by the upper Triassic Shublik Formation (Fig. 1-3). The Ivishak Sandstone represents the only fluviodeltaic deposits within a marine-dominated Permian through Early Jurassic succession.

Field Characteristics and Production History

The Prudhoe Bay Field underlies an area of more than 225 square miles (585 km²) at an average depth of 8,500 feet (2,590 m) subsea. The Ivishak reservoir has a maximum light oil column thickness of 425 feet (130 m) and has an average net-to-gross ratio of 0.87. Average porosity, permeability, and water saturation are 22%, 400 md, and 35%, respectively. Field production began in 1977, and presently the field contains more than 900 wells (846 active wells as of February, 1989) ranging in spacing from 1,867 to 2,640 feet (569–805 m). The field is managed by BP Exploration (Alaska) Incorporated, who operate the western part, and ARCO Alaska Incorporated, who operate the eastern part (Fig. 1-4).

Initial production was facilitated by gravity drainage combined with gas-cap expansion. To enhance production and maintain reservoir pressure, a produced-water and seawater injection program was initiated in 1984 in areas with limited aquifer support. Pilot EOR studies involving the use of combined water and miscible gas (WAG) have been in operation since 1983. At present, all three modes of recovery, primary, secondary, and tertiary, are being used concurrently in different parts of the field. This efficient development plan has resulted in the net production of more than 6 billion barrels (9.5 × 10⁸ m³) of oil.

Comprehensive accounts of the discovery and subsequent development planning of the field can be found in Jamison and others (1980) and Alwin and others (in press).

Reservoir Characteristics

Structure, Trapping, and Oil Type

The Prudhoe Bay Field is a combination structural and unconformity truncation trap. The structure is an anticline with a gently dipping southern flank and a highly faulted northern flank (Figs. 1-2, 1-4). To the north, the accumulation is bounded by northward-dipping normal faults, to the east by truncation along the LCU and overlying unconformable Cretaceous shales, to the south by the oil/water contact, and to the west by another series of normal faults (Fig. 1-4). The truncating unconformity is Early Cretaceous in age, and the overlying Cretaceous shale forms the main seal over the eastern flank of the field. Seismic (Fig. 1-5) and well data clearly show the Cretaceous unconformity truncating progressively older strata in a northeasterly direction. The top of the Ivishak reservoir within the field ranges in depth from 8,000 feet (2,440 m) to 9,200 feet (2,800 m) subsea and dips to the south and west at approximately 1° to 2° (Fig. 1-4). The field has a total hydrocarbon column of approximately 1,200 feet (365 m) from the top of the gas accumulation to the oil/water contact. This contact is tilted and ranges from 8,925 feet (2,720 m) to 9,061 feet (2,762 m) subsea.

Basin analysis suggests that the structural development of the Prudhoe Bay region focused migrating fluids toward the field during the Late Cretaceous and Early Tertiary. Oil generation and migration began during the deposition of the Colville Group and continued during the deposition of the Sagavanirktok Formation. As the Colville Trough subsided, oil and gas were generated and migrated updip, generally in a northward direction. Hydrocarbon generation and migration were probably complete by the end of the Eocene. A tilting event during the late Eocene (ca. 40 Ma) is interpreted as having caused spillage of some of the original oil column into the western part of the field (Fig. 1-4). High oil saturations within the micropores of rocks of the current gas cap and the occurrence of residual oil far below the present-day oil/water contact suggest that the original oil column was about 2,000 feet (610 m) thick. A heavy oil/tar mat at the base of the present-day oil column is probably the result of a deasphalting process caused by the later introduction of gas into the Ivishak reservoir.

Prudhoe Bay Field oils have an API gravity range of 24.9° to 32.4° (average of 27.9°) and an average sulfur content of 1.01% (Sedivy et al., 1987). The oil is a mixture generated from several source formations. Identified co-sources include the Triassic age Shublik Formation, Jurassic age Kingak Shale, and the organic-rich part of the Lower Cretaceous Pebble Shale (part of the "Unnamed Shale" of Fig. 1-3) (Seifert et al., 1979). Other North Slope reservoirs in the vicinity of the Prudhoe Bay Field (e.g., Kuparuk, West Sak) contain the same oil type as the Ivishak

C.D. Atkinson et al.

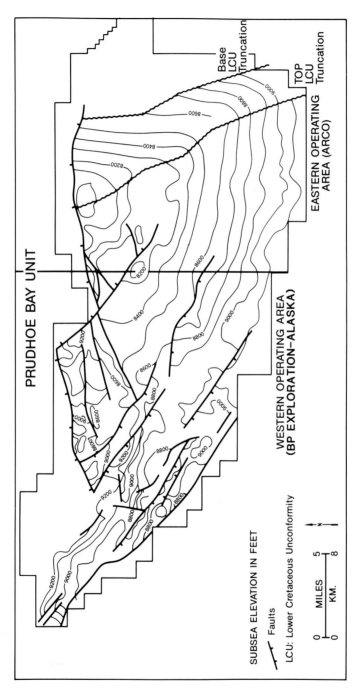

Fig. 1-4. Structure map of the Top Ivishak Sandstone Formation in the Prudhoe Bay Field. Contour interval is 100 feet (30.5 m).

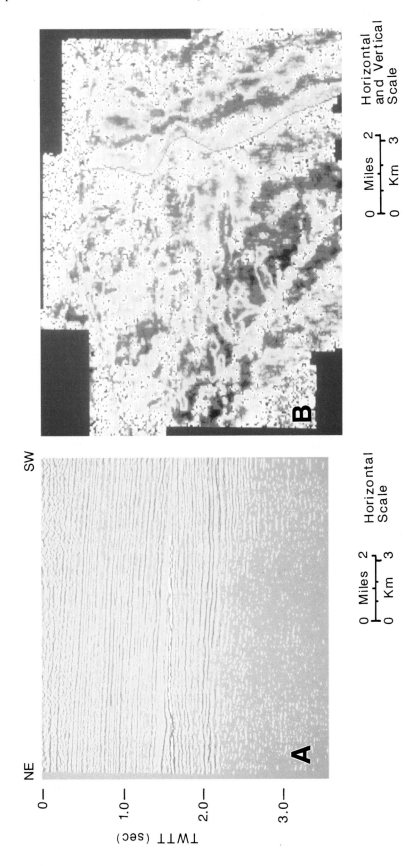

Figure 1-5. (A) Seismic traverse from SW to NE through the northeastern portion of the field. The lower Cretaceous unconformity (LCU) is conspicuously discordant with the underlying formations and truncates the Ivishak (top highlighted) in the NE part of the line. Section displayed in variable intensity. TWTT = Two-way travel time in seconds. (B) Amplitude time slice at approximately 1.8 seconds showing the areal extent of the 3-D seismic survey in the northeastern part of the field. The structural texture of the Ivishak reservoir is evidenced by the coloration of the time slice and shows the trend and dimension of various structural features. The prominent red line denotes the beginning of the truncation of the Ivishak by the LCU to the east.

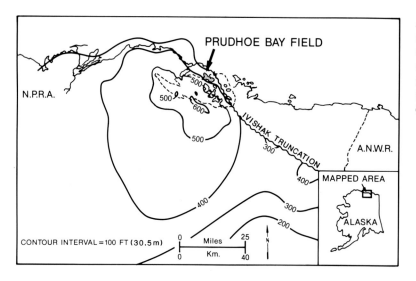

Fig. 1-6. Sandstone thickness map (in feet) of the Ivishak Sandstone in the North Slope region (from Alwin et al., in press, and reprinted by permission of American Association of Petroleum Geologists).

Sandstone, indicating a common co-source for all the oils (Jones and Speers, 1976; Sedivy et al., 1987).

Depositional Environment and Facies

The gross thickness of the Ivishak Sandstone in the field ranges from zero at the unconformity contact to 650 feet (200 m) and averages 550 feet (170 m). Net sandstone thickness averages 484 feet (148 m), with a maximum of 572 feet (174 m). Outside the field, sandstone thickness and the percentage of sandstone and conglomerate within the formation decrease toward the south and southwest (Figs. 1-6, 1-7). Both these factors indicate that the source for the Ivishak sediment was to the north of the present-day coastline. Most authors agree that the Ivishak Sandstone was deposited in fluviodeltaic complexes (Fig. 1-8) which prograded southward into a marine basin (Jamison et al., 1980; Melvin and Knight, 1984; McGowen and Bloch, 1985; Lawton et al., 1987; Atkinson et al., 1988). Stratigraphically, the Ivishak comprises two main depositional megacycles (Fig. 1-9): (1) a lower, upward-coarsening "fluvial progradation" (overall regressive) sequence involving a vertical transition from predominantly interbedded sandstone and marine shale to amalgamated sandstone and conglomerate, and (2) an upper, finer-grained interval of fluvial sandstone and shale which is interpreted to represent a period of "fluvial retreat" (overall transgressive). The distinctive

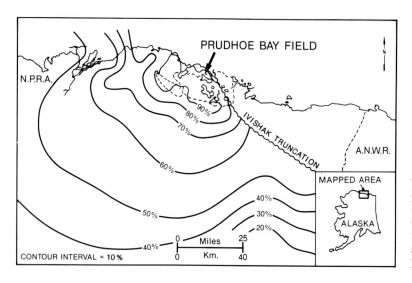

Fig. 1-7. Map of percentage conglomerate/sandstone within the Ivishak Sandstone in the North Slope region (from Alwin et al., in press, and reprinted by permission of American Association of Petroleum Geologists).

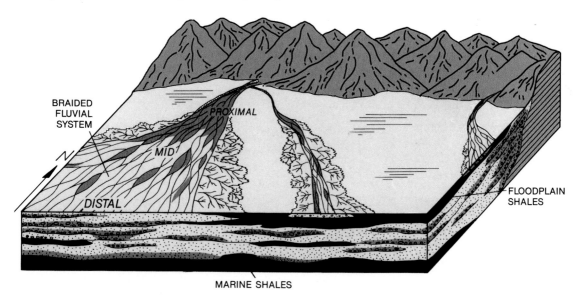

Fig. 1-8. Block diagram illustrating conceptual, braided-river dominated, fluviodeltaic depositional model for the Ivishak Sandstone (modified from Atkinson et al., 1988, and reprinted by permission of the Society of Economic Paleontologists and Mineralogists).

lithologies of the Ivishak Sandstone facilitate a simple, yet effective, petrophysical zonation of the reservoir (zones 1-4, Fig. 1-9). In general, these zones and their subdivisions are stratigraphically continuous throughout the field and can be easily correlated from well logs. Although there is general agreement between these zones/subzones and depositional environment, the relationship is by no means ubiquitous fieldwide (see later discussion).

The Ivishak Sandstone accumulated in depositional environments ranging from delta front to braided stream. Braided-stream processes distributed chert-rich gravel and quartz- and chert-rich sand radially across a coastal plain to construct the subaerial part of the system. Subaerial facies can be broadly categorized as mid-braided stream (alternating conglomerate and sandstone), distal braided/ meandering stream (chiefly sandstone), and abandoned channel-fill, overbank, and pond facies (mudstone, siltstone, and fine-grained sandstone). The fluvial deposits generally comprise multistoried arrangements of erosive-based, upward-fining, channel- and bar-fill sequences. Clast- to matrix-supported conglomerates (Fig. 1-10A), massive (unstratified) to cross-stratified conglomeratic sandstones (Figs. 1-10B, 1-10C), and cross-stratified to parallel-laminated fine- to coarse-grained sandstones (Fig. 1-10D) are the main lithofacies. Marine time-equivalents of the fluvial system are delta-front sandstones (including river-mouth bars and delta fringe/distal bar) and prodelta sandstones, siltstones, and mudstones. Delta-front sandstones were influenced by both fluvial and marine processes and are finer-grained and better sorted than are the contemporaneous fluvial deposits. Dominant lithofacies include parallel- to ripple-laminated sandstone (Fig. 1-11A) and siltstone and bioturbated siltstone and mudstone (Fig. 1-11B). The prodelta sediments comprise interbedded very fine-grained sandstones, siltstones, and mudstones and belong, for the most part, to the underlying Kavik Shale. Grain size decreases and sorting improves from the braided stream through delta front and into the marine-dominated deposits of the Ivishak Sandstone. The scale of sedimentation units decreases from the proximal to the distal facies.

Facies and interpreted environmental relationships are illustrated on north-south (parallel to depositional dip) and west-east (normal to depositional dip) cross sections (Figs. 1-12, 1-13). From north to south, a decrease in the thickness of the conglomeratic mid-braided stream deposits is accompanied by a corresponding increase in thickness in the more sandy, distal-fluvial deposits (Fig. 1-12). The cross section demonstrates a pulsed southward outbuilding of the Ivishak fluviodeltaic system, followed by a northward retreat of the system tract. The east-west cross section (Fig. 1-13) is approximately transverse to the overall sediment transport direction

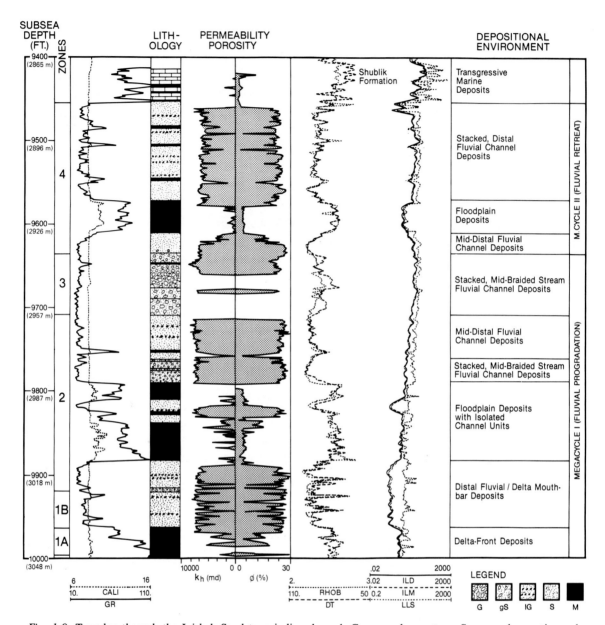

Fig. 1-9. Type log through the Ivishak Sandstone indicating lithology, reservoir quality, depositional environments, and petrophysical zonation (1-4). Lithologic legend: G = conglomerates, gS = conglomeratic sandstones, IG = intraformational conglomerates, S = sandstone, M = shale.

Fig. 1-10. Core photographs representative of the Ivishak fluvial lithofacies: (A) massive, clast- to matrix-supported conglomerate (white and black clasts are microporous and dense chert, respectively), (B) cross-stratified conglomeratic sandstone, (C) cross-stratified pebbly sandstones, and, (D) cross-stratified medium- to coarse-grained sandstone. Scale bar is 1 inch (2.5 cm).

Fig. 1-11. Delta-front lithofacies: (A) low-angle to parallel-laminated, very fine-grained sandstone containing mud drapes and intraformational conglomerate, (B) *Planolites* burrowed delta-front mudstone. Scale bar is 1 inch (2.5 cm).

and illustrates the tendency for more proximal facies to grade laterally at approximately the same stratigraphic level into more distal deposits. This suggests that the Ivishak coastal braidplain consisted of major, braided-river channel belts separated by interfluvial areas comprising overbank deposits and smaller, finer-grained fluvial channels (see Fig. 1-8). The overall vertical succession (prodelta, delta front, distal- to mid-fluvial) in some wells indicates a southward progradation by the Ivishak fluviodeltaic system. Outbuilding was produced by river systems that not only transported coarse-grained sediment to the sea but also spread sediment laterally as the braided channels continually shifted their courses in order to adjust to sediment load, discharge, and slope.

Petrography and Diagenesis

Monocrystalline quartz and chert are the most abundant detrital components in the Ivishak clastic-

rocks. The Ivishak sandstones are litharenites and sublitharenites (terminology of Folk et al., 1970). The quartz-chert ratio is a function of grain size and increases with increasing distance from the sediment source area. Two types of chert are present: (1) nonporous ("dense") chert, and (2) microporous chert (Fig. 1-14). Other detrital components, typically present in minor amounts, consist of polycrystalline quartz, sedimentary (other than chert) and metasedimentary rock fragments, and trace amounts of feldspar. Detrital clay matrix is a minor constituent except in very fine- grained sandstones, siltstones, and mudstones. The advanced stage of mineralogical maturity suggests that Ivishak rocks originated as a recycled sediment accumulation.

Diagenetic effects were gradually superimposed on the component facies of the fluviodeltaic system. Porosity-reducing diagenesis most commonly consists of partial cementation by quartz, siderite, kaolinite, pyrite, and ferroan carbonate, and compac-

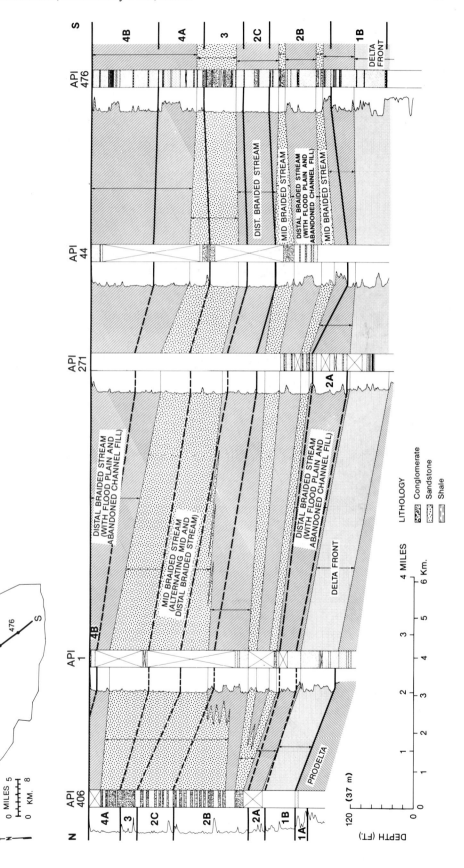

Figure 1-12. North-south stratigraphic cross section through the Ivishak Sandstone illustrating intrareservoir facies distribution and detailed petrophysical zonation (shown as 1A–4A). Note cross-cutting relationship between petrophysical zonation and interpreted depositional environment in the northern part of the section.

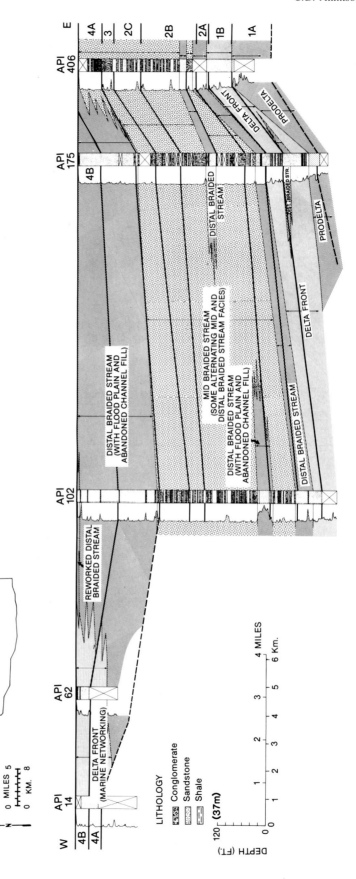

Fig. 1-13. East-west stratigraphic cross section through the Ivishak Sandstone illustrating intrareservoir facies distribution and detailed petrophysical zonation (marked as 1A-4A). Note cross-cutting relationship between petrophysical zonation and interpreted depositional environment in the western and eastern parts of the section.

Fig. 1-14. Photomicrograph of typical medium-grained Ivishak Sandstone consisting of detrital quartz with well-developed overgrowths and dense and microporous chert (arrowed). Scale bar is 1 mm. Plane polarized light.

Fig. 1-15. Photomicrograph of typical coarse-grained Ivishak Sandstone displaying limited mechanical compaction. Scale bar is 1 mm. Plane polarized light.

tion (including formation of stylolites and pseudo-matrix). The fabric of most samples is characterized by tangential and long grain-to-grain contacts (Fig. 1-15). Pore types include primary intergranular macropores, relatively sparse secondary intragranular macropores, and micropores. Secondary macroporosity is the result of partial to complete dissolution of siderite cement and sedimentary rock fragments. Micropores, an important form of microscopic heterogeneity in the reservoir, occur in partially leached chert fragments and in kaolinite cement.

Reservoir Quality

Reservoir quality is, to a large extent, controlled by sedimentary facies (Table 1-1; Figs. 1-16, 1-17) and, in particular, by their textural characteristics. At the reservoir scale, grain size is the primary control of permeability, and a decrease in grain size from more proximal to more distal facies is generally accompanied by decreasing permeabilities. However, sorting also significantly influences reservoir quality of the sandstones. In sandy conglomerates and conglomeratic sandstones, grain size bimodality is a major control of porosity and permeability (Fig. 1-18). For example, sandy conglomerates of lower mid-braided stream affinity have lower porosities but higher permeabilities than do moderately sorted, medium- to coarse-grained sandstones of distal braided-

stream origin (Table 1-1; Figs. 1-16, 1-17). This observation is in agreement with the conclusion of Clarke (1979) that "the effects of bimodality are more important than diagenesis in determining the quality of some oil-field reservoirs."

Comparison of the overall permeability trends among facies of the Ivishak Sandstone with permeability patterns displayed by unconsolidated sands with analogous grain size and sorting (Beard and Weyl, 1973) indicates that the general trends which existed in the unconsolidated original sediments are still recognizable in spite of the diagenetic overprint. Secondary porosity appears only to have enhanced trends in reservoir quality that existed in the Ivishak sediments prior to burial.

Reservoir Heterogeneity and Production Behavior

Although the Ivishak succession is characterized by a high percentage of sandstone and conglomerate, and in comparison to many other reservoirs can be regarded as relatively homogeneous, it does exhibit significant degrees of internal heterogeneity. Heterogeneity exists at scales ranging from the macroscopic (vertical scale in tens of feet, tens of meters; horizontal scale in hundreds of feet, hundreds of meters) to the mesoscopic (vertical and horizontal scales in ones to tens of feet; decimeters to meters)

Table 1-1. Relationship between depositional facies and average core-plug-measured porosities and permeabilities in a well in the eastern part of the field. This well had the most complete core coverage and the largest number of samples analyzed for porosity and permeability of all the wells available in the field. Reservoir quality trends observed in this well are representative of the field.

Facies	Total thickness ft (m)	$\bar{k}$ (md)	ϕ (%)	Comments
Mid-braided stream	13 (4.0)	639.0	22.5	* No conglomerate
Lower mid-braided stream	35 (10.7)	512.0	18.9	* 16 conglomerate samples, $\bar{k}$ = 359 md; 19 sandstone samples, $\bar{k}$ = 640 md
Mid-braided stream to distal-braided stream	27 (8.2)	495.0 495.0	23.9 23.9	* 1 conglomerate sample, $\bar{k}$ = 71 md; 26 sandstone samples
Upper distal-braided stream	64 (19.5)	685.0	25.2	* No conglomerate
Distal-braided stream	248 (75.6)	349.0	22.6	
Distal-braided stream to delta front	18 (5.5)	73.0	20.3	
Delta front	44 (13.4)	13.0	14.9	13 samples coarser than fine-grained, $\bar{k}$ = 33 md; 30 samples fine-grained and finer, $\bar{k}$ = 7 md
Floodplain	63 (19.2)	0.8	5.0	

* Measured permeability in conglomerates is low owing to the inability to obtain accurate measurements in this facies using core plug samples. Whole core and selective pressure test data suggests significantly higher permeability values.

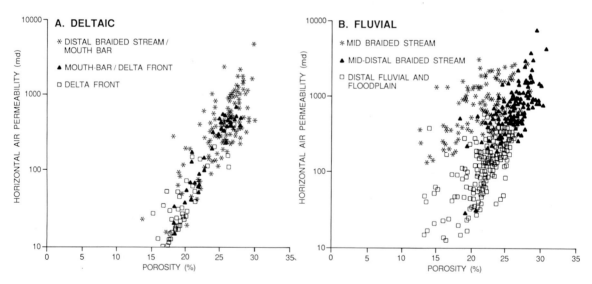

Fig. 1-16. Relationship between depositional facies and core plug measured porosity/permeability for the deltaic and fluvial components of the Ivishak reservoir. Note relatively low permeabilities associated with mid-braided stream facies owing to inability to obtain representative plug samples in this predominantly conglomeratic lithology. Whole core and production test measurements from this facies indicate much higher permeabilities than those suggested here.

ENVIRONMENT TYPE	PREDOMINANT LITHOLOGY	GRAIN SIZE TREND	SORTING TREND	PERMEABILITY TREND LOW HIGH
MID-BRAIDED STREAM	SANDY CONGLOMERATE CONGLOMERATIC SS COARSE-GRAINED SS	DECREASE	DECREASE	
DISTAL BRAIDED STREAM	MEDIUM-GRAINED SANDSTONE			
DELTA FRONT	FINE- TO VERY FINE-GRAINED SS			

Fig. 1-17. Qualitative relationship among depositional facies, lithology, sediment texture, and permeability.

and ultimately to the microscopic (vertical and horizontal scales in tenths to hundreds of inches; millimeters).

One of the most important forms of macro- to mesoscopic heterogeneities are the shale intervals, which occur throughout the reservoir (Fig. 1-19). Previous studies have shown that the relative extent of a shale interval in the reservoir is linked to its depositional environment (Geehan et al., 1986). Continuous shales, those which extend over two or more well spacings, were deposited in floodplain, prodelta, and marsh/bay environments (Fig. 1-19). They occur throughout the Ivishak succession but are most common in the lower parts of the reservoir in the eastern part of the field. Of these shales, the thick floodplain deposits act as the most effective vertical permeability barriers and commonly divide the oil column into isolated production intervals. Depending upon their location within the field, "continuous" shales may be either advantageous or disadvantageous to production (Fig. 1-20). Locally, shale barriers can assist in increasing production rates by preventing both gas coning and water influx. Elsewhere, their presence can be detrimental. First, they may reduce pressure support from the gas cap if production takes place below these shales. Second, shales can promote gas underrunning where production-induced pressure sinks develop below them when they are continuous updip into the gas cap (Geehan et al., 1986; Haldorsen and Chang, 1986). Discontinuous shales, those smaller than the interwell distance, comprise abandoned channel fills and thin intrachannel drape deposits (Fig. 1-19). They are widespread in their distribution and occur

throughout the fluvially dominated parts of the reservoir. Where present, the shales form partial baffles to vertical fluid flow and act to reduce effective vertical permeability in the reservoir. In areas where gravity drainage occurs, oil may accumulate on top of the shales and be bypassed as the gas cap expands downward (Geehan et al., 1986; Haldorsen and Chang, 1986).

Recent studies have shown that mesoscale heterogeneities resulting from permeability variation within the sandstones and conglomerates also exert a significant control on reservoir performance. Dur-

Fig. 1-18. Photomicrograph of quartz and chert grains filling interstices between a framework of chert granules and pebbles. Scale bar is 1 mm. Plane polarized light.

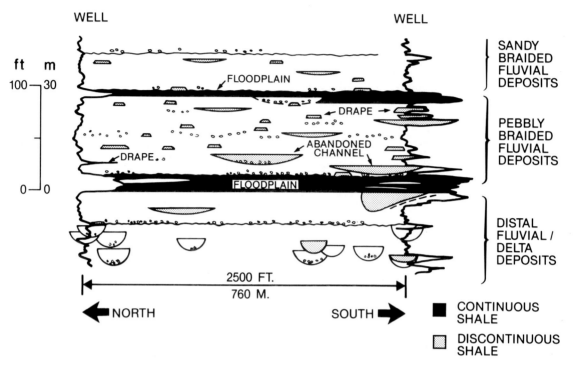

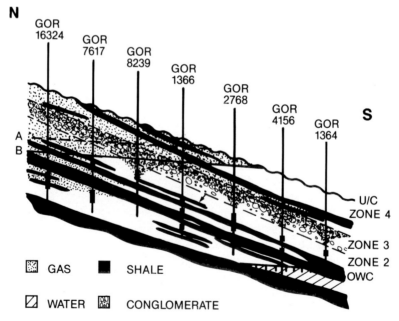

Fig. 1-19. Interpreted depositional environments and associated genetic shale facies of the zone 2 interval of the Ivishak Sandstone in the eastern part of the field. Interpretations based upon log response and core data (modified from Geehan et al., 1986, and reprinted by permission of Academic Press).

Fig. 1-20. Schematic north-south cross section through the eastern part of the field illustrating gas underrunning shales. Arrow indicates a "continuous" shale which inhibits gas coning but which does not intersect the gas cap and promote underrunning. GOR = gas:oil ratio for each
well, A = position of original gas/oil contact (1977), B = position of gas/oil contact as of March, 1983 (modified from Geehan et al., 1986, and reprinted by permission of Academic Press).

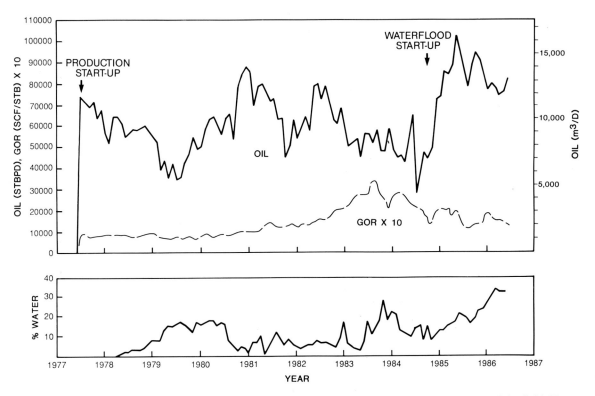

Fig. 1-21. Typical production-history curves for a single drill-site area (34 wells) in the eastern part of the field. Note the increase in oil production and decrease in gas:oil ratio (GOR) associated with late 1984 waterflood start-up.

ing primary production from the field, the presence of high-permeability intervals within the coarse-grained, fluvial deposits facilitated high oil production rates from many wells. As emphasis switches to secondary and tertiary recovery techniques, these same high permeabilities are presenting difficulties in optimizing recovery from the field. Field development problems result primarily because the permeabilities are not uniform within the reservoir. In many wells, permeability profiles are highly irregular, with thick intervals of good overall horizontal permeability (100s of md) being punctuated by thinner horizons of extremely high permeability (1,000s of md). This irregularity is the result of the facies distribution within the Ivishak Sandstone where coarser (high permeability)- and finer (lower permeability)-grained channel fills alternate.

The control exerted on reservoir performance by this facies variation is illustrated by an example from the production/injection characteristics of a single drill site in the eastern part of the field. This drill site comprises a total of 34 wells, 10 of which are cur-

rently used for water injection. A production summary for the area is illustrated in Figure 1-21. In this area, detailed geological studies supported by engineering data have been used to subdivide the reservoir into a series of major mappable "flow units" (Fig. 1-22). A flow unit is here defined in the sense of Ebanks (1987) as "a volume of the total reservoir rock within which geological and petrophysical properties that affect fluid flow are internally consistent and predictably different from the properties of other rock volumes, i.e., flow units". There is a close relationship between the reservoir flow unit subdivision and gross depositional facies (Fig. 1-22). Furthermore, as discussed in the previous section, the production/injection characteristics of each flow unit are strongly influenced by the relative reservoir properties of the component facies. In wells perforated throughout the entire Ivishak interval, injected water more easily enters flow units V1, V2, and V3. These flow units are composed predominantly of coarser-grained, braided-stream conglomerates and conglomeratic sandstones. In contrast, it is relatively

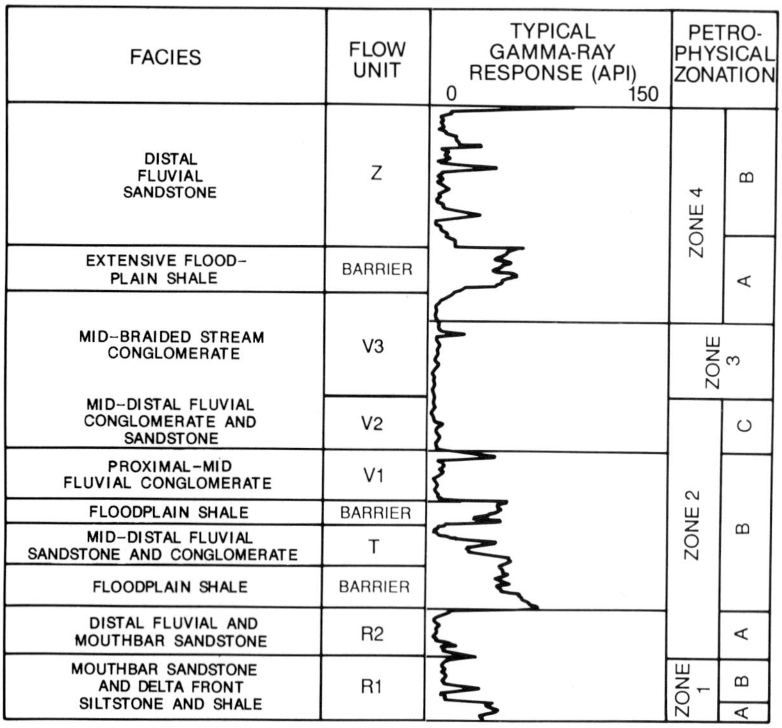

FACIES	FLOW UNIT	TYPICAL GAMMA-RAY RESPONSE (API) 0 150	PETRO-PHYSICAL ZONATION	
DISTAL FLUVIAL SANDSTONE	Z		ZONE 4	B
EXTENSIVE FLOOD-PLAIN SHALE	BARRIER			A
MID–BRAIDED STREAM CONGLOMERATE	V3		ZONE 3	
MID–DISTAL FLUVIAL CONGLOMERATE AND SANDSTONE	V2			C
PROXIMAL–MID FLUVIAL CONGLOMERATE	V1		ZONE 2	
FLOODPLAIN SHALE	BARRIER			B
MID–DISTAL FLUVIAL SANDSTONE AND CONGLOMERATE	T			
FLOODPLAIN SHALE	BARRIER			
DISTAL FLUVIAL AND MOUTHBAR SANDSTONE	R2			A
MOUTHBAR SANDSTONE AND DELTA FRONT SILTSTONE AND SHALE	R1		ZONE 1	B / A

Fig. 1-22. Type log illustrating the relationship between depositional facies, petrophysical zonation, and flow-unit subdivision in the drill site of Figure 1-21. Flow units Z, V, T, and R are named after shale mapping units.

more difficult to inject water into flow units Z, R1, and R2, which are dominated by finer-grained, distal-fluvial and delta-front deposits (Fig. 1-22). The ability to inject water is a reflection of the effective permeabilities of each of the flow units, which are governed by the grain size and sorting characteristics of the component depositional facies.

Flow-unit mapping in the drill site has helped recognize and predict several high-permeability "thief zones" within the reservoir. These thief zones are typically clast-supported conglomerates with exceedingly high permeabilities and lateral extents of several well spacings (thousands of feet; hundreds of meters). A more detailed examination of the injectivity profile of a particular well highlights this thief-zone problem (Fig. 1-23). In this well, which was perforated selectively throughout the V2-Z interval, spinner data indicate that very high flow rates (95% of the total injected water) occurred initially in a relatively thin, 10- foot (3.0-m) thick conglomeratic zone within the V3 interval (8,915–8,925 ft, 2,717–2,720 m, Fig. 1-23). This conglomeratic interval, with an estimated permeability of more than 4,000 md, acts as a major thief zone. Although several other zones were open to flow in this well, no water entered at these levels despite

measured air permeabilities from core of more than 200 md. Following well-profile modification and the isolation of this thief zone, water injection into the overlying more distal fluvial sandstones has been fairly uniform. This is a product of the similar grain-size distributions and sorting characteristics that these sandstones possess, in contrast to the underlying conglomerates (Fig. 1-23).

Microscopic-scale heterogeneities also exist within the reservoir. These further complicate the saturation distribution and recovery efficiency across the field. The most important of these are the micropores present within chert grains and clasts in the sandstones and conglomerates. The amount of microporous chert varies throughout the field, but in places it can be as high as 35% of the total rock volume. Measured intragranular porosity in microporous chert averages 40% and may be more than 60% in some instances (Alwin et al., in press). Where present, the chert makes formation evaluation difficult, since the micropores may be filled with either water or oil depending upon location within the field. Where these micropores occur in the oil column, they are often water saturated. In such cases, calculated water saturations from resistivity logs may indicate values as high as 60%, but the zones produce water-free oil.

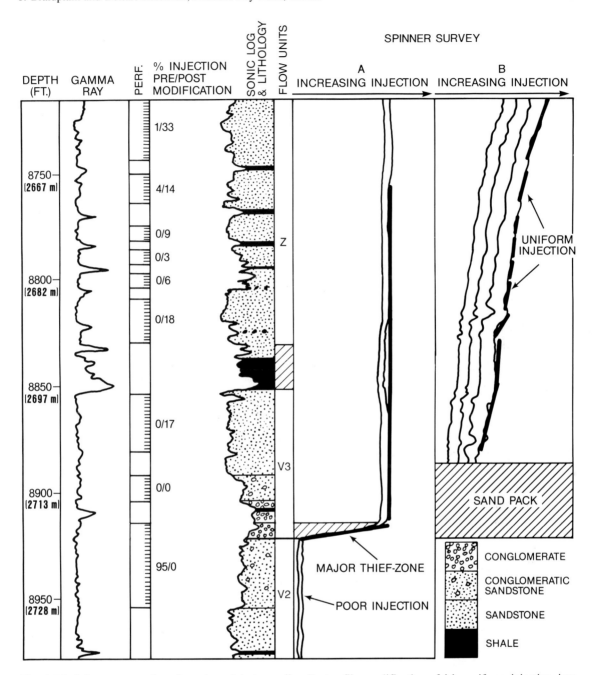

Fig. 1-23. Spinner survey data through an injector well in the drill site of Figure 1-21. (A) Preprofile modification: high-permeability conglomeratic thief zone within the V3 flow unit takes 95% of total injected water. (B) Postprofile modification: fairly uniform injection into sandstones with similar grain-size and sorting comprising the upper V3 and Z flow units. PERF. on the figure refers to perforations.

Exploration and Production Strategy

The location of the giant Ivishak accumulation is the result of a delicate relationship among several key factors. These factors are the presence of a thick and extensive section of excellent reservoir rocks within a structural/stratigraphic trap, favorable burial history, presence of rich source rocks and a geothermal gradient sufficient to generate hydrocarbons, and the proper timing of focused migration of hydrocarbons into the trap.

Exploration outside the field has shown that the Ivishak Sandstone is laterally continuous across the North Slope region on the southern flank of the Barrow arch. To the south, in the Colville Trough, the Ivishak succession becomes increasingly finer grained and eventually passes distally into marine shales. The Ivishak thus forms a clastic wedge-like interval on the flank of a major basement uplift. Exploratory drilling has followed the Ivishak trend north and west of Prudhoe Bay. Several favorable structures with good reservoir quality have been encountered, but only one other significant hydrocarbon accumulation, Seal Island, has been found. Despite the presence of more than 300 million barrels (4.8×10^{10} m³) of oil in place, this field is currently uneconomic. Although good hydrocarbon shows have been encountered in other wells drilled along the Ivishak trend, leaky seals combined with unfavorable tectonic histories appear to have prevented any economic hydrocarbon accumulation. The laterally extensive nature, high net-to-gross ratio, and overall good reservoir quality of the sandstones and conglomerates comprising the reservoir have resulted in a relatively straightforward development strategy for the field. Development wells have been drilled on a standard 160/80-acre spacing from a series of drill sites throughout the field. The waterflood and water-alternating gas (WAG) programs are utilizing an inverted nine-spot pattern in the central part of the field whereas on the periphery, where the oil leg is thinnest, a five-spot pattern has been adopted. Future modifications to this strategy are dependent upon integrating field economics with the results of continued reservoir monitoring and description efforts.

Summary

The Ivishak Sandstone reservoir at Prudhoe Bay comprises an extensive, sheet-like body of amalgamated sandstones, conglomerates, and interbedded shales of deltaic and braidplain origin. Reservoir quality is primarily controlled by sediment grain size and sorting and, hence, by depositional facies and environment. The highest permeability intervals with the greatest flow and injection rates are associated with coarser-grained, mid-braided stream conglomerates and conglomeratic sandstones. In contrast, lower permeabilities and poorer production/ injectivity are characteristic of the finer-grained, distal fluvial and deltaic sandstones.

Despite the predominance within the reservoir of high permeability sandstones and conglomerates, the presence of numerous intrareservoir shales, together with a complicated facies distribution, has combined to produce a reservoir with a variable, heterogeneous, and layered fabric. As development strategy has changed from primary to secondary and tertiary recovery, the significance of this reservoir heterogeneity for ultimate field production has become increasingly evident. The current goal of combined geological and engineering reservoir description in the field is to quantify and map heterogeneity using the flow unit concept. Through this methodology, an efficient reservoir management program is being formulated to maximize the ultimate recovery from this super giant field.

Acknowledgments. We thank ARCO Alaska, Inc. and the other co-owners of the Prudhoe Bay Field for permission to publish this work. The views expressed here are those of the authors and do not necessarily represent the views or opinions of any of the working-interest owners. Since the field was discovered in 1968, many people from the exploration and development groups of ARCO, SAPC (now BP Exploration, Alaska, Inc.) and Exxon have been involved in documenting and understanding the geological/engineering complexities which exist within the Ivishak reservoir. This review article utilizes information from many of these sources. The following people are acknowledged for their help: Fritz Christie, Carol Baker, Meg Kremer, Wayne Zeck (all ARCO Alaska Inc.), Roger Slatt and Jim Ebanks (both ARCO Oil and Gas, Plano). The manuscript has benefited from critical reviews by Peter Barker, Naresh Kumar, Tim Lawton, John McPherson, and Michael Wilson. Finally, we thank the ARCO Plano Graphic Services department for its production of all figures and photographs.

Reservoir Summary

Field: Prudhoe Bay
Location: North Slope, Alaska
Operators: ARCO Alaska, Inc., BP Exploration (Alaska) Inc.
Discovery: 1968
Basin: Colville Trough
Tectonic/Regional Paleosetting: Combination extensional rift and later compressional foreland basin
Geologic Structure: Faulted, truncated anticline
Trap Type: Combination anticline/subunconformity with three-way dip closure
Reservoir Drive Mechanism: Gravity drainage with gas-cap expansion (with limited aquifer support)
 • **Original Reservoir Pressure:** 4,335 psi (3.0×10^4 kPa) at 8,575 feet (2,614 m) subsea
 • **Present Pressure:** 3,750 psi (2.6×10^4 kPa) at 8,800 feet (2,682 m) subsea
 • **Pressure Gradient:** 0.50 psi/ft (11.3 kPa/m)
Reservoir Rocks
 • **Age:** Early Triassic (?), early-middle Scythian
 • **Stratigraphic Unit:** Ivishak Sandstone
 • **Lithology:** Very fine- to very coarse-grained sandstones (litharenites and sublitharenites) and chert-rich, pebble- to cobble-grade conglomerates
 • **Depositional Environment:** Fluviodeltaic
 • **Productive Facies:** Fluvial channel-fill sandstones and conglomerates and delta-front/mouth-bar sandstones
 • **Petrophysics**
   • **Porosity Type:** Intergranular; secondary microporosity
   • **ϕ:** Average 22%, range 10 to 30%, cutoff 12% (cores)
   • **k:** Average 400 md, range 20 to >4,000 md, cutoff 20 md (cores)
   • **S_w:** Average 35%, range 5 to 60% (cores)
   • **S_o:** Average 18%, range 25 to 30% (relative to water; cores)
Reservoir Geometry
 • **Depth:** 8,000 to 9,200 feet (2,440–2,800 m)
 • **Areal Dimensions:** 37 by 12 miles (60×19 km)
 • **Productive Area:** 150,500 acres (6.1×10^4 ha.)
 • **Number of Reservoirs:** 1
 • **Hydrocarbon Column Height:** 425 feet (130 m)
 • **Fluid Contacts:** Oil/water at 8,925 to 9,061 feet (2,720–2,762 m) subsea, tilted; gas/oil at 8,575 feet (2,614 m) subsea
 • **Gross Sandstone Thickness:** 0 to 650 feet (0–200 m)
 • **Net Sandstone Thickness:** Average 484 feet (148 m), maximum 572 feet (174 m)
 • **Net/Gross:** 0.87
Hydrocarbon Source, Migration
 • **Lithologies and Stratigraphic Units:** Marine shales (organic, pyritic & phosphatic), Shublik Formation (Triassic), Kingak Shale (Jurassic), and Pebble Shale (Lower Cretaceous)
 • **Average TOC:** Pebble Shale 5.0%, Kingak 1.7%, Shublik 2.1%
 • **Kerogen Type:** Pebble Shale Type II/III, Kingak Type II/III with minor I, Shublik Type II with minor Type I & III
 • **Time of Hydrocarbon Maturation:** Late Cretaceous-Tertiary
 • **Time of Trap Formation:** Late Cretaceous
 • **Time of Migration:** Late Cretaceous to late Eocene
Hydrocarbons
 • **Type:** Naphthenic to aromatic intermediate crude
 • **GOR:** 745 SCF/bbl (131 m³/m³)
 • **API Gravity:** 27.9°
 • **FVF:** 1.36
 • **Viscosity:** 0.8 cP (0.8×10^3 Pa·s) at 200°F (93°C)
Volumetrics (Ivishak reservoir, primary and EOR)
 • **In-Place:** 21,500 MMBO (3.4×10^9 m³), 46.5 TCFG (1.3×10^{12} m³)
 • **Cumulative Production:** 5,500 MMBO (8.8×10^8 m³), 7 TCFG (2.0×10^{11} m³)

Volumetrics (Ivishak reservoir, primary and EOR) (*cont.*)
  • **Ultimate Recovery:** NA
  • **Recovery Efficiency:** NA
Wells
  • **Spacing:** 1,867 to 2,640 feet (570–805 m), nominal 80 to 160 acres (32.4–64.8 ha.)
  • **Total:** 846 active (December 31, 1988)
  • **Projected Number of Wells:** 1,346 through December 31, 1992
  • **Types:** Vertical, high angle, and horizontal
  • **Drilling Mud:** Lightly dispersed freshwater system
  • **Well Treatment:** Perforated underbalanced; later treatments may include acid stimulation
  • **Testing Practice:** Wells brought on slowly to avoid formation damage
Typical Well Production
  • **Average Daily:** 2,400 BO (382 m³)
  • **Cumulative:** 10 MMBO (1.6 × 10⁶ m³)
Other
  • **Water Salinity (TDS):** Variable, mean 20,000 ppm
  • **Resistivity of Water:** Variable, mean 0.344 ohm-m at 68°F (20°C)
  • **BH Temperature:** 175° to 230°F (79 °–110°C)
  • **Geothermal Gradient:** 2.3°F/100 feet (4.1°C/100 m)
  • **EOR Techniques:** Waterflood and miscible gas (WAG)

References

Alwin, B. W., Davidson, M. C., and Zeck, W., in press, Prudhoe Bay Field – USA, Alaska, North Slope: American Association of Petroleum Geologists Atlas of Oil and Gas Fields, American Association of Petroleum Geologists Treatise of Petroleum Geology.

Atkinson, C. D., Trumbly, P. N., and Kremer, M. C., 1988, Sedimentology and depositional environments of the Ivishak Sandstone, Prudhoe Bay field, North Slope, Alaska: Society of Economic Paleontologists and Mineralogists Core Workshop No. 12, v. 2, p. 561–614.

Beard, D. C., and Weyl, P. K., 1973, Influence of texture on porosity and permeability of unconsolidated sand: American Association of Petroleum Geologists Bulletin, v. 57, p. 349–369.

Carman, G. J., and Hardwick, P., 1983, Geology and regional setting of the Kuparuk oil field, Alaska: American Association of Petroleum Geologists Bulletin, v. 67, p. 1014–1031.

Clarke, R. H., 1979, Reservoir properties of conglomerates and conglomeratic sandstones: American Association of Petroleum Geologists Bulletin, v. 63, p. 799–809.

Ebanks, W. J., Jr., 1987, Flow unit concept – Integrated approach to reservoir description for engineering projects: American Association of Petroleum Geologists Bulletin, v. 71, p. 551–552.

Folk, R. L., Andrews, P. B., and Lewis, D. W., 1970, Detrital sedimentary rock classification and nomenclature for use in New Zealand: New Zealand Journal of Geology and Geophysics, v. 13, p. 937–968.

Geehan, G. W., Lawton, T. F., Sakurai, S., Klob, H., Clifton, T. R., Inman, K. F., and Nitzberg, K. E., 1986, Geologic prediction of shale continuity, Prudhoe Bay field, *in* Lake, L. W. and Carroll, H. B., Jr., eds., Reservoir Characterization: New York, Academic Press, p. 63–82.

Haldorsen, H. H., and Chang, D. M., 1986, Notes on stochastic shales; from outcrop to simulation model, *in* Lake, L. W. and Carroll, H. B., Jr., eds., Reservoir Characterization: New York, Academic Press, p. 445–486.

Jamison, H. C., Brockett, L. D., and McIntosh, R. A., 1980, Prudhoe Bay – A 10-year perspective, *in* Halbouty, M. T., ed., Giant Oil and Gas Fields of the Decade: American Association of Petroleum Geologists Memoir 30, p. 289–310.

Jones, H. P., and Speers, R. G., 1976, Permo-Triassic reservoirs of Prudhoe Bay Field, North Slope, Alaska, *in* Braunstein, J., ed., North American Oil and Gas Fields: American Association of Petroleum Geologists Memoir 24, p. 23–50.

Lawton, T. F., Geehan, G. W., and Vorhees, B. J., 1987, Lithofacies and depositional environments of the Ivishak Formation, Prudhoe Bay field, *in* Tailleur, I. and Weimer, P., eds., Alaskan North Slope Geology: Society of Economic Paleontologists and Mineralogists Pacific Section, Bakersfield, California, and Alaskan Geological Society, Anchorage, Alaska, v. 1, p. 61–76.

Leffingwell, E. de K., 1919, The Canning River Region, Northern Alaska: U.S. Geological Survey Professional Paper 109, 251 p.

McGowen, J. H., and Bloch, S., 1985, Depositional facies, diagenesis and reservoir quality of Ivishak Sandstone (Sadlerochit Group), Prudhoe Bay field [abst.]: American Association of Petroleum Geologists Bulletin, v. 69, p. 286.

Melvin, J., and Knight, A. S., 1984, Lithofacies, diagenesis and porosity of the Ivishak Formation, Prudhoe Bay area, Alaska, *in* McDonald, D. A. and Surdam, R. C., eds., Clastic Diagenesis: American Association of Petroleum Geologists Memoir 37, p. 347–365.

Sedivy, R. A., Penfield, I. E., Halpern, H. I., Drozd, R. J., Cole, G. A., and Burwood, R., 1987, Investigation of source rock-crude oil relationships in the northern Alaska hydrocarbon habitat, *in* Tailleur, I. and Weimer, P., eds., Alaskan North Slope Geology: Society of Economic Paleontologists and Mineralogists Pacific Section, Bakersfield, California, and Alaskan Geological Society, Anchorage, Alaska, v. 1, p. 169–179.

Seifert, W. K., Moldowan, J. M., and Jones, R. W., 1979, Application of biological marker chemistry to petroleum exploration: Proceedings of Tenth World Petroleum Congress Special Paper 8, p. 425–440.

Key Words

Prudhoe Bay Field, Alaska, Colville trough, Ivishak Formation, Triassic, Scythian, fluviodeltaic, braided-stream deposits, super giant field, North Slope, Sadlerochit Group, EOR, unconformity truncation trap, reservoir heterogeneity, flow units, braidplain reservoir.

2

A Braided Fluvial Reservoir, Peco Field, Alberta, Canada

Scott Gardiner, David V. Thomas, E. Dale Bowering, and Larry S. McMinn

Esso Resources Canada Limited, Calgary, Alberta, Canada T2P 0H6

Introduction

Although the deposits of braided fluvial systems commonly have excellent reservoir characteristics, as demonstrated by other chapters in this volume, diagenesis may nevertheless greatly impair their reservoir quality. The ability to predict reservoir quality successfully where diagenesis may have overprinted the depositional controls lies with understanding the diagenesis itself. This chapter concerning the Belly River Formation shows what is probably a very common case for reservoirs, in which diagenetic alterations that determine the final reservoir quality are directly controlled by depositional factors, principally grain size. The finer-grained sandstones have been compacted and cemented to the point of almost totally eliminating porosity, whereas the conglomeratic sandstones retained their depositional porosity and permeability and now are the main reservoir facies.

Since the mid-1960s oil has been produced from reservoirs in the Belly River Formation (Late Cretaceous) in Alberta. Serendipity led to many of the early discoveries in this formation, uphole from deeper targets. In recent years, this overlooked objective has gained industry attention. The Peco Belly River pools, one of the most recent Belly River oil discoveries in Alberta, are the deepest (7,200 ft; 2,195 m) and westernmost in this formation, which is the youngest stratigraphic conventional oil-producing horizon in Alberta. This chapter is the first published interpretation of the Belly River Formation as a braided fluvial system. In this braided fluvial system, predicting reservoir quality and geometry and understanding the continuity of the oil-productive lithofacies at these depths are new challenges for the industry to explore and develop this reservoir economically. Therefore, the experience presented here should be valuable as exploration and development continues in deeper parts of the western Canada basin.

The Peco Field area is located in Townships 46 and 47, Ranges 14, 15, and 16 west of the Fifth Meridian in west-central Alberta in the western Canada basin (Fig. 2-1). This field area contains many oil and gas pools that produce from several reservoirs of different geologic ages. This chapter describes two Late Cretaceous (Campanian) Belly River oil pools, called "A&N" and "C," which together form one of the ten major Belly River oil accumulations in the basin (Fig. 2-1). Production from these two Belly River Peco oil pools is primarily from fluvial channel reservoirs of the lowermost (basal) Belly River Formation. The initial oil reserves from all the Belly River Formation pools in Alberta are 191 MMBO (3.0×10^7 m^3), which represents approximately 1.4% of conventional recoverable oil reserves in Alberta.

Discovery and Production History

The Peco "A&N" and "C" pools were discovered in 1964 and 1983, respectively. The "C" oil pool is separated from the "A&N" oil pool by the presence of a

32   S. Gardiner et al.

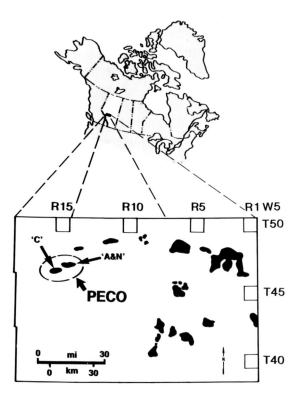

Fig. 2-1. Location map of study area, Peco Belly River "A&N" and "C" pools and other nearby Belly River Formation pools. Pool areas not to scale.

gas cap in the "C" pool. Fifty-four wells are currently on production in the Peco "A&N" and "C" pools at 160-acre (64.8-ha.) spacing. Of 51 Belly River Formation cores currently available (Fig. 2-1), 47 cores are from the basal Belly River interval, and 40 of these 47 cores were examined for this study.

As of December 1987, the "A&N" and "C" pools have been delineated to areas of 2,730 acres (1,105 ha.) and 1,900 acres (770 ha.), respectively. The aggregate length of these pools is approximately 12 miles (19 km) (Fig. 2-1). These pools are currently being extended by drilling to the east, but their extent to the west has been completely delineated. Both pools are operated by several companies (Amoco, Esso, Pan Continental, PetroCanada, Sceptre, and Texaco); however, neither pool has been unitized.

Cumulative production in the Peco "A&N" is 1.3 MMB (2.1×10^5 m³), on primary recovery, of oil with a gravity of 42° API, about 76% of the total recoverable reserves of 1.7 MMBO (2.7×10^5 m³). The 4-21-47-15 well, on production since 1965, is

still producing and has the highest cumulative production (352 MBO; 5.6×10^4 m³) from 160-acre (64.8-ha.) spacing in the "A&N" pool. The Peco "C" pool is located approximately three miles (4.8 km) downdip from the older Peco "A&N" pool in the basal Belly River Formation. The "C" pool has produced 410 MBO (6.5×10^4 m³) of the total recoverable. Separate gas caps are present in each of the Peco pools, with an aggregate recoverable reserve of 11 BCF (3.1×10^8 m³). The GOR of the oil ranges between 450 and 560 SCF/bbl (80-99 m³/m³) at pressures of 1,763 to 1,874 psi (12.2-12.9 $\times 10^3$ kPa). Minor oil production occurs uphole in younger Belly River sandstones above the "A&N" and "C" pools. These younger producing horizons will not be discussed in this chapter.

Regional Data

Regional Setting and Stratigraphy

The stratigraphy of the Upper Cretaceous Belly River Formation and equivalent rocks in Alberta has been defined from studies of foothill outcrops and river exposures by Gleddie (1949), Stott (1959), McLean (1971), McCrory and Walker (1986), and Rosenthal and Walker (1987). Putnam (1989) discusses the petroleum potential of the Belly River Formation and equivalent rocks in west-central Alberta where it consists of an eastward thinning and prograding continental clastic wedge along the western shoreline of the Cretaceous epeiric sea (cf. Williams and Stelk, 1975). The Belly River Formation in the Peco area consists of sharp-based fluvial channel sandstones, overlain by thinner interbedded, continental sandstones, shales, and coal deposits. In the Peco area, the total thickness of the Belly River Formation is approximately 1,200 feet (365 m), is underlain by marine shale and interbedded fine-grained sandstone of the Lea Park Formation, and is overlain by marine shale of the Bearpaw Formation (Fig. 2-2). A local stratigraphic section of the basal Belly River Formation across the Peco "A&N" pool indicates the cross-channel reservoir geometry and sandstone thickness (Fig. 2-3).

No interpretations of the sequence stratigraphy of the Belly River Formation have been published. From outcrop studies, Rosenthal and Walker (1987) observed that the basal fluvial Belly River sandstones "appear to rest directly on bioturbated off-

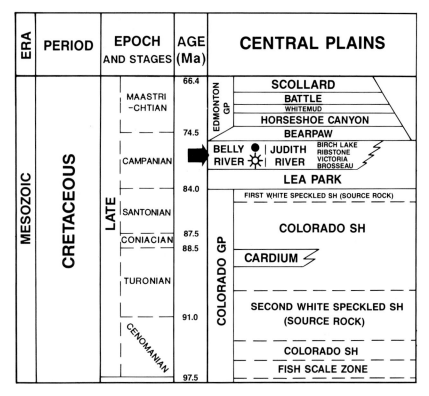

Fig. 2-2. Stratigraphic chart showing the position of the Belly River Formation (see arrow) within the Upper Cretaceous. The stage boundaries are approximate only. SH is an abbreviation for shale. (Reprinted by permission of ERCB, 1984.)

shore mudstones"; however, no interpretation was made regarding the potential significance of this contact. In the Peco area, the interpretation of the depositional environment of the shale underlying the sharp-based fluvial sandstones is uncertain. If it is a marine shale, this would suggest that a relative sea-level drop occurred. However, the low confidence in the interpretation of the depositional environment of the shale prohibits the conclusion that a basin-wide erosional surface exists. A regional stratigraphic correlation of this contact is needed to test the hypothesis that this contact may represent a sequence boundary, perhaps at the base of an incised valley system.

Structural Setting

Peco Field is located in the western portion of the western Canada basin in the eastward extension of the foothills thrust belt of the Canadian Cordillera. In this area, subtle postdepositional structural deforma-

tion has occurred and plays a significant role in the petroleum accumulation of both pools. The Peco reservoirs have undergone a history of thrust faulting as observed on seismic lines (Fig. 2-4) and shown schematically in Figure 2-5. The presence of these faults and their control on hydrocarbon trapping were not known when the discovery wells were drilled. Seismic and well-log data from subsequent reservoir delineation established the significance of the faulting for hydrocarbon entrapment. Some thrust faults intersect the Belly River Formation such as at the western boundary of the pool. More commonly, they terminate just above the stratigraphic level of the underlying Cardium Formation and below the basal Belly River Formation (Figs. 2-4, 2-5). These thrust faults at the Cardium Formation level created an anticline in the overlying Belly River Formation (Fig. 2-6). A normal fault at the eastern portion of Peco probably resulted from the internal rotation of the compressional forces that caused the thrust faults (Etchecopar, 1974).

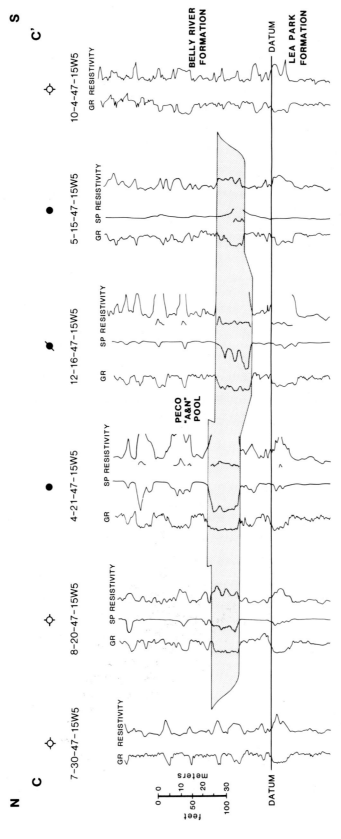

Figure 2-3. Stratigraphic cross section C-C' through the Peco "A&N" pool which indicates the thickness and lateral extent of the reservoir. See Figure 2-7 for the location of cross section.

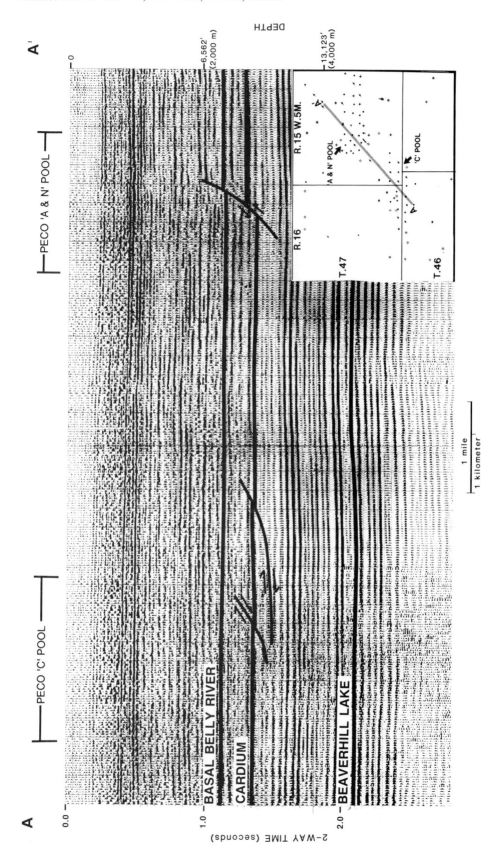

Figure 2-4. Seismic line A-A' indicates the presence of thrust faulting at Peco. The major thrust fault indicated on the west side of the "C" pool in Figures 2-5 and 2-6 is not on this seismic line and is located approximately one mile west of the southwestern limit of line A-A'.

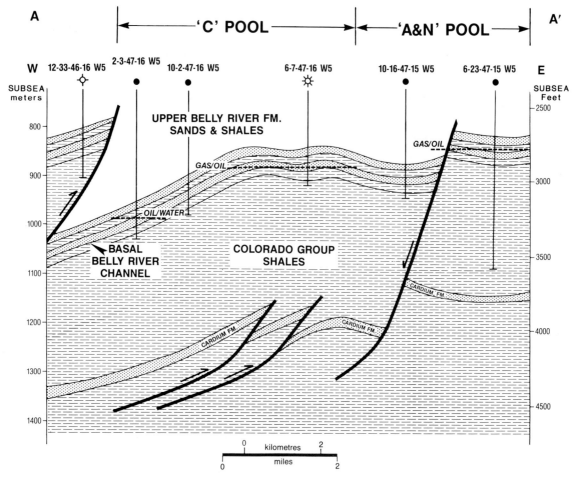

Fig. 2-5. Schematic structural cross section along the line A-A' shown in Figure 2-6 indicating the nature of the faulting and the anticlinal trapping at Peco. Fluid contacts are also indicated. Refer to Figure 2-2 for stratigraphic nomenclature.

Reservoir Characterization

Geometry

The gross sandstone isopach map (Fig. 2-7) of the basal Belly River reservoir is based on the total sandstone thickness indicated on the gamma-ray log. The maximum gross sandstone thickness is 69 feet (21 m) in the central longitudinal axis of the reservoir in an overall west-east trend. This interval consists of vertically continuous sandstone of varying texture. Any well which contains less than 16 feet (5.0 m) of gross sandstone is considered to define the lateral limit of the reservoir. The greatest porous sandstone thickness (net sandstone) is 46 feet (14 m) and lies

predictably along the central longitudinal axis, mirroring the distribution of the gross sandstone thickness (Fig. 2-8). The criterion for net sandstone was a 6% cutoff on the density-porosity log, as discussed later in the petrophysics section of this chapter. Average net sandstone thickness is 16 feet (5.0 m).

Trapping Mechanism and Hydrocarbon Migration

The geologic structure in the Peco area and its control on hydrocarbon entrapment of the "A&N" and "C" pools were determined from well logs and seismic data. The association of gas and oil in the Peco pools in relation to structure is shown in Figures 2-5

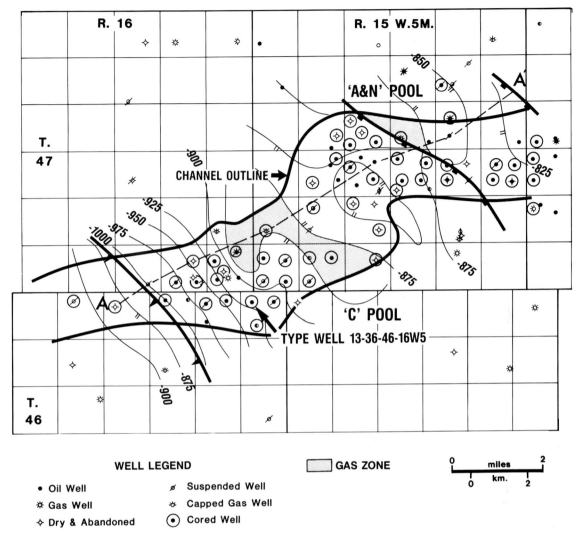

WELL LEGEND

- Oil Well
- ☀ Gas Well
- ✦ Dry & Abandoned
- ⌀ Suspended Well
- ☼ Capped Gas Well
- ⊙ Cored Well

▨ GAS ZONE

Fig. 2-6. Structure map of a shale marker in the Lea Park Formation immediately below the basal Belly River Formation. Contour interval is 25 meters (82 ft) and depths are subsea. Note the two gas caps. The schematic structural cross section in Figure 2-5 occurs along the line A-A′.

and 2-6. Gas in the "A&N" pool is trapped in the high side of a normal fault block. The crest of the anticline associated with a thrust fault locally traps gas in the Peco "C" pool. The gas-oil contact in the "C" pool occurs at 2,870 feet (875 m) subsea. A water leg at approximately 3,200 feet (975 m) subsea defines the western downdip limit of the "C" pool. Lateral seal is provided by the shale which surrounds the sandstone reservoir on the north and south. The top seal is provided by the overlying shale and fine-grained sandstone within the Belly River Formation.

Allan and Creaney (1988) determined that the source rocks for the Belly River oil of west-central Alberta are the organic-rich First and Second White Speckled shales within the underlying Colorado Group (Turonian to Campanian). Commercial quantities of oil have also been generated from portions of leaner shales between these condensed intervals. The thrust faults which cause the trap at Peco Field sole out in the underlying shale of the Colorado Group, probably in the organic-rich source rock of the Second White Specks Formation. The Second

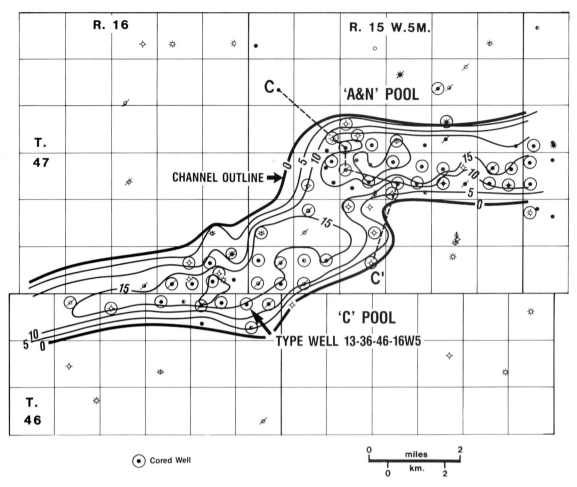

Fig. 2-7. Gross sandstone isopach map contoured in meters of the oil-producing basal Belly River Formation at Peco. Note the location of the Peco "A&N" and "C" pools.

The location of the type log in Figure 2-9 is indicated (see arrow). The location of the local stratigraphic cross section C-C' in Figure 2-3 is also shown.

White Specks Formation is interpreted to have been the glide plane for fault movement. The timing and pathway of oil migration from the source rock to the reservoir are not well understood. Oil may have migrated directly from the Second White Specks source up a conduit provided by the thrust faults to the basal Belly River Formation.

Lithofacies

Three lithofacies (A, B, and C) in the basal Belly River Formation and three lithofacies (D, E, and F) in the underlying Lea Park Formation have been recognized at Peco. The lithofacies have been identi-

fied by lithology, grain size, bed thickness, sedimentary structures, and trace fossils. After lithofacies descriptions, interpretations of the depositional environment and a facies model for the reservoir (lithofacies A, B, and C) will be presented. The lithofacies descriptions, interpretation of depositional environment, and the associated gamma-ray log response for a core taken in a typical Peco well are shown in Figure 2-9. Because lithofacies A, B, and C comprise the main reservoir sandstone at Peco, their interpretation is more rigorous than that for lithofacies D, E, and F, which comprise the underlying nonreservoir Lea Park Formation. Although lithofacies D, E, and F have no direct

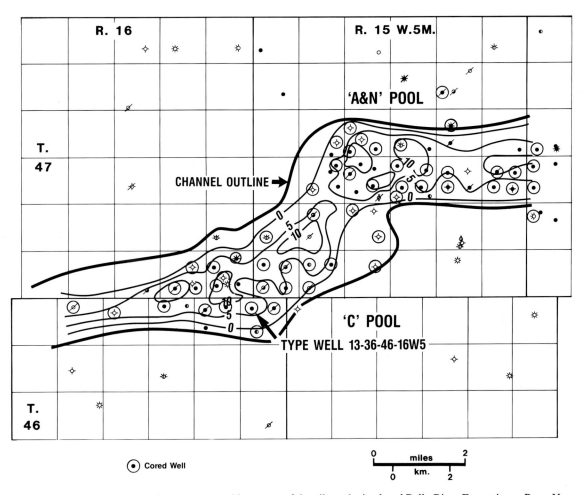

Fig. 2-8. Net sandstone isopach map contoured in meters of the oil-producing basal Belly River Formation at Peco. Net sandstone has porosity ≥ 6%.

impact on Peco oil production, an understanding of these lithofacies puts the reservoir lithofacies in the proper context. East of the Peco pools, at the Wilson Creek and Ferrybank pools, Belly River oil production occurs from reservoirs which were deposited in marine depositional settings similar to those described here in lithofacies D, E, and F.

Rippled, Very Fine-Grained Sandstone, Siltstone, and Shale (Lithofacies A). Lithofacies A consists of very fine-grained sandstone interbedded with siltstone and shale. The beds average 0.2 to 1.0 inch (0.5-2.5 cm) thick (Figs. 2-10A, 2-10B). The sandstone displays abundant ripple cross-lamination. Carbonaceous laminae and thin (0.5-1.0 in.; 1.3-2.5 cm) coaly layers are present. No trace fossils were observed.

Fine- to Medium-Grained Sandstone (Lithofacies B). Lithofacies B consists of fine- to medium-grained sandstone (Fig. 2-10C). It is typically calcite cemented and has a clayey matrix. High-angle (20–30°) cross-beds (Fig. 2-10C) are common, as are unstratified beds. Parallel laminations and cross-laminations generated by current ripples and climbing ripples are present in this lithofacies. Lithofacies B also contains carbonaceous fragments and laminae as well as shaly parallel laminae separated on a scale of 1 to 6 inches (2.5-15 cm) between laminae.

Sandy Conglomerate to Pebbly Sandstone (Lithofacies C). This lithofacies consists of poorly sorted, sand-matrix-supported conglomerate and pebbly sandstone (Figs. 2-10D through 2-10I). The compo-

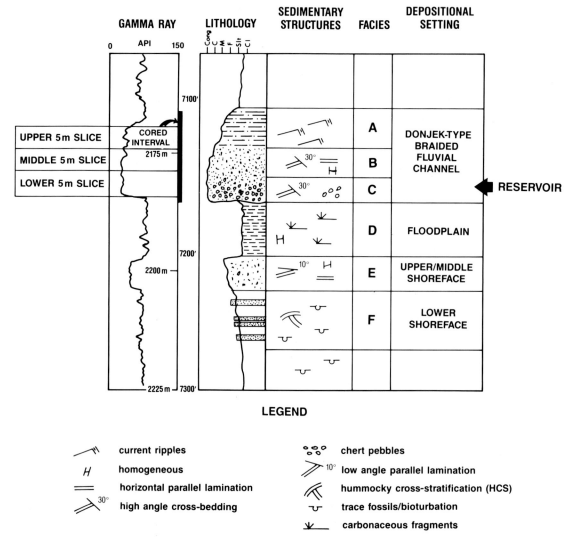

Fig. 2-9. Peco pool type well log of the basal Belly River Formation from the Esso Canterra 13-36-46-16 W5M, lithofacies observed, interpretation of depositional environment, and location of the 16-foot (5.0-m) slice intervals. Refer to Figure 2-7 for the location of this type well log.

sition of the pebbles is dominantly chert with some quartzose, granitic, or feldspathic pebbles (Figs. 2-10D through 2-10F). The grain size of the conglomerate ranges from granules up to small pebbles (Figs. 2-10D through 2-10F), with most pebbles subangular to subrounded. The matrix consists of medium- to coarse-grained sand (Fig. 2-10F, 2-10G). Shale rip-up clasts with long axes up to 4 inches (10 cm) in size are common (Fig. 2-10G).

The dominant sedimentary structure is high angle (20-30°) cross-bedding (Figs. 10D, 10F, 10H),

although unstratified sequences are common. The pebbles exhibit some imbrication by long axes alignment parallel to the cross-bed laminations. This lithofacies is commonly 16 feet (5.0 m) thick with a maximum thickness of 40 feet (12 m). It usually occurs at the bottom of the Peco reservoir but does occur elsewhere in the vertical profile. The amount of sand matrix between pebbles is highly variable and, overall, the amount of matrix increases upward. This variation makes it difficult to define lithofacies C as either a conglomerate or a pebbly sandstone. Several amalgamated, upward-fining trends have been recognized based on an upward decrease in the concentration of pebbles and an increase in the concentration of sand matrix, punctuated by grain size reversals. Typically, two to three of these erosive, amalgamated beds or cycles exist in this lithofacies at any one location. However, no predictable vertical cyclicity of these pebbly sandstones exists within a wellbore and between wells. Confidence in correlations of individual amalgamated pebbly sandstone beds from one well to another is low, especially when correlation distances exceed more than one well-spacing unit (1,310 feet; 400 m). A statistical technique was used to better understand intrareservoir vertical and horizontal continuity and heterogeneity in this lithofacies, as described later in this chapter.

Gray to Black Mudstone (Lithofacies D). Lithofacies D is dominantly composed of gray to black mudstone with very minor thin siltstone laminae (Fig. 2-10I). Typically, lithofacies D is unstratified and does not exhibit sedimentary structures. It does not exhibit the fissility that is typical of shales in lithofacies F, and as a result, it shows no splitting tendency in core. Carbonaceous fragments are present but rare (Fig. 2-10I). This lithofacies occurs below lithofacies A and above lithofacies E (Fig. 2-9).

Hummocky Cross-Stratified Sandstone (Lithofacies E). Lithofacies E consists of very fine- to fine-grained sandstone which is unstratified or contains parallel lamination and some low- to medium-angle (5-15°) hummocky cross-stratification (HCS) (Fig. 2-10J). Shale partings or interbeds are rare to absent. The thickness of this lithofacies is commonly about 10 feet (3.0 m).

Fine-Grained Sandstone and Interbedded Shale (Lithofacies F). Lithofacies F consists of very fine- to fine-grained sandstones interbedded with shales and bioturbated sandy shales (Figs. 2-10K, 2-10L). Interbedding occurs on a scale of 1 to 6 inches (2.5-15 cm). The thickness of the lithofacies based on well logs varies from 10 to 50 feet (3.0-15 m).

The sandstone beds of lithofacies F have sharp bases and either exhibit parallel to low-angle (≥ 10°) laminations, exhibit graded bedding, or are unstratified. The interbedded mudstones are slightly to extensively bioturbated by *Skolithos* and *Planolites*. Sideritic concretions, 0.5 to 2 inches (1.3-5.1 cm) in size, are common.

Depositional Setting

The only published interpretation of the depositional environment of the basal Belly River Formation in west-central Alberta is that of Shouldice (1979), who determined that it was deposited "in a fluvial setting with associated marine shorelines." This general interpretation needs to be refined to account for different paleogeographic and stratigraphic settings in the basin.

At Peco, lithofacies A, B, and C are interpreted to have been deposited in a braided fluvial system similar to the Donjek model proposed by Miall (1977, 1978). Lithofacies C is very similar to facies Gt, St, and Sp of Miall's (1977) classification of braided rivers. The upward-fining trend from lithofacies C to B and A at Peco correlates to the threefold vertical division from cross-stratified gravels to cross-stratified gravelly sandstone to fine-grained rippled sands and muds observed in the Donjek river (Williams and Rust, 1969). This upward-fining trend and the coarse-grained nature of the channel fill, internal grading, and cyclicity support the Donjek model interpretation.

The braided-channel system at Peco was probably distal from its sedimentary source as evidenced by its sand and gravel elements. The Donjek model was based largely on the gravel-dominated proximal parts of the Donjek river. However, Miall (1978) proposed that the Donjek river model can contain between 10 and 90% conglomerate/pebbly sandstone, which allows for a great deal of variation of internal characteristics. This variation makes it difficult to unequivocally distinguish the Donjek model for Peco from the south Saskatchewan braided model of Cant (1978).

The cross-bedded pebbly sandstone and sandy conglomerate of lithofacies C probably represents sedimentation as linguoid and transverse bars and scour fills. The lateral extent of these bars is unknown due to a lack of sufficient well control. Based on modern studies by Collinson (1970) and Smith (1970, 1971, 1972, 1974), linguoid and transverse bars have great variability in width, from 6 to 10 feet (1.8-3.0 m) to 3,440 feet (1,050 m), and in length, up to 6,880 feet (2,100 m). The maximum height is typically 1.5 to 3.0 feet (0.5-1.0 m). These two bar types are probably indistinguishable from each other in the ancient record (Miall, 1977). Although the lateral extent of these bars is impossible to determine with limited well spacing

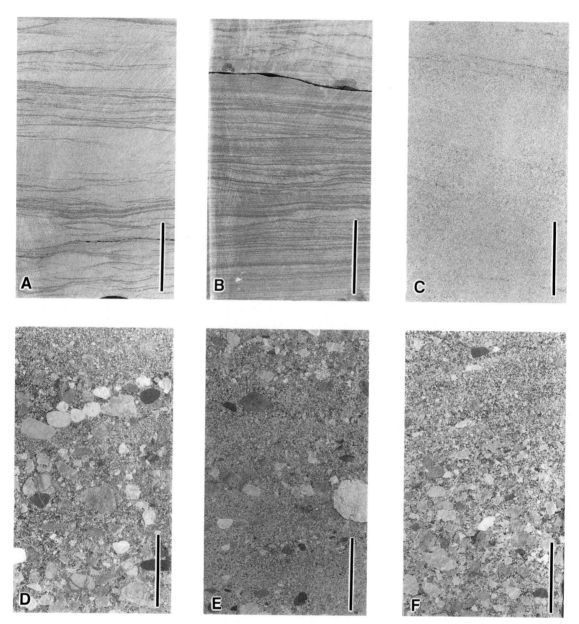

Fig. 2-10. Core photographs representative of lithofacies of the Belly River Formation. Scale bar is 1 inch (2.5 cm). (A,B) Lithofacies A. Current-rippled, very fine-grained sandstone and siltstone. Well location is 10-6-47-15 W5. (C) Lithofacies B. Nonreservoir, medium-grained cross-bedded sandstone. No pebbles are present. Well location is 13-36-46-16 W5. (D) Lithofacies C. This pebbly sandstone which is the net pay at Peco consists of poorly sorted, subrounded chert pebbles in a medium-grained sandstone matrix. Some evidence for pebbles aligned along cross-bedding planes. Well location is 13-36-46-16 W5. (E) Lithofacies C. Poorly sorted pebbly sandstone with pebbles up to 0.75 inch (2 cm) in size. Well location is 13-36-46-16 W5. (F) Lithofacies C. Note the gradual grain-size decrease upward from pebbles at the bottom to medium- to coarse-grained sandstone with scattered pebbles at the top of the photograph. Well location is 13-36-46-16 W5. (G) Lithofacies C. Shale rip-up clast and abrupt grain size changes and reversals. Well location is 10-6-47-15 W5. (H) Lithofacies C. Coarse-grained cross-bedded sandstone. A few scattered small pebbles are present. Well location is 13-36-46-16 W5. (I) Lithofacies D. Gray to black shale contains comminuted and larger whole carbonaceous fragments (see arrows). This perspective is end-on where the core is split open. Well location is 10-6-47-15 W5. (J) Lithofacies E. Thickly bedded hummocky sandstones. Note the sharp base of the fine-grained sandstone overlying the shale and the decreasing-upward angle of the cross-bedding. This cross-bedding is interpreted to be hummocky cross-stratification (HCS).

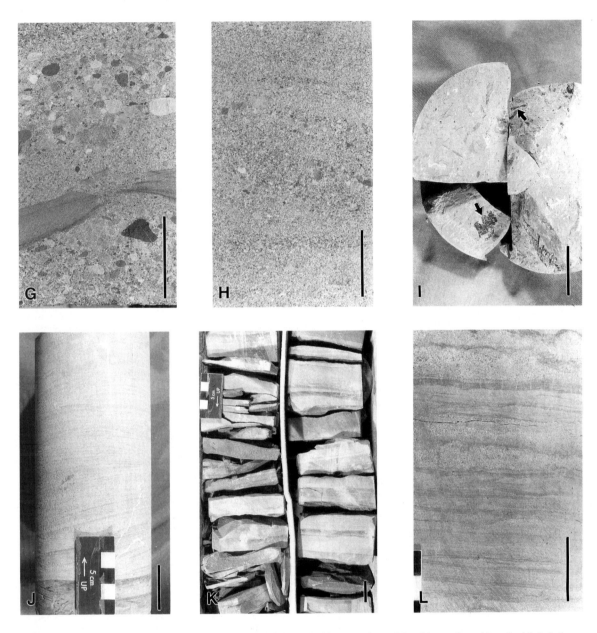

Well location is 4-22-47-15 W5. (K) Lithofacies F. Thinly bedded very fine-grained sandstones and interbedded shales. Well location is 16-17-47-15 W5. (L) Lithofacies F. Close up view of the thin sands and interbedded shales. Note the relatively sharp base of the thin sandstones. Well location is 16-17-47-15 W5.

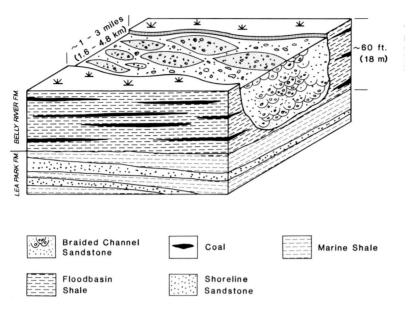

Fig. 2-11. A schematic diagram showing the deposition of the Belly River Formation as found in the Peco reservoirs.

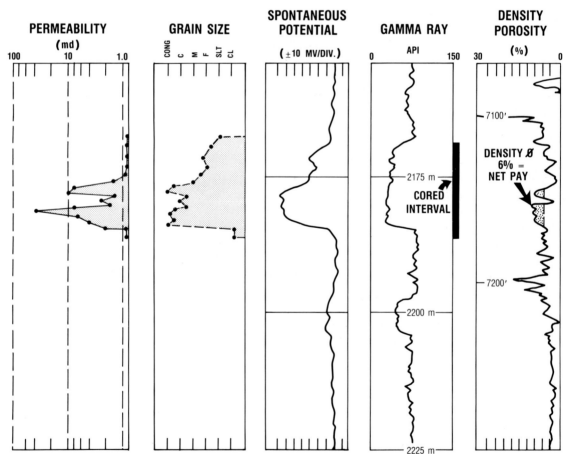

Fig. 2-12. Well-log character, lithofacies, grain size, and core permeability of the reservoir and associated facies in well 13-36-47-15 W5M.

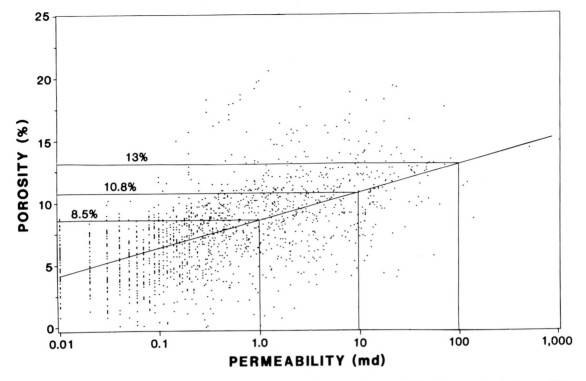

Fig. 2-13. Porosity versus permeability cross-plot based on data from 47 Peco Basal Belly River cores. The least-squares-fit straight line indicates a core porosity of 8.5% is equal to 1 md permeability. Log porosity indicates 6% is approximately equal to 1 md permeability.

and cores, it has a large impact on the reservoir continuity and heterogeneity, which will be discussed later in this chapter.

Figure 2-11 illustrates the facies model and resulting vertical profile envisioned for the basal Belly River at Peco. Farther east in the basin 50 to 100 miles (80-160 km), where the depositional slope was gentler, it is likely that a different fluvial facies model should be applied to reflect the anticipated decrease in the conglomerate/sandstone ratio and the relatively more meandering river profile.

Lithofacies E and F comprise an upward-coarsening sequence and are collectively interpreted here to represent a sandy marine shoreline. Lithofacies E represents upper to middle shoreface deposition, while lithofacies F was deposited in the lower shoreface to offshore transition setting. The presence of carbonaceous fragments and lack of any diagnostic marine fauna suggest that lithofacies D was deposited in a continental floodplain adjacent to the channel system. However, because few cores of lithofacies D have been taken and because those

cores which are available exhibit nondiagnostic features, the interpretation of this facies as nonmarine is equivocal.

Petrophysics

Lithofacies C, which contains the coarsest grain size, is the only lithofacies capable of oil production at Peco. Therefore, the well-log characteristics of the other lithofacies (B-F) will not be discussed, except to say that they can be recognized on gamma-ray (GR) and sonic- or density-porosity well logs. Figure 2-12 indicates the well-log response of the gamma-ray, density-porosity, and spontaneous potential (SP) curves for the reservoir and compares these to grain size and core permeability. The log porosity in lithofacies A ranges from 4 to 12% with an average of 8 to 9%. Based on a cross-plot of core porosity versus permeability (Fig. 2-13), a core porosity of 8.5% equates to 1 md permeability, which is arbitrarily considered to be the lower productive limit for this reservoir. The density- and sonic-log

Table 2-1. Compositional point-count data of sandstones from the Peco reservoir.

Well	KB depth (m)	Litho-facies	Quartz (%)	Feldspar (%)	Chert (%)	Sand-stone/shale (%)	Silt-stone (%)	Meta-morphics (%)	Vol-canics (%)	Other (%)	Grain size (mm)	Sorting	Core porosity (%)	Core perme-ability (md)
4-6-47-15W5	2126.80	C	11	5	70	2	4	1	6	1	0.70*	Well	7.5	1.1
"	2129.10	B	23	20	29	6	4	3	13	2	0.30	Well	9.7	0.09
10-6-47-15W5	2138.20	C	20	10	51	2	5	2	10	—	0.50	Mod. Well	9.4	1.13
13-36-47-15W5	2171.70	B	23	16	33	5	5	3	14	1	0.35	Well	6.3	0.73
2-6-47-15W5	2151.05	B	31	12	37	2	4	—	14	—	0.30	Well	9.0	0.29
"	2153.24	B	33	19	33	5	—	TR	10	TR	0.30	Well	5.0	0.05
"	2154.65	C	16	4	64	3	4	1	8	—	0.50	Well	11.0	12.9
"	2155.84	C	23	2	68	3	3	TR	1	—	0.60*	Bimodal	4.0	0.2
"	2156.55	C	23	2	67	2	TR	—	6	—	0.60*	Mod. Well	12.0	40.7
"	2156.83	C	24	8	54	2	2	—	10	—	0.35*	Well	13.0	5.12
"	2157.43	C	28	6	48	1	3	1	13	—	0.45*	Well	11.0	1.22
"	2158.26	C	17	TR	75	2	1	—	5	—	0.90*	Mod. Well	12.0	122.
"	2158.59	C	19	TR	70	3	—	—	8	—	0.90*	Bimodal	11.0	111.
"	2159.53	C	23	5	60	2	4	—	6	—	0.60*	Moderate	11.0	13.1

*Sandstone contains floating pebbles.
TR = Trace

porosity values are typically 2% lower than the core porosity. This difference in porosity measurements between core and log data is probably due to the effect of authigenic clays which reduce effective porosity at reservoir conditions but cause an overestimate of porosity in air-dried cores measured in laboratory conditions at surface. Net pay is determined using the porosity logs and cores where they are available. In addition, empirical results indicate that an SP log cutoff of ±40 millivolts corresponds to permeability greater than or equal to 1 md (Fig. 2-12). Figure 2-3 indicates the variation in the response of the SP log across the reservoir. The greater the magnitude of the SP response, the greater the reservoir permeability. The permeability variation is due to the presence or absence of lithofacies C (net pay) in each well.

Dipmeter logs have been used in four Peco wells to determine the dip of the high-angle cross-beds within the channel. The dominant paleoflow direction in lithofacies C is eastward, with slight deviations to the northeast and southeast. The paleoflow direction aids in choosing development drilling locations in the reservoir.

Determining the reservoir fluid in Peco development wells is usually achieved through core examination. Well logs at Peco typically indicate a reservoir resistivity of 40 to 50 ohm-meters in the oil-productive zones. However, the resistivity values are highly variable due to the invasion of drilling fluids, as described later in the well and reservoir performance section of this chapter.

Diagenesis

The basal Belly River sandstones and conglomerates at Peco are chert-rich litharenites (classification of Folk et al., 1970) composed of chert (29-75%), with minor quartz (11-33%), feldspar (2-20%), and volcanic, sedimentary, and metamorphic rock fragments (7-18%) (Table 2-1; Fig. 2-14). The percentage of chert increases with increasing grain size (Fig. 2-15). Coarse-grained to pebbly sandstones and sandy conglomerates (lithofacies C) are typically composed of 50% chert. Calcite and minor quartz cement and authigenic chlorite and kaolinite clays are the main pore-filling constituents.

At Peco Field, reservoir quality of the Belly River Formation is largely controlled by depositional texture, principally grain size, and diagenetic alterations (i.e., compaction, cementation, and leaching). Figure 2-16 shows the relationship between permeability and grain size. Effective reservoir permeability greater than or equal to 1 md is preserved in virtually all coarse-grained samples with an average grain size coarser than 0.02 inch (0.5 mm)(1.0 phi). Similarly,

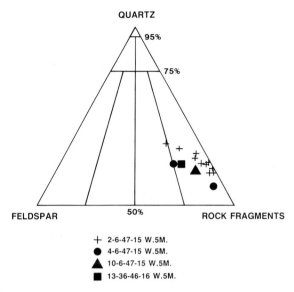

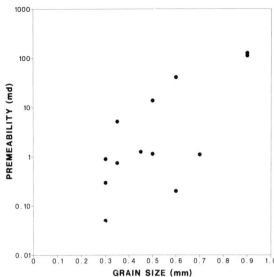

Fig. 2-14. Composition of sandstones from the Peco Belly River "C" pool reservoir. Classification after Folk et al. (1970).

Fig. 2-16. Plot of permeability versus grain size for the Peco pool. The general trend shown is that oil-productive lithofacies containing coarser grain sizes have higher permeability.

all fine-grained samples with a grain size less than 0.014 inch (0.35 mm)(1.5 phi) contain dominantly noneffective reservoir (<1 md). A generalized diagenetic sequence (Fig. 2-17) based on petrographic analyses of thin sections from the Peco reservoir summarizes the diagenetic changes observed.

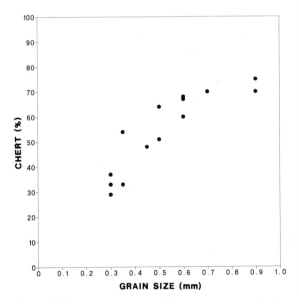

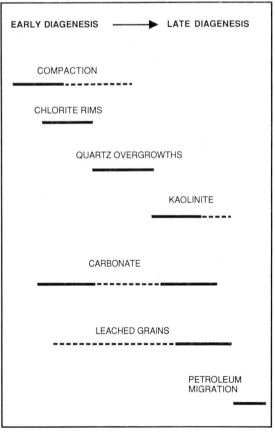

Fig. 2-15. Plot of percentage of chert versus grain size for the Peco pool. The general trend shown is that coarser grain sizes contain a higher percentage of chert.

Fig. 2-17. Relative timing of the diagenetic sequence of events in the Belly River Formation at the Peco "A&N" and "C" pools. No absolute time scale is implied.

Compaction is largely responsible for the difference in reservoir potential exhibited between lithofacies C (Fig. 2-18A) and lithofacies B and A (Fig. 2-18B). The interstices of fine-grained sandstones of lithofacies B and A are initially smaller and, therefore, the pore throats are smaller than those of coarse-grained sandstones of lithofacies C. Therefore, lithofacies A and B are more prone to porosity/permeability reductions through compaction. In addition, lithofacies B and A typically contain more volcanic and argillaceous rock fragments than does lithofacies C (Table 2-1). These rock fragments are more prone to deformation and welding around more competent grains such as quartz. This deformation and welding tends to reduce pore throat sizes further (Fig. 2-18B).

Reservoir quality has also been reduced by diagenetic pore-filling clays and cements that were precipitated under meteoric water conditions (Ayalon and Longstaffe, 1988). Coarse-grained sediments of lithofacies C, which contain primary porosity that survived compaction, have been particularly affected by secondary cements and clays. In order of emplacement, chlorite, quartz overgrowths, calcite cement, and authigenic kaolinite are the principal diagenetic minerals present.

Chlorite is the most abundant of the authigenic clays at Peco. It occurs as grain-rimming cements that formed early in the diagenetic sequence (Fig. 2-18C). The presence of chlorite rims further restricts pore throats and consequently reduces permeability (Iwuagwa and Leberkmo, 1981).

Authigenic quartz cement is present in minor amounts at Peco. It occurs as syntaxial overgrowths on quartz grains (Fig. 2-18D) and as small crystals on chert and on quartzose grains. Precipitation of quartz is inhibited by the presence of chlorite rims. Minor quartz cement is observed on thin chlorite coatings (Fig. 2-18E), whereas thick chlorite coatings may totally preclude precipitation of quartz cement.

Calcite cement is a minor constituent in the Peco reservoir. Calcite occurs either as a pervasive poikilotopic cement (Fig. 2-18E) or in patchy clusters (Fig. 2-18F). Occasionally, calcite cement can form early and completely preclude all other authigenic minerals. Calcite cement, however, as can be observed in Figure 2-18E, can also postdate chlorite and quartz overgrowths.

Authigenic kaolinite is very common in the Peco reservoir. It is commonly distributed in patches or as pervasive pore-fillings and is the last authigenic mineral to form. Figure 2-18G shows kaolinite postdating quartz overgrowths, and Figure 2-18F shows the corrosive outline of calcite cement in contact with kaolinite, indicating that kaolinite preceded calcite cementation.

Secondary leaching of the Peco Belly River reservoirs is common and contributes approximately 50% to the porosity and permeability of these

Fig. 2-18. Photomicrographs of Peco Belly River Formation reservoirs. The scale bar is 0.50 mm. (A) Coarse-grained, chert-rich litharenite. Note open grain framework. Quartz (q), chert (ch), volcanic rock fragment (v), pore space (p). Esso Canterra Peco 4-6-47-15 W5. 7,005 feet (2135.2 m). Plane polarized light. (B) Tight fabric characteristic of fine-grained interbeds. Deformation of argillites (a) and welded grain contacts have destroyed all effective permeability. Esso Canterra Peco 4-6-47-15 W5. 7116.5 feet (2169.1 m). Plane polarized light. (C) Chlorite (cl) coatings contribute to the permeability reductions of reservoir sands at Peco. Remnants of partially dissolved volcanic fragments (v) can be found in the pore spaces (p). Note the absence of quartz overgrowths. Chert (ch). Esso Canterra Peco 13-35-46-16 W5. 7149.1 feet (2179.0 m). Plane polarized light. (D) This coarse-grained sandstone has excellent reservoir characteristics. Note quartz overgrowths (o) that have strengthened the grain framework. Esso Canterra Peco 10-6-47-15 W5. 6867.5 feet (2093.2 m). Plane polarized light. (E) Pore network in sandstone is affected successively by (1) chlorite rims (cl), (2) local quartz overgrowths (o) (left center), and (3) coarse sparry calcite cement (cc). Note the absence of a chlorite rim due to leaching in chert grain bordering pore space (bottom center). Esso Canterra Peco 4-6-47-15 W5. 6996.4 feet (2132.5 m). Plane polarized light. (F) Patchy cluster of calcite cement (cc). Note the corrosive outline of the calcite cement with the kaolinite (k) indicating that kaolinite postdates calcite. Rock fragment (rf), volcanic rock fragment (v). Esso Canterra 10-6-47-15W5. 7033.5 feet (2143.8 m). Crossed polarizers. (G) Kaolinite (k) postdates quartz overgrowths (o). Quartz (q), pore (p). Esso Canterra Peco 10-6-47-15 W5. 6870.0 feet (2094.0 m). Plane polarized light. (H) Corroded grain surfaces due to leaching (upper left). Secondary porosity caused by leaching considerably enhances reservoir quality in the coarse-grained basal channel sandstone beds. Chert (ch), quartz (q), rock fragment (rf), pore (p). Esso Canterra Peco 10-6-47-15 W5. 7034.4 feet (2144.1 m). Plane polarized light.

49

Table 2-2. Point-count porosity data showing the types of porosity in sandstones from the Peco reservoir in Well 2-6-47-16 W5.

KB depth (m) (ft)	Point-count porosities		
	Intragranular (%)	Secondary (%)	Total (%)
2152.05 (7060.5)	1	1	2
2153.24 (7064.4)	–	–	–
2154.65 (7069.1)	9	6	15
2155.84 (7073.0)	TR	–	TR
2156.55 (7075.3)	6	12	18
2156.83 (7076.2)	3	3	6
2157.43 (7078.2)	3	2	5
2158.26 (7080.9)	7	6	13
2158.59 (7082.0)	9	6	15
2159.53 (7085.1)	3	–	3

TR = Trace

sediments (Table 2-2). Feldspar and volcanic rock fragments typically undergo dissolution, creating oversized porosity. However, in lithofacies C, leaching has been more extensive and has produced widespread corrosion of framework quartz and chert grains. The higher original permeabilities preserved in lithofacies C (Table 2-1) probably allowed diagenetic waters to better access and leach this reservoir. Figure 2-18H shows an example of this widespread leaching as exhibited by corroded chert grains.

Production Characteristics

Well and Reservoir Performance

The Peco Basal Belly River "A&N" and "C" pools are currently drilled on 160-acre (64.8-ha.) spacing and produce under primary production. A secondary recovery scheme has not yet been initiated by any operator. The primary drive mechanism is solution gas and gas-cap expansion.

No known active water drive exists in the reservoir. Water has been encountered in the structurally low, downdip western portion of the "C" pool (Fig. 2-8). However, because the reservoir has been isolated through faulting from its westernmost extension, no conduit for water migration to the main reservoir appears to exist.

All wells in the Peco "A&N" and "C" pools have been hydraulically fracture-stimulated to enhance productivity by overcoming formation damage and increasing the permeability height (k_h) of the reservoir. Because Belly River reservoirs are underpressured (0.25 psi/ft; 5.7 kPa/m), they are typically damaged during drilling operations as a result of the invasion of drilling filtrate and drilling mud solids. Core flood studies conducted by Esso have shown that freshwater drilling filtrate destabilizes chlorite and kaolinite clays in the reservoir. The high pressure differential causes these destabilized clays to dislodge and migrate, along with the drilling solids, into the reservoir, resulting in blocked pore throats. The resulting loss of permeability near the wellbore requires fracture stimulation to alleviate this formation damage.

Producing oil wells at Peco can be categorized into two general types: Type 1 exhibits a relatively high initial production rate, 63 to 220 BOPD (10-35 m³/D), that decreases relatively slowly over the life of the well. Type 2 exhibits a rate of up to 125 BOPD (20 m³/D) that declines very rapidly relative to Type 1 and stabilizes at rates of between 12 and 31 BOPD (2-5 m³/D).

The initial higher production rates observed in the Type 1 and Type 2 wells are interpreted to result from fracture stimulation and the associated introduction of artificially high formation pressures adjacent to the wellbore. With production, the artificially high pressures are reduced, and the resultant production rates become a function of the ability of the formation to replenish the produced oil. This ability is greater in Type 1 wells and, as a result, production declines less rapidly than for Type 2 wells.

Reservoir Continuity/Heterogeneity

Productivity of a well is a function not only of the thickness and quality of lithofacies C but also of the distribution or continuity of this lithofacies throughout the pool. For example, both Type 1 and 2 wells may produce from reservoirs exhibiting similar thicknesses of lithofacies C. However, Type 1 wells are generally in the thicker, more continuous net-pay areas of the reservoir (i.e., down the channel

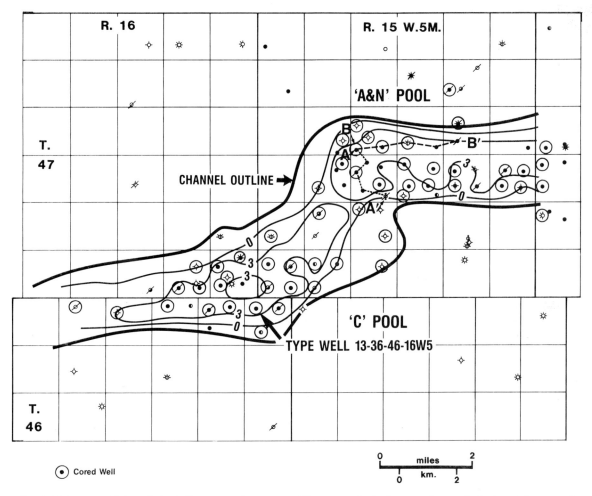

Fig. 2-19. Bottom 16-foot (5.0-m) slice net sandstone map. The net pay in the bottom slice is relatively continuous and has a maximum of 10 feet (3.0 m) of net pay. Contoured in meters.

axis). Type 2 wells are generally in the thinner, less continuous net-pay areas of the reservoir (i.e., in the peripheral portions of the channel). These more peripheral wells are interpreted to drain smaller, relatively less continuous reserves of oil, whereas Type 1 wells are interpreted to drain larger, more continuous reserves.

In order to quantify the reservoir continuity, a modified version of the statistical technique of Delaney and Tsang (1982), Stiles (1976), Stiles and George (1977), and Ghauri et al. (1974) was used in this study. This technique can be used between any two wells where any net pay is considered to be "continuous" if it is stratigraphically correlatable and "discontinuous" if it is not stratigraphically correlatable (Delaney and Tsang, 1982).

The "continuity" between two wells is quantified by the ratio of the summation of all the correlatable (effective) net pay to the total net pay.

The Peco Basal Belly River reservoir was horizontally "sliced" into 16-foot (5.0-m) net-sandstone increments using well logs from the "A&N" and "C" pools (Fig. 2-9). The net pay from each slice was determined using a density-log porosity cutoff of 6%. The thickness of net pay in each of the three slices was plotted on a map and contoured (Figs. 2-19, 2-20, 2-21). The contoured slice maps suggest that the net pay in the basal slice has the greatest continuity and that continuity decreases stratigraphically upward in the channel. The continuity of each slice was determined on two cross sections: one across the channel reservoir and the other parallel to

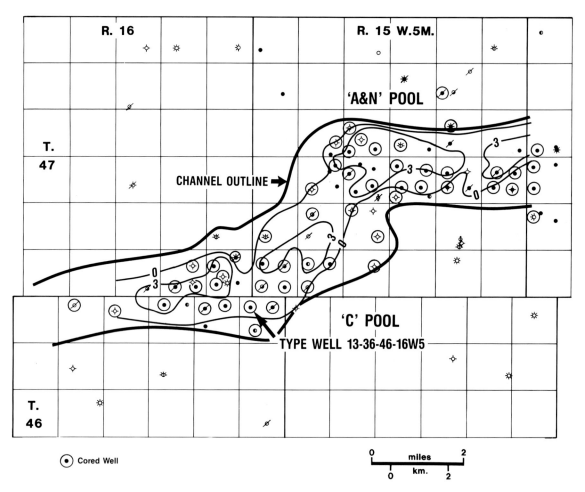

Fig. 2-20. Middle 16-foot (5.0-m) slice net sandstone map. The net pay in the middle slice is less continuous than the basal slice but also has a maximum of 10 feet (3.0 m) of net pay. Contoured in meters.

the channel axis, using the technique of Delaney and Tsang (1982). These continuity values were then plotted graphically versus the interwell distance for the lowermost of the three slices (Fig. 2-22). Because the majority of the net pay is concentrated in the basal slice, no continuity plot was attempted for the middle and upper net-sandstone slices. Figure 2-22 indicates that the wells down the channel axis exhibit good continuity (up to 80%) for more than 9,850 feet (3,000 m). The cross-channel plot shows good continuity for approximately 3,940 feet (1,200 m), which then deteriorates rapidly toward the channel margins. This relationship suggests reservoir continuity is highest down depositional dip parallel to the channel axis and decreases perpendicular to the channel axis. Eighty-three percent of wells with less than 60% continuity were nonpro-

ductive (either shut-in or abandoned). These observations about reservoir continuity suggest that the basal 16-foot (5.0-m) slice is in contact with an areally extensive connected reservoir and that wells with poor continuity are probably in contact with a relatively smaller drainage area within the reservoir.

Exploration And Production Strategy

The Peco pools are the deepest and westernmost Belly River pools in the western Canada basin. Historically, structurally trapped Belly River targets in this part of the basin have not been considered as a primary exploration objective. Assuming abundant mature source rock and thrust-fault traps

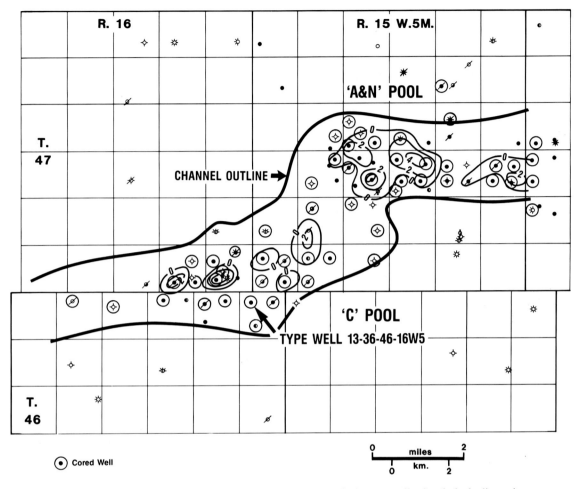

Fig. 2-21. Upper 16-foot (5.0-m) slice net sandstone map. The net pay in the upper slice is relatively discontinuous compared to the bottom and middle slices. Contoured in meters.

parallel to the foothills, the potential for many more basal Belly River oil discoveries in western Alberta is vast.

Two critical factors are necessary to optimize exploration success for these reservoirs. Seismic identification of a suitable structural trap configuration is the first requirement. To complement this seismic data, structure or residual structure maps from well log data are often valuable. Second, mapping the presence of basal Belly River channel sandstones requires the examination of many well logs, mostly from old wells drilled for much deeper targets. Often only one or two widely spaced wells, logged using outdated technology, are the only data available to map potential channel trends in the sparsely drilled parts of the basin.

Production efficiency depends on the relationships between grain size, porosity, and permeability,

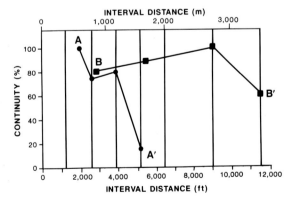

Fig. 2-22. Plot illustrating the continuity of the net sandstone in the basal slice of the "A&N" pool in two directions: across the channel (A-A') and down the channel axis (B-B'). Continuity is higher in a direction down the channel axis. Refer to Figure 2-19 for the location of A-A' and B-B'.

as well as the role of burial depth and diagenesis is modifying or overprinting these primary rock properties is needed. For example, in two siliciclastic reservoirs which have the same porosity, the reservoir with the coarser grain size will have higher permeabilities, if all other factors are equal in water-wet reservoirs. Production optimization in braided fluvial channel reservoir also depends on understanding the vertical and lateral continuity of the net pay. Many times, however, these critical factors are often never fully understood until the reservoir development is mature with closely spaced wells, and the reservoir is largely depleted. It is hoped that analog studies like this one can provide data for more efficient reservoir development at an early stage.

Conclusions

The Peco "A&N" and "C" pools occur in a faulted and folded braided fluvial channel deposit of the Upper Cretaceous (Campanian) Belly River Formation. Production is primarily from the lowermost (basal) Belly River fluvial channel-fill sandstones. Minor production occurs from stratigraphically higher channel sandstones. Cumulative recoverable reserves for the two pools are 3.4 MMBO (5.4×10^5 m³) and 11 BCF (3.1×10^8 m³) of gas.

The productive reservoir is a sandy conglomerate to pebbly sandstone (lithofacies C). Associated fine- and medium-grained sandstones are nonreservoir because of the diagenetic effects of compaction and

cementation. Authigenic chlorite is the main permeability-reducing component and is followed in decreasing order of importance by kaolinite, quartz overgrowths, and calcite. Significant leaching of lithofacies C has generated some secondary porosity.

The thickness and continuity of the lithofacies C (net pay sandstone) play a key role in determining both the producibility and the expected primary recovery of oil from each well. Lithofacies C exhibits the highest lateral continuity in the basal portions of the reservoir in wells located near the center of the channel. These wells typically exhibit exponential decline and recovery factors of 8% to 10%. Wells located away from the center of the channel, where net pay in lithofacies C is less continuous, typically exhibit harmonic decline and recovery factors of 10% to 12%.

Acknowledgments. Permission to publish from Esso Resources Canada Limited is gratefully acknowledged. We thank F. H. Lane, W. D. Evans, J. T. Code, and C. G. Zinkan for their encouragement and helpful comments in preparing the manuscript. Petrographic information was obtained from various Esso technical reports authored by H. W. Nelson, A. Cochran, and R. S. Dean. Also from Esso, we thank K. H. Mueller for contributing to the understanding of the reservoir engineering and production characteristics, D. Taylor for typing the manuscript, and W. Ritco for drafting the figures. Many valuable improvements to the manuscript were suggested by three reviewers.

Reservoir Summary

Field: Peco "A&N" and "C" Pools
Location: West-central Alberta
Operators: Amoco, Esso, Pan Continental, PetroCanada, Sceptre, Texaco
Discovery: 1964, 1983
Basin: Western Interior (Alberta) Basin
Tectonic/Regional Paleosetting: Marginal to a shallow epicontinental seaway (foreland basin)
Geological Structure: Thrust faults, normal faults, and associated anticlines
Trap Type: Stratigraphic pinch-out on anticlinal and fault closures.
Reservoir Drive Mechanism: Solution gas
  • **Original Reservoir Pressure:** 1,763 to 1,874 psia (12.2-12.9 × 10³ kPa) at approximately 3,000 feet (915 m) subsea
Reservoir Rocks:
  • **Age:** Late Cretaceous (Campanian)
  • **Stratigraphic Unit:** Belly River Formation (basal portion)
  • **Lithology:** Sandy conglomerate to pebbly sandstone, fining-upward to medium- to fine-grained sandstone to silty shale
  • **Depositional Environment:** Braided river (Donjek type)

- **Productive Facies:** Sandy conglomerate to pebbly sandstone
- **Petrophysics:**
  - **Porosity Type:** Intergranular
  - **ϕ:** Average 10%, range 6 to 13%, cutoff 8.5% (cores), 6.5% (density logs)
  - **k:** Average 5 md, range 1 to 150 md, cutoff 1 md (cores)
  - **S_w:** Average 43%, range 35 to 50%, cutoff 50% (logs)

Reservoir Dimensions
- **Depth:** 7,200 feet (2,195 m)
- **Areal Dimensions:** 1.5 to 3.0 miles (2.4-4.8 km) wide by 12 miles (19 km) long
- **Productive Area:** 4,630 acres (1,875 ha.)
- **Number of Pay Zones:** Two; only the basal, main pay zone presented in this chapter
- **Hydrocarbon Column Height:** 1,791 feet (546 m)
- **Fluid Contacts:** Gas/oil at 2,870 feet (875 m) subsea; oil/water at 3,200 feet (975 m) subsea for "C" Pool
- **Gross Sandstone Thickness:** 60 feet (18 m)
- **Net Sandstone Thickness:** 18 feet (5.5 m)
- **Net/Gross:** 0.30

Source Rocks
- **Lithology & Stratigraphic Unit:** Organic-rich marine shales, Second White Speckled Shale Formation, Colorado Group, Late Cretaceous (Cenomanian-Turonian)
- **Time of Hydrocarbon Maturation:** Unknown
- **Time of Trap Formation:** Unknown
- **Time of Migration:** Unknown

Hydrocarbons
- **Type:** Oil with gas cap
- **GOR:** 450 to 560 SCF/bbl (80-99 m³/m³)
- **API Gravity:** 42°
- **FVF:** 0.78
- **Viscosity:** 2.14 cP (2.1 × 10³ Pa·s) at 86°F (30°C)

Volumetrics
- **In-Place:** 34 MMBO (5.4 × 10⁶ m³)
- **Cumulative Production:** 1.7 MMBO (2.7 × 10⁵ m³)
- **Ultimate Recovery:** 3.4 MMBO (5.4 × 10⁵ m³)
- **Recovery Efficiency:** 10%

Wells
- **Spacing:** 1,320 feet (400 m), 160 acres (64.8 ha.)
- **Total:** 61
- **Dry Holes:** 7

Typical Well Production
- **Average Daily:** 60 to 90 bbl (9.5-14.3 m³)
- **Cumulative:** 200 to 300 MBO (3.2-4.8 × m⁴)

Stimulation: Various types of frac (polyemulsion, gelled diesel) using 44,000 to 88,000 pounds (2.0-4.0 × 10⁴ kg) of sand

References

Allan, J., and Creaney, S., 1988, Sequence stratigraphic control of source rocks: Viking-Belly River System [abst.]: Proceedings Canadian Society of Petroleum Geologists Technical Meeting, Sequences, Stratigraphy, Sedimentology: Surface and Subsurface, p. 575.

Ayalon, A., and Longstaffe, F.J., 1988, Oxygen isotope studies of diagenesis and pore-water evolution in the western Canada sedimentary basin: Evidence from the Upper Cretaceous basal Belly River sandstone, Alberta: Journal of Sedimentary Petrology, v. 58, p. 489–505.

Cant, D.J., 1978, Development of a facies model for sandy braided river sedimentation: Comparison of the South Saskatchewan River and the Battery Point Formation, in Miall, A.D., ed., Fluvial Sedimentology: Canadian Society of Petroleum Geology Memoir 5, p. 627–639.

Collinson, J.D., 1970, Bed forms of the Tana River, Norway: Geografiska Annaler, v. 52-A, p. 31–56.

Delaney, R.P., and Tsang, P.B., 1982, Computer reservoir continuity study at Judy Creek: Journal of Canadian Petroleum Technology, Jan.–Feb., p. 38–44.

Etchecopar, A., 1974, Simulation par ordinateur de la déformation progressive d'un agrégat polycristalline: Faculté des Sciences de Nantes, Nantes, France, 135 p.

Folk, R.L., Andrews, P.B., and Lewis, D.W., 1970, Detrital sedimentary rock classification and nomenclature for use in New Zealand: New Zealand Journal of Geology and Geophysics, v. 13, p. 937–968.

Ghauri, W.K., Osborne, A.F., and Magnuson, W.L., 1974, Changing concepts in carbonate waterflooding, west Texas Denver Unit project—An illustrative example: Journal of Petroleum Technology, June, p. 595–606.

Gleddie, J., 1949, Upper Cretaceous in western Peace River plains, Alberta: American Association of Petroleum Geologists Bulletin, v. 33, p. 511–532.

Iwuagwa, C.J., and Leberkmo, J.F., 1981, The role of authigenic clays in some reservoir characteristics of the basal Belly River sandstone, Pembina Field, Alberta: Bulletin of Canadian Petroleum Geology, v. 29, p. 1–62.

McCrory, V.L.C., and Walker, R.G., 1986, A storm and tidally influenced prograding shoreline—Upper Cretaceous Milk River Formation of southern Alberta, Canada: Sedimentology, v. 33, p. 47–60.

McLean, J.R., 1971, Stratigraphy of the Upper Cretaceous Judith River Formation in the Canadian Great Plains: Saskatchewan Research Council, Geology Division, Report 11, 96 p.

Miall, A.D., 1977, A review of the braided river depositional environment: Earth-Science Reviews, v. 13, p. 1–62.

Miall, A.D., 1978, Lithofacies types and vertical profile models in braided river deposits: A summary, in Fluvial Sedimentology, A. D. Miall, ed.: Canadian Society of Petroleum Geologists, Memoir 5, p. 597–625.

Putnam, P., 1989, The Belly River Formation of west-central Alberta: Anatomy of an emerging deep basin oil play [abst.]: American Association of Petroleum Geologists Bulletin, v. 73, p. 402.

Rosenthal, L.R.P., and Walker, R.G., 1987, Lateral and vertical facies sequences in the Upper Cretaceous Chungo Member, Wapiabi Formation, southern Alberta: Canadian Journal of Earth Sciences, v. 24, p. 771–783.

Shouldice, J.R., 1979, Nature and potential of Belly River gas sand traps and reservoirs in western Canada: Bulletin of Canadian Petroleum Geology, v. 27, p. 229–241.

Smith, N.D., 1970, The braided stream depositional environment: Comparison of the Platte River with some Silurian clastic rocks, north-central Appalachians: Geological Society of America Bulletin, v. 81, p.2993–3014.

Smith, N.D., 1971, Transverse bars and braiding in the lower Platte River, Nebraska: Geological Society of America Bulletin, v. 82, p. 3407–3420.

Smith, N.D., 1972, Some sedimentological aspects of planar cross-stratification in a sandy braided river: Journal of Sedimentary Petrology, v. 42, p. 624–634.

Smith, N.D., 1974, Sedimentology and bar formation in the upper Kicking Horse River, a braided outwash stream: Journal of Geology, v. 82, p. 205–223.

Stiles, L.H., 1976, Optimizing waterflood recovery in a mature waterflood—The Fullerton Clearfork Unit: Society of Petroleum Engineers Paper 6198, p. 1–9.

Stiles, L.H., and George, C.J., 1977, Improved techniques for evaluating carbonate waterfloods in west Texas: Society of Petroleum Engineers Paper 6739, p. 1–16.

Stott, D.F., 1959, The Cretaceous Alberta Group and equivalent rocks, Rocky Mountain Foothills, Alberta: Geological Survey of Canada Memoir 317, 306 p.

Williams, P.F., and Rust, B.R., 1969, The sedimentology of a braided river: Journal of Sedimentary Petrology, v. 39, p. 649–679.

Williams, G.D., and Stelk, C.R., 1975, Speculation in the Cretaceous Paleogeography of North America, in Caldwell, W.G.E., ed., The Cretaceous System in the Western Interior of North America: Canadian Society of Petroleum Geologists Special Paper 13, p. 1–20.

Key Words

Peco Field, Alberta, Canada, Western Interior (Alberta) Foreland basin, Belly River Formation, Cretaceous, Campanian, braided river, dipmeters, diagenesis, chlorite, formation damage, compaction, reservoir continuity.

3

Coarse-Grained Meander-Belt Reservoirs, Rocky Ridge Field, North Dakota

John O. Hastings, Jr.

Shell Western E&P Inc., Houston, Texas 77001

Introduction

Fluvial sandstones are among the most extensively studied and documented terrigenous clastic sequences. Numerous depositional models have been developed to generalize fluvial processes and sedimentary structures, and countless papers describe various aspects of modern fluvial environments. However, few studies have integrated the characteristics of fluvial sandstone reservoirs at all scales, and none have done so for coarse-grained meander-belt systems.

The Rocky Ridge area in North Dakota has been extensively explored and produced. Thus, the area is ideal for documenting reservoir characteristics at all scales (see also Barwis, this volume). This chapter presents an integrated evaluation of the depositional controls and reservoir characteristics of a coarse-grained meander-belt reservoir system from a seismic to a pore scale.

Field Location

Rocky Ridge Field is located in southwestern North Dakota along the southern flank of the intracratonic Williston basin (Fig. 3-1) and produces high-gravity oil from Lower Pennsylvanian sandstones. The field lies within the Rocky Ridge, Medora-Dickinson area, which has yielded about 60 MMBO (9.6×10^6 m³) from the nonmarine and marginal marine Tyler Formation. Total proven recoverable Tyler reserves for the Williston basin are about 170 MMBO (2.7×10^7 m³).

Production History

Rocky Ridge Field (Main Pool) was discovered in 1957 through serendipity when the Northern Pump-Lucy Fritz No. 1 well encountered oil-bearing sandstones in the Tyler Formation (Fig. 3-2). The prospect was a seismically defined structural test for a deeper objective. That objective was tested as water-productive, but the uphole Tyler Formation was tested at an initial potential of 1,224 BOPD (195 m³/D). Subsequent drilling has resulted in nine other producing wells and 19 dry holes.

Another pool (Southeast Pool) a few miles to the southeast was discovered in 1969 when the Shell State No. 41-36 well encountered oil-bearing Tyler sandstones on a seismically defined stratigraphic prospect. Subsequent drilling of this pool resulted in seven other oil wells and 15 dry holes.

Until 1983, Shell Oil Company was the major operator of the field. Other operators included Pan American (Amoco), Northern Pump, and Mule Creek Oil. Apache Corporation is currently the major operator. The field has a conspicuously linear northwest-southeast trend and is 8 miles (13 km) long and 1 mile (1.6 km) wide. Production is from a complex network of coarse-grained meander-belt channel sandstones of the Tyler Formation. Cumulative production is 5.04 MMBO (8.0×10^5 m³)

Casebooks in Earth Science
Sandstone Petroleum Reservoirs
Eds.: Barwis/McPherson/Studlick

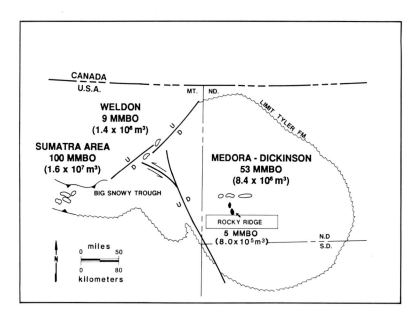

Fig. 3-1. Index map of the Williston basin showing Rocky Ridge Field and other key areas with Tyler production. The map shows the erosional limit of the Tyler Formation and Big Snowy trough, which connected the basin with the ocean to the west. The large northwest-trending fault is the Cedar Creek fault.

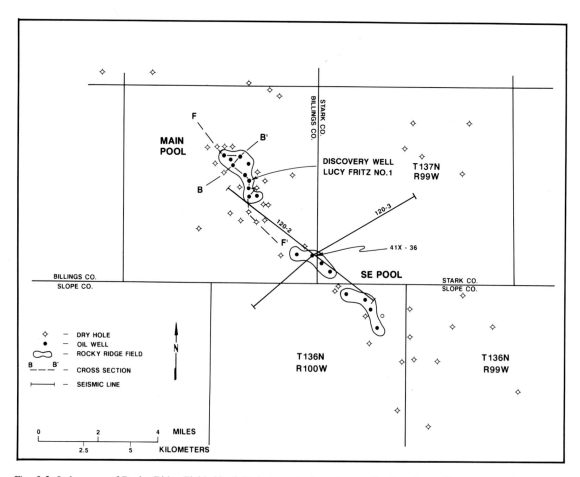

Fig. 3-2. Index map of Rocky Ridge Field, North Dakota. Wells shown are all Tyler Formation penetrations in the area. The map shows field outlines and discovery wells for both pools as well as locations of key cross sections and two seismic lines (120-2 and 120-3). 120-2, F-F', and B-B' are the sections of Figures 3-10, 3-12, and 3-15, respectively.

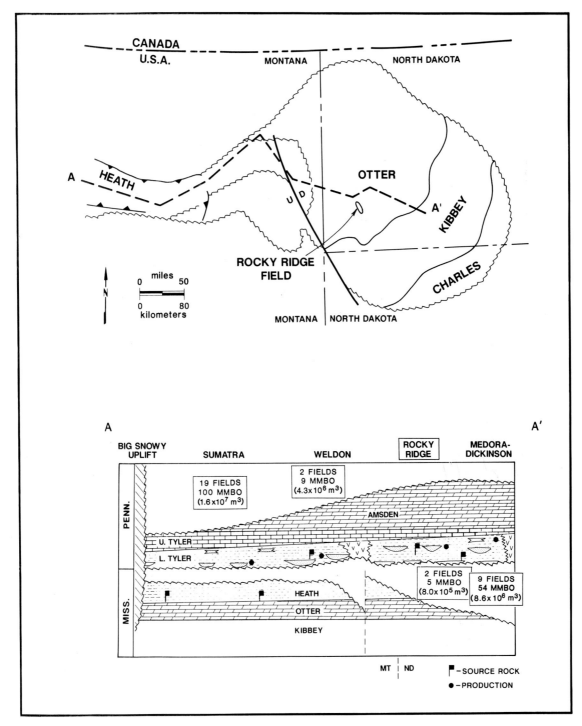

Fig. 3-3. Tyler subcrop map and regional time-stratigraphic section (A-A') across the Williston basin. The figure shows the formations below the erosional unconformity at the base of the Tyler Formation and illustrates the stratigraphic relationships of the Tyler Formation and surrounding units. Note the stratigraphic position of production and major source rocks. Datum for the section is top Tyler Formation.

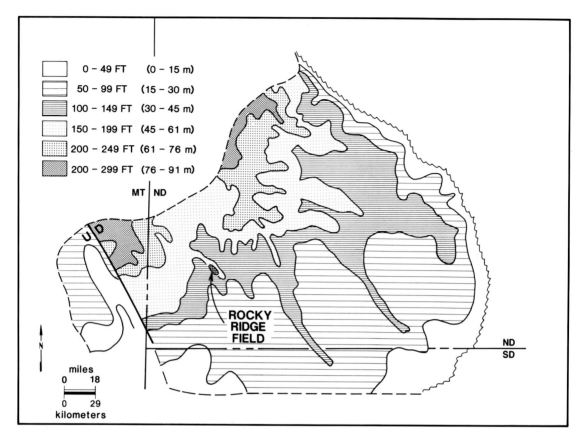

Fig. 3-4. Regional Tyler interval-isopach map, southern Williston basin, based on approximately 1,200 wells; data from Barwis, this volume. The isopach thicks represent paleotopographic lows and demonstrate a radially coalescing drainage pattern on the Otter surface. Note the location of Rocky Ridge Field in the Tyler isopach thick.

and 2.12 MMBW (3.4×10^5 m³). Estimated ultimate recovery with current waterflood technology is about 5.1 MMBO (8.1×10^5 m³) with the field production presently in steep decline.

Regional Geology

Geologic Setting

Paleozoic rocks in the Williston basin are mostly carbonates with the Tyler Formation being one of the few clastic sequences. Throughout the Carboniferous, the Williston basin was a very shallow, intracratonic basin which bordered the Canadian Shield to the northeast and opened westward to the Cordilleran Miogeosyncline through the narrow Big Snowy trough (Maughan, 1984).

Important orogenies, including the Alleghenian, Ouachita, Arbuckle, and Ancestral Rocky Mountain events, occurred during latest Mississippian time. The plate collisions responsible for these structural events (Kluth and Coney, 1981) resulted in cratonic uplift and subaerial exposure of the Williston basin (Gerhard et al., 1982). Mississippian carbonates and clastics were deeply eroded during this time, particularly in the southeastern part of the basin (Fig. 3-3). Regional Tyler interval isopach data demonstrate that, during this relative lowstand, Mississippian strata were eroded by an incised dendritic drainage system which coalesced radially to feed deltas in the Big Snowy trough in eastern Montana (Fig. 3-4). Sandstone deposition was focused into these incised valleys during most of Tyler Formation deposition.

Regional Stratigraphy

Depositional sequences of the Tyler Formation reflect the influences of Mississippian tectonics and relative sea-level changes. The Tyler Formation lies

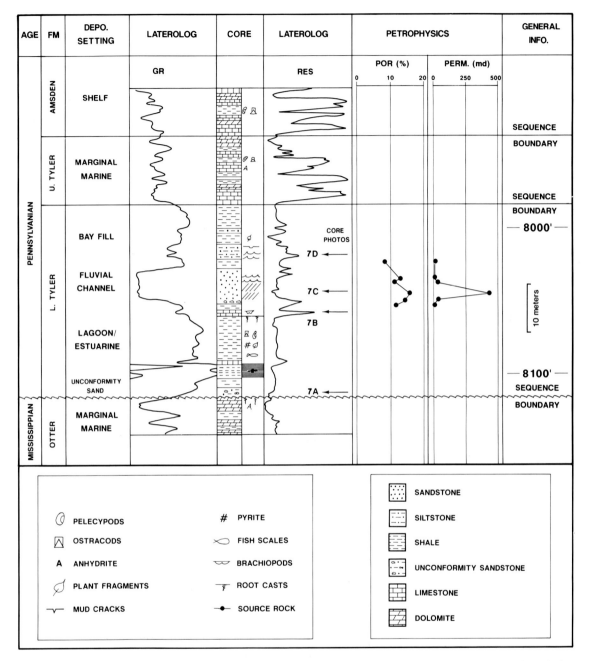

Fig. 3-5. Type log showing the principal lithofacies of the Tyler Formation and surrounding units in Rocky Ridge Field and the representative petrophysical properties of the reservoir sandstone. Note the sequence boundaries and locations of core photos. The well used is the Shell State No. 41-36.

with an angular unconformity above an erosional surface which developed on Mississippian marine sediments. This surface is a major sequence boundary throughout the Williston basin. Tyler Formation depositional environments grade upward from nonmarine to marginal marine to shelf, indicating an overall relative rise in sea level. Facies transitions are relatively sharp on well logs and are regionally correlatable. Fluvial, deltaic, and barrier-island sandstones are developed in the Lower Tyler section.

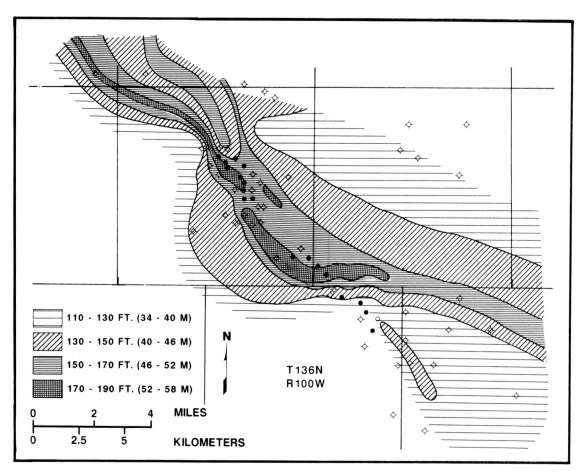

Fig. 3-6. Tyler interval-isopach map, Rocky Ridge Field area. Isopach thicks represent paleotopographic lows on the Otter surface. Note the isopach thinning in the Southeast Pool area and the thick trend toward the east-southeast.

Local Stratigraphy

The Tyler Formation in the Rocky Ridge area unconformably overlies marine shales of the Upper Mississippian Otter Formation and is conformably overlain by shelf carbonates of the Lower Pennsylvanian Amsden Formation (Fig. 3-5).

The Tyler Formation can be divided into two distinct stratigraphic intervals. The lower unit is highly variable in thickness and lithology, primarily due to paleotopography below the basal contact (i.e., the Otter surface). The upper unit is more uniform in thickness and lithology, suggesting a more stable and aerially extensive depositional setting. Thick-

▷

Fig. 3-7. Core photographs representative of the principal lithofacies in the Lower Tyler Formation. The core is from the Shell State No. 41-36. See the type log (Fig. 3-5) for photograph locations. Scale bar is 1 inch (2.5 cm). (A) Unconformity sandstone. The sandstone is composed of large exotic Mississippian and older clasts (arrows) derived from highlands to the southeast during the Mississippian erosional event. (B) Transgressive estuarine shale.

The shale represents the first marine incursion after the Mississippian erosional event. Abundant freshwater ostracods (arrows) indicate an estuarine environment. (C) Fluvial channel sandstone near a channel base. Oil staining is evident from this sample. (D) Bay fill or levee siltstone. The strata are micrograded with each bedset representing one flooding or tidal event. Reddish-brown coloration is due to siderite (si) staining.

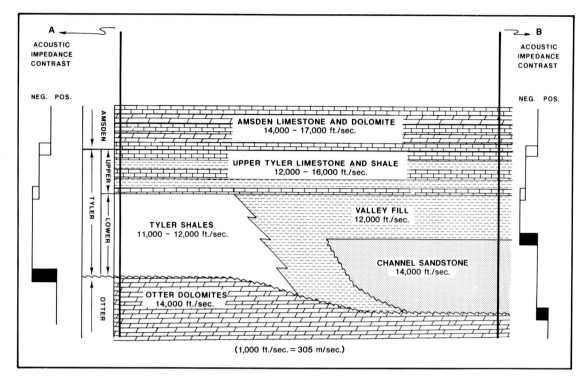

Fig. 3-8. Geophysical model showing the lateral and vertical velocity relationships and acoustic impedance contrasts of the Tyler section in the Rocky Ridge Field area. Note the 2,000 feet/second (610 m/s) vertical and lateral velocity contrast between the Tyler valley fill and the channel sandstone, and the additional positive impedance horizon on top of the Tyler sandstone body.

nesses in the Lower Tyler Formation range from 80 to 190 feet (24-60 m). The interval isopach map of the Tyler Formation suggests that as much as 70 feet (21 m) of topographic relief existed before Tyler deposition (assuming that the uppermost Tyler transgressive shelf was a relatively flat surface) (Fig. 3-6).

A thin unconformity-draping sandstone representing the Mississippian erosional event is present in the lowermost Tyler Formation (Fig. 3-7A). Following this erosional event, Otter paleotopographic lows were flooded by a sea transgressing from the present-day northwest. River valleys became estuaries and received silts and mudstones containing brackish to freshwater fauna (Grenda, 1977) (Fig. 3-7B). Marsh environments lined the estuaries, as evidenced by abundant grassy plant fragments, rooted zones, and occasional very thin lignites. Supratidal flats and subaerial exposure surfaces also existed lateral to the estuaries as indicated by vari-colored, anhydritic shales and various weathered lithologies observed in cores.

The transgressive event was followed by a relative lowstand in which the estuaries were reactivated as fluvial channels, supplied by sediment from the southeast. The system developed as a complex network of four channel-fill sequences and deposited the sandstones that now form the reservoir in Rocky Ridge Field (Fig. 3-7C). Levee, overbank swamp, and brackish bay-fill sediments were deposited adjacent to the channels (Fig. 3-7D). These channels were also the likely feeders for the transgressive sandstones in the Medora-Dickinson barrier-island trend downdip to the north (see Barwis, this volume).

The Upper Tyler Formation is 50 to 60 feet (15-18 m) thick and consists of dark gray to black, dense lime mudstones and wackestones interbedded with fissile black shales. These carbonates and shales are interpreted as marginal-marine deposits and represent a transgression from the northwest (Grenda, 1977).

Reservoir Characteristics

Seismic Attributes

The technique of using seismic methods to detect high-velocity sedimentary units (e.g., channel-fill sandstone reservoirs) in the subsurface requires

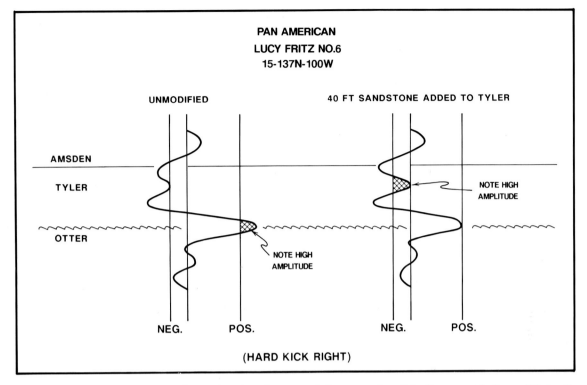

Fig. 3-9. Synthetic seismogram of Pan American-Lucy Fritz No. 6, Main Pool, Rocky Ridge Field. Note the increase in amplitude character of the intra-Tyler section and the attenuation of the Otter event in the modified well synthetic (40-foot (12-m) thick sandstone added).

significant lateral velocity contrasts. The contrasts should be about 2,000 feet per second (610 m/s) or greater, assuming a minimum vertical interval of 50 feet (15 m) (more recently acquired or reprocessed seismic data with higher frequency content can detect units 20 to 30 feet (6-9 m) thick assuming similar velocity contrasts). Lateral velocity changes of this magnitude occur within the lower unit of the Tyler Formation at Rocky Ridge Field and are the result of meander-belt facies architecture.

Tyler channel-fill sandstones, with relatively high velocities (14,000 ft/s, 4,270 m/s) and gross sandstone thicknesses greater than 50 feet (15 m), pass abruptly laterally to mudstones with significantly lower velocities (11,000-12,000 ft/s, 3,350-3,660 m/s) (Fig. 3-8). The effect of this lithofacies change is a noticeable change in the seismic character of the Tyler section. The change is illustrated on synthetic seismograms (Fig. 3-9). The modified synthetic seismogram with an additional 40 feet (12 m) of sandstone shows a higher positive amplitude response within the Tyler section and a lower positive amplitude response at the top of the Otter Formation. Thus, a seismic wavelet passing through the Tyler section containing a thick sandstone body (as opposed to no sandstone) will encounter the additional shale-sandstone interface which is characterized by a large positive acoustic impedance contrast. This affects the entire impedance scale of the Tyler section by (1) strengthening the positive amplitude event within the Tyler, and (2) attenuating the positive amplitude event at the top of the Otter. A dip-oriented seismic line across Rocky Ridge Field illustrates this seismic model (Fig. 3-10). It shows higher amplitudes within the Tyler sequence where channel sandstones are present and attenuation of the Otter amplitude event below channel sandstones.

Geometry

Four major channel sequences are developed within the Tyler Formation in Rocky Ridge Field. A gross-sandstone isopach map of all four channel-fill sandstones shows a dip-oriented linear morphology with all four channel sequences merging at the Main Pool (Fig. 3-11). The area of thickest sandstone coincides with the most productive part of the field. Furthermore, there is close correlation between the gross-

Fig. 3-10. Dip-oriented CDP seismic line (120-2 from Fig. 3-2) across Rocky Ridge Field. The line shows the high amplitudes within the Tyler sequence where fluvial channel sandstones are present. Also note the attenuation of the Otter event below the documented channel sandstones, especially at the Bell No. 23-31 well. The line was shot in 1970 and reprocessed in 1980.

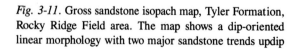

Fig. 3-11. Gross sandstone isopach map, Tyler Formation, Rocky Ridge Field area. The map shows a dip-oriented linear morphology with two major sandstone trends updip (southeast). The two trends merge at the Main Pool, where the isopach is the thickest. The northernmost trend updip has been inferred from the interval-isopach data (Fig. 3-6).

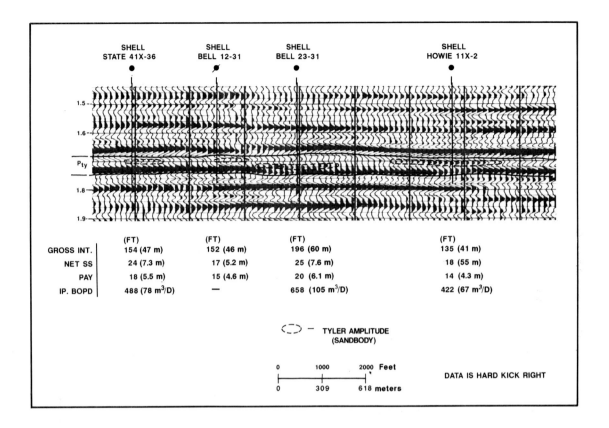

	(FT)	(FT)	(FT)	(FT)
GROSS INT.	154 (47 m)	152 (46 m)	196 (60 m)	135 (41 m)
NET SS	24 (7.3 m)	17 (5.2 m)	25 (7.6 m)	18 (55 m)
PAY	18 (5.5 m)	15 (4.6 m)	20 (6.1 m)	14 (4.3 m)
IP. BOPD	488 (78 m³/D)	—	658 (105 m³/D)	422 (67 m³/D)

$\langle\;\supset\;\rangle$ — TYLER AMPLITUDE
(SANDBODY)

```
0        1000       2000  Feet
|----------|----------|
0         309        618  meters
```

DATA IS HARD KICK RIGHT

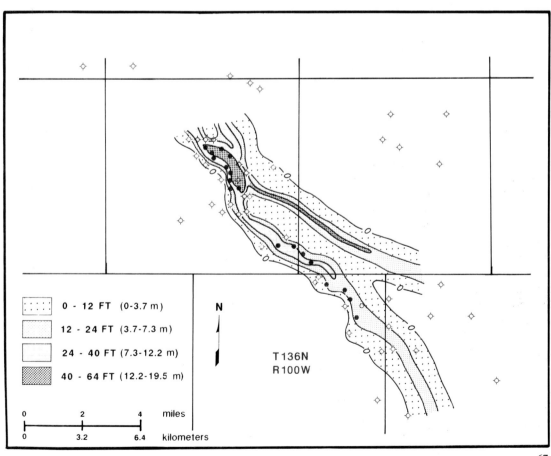

∷∷∷	0 - 12 FT (0-3.7 m)
▨	12 - 24 FT (3.7-7.3 m)
☐	24 - 40 FT (7.3-12.2 m)
▦	40 - 64 FT (12.2-19.5 m)

N

T 136N
R 100W

```
0         2          4   miles
|----------|----------|
0        3.2        6.4  kilometers
```

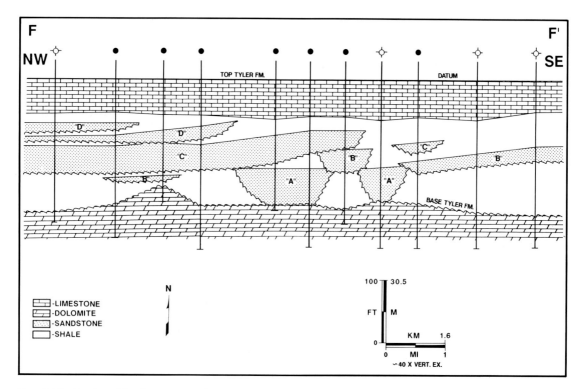

Fig. 3-12. Stratigraphic dip cross section (F-F′) across Rocky Ridge Field, Main Pool. The section shows the four major fluvial channels which are developed in the field area and their stratigraphic relationships. Note the offlapping, en echelon geometry of the channels. The chan- nels "disappear" along the section because several wells shown are outside the various channel boundaries. The datum used is the top of the Tyler Formation. See Figure 3-2 for the section location.

interval and the gross-sandstone isopach thicks in the Tyler Formation; thick Tyler sections are coincident with or proximal to Tyler channel-fill sandstones. Since thickest sections of the Tyler Formation represent areas of greatest pre-Tyler erosional relief, the Otter Formation paleotopography controlled deposition of Tyler channel-fill sandstone sequences by controlling drainage within the paleovalley. This concept is important to exploration for Tyler sandstone bodies.

The four channel-fill sandstones are variable in size and stratigraphic position within the Lower Tyler Formation. They are also highly discontinuous along strike. A dip-oriented stratigraphic cross section (i.e., parallel to the valley axis) across the field illustrates the timing relationships of all four channel sandstones as well as their en echelon stacking pattern (Fig. 3-12). A series of gross-sandstone isopach maps of the channel-fill sequences shows the geometry and trend of each sandstone (Fig. 3-13). These units are characterized as "channels" A through D:

Channel A is stratigraphically lowest and attains thicknesses of more than 60 feet (18 m). Channel A sandstone is present in only five wells and is water-bearing (Fig. 3-13A).

Channel B is stratigraphically higher and forms the reservoir for the Southeast Pool. The Channel B sandstone is also productive in the Main Pool but is largely below the oil/water contact. This sandstone has a maximum thickness of more than 30 feet (9 m) (Fig. 3-13B).

Channel C is the primary producing sandstone in the Main Pool, with gross sandstone thicknesses of more than 30 feet (9 m) (Fig. 3-13C).

Channel D is the highest stratigraphically and consists of two composite sequences, each of which ranges from 5 to 15 feet (1.5-4.5 m) thick. Channel D sandstone is productive in the Main Pool (Fig. 3-13D).

Although these sequences are mappable as four different units, they are probably the depositional

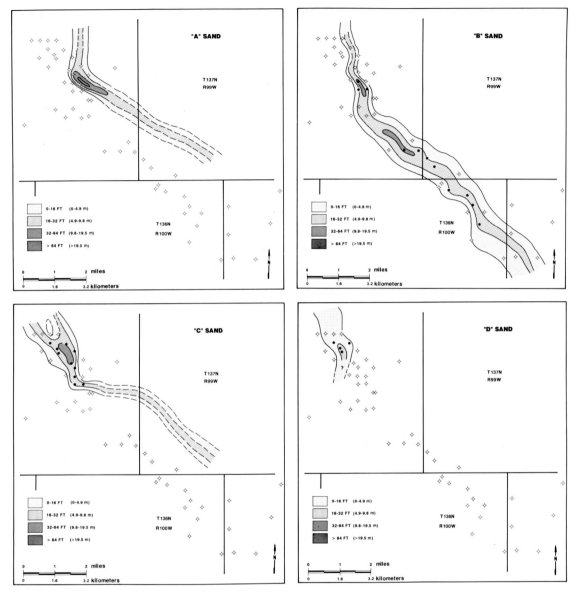

Fig. 3-13. Gross sandstone isopach maps and production of the four individual fluvial channel sandstones in Rocky Ridge Field. Note that the production is confined almost exclusively to the channel axes. All maps are at the same scale and contour interval. (A) A Sandstone. Stratigraphically lowest and water-bearing and probably well-developed east of Rocky Ridge Field. (B) B Sandstone. Reservoir in Southeast Pool but largely below the oil/water contact in Main Pool. (C) C Sandstone. Primary reservoir in Main Pool and also probably developed east of Rocky Ridge Field. (D) D Sandstone. Stratigraphically highest and productive in Main Pool.

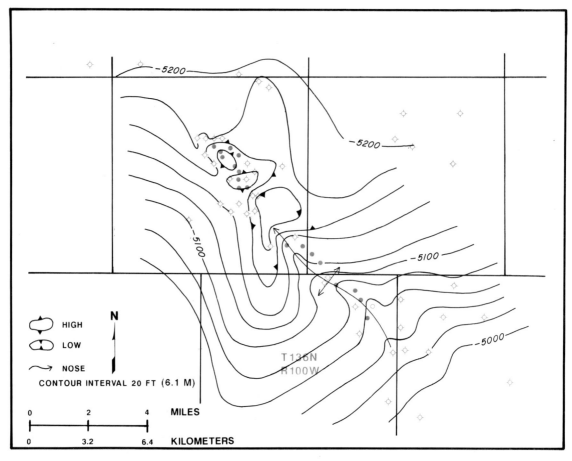

Fig. 3-14. Structure map on the top of the Tyler Formation, Rocky Ridge Field area. The map shows a dome in the Main Pool and a northwest-plunging nose in the Southeast Pool. The structure shown is representative of that in the lower Tyler Formation.

result of avulsion processes in one river system. Production from these channel-fill sandstones is almost exclusively confined to the thickest part of the channel.

Trap and Hydrocarbon Source

The trapping mechanism at Rocky Ridge Field combines both structural and stratigraphic elements. The structure at the top of the Tyler Formation is characterized by a dome with about 30 feet (9 m) of closure at the Main Pool and a northwest-plunging nose at the Southeast Pool (Fig. 3-14). Channel sandstone discontinuity limits the trap along strike (Fig. 3-15). High concentrations of anhydrite and clays in reservoir sandstones in updip wells in the Southeast Pool suggest a diagenetic trapping component. The reservoirs are effectively sealed above and laterally by surrounding estuarine and marine mudstone and shale. Trap analysis indicates that the Main Pool structure is filled with oil to the structural spill point.

The structure of the field is attributable mainly to differential compaction, because sand compacts about one third less than mud, areas with thick sand accumulations compact less than sand-poor areas, and the sand-rich areas remain structurally higher. This compactional phenomenon is illustrated by an overlay of the gross-sandstone isopach map and the top Tyler structure map (Fig. 3-16). The thickest gross-sandstone area is present at the apex of the dome. Also, the nose at the Southeast Pool essentially follows the axis of the gross-sandstone map (channel B) in the area.

Deep-seated faulting may also have affected the structure of the Tyler Formation. Several seismic

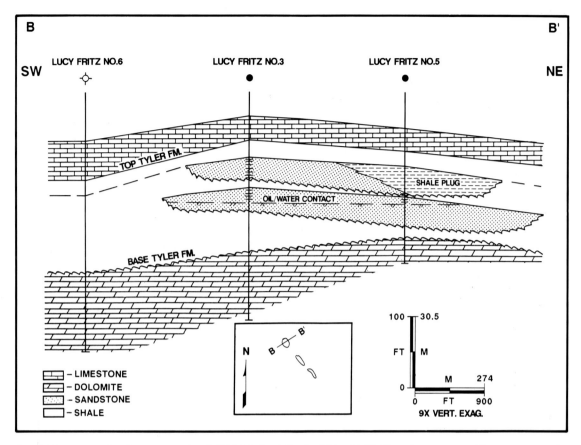

Fig. 3-15. Structural strike cross section B-B', Main Pool, Rocky Ridge Field. Note the common oil/water contact, the extreme discontinuity of the channel sandstones, and the effective vertical and lateral seals.

lines show similar or greater structural relief on deeper horizons. These faults probably influenced Tyler drainage patterns (Thomas, 1974; Brown, 1978; Sturm, 1983) by creating fracture zones which were relatively more erodible during the Mississippian lowstand.

Hydrocarbons in the Tyler fluvial channel sandstones were sourced from within the formation. Dark gray to black, marginal-marine to estuarine shale and limestone in the Upper and Lower Tyler Formation were the major source of Tyler oil, as evidenced by geochemical matching of oils and source rocks (unpublished Shell Oil Company proprietary data). These rocks are found both above and below the reservoir sandstones. Pyrolysis and fluorescence analyses of these potential source rocks as well as visual kerogen analysis establish that they are organically rich and thermally mature. Burial and temperature history modeling show that the source rocks began to expel oil in the Oligocene (J. H. Barwis, personal communication, 1988).

Depositional Setting

A block diagram summarizing the depositional setting and lithofacies associations within the Lower Tyler Formation in the Rocky Ridge area is shown in Figure 3-17. The channel-fill sandstones are interpreted to be fluvial on the combined evidence from wireline logs and cores which display (1) basal erosional surfaces with pebble-lag deposits and mudstone intraclasts, (2) upward-fining sequences, commonly capped by lignites and mud-cracked exposure surfaces, (3) lack of bioturbation, and (4) freshwater ostracods. A fluvial interpretation is further evidenced by the dip-oriented, linear morphology of the sandstone bodies and by their paleotopographic position updip from a documented Tyler shoreline (see Barwis, this volume). The fluvial system in the area appears to have had two major tributaries which merged at the Main Pool in Rocky Ridge Field. Gross-interval and sandstone isopach trends delineate these tributaries.

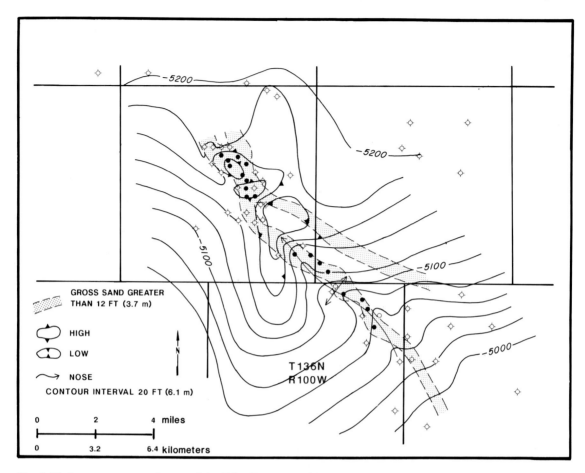

Fig. 3-16. Structure map on the top of the Tyler Formation, with the Tyler gross-sandstone isopach map superimposed. The coincidence of the structural highs with the sandstone thicks suggests that high areas are created by differential compaction of the strata.

The fluvial sandstones are interpreted as active channel fill in a coarse-grained meander-belt system. The abundance of erosional surfaces, rip-up clasts, upward-fining sequences, thin coals and mud-cracked surfaces, and especially the mapped sinuosity of individual channel-fill sandstones are all supportive evidence for a meandering system. Furthermore, the near-coastal setting is conducive to development of meandering fluvial systems. The low gradients and high suspended/bedload ratios provide for bank stability and promote meandering. The low sinuosity of the gross channel network is probably a function of the amalgamation and stacking of individual channels within the meander-belt network, which was in turn the result of fault-controlled, linear valley shape.

Other lithologies in the Lower Tyler Formation include anhydritic shales, lignites, and paleosol hori-

zons, suggesting that marshes and supratidal flats existed adjacent to the channels. Interbedded thin sandstones, siltstones, and mudstones adjacent to the sandstones represent levee, crevasse-splay, and bay-fill deposits. Thin limestones represent minor marine transgressions. These finer-grained lithofacies encase the long and narrow channel-fill sequences.

The relative absence of mudstone in the upper portions of channel sandstones, along with relatively linear isopach patterns, suggests appreciable chute-bar modification of the point bars. Chute-modified point bars are common in coarse-grained meandering fluvial systems (Galloway and Hobday, 1973). In these systems, sandbelts are vertically and horizontally amalgamated and generally dip-oriented. The width-to-thickness ratio is typically moderate to high (10-40:1), and development of coalesced multilateral channel fills is common.

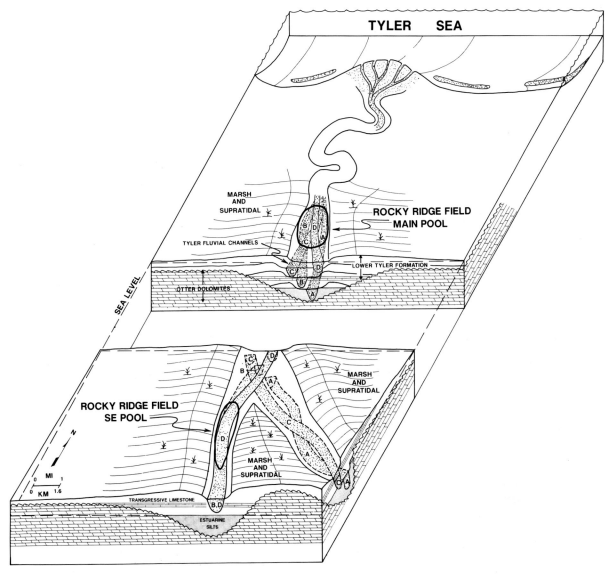

Fig. 3-17. Block diagram illustrating the depositional setting of the Lower Tyler Formation, Rocky Ridge Field area. The diagram shows the four major fluvial channels which merge at the Main Pool. This fluvial system probably fed the fluvial-deltaic and barrier island sandstones of the Tyler Formation downdip from the Medora-Dickinson trend (see Barwis, this volume, Chapter 16).

Lithofacies

Chute-bar modification of Tyler point bars has resulted in a somewhat less heterolithic lithofacies assemblage than that found in finer-grained point bars of high-sinuosity meandering systems. Lithofacies representing three major environments are developed within the Tyler reservoir section at Rocky Ridge Field: (1) channel-fill, (2) channel-margin, and (3) interchannel-floodbasin.

The channel-fill sequences, the reservoir for Rocky Ridge Field, consist of individual meander-belt units of fine- to coarse-grained, poorly to well-sorted lower point-bar and chute-modified upper point-bar sandstones. Bed thicknesses range from 1 to 3 feet (0.3-1.0 m). The lower point-bar deposits usually contain medium-scale trough cross-stratification, whereas chute-modified upper point-bar sandstones contain medium- to large-scale planar/tabular cross-stratification.

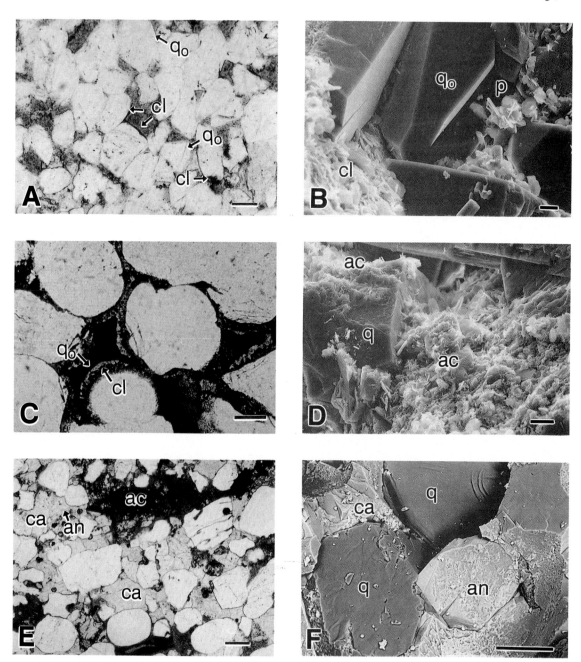

Fig. 3-18. Thin sections and SEM photomicrographs showing diagenetic components of the Tyler sandstones at Rocky Ridge Field. Samples are from the Shell State No. 41X-36 well (see Fig. 3-5) except for part E, which is from the Shell Gov't 21-18 well. (A) High-quality reservoir sandstone (channel-fill facies) (ϕ = 17%, k = 450 md) showing chlorite grain coatings (cl) and quartz overgrowths (q_o). Note the well-developed sorting of the grains and the absence of authigenic cements. 8,045 feet (2,452 m); plane light. Scale bar is 0.10 mm. (B) Equivalent sandstone to part A, showing quartz grains with euhedral quartz overgrowths (q_o) and grain-coating chlo- rite (cl). Note the good porosity (p). Top of 8,045 feet (2,452 m). Scale bar is 0.01 mm. (C) Reservoir sandstone showing extensive grain coatings of chlorite (cl). Also note the quartz overgrowth (q_o). The pore space is black because of oil staining. 8,035 feet (2,449 m); plane light. Scale bar is 0.10 mm. (D) Similar sandstone to part C, showing quartz grain (q) surrounded by complex authi- genic mixed-layer clays (ac) which clog pores and reduce reservoir quality. Elemental analysis indicates that the clays are primarily smectite, illite, and kaolinite. Base of 8,045 feet (2,452 m). Scale bar is 0.01 mm. (E) Non- reservoir sandstone (channel-margin facies) (ϕ = 7.5%,

The channel-margin facies comprises finer-grained rocks interpreted to be crevasse-splay and levee deposits. Both deposits are characterized by ripple and climbing-ripple cross-laminations and by wavy and planar laminations, clay drapes, laminated mud layers, and root-disturbed zones. This facies is relatively impermeable, and its position overlying and adjacent to the channel-fill sandstones forms partial top and lateral seals.

The interchannel-floodbasin facies contains very fine-grained material deposited from suspension between and lateral to active channels. This facies consists largely of red, green, and gray mudstones deposited as channel plugs, bay fill, and soil deposits. These strata are typically horizontally laminated but have been extensively reworked by plant growth and pedogenic processes. This facies is commonly evaporitic and lignitic. Interbedded algally laminated dark shales represent transgressive episodes and provide both the source of oil and the ultimate lateral and vertical seals for the productive Tyler sandstones at Rocky Ridge Field.

This assemblage of lithofacies is similar to those described in modern coarse-grained point-bar deposits of the Amite and Colorado rivers (McGowen and Garner, 1970). The main similarities to Rocky Ridge sandstones are linear, dip-oriented morphology, "blocky" vertical sequence, medium-to large-scale cross beds near the top of the sequence, and the relative absence of mud.

Petrography and Diagenesis

The reservoir sandstones of the Tyler Formation in Rocky Ridge Field are quartzarenites. Other primary minerals include chert and rock fragments and rare carbonate grains, feldspar, and assorted heavy minerals. The fluvial channel sandstones of the Tyler Formation are predominantly medium-grained but vary from fine- to coarse-grained. Reservoir quality rocks are usually medium-grained and are typically well-sorted and rounded.

The first diagenetic event affecting reservoir quality was early grain-coating chlorite, followed by quartz overgrowths (Figs. 3-18A, 3-18B, 3-18C). The chlorite was probably derived from a detrital clay matrix. Further reduction in reservoir quality occurred with formation of complex authigenic mixed-layer clays in reservoir pores (D. Beard, personal communication, 1988) (Figs. 3-18D, 3-18E, 3-18F). The clays are primarily smectite, illite, and kaolinite. Calcite, ankerite, anhydrite, and hematite cements are authigenic components that probably formed late in the burial history of the reservoir sandstones (Figs. 3-18E, 3-18F). Although the calcite, ankerite, and anhydrite cements totally occlude porosity locally, they are not widespread cements in the formation. Porosity and permeability have been slightly enhanced by the late-stage dissolution of calcite and anhydrite cements (Sturm, 1983).

The chlorite grain coatings appear to have prevented later porosity and permeability reduction caused by quartz overgrowths and/or the calcite, ankerite, and anhydrite cements (Fig. 3-18C). These cements are largely absent from the productive sandstones; however, where chlorite is not present, these cements may totally occlude porosity. The cements may provide the updip seal in the Southeast Pool. There does not appear to be any lithofacies-dependent diagenesis.

Petrophysics

Porosity in the reservoir sandstones at Rocky Ridge Field is primarily intergranular (Figs. 3-18A, 3-18B). Porosities average 13 to 17% but may be as high as 25%. Permeabilities average 60 to 170 md and are as high as 780 md. Reservoir quality is directly related to lithofacies as indicated by the porosity-permeability relationships of three different groupings of facies in Figure 3-19. The fluvial channel sandstones are the best reservoirs and contain the most oil. Channel-margin sandstones have low permeabilities and are rarely oil-bearing.

---

k = 0.1 md) showing abundant authigenic mixed-layer clays (ac) with some possible matrix and calcite (ca) and ankerite (an) cements. The authigenic cements have completely occluded porosity. Note the poorly developed sorting versus that found in the reservoir sandstone. 7,809 feet (2,380 m); plane light. Scale bar is 0.01 mm. (F) Equivalent nonreservoir sandstone to part E, showing quartz grains (q) surrounded by authigenic calcite (ca) and ankerite (an) cements which have completely occluded porosity. Photo taken in backscatter mode. Base of 8,045 feet (2,452 m). Scale bar is 0.01 mm.

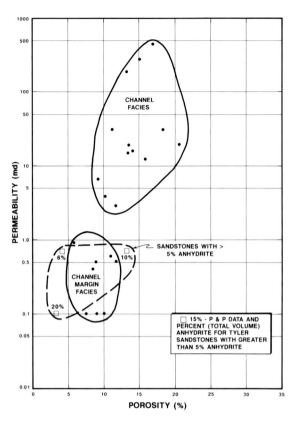

Fig. 3-19. Plot of porosity versus permeability with the two major sandstone lithofacies distinguished. Note the wide range of permeability for a given porosity (sandstones with 12% porosity have permeabilities which range from 0.5 to 200 md). Clearly, the fluvial channel-fill facies is the reservoir at Rocky Ridge Field.

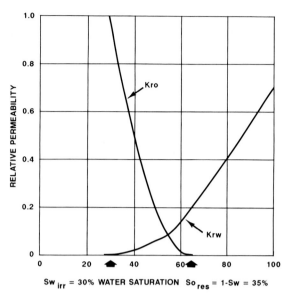

Fig. 3-21. Relative permeability plot of typical fluvial channel reservoir sandstones at Rocky Ridge Field. The plot shows average irreducible water saturations (Sw_{irr}) of about 30% and average residual oil saturations (So_{res}) of about 35%. $K_{ro} = (1 - Sw*)^2$ (relative permeability to oil); $K_{rw} = (Sw*)^{1/2}(Sw)^4$ (relative permeability to water); $Sw* = (Sw - Swi)/(1 - Swi - Sor)$ (effective water saturation − ratio of movable water over pore space available for movable liquids).

Porosities and permeabilities in anhydritic fluvial-channel sandstones mimic those in the channel-margin facies because high concentrations of anhydrite cement have occluded porosity.

Capillary-pressure data of a typical reservoir sandstone show low entry pressures and irreducible water saturations of about 20 to 30% (Fig. 3-20). Nonreservoir sandstones have high entry pressures and effectively act as seals. The pore-throat analysis for the reservoir quality sandstone shows relatively high uniformity of pore and pore throat sizes and suggests high continuity between pore space. Absolute average pore throat sizes are about 20 microns and 0.4 microns for reservoir and nonreservoir sandstones, respectively.

The relatively large pore throats and even porosity distribution in reservoir sandstones result in short transition zones above the oil-water contact in both

pools. Oil saturations of 40% and 60% are calculated to be encountered about 11 feet (3.4 m) and 15 feet (4.6 m), respectively, above the oil-water contact (Fig. 3-20). Typical productive sandstones exhibit residual oil saturations of about 35% (Fig. 3-21). Relative permeability of the sandstones to oil increases dramatically below water saturations of about 55%. Pay thicknesses average 30 feet (9.1 m) for the Main Pool and 20 feet (6.1 m) for the Southeast Pool.

Production Characteristics

Well Performance

Rocky Ridge Field was developed on an 80-acre (32.4-ha.) spacing. The discovery well (Lucy Fritz No. 1) has been by far the most prolific with an initial potential of 1,224 BOPD (195 m³/D) and a cumulative production of more than 1.0 MMBO (1.6 × 10⁵ m³). Initial potentials in the field ranged from 25 to 1,224 BOPD (4.0-195 m³/D) and averaged 260 BOPD (41 m³/D) (Fig. 3-22). Average well

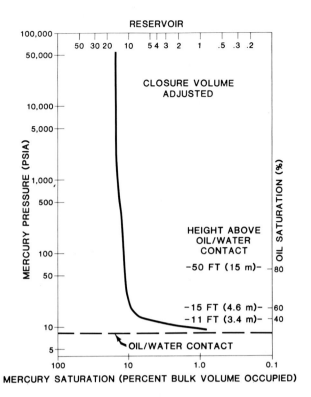

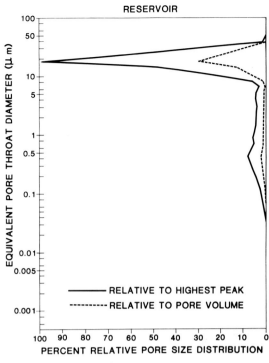

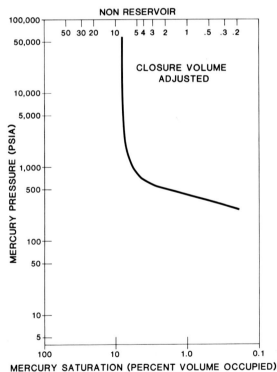

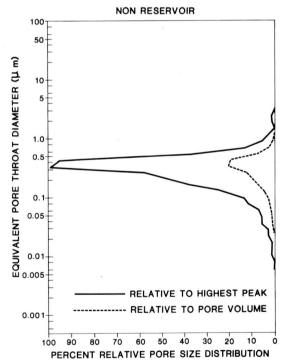

Fig. 3-20. Capillary-pressure curves and pore-throat diameter distribution curves for typical reservoir and nonreservoir quality sandstones in Rocky Ridge Field. (A) and (B) are from a reservoir quality sandstone in the Shell State No. 41X-36 and have low entry pressures of about 8 psi (55 kPa) and relatively large and uniform pore-throat diameters of about 20 microns. Note the calculated oil saturations for various heights above the oil/water con- tact. The reservoir quality sandstones have relatively short transition zones. 8,045 feet (2,452 m). (C) and (D) are from a nonreservoir quality sandstone in the Shell Gov't No. 21-18 and have high entry pressures of about 250 psi (1.7 × 10³ kPa) and small and uniform pore-throat diameters of about 0.4 microns. These sandstones act effectively as seals. 7,809 feet (2,380 m).

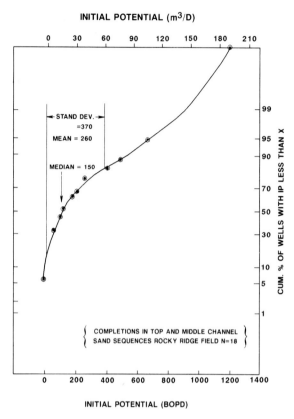

Fig. 3-22. Graph showing the range in initial production potential of wells in Rocky Ridge Field.

production in the Main Pool for the first 10 years ranged from 16 to 230 BOPD (2.5-37 m³/D) and averaged 110 BOPD (17.5 m³/D). For the entire field, cumulative production per well is also quite variable, ranging from 2.6 MBO to 1.05 MMBO (0.4-167 × 10³ m³) and averaging about 250 MBO (4.0 × 10⁴ m³) (Fig. 3-23). About 40% of the wells have produced less than 50 MBO (8.0 × 10³ m³).

Reservoir Performance

The point-bar sandstones at Rocky Ridge Field produce oil by a depletion-drive system. Typical reservoir performance is illustrated by the decline curve for the discovery well (Fig. 3-24). Wells maintain relatively high production rates for the first two to five years of their life, followed by gradual production declines. Reservoir pressures gradually decrease with time and little water production occurs during primary recovery, both characteristic of depletion-drive systems. Stimulation procedures are usually unnecessary for well completion unless initial production rates are less than 100 BOPD (16 m³/D), in which case the reservoir may be acidized and sometimes hydraulically fractured.

A waterflood was initiated in the Main Pool in 1967 using one injection well. The program was marginally successful, as it increased production about 50% for more than a year. However, this waterflood was initiated ten years after field discovery and

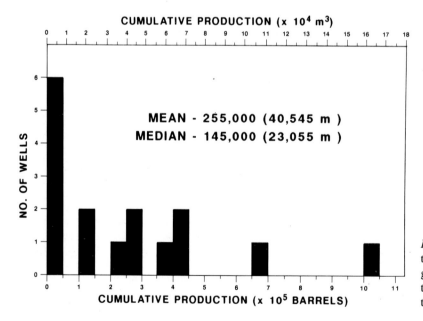

Fig. 3-23. Cumulative production graph showing wells with given cumulative production and the range of cumulative production.

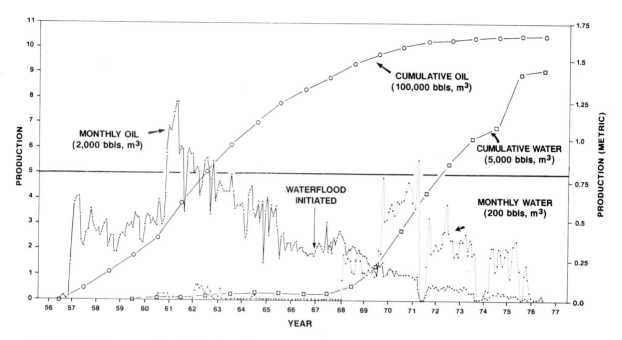

Fig. 3-24. Decline curve for the Northern Pump-Lucy Fritz No. 1, Main Pool, Rocky Ridge Field. The curve is representative of well performance in the field. Note the high initial monthly production rates followed by steady declines. The curve also demonstrates the minimal effects of the waterflood on production. Note the absence of water production prior to the waterflood. For conversion to metric, 1 barrel = 0.159 m³.

development, when production rates and reservoir pressures had significantly declined, and therefore the waterflood did not result in significant additions to ultimate recovery. In addition, poor placement of the injection well may have also been a factor in limiting the waterflood results, both in the Main Pool and in the Southeast Pool, where a similar waterflood was initiated in 1970.

Pressure data from the Main Pool indicate that the three productive channel sandstones are in communication, with little if any effective interreservoir boundaries. This reflects the chute-bar modification of the point-bar sandstone bodies and explains the common oil/water contact. Pressure data suggest that the Southeast Pool is not in communication with the Main Pool and that several of the wells in the Southeast Pool are isolated.

A map of "net feet of pay" shows the distribution of productive reservoir within the two pools (Fig. 3-25). Reservoir richness ranges from 6,355 BO/acre (2,495 m³/ha.) in the Main Pool to 2,930 BO/acre (1,150 m³/ha.) in the Southeast Pool. The relatively high richness of the Main Pool results from the greater pay thickness. Ultimate production per well is also higher in the Main Pool (Fig. 3-26). The thicker pay section

and ultimate production values reflect multiple pay zones due to the stacking of productive channel sandstones at the Main Pool and result in recovery efficiencies of about 30% following primary and secondary (waterflood) recovery. The lower richness values in the Southeast Pool may result from intrareservoir boundaries. The boundaries reduce drainage area per well and could be caused by shale channel plugs or mudstone drapes, which are present but uncommon in low- sinuosity meander belts. Local zones of authigenic cement may have also created intrareservoir barriers.

Thickest net pay sections and high ultimate production values do not perfectly coincide, and they cannot be closely related to structural position. Data from the Main Pool show a poor relationship between ultimate well production and top-of-pay distance above the oil/water contact (Fig. 3-27). The five wells which produce from 20 to 25 feet (6.1-7.6 m) above the oil/water contact show widely varying ultimate production. Well performance, although clearly influenced by pay thickness, is more importantly affected by variation in average reservoir permeability. Thus, the ability to predict reservoir quality, and not just thickness, is the key to optimum

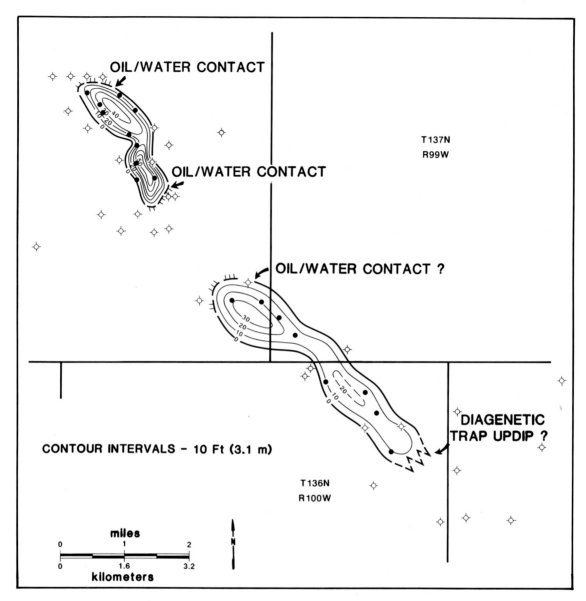

Fig. 3-25. Map of the "net feet of pay" at Rocky Ridge Field contoured in feet. The higher "net feet of pay" values in the Main Pool reflect multiple pay sandstones due to stacked channels, explaining the higher richness values there.

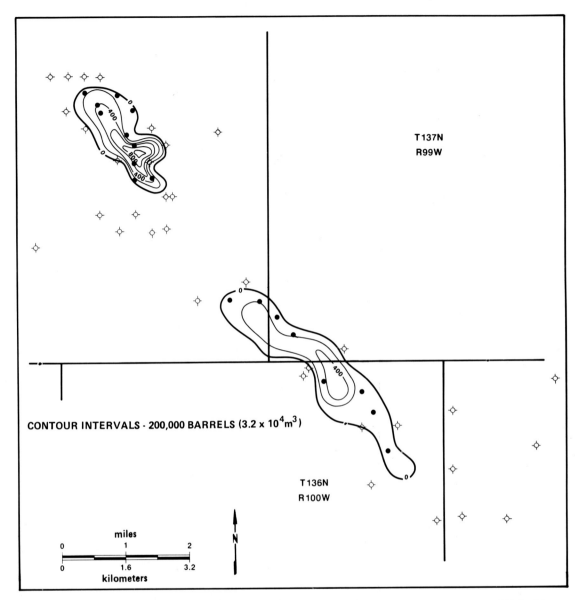

Fig. 3-26. Map of ultimate production at Rocky Ridge Field contoured in barrels. Note the higher ultimate production from wells in the Main Pool. Note also that the highest ultimates are not always at the the crest of the structure, and the high net feet of pay and high ultimate production do not perfectly coincide. Both are due to the depletion-drive mechanism and variations in reservoir permeability.

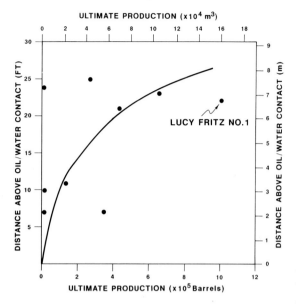

ULTIMATE PRODUCTION (x10⁴ m³)

LUCY FRITZ NO.1

DISTANCE ABOVE OIL/WATER CONTACT (FT)

DISTANCE ABOVE OIL/WATER CONTACT (m)

ULTIMATE PRODUCTION (x10⁵ Barrels)

Fig. 3-27. Cumulative production plotted against height of "top of pay" (in feet and meters) above the oil/water contact. The plot shows that neither ultimate production nor "net feet of pay" closely relates to structural position because of variations in reservoir quality within the producing sandstones.

field development. A knowledge of the lithofacies relationships is a key to this prediction.

Exploration and Production Strategy

Regional interval- and gross-sandstone isopach data suggest that additional dip-trending Tyler fluvial systems should exist along strike from the Rocky Ridge fluvial system. These trends can be identified by incorporating regional seismic data with well control. Lead areas which have high potential for sandstone development must then be scrutinized for structural character and seismic attributes. New geophysical technology, such as shear seismic and seismic amplitude versus offset, should help delineate these major channel-sandstone bodies and distinguish between various Tyler lithologies. An integrated geological/geophysical approach to exploration should result in future discoveries in the Tyler.

Rocky Ridge Field illustrates the risk that can be involved in developing Tyler fluvial sandstone bodies once they have been discovered. Lateral discontinuity of sandstone bodies makes reservoir prediction precarious and explains the poor Rocky

Ridge Field development program success of 32%. Furthermore, intrareservoir inhomogeneities, such as authigenic clays and cements or shale plugs, make predictions even more difficult. However, these problems can be overcome by developing a geologic model to predict both the geometry and the reservoir quality of the channel sandstone bodies and by defining the geophysical and geological parameters of the productive reservoir early in the field development stage.

Conclusions

Channel sandstones in the Tyler Formation at Rocky Ridge Field produce oil from 8,100 feet (2,470 m) in two pools. The sandstones are chute-modified point bars deposited in a coarse-grained meander-belt system which transported sand from south to north. The system deposited four major low-sinuosity, dip-trending, laterally discontinuous channel-fill sandstones. Sandstone deposition was focused in Otter Formation paleotopographic lows. Trapping is due primarily to differential compaction over amalgamated channel sandstones.

Reservoir quality is best developed in the channel-fill facies. Chute modification of point-bar sands created a relatively homogeneous sequence of sandstone with few vertical barriers to flow among stacked channel-fill units. Lateral flow barriers, however, are important, especially in the Southeast Pool, and are caused by local shale channel plugs, mudstone drapes, and authigenic cements. These barriers limit the drainage areas of individual wells.

An integrated geological/geophysical model permits the exploration geologist to prioritize specific prospective lead areas where fluvial channel development is likely and helps the production geologist categorize reservoir controls and parameters for optimum well placement in development programs.

Acknowledgments. The author thanks Shell Oil Company for permission to publish the material and for project funding. The author is especially indebted to John Barwis for his advice during the Rocky Ridge project and contributions to the final report and to Don Beard for his help with petrography and reservoir description.

Reservoir Summary

Field: Rocky Ridge (Main Pool and Southeast Pool)
Location: Southwest North Dakota
Operators: Shell, Amoco, Northern Pump, Mule Creek Oil
Discovery: 1957, 1969
Basin: Williston basin
Tectonic/Regional Paleosetting: Intracratonic basin
Geologic Structure: Dome (Main Pool); NW-plunging anticlinal nose (Southeast Pool)
Trap Type: Stratigraphic pinch-out on compaction-generated closures
Reservoir Drive Mechanism: Depletion
  • Original Reservoir Pressure: 4,200 psi (2.9×10^4 kPa) at 5,270 feet (1,606 m) subsea
Reservoir Rocks
  • **Age:** Early Pennsylvanian, Bashkirian
  • **Stratigraphic Unit:** Tyler Formation
  • **Lithology:** Medium-grained quartzarenite
  • **Depositional Environment:** Fluvial (coarse-grained meander belt)
  • **Productive Facies:** Point-bar sandstones
  • **Petrophysics:**
    • Porosity Type: Total 16%; primary >90%, secondary intergranular
    • ϕ: Average 16% (17% Main Pool, 15% Southeast Pool), range 13 to 25% (cores)
    • k: Average 115 md, range 60 to 780 md (cores)
    • S_w: Average 30%, range 20 to 60%, cutoff 60% (cores, logs)
    • S_o: Average 35% (cores, logs)
Reservoir Dimensions
  • **Depth:** 8,100 feet (2,470 m)
  • **Areal Dimensions:** 8 miles by 1 mile (13×1.6 km)
  • **Productive Area:** Main Pool: 535 acres (217 ha.); Southeast Pool: 580 acres (235 ha.)
  • **Number of Reservoirs:** 4 (Main Pool), 1 (Southeast Pool)
  • **Hydrocarbon Column Height:** 50 feet (15 m) (Main Pool), 30 feet (9 m) (Southeast Pool)
  • **Fluid Contacts:** Oil/water contact at 5,250 feet (1,600 m) subsea (Main Pool)
  • **Number of Pay Zones:** 3 (Main Pool), 1 (Southeast Pool)
  • **Gross Sandstone Thickness:** 30 feet (9.1 m)
  • **Net Sandstone Thickness:** 25 feet (7.6 m)
  • **Net/Gross:** 0.83
Source Rocks:
  • **Lithologies and Stratigraphic Units:** Shale and limestone, Tyler Formation
  • **Time of Hydrocarbon Maturation:** Oligocene
  • **Time of Trap Formation:** Soon after deposition, during compaction; Paleocene structuring
  • **Time of Migration:** Oligocene and Miocene
Hydrocarbons
  • **Type:** Oil
  • **GOR:** 200:1
  • **API Gravity:** 36°
  • **FVF:** 1.16 at 218°F (103°C)
  • **Viscosity:** 2.43 cP (2.4×10^3 Pa·s) at 218°F (103°C)
Volumetrics:
  • **In-Place:** 11.4 MMBO (1.8×10^6 m³) Main Pool; 8.7 MMBO (1.4×10^6 m³) Southeast Pool
  • **Cumulative Production:** 3.4 MMBO (5.4×10^5 m³), 1.4 MMBW (2.2×10^5 m³) Main Pool; 1.6 MMBO (2.5×10^5 m³), 0.7 MMBW (1.1×10^5 m³) Southeast Pool
  • **Ultimate Recovery:** 3.4 MMBO (5.4×10^5 m³) Main Pool; 1.7 MMBO (2.7×10^5 m³) Southeast Pool
  • **Recovery Efficiency:** 30% (Main Pool); 20% (Southeast Pool)
Wells
  • **Spacing:** 80 acres (32.4 ha.)

Typical Well Production:
- **Total:** 52 (29 Main Pool, 23 Southeast Pool)
- **Dry Holes:** 34 (19 Main Pool, 15 Southeast Pool)
- **Average Daily:** 110 BO (17.5 m³) (Main Pool)
- **Cumulative:** 250 MBO (4.0×10^4 m³)

References

Brown, D. L., 1978, Wrench-style deformation patterns associated with a meridional stress axis recognized in Paleozoic rocks in parts of Montana, South Dakota and Wyoming, *in* Williston Basin Symposium: Montana Geological Society, 24th Annual Conference, p. 17–34.

Galloway, W. E., and Hobday, D. K., 1973, Terrigenous Clastic Depositional Systems: Applications to Petroleum, Coal, and Uranium Exploration: New York, Springer-Verlag, 423 p.

Gerhard, L. C., Anderson, S. B., LeFever, J. A., and Carlson, C. G., 1982, Geological development, origin, and energy mineral resources of the Williston Basin, North Dakota: American Association of Petroleum Geologists Bulletin, v. 66, p. 989–1020.

Grenda, J. C., 1977, Paleozoology of cores from the Tyler Formation (Pennsylvanian) in North Dakota, U.S.A. [Ph.D. thesis]: Grand Forks, North Dakota, University of North Dakota, 338 p.

Kluth, C. F., and Coney, P. J., 1981, Plate tectonics of the ancestral Rocky Mountains: Geology, v. 9, p. 10–15.

Maughan, E. K., 1984, Paleogeographic setting of Pennsylvanian Tyler Formation and relation to underlying Mississippian rocks in Montana and North Dakota: American Association of Petroleum Geologists Bulletin, v. 68, p. 178–195.

McGowen, J. H., and Garner, L. E., 1970, Physiographic features and stratification types of coarse-grained point bars: Modern and ancient examples: Sedimentology, v. 14, p. 77–111.

Sturm, S. D., 1983, Depositional environments and sandstone diagenesis in the Tyler Formation (Pennsylvanian), southwestern North Dakota: North Dakota Geol. Survey, Report Investigations No. 76, 48 p.

Thomas, G. E., 1974, Lineament-block tectonics: Williston-Blood Creek Basin: American Association of Petroleum Geologists Bulletin, v. 58, p. 1305–1322.

Key Words

Rocky Ridge Field, North Dakota, Williston basin, Tyler Formation, Pennsylvanian, Bashkirian, coarse-grained meander belt, differential compaction, diagenesis, chlorite, synthetic seismogram, channel sandstones, paleotopography, chute-modified point bar.

4

Meander-Belt Reservoir Geology, Mid-Dip Tuscaloosa, Little Creek Field, Mississippi

Ernie G. Werren, Roger D. Shew, Emmet R. Adams, and Richard J. Stancliffe

BP Exploration Inc., Houston, Texas 77074; Shell Development Company, Houston, Texas 77001; Consultant, New Orleans, Louisiana 70124; Shell Offshore Inc., New Orleans, Louisiana 70161

Introduction

General

Meander belts represent the first genetic sequences that were understood by geologists. Research on modern point bars in the Brazos River of South Texas in the period 1953 to 1960 led to the formulation of a facies model that was directly applicable to subsurface exploration (Bernard and Major, 1963). It is this model which, in the late 1950s and early 1960s, guided the exploration and development of Little Creek Field, the case study of this chapter. This chapter represents one of the first uses of facies models in the oil industry and is a documentation of the first point-bar reservoir to be explored and produced with geologic forethought. The success of the point-bar facies model in predicting reservoir distribution can be measured by the high success rate of drilling by the main operators as compared with that of the surrounding farmed-out acreage.

Little Creek Field is located within the Upper Cretaceous Mid-Dip Tuscaloosa trend of southwestern Mississippi and northeastern Louisiana. This mature exploration and producing trend has initial estimated oil in place of greater than 1 billion barrel equivalents (1.6×10^8 m³). Little Creek was one of the larger fields discovered in the trend with more than 102 MMBO (1.6×10^7 m³) in place. The depositional system consists of composite point bars, and trapping is by a combination of structural and stratigraphic closure. A large number of wells and cores and a long history of production, including primary, waterflood, and enhanced oil recovery using carbon dioxide (CO_2), provide important information concerning the controls on and productivity of this well-defined point-bar complex.

Field and Reservoir Development

Little Creek was discovered in 1958 by Shell Oil Company. The discovery well, Lemann No. 1 in Section 35, T5N, R8E, was completed flowing 588 BOPD (93.5 m³/D), 39° API, and 260 MCFGPD (7.4×10^3 m³/D) on a 15/64-inch (0.6-cm) choke at a FTP of 730 psig (5.0×10^3 kPa). The open-hole interval was 10,770 to 10,790 feet (3,283–3,289 m). The well was drilled based on a knowledge of oil production from correlative stratigraphic intervals in nearby areas and on the seismic interpretation of minor simple closure at the Lower Tuscaloosa horizon. Following discovery, development drilling on 40-acre (16.2-ha.) spacing was rapid. A total of 208 wells were drilled with 162 being successful. Total oil in place was estimated at 102 MMBO (1.6×10^7 m³) in the 6,200-acre (2,510-ha.) productive area.

The early reservoir-pressure performance indicated a solution-gas drive with only limited water influx. The field was voluntarily unitized, and one of the industry's deepest waterfloods at the time was initiated in 1962 after primary production of 23 MMBO (3.7×10^6 m³). As Mississippi was without a force-pooling law, it was a significant achievement to get 1,525 working interest and royalty owners to

Fig. 4-1. Location map of Little Creek Field within the Mid-Dip Tuscaloosa trend. Also shown are Olive and Liberty fields, two of the more recent discoveries within the trend.

ratify the unitization agreement. An additional 25 MMBO (4.0×10^6 m³) was recovered by the waterflood (Cronquist, 1968), which was finished in 1970. Primary and secondary (waterflood) methods recovered 47% of the OOIP.

The large amount of remaining oil (54 MMBO; 8.6×10^6 m³) led Shell to conduct a CO_2 pilot flood from 1974 to 1977 in a small area in the eastern part of the field (Hansen, 1977). The results were encouraging and, after producing 125 MBO (2.0×10^4 m³), plans for a phased full-scale CO_2 flood were initiated. Following construction of a pipeline from the CO_2 source area (Northeast Jackson Dome), near Jackson, Mississippi, and field facilities construction, CO_2 injection began in December, 1985.

Regional Data

Location and Regional Geology

Little Creek Field is located in Lincoln and Pike Counties, southwestern Mississippi (Fig. 4-1), on the south rim of the Mississippi Salt basin. It is

within the Upper Cretaceous Mid-Dip Tuscaloosa trend which occurs updip of the Lower Cretaceous shelf margin. In Mississippi, production from the Mid-Dip trend extends 150 miles (240 km) to the east-southeast from the Mississippi River in a belt 30 to 60 miles (50-100 km) wide.

The Upper Cretaceous (Cenomanian) Lower Tuscaloosa Formation is composed primarily of terrigenous clastics derived from the Ouachita orogenic belt. An initial progradation, represented by Lower Tuscaloosa sediments unconformably overlying the Lower Cretaceous Washita-Fredericksburg group, led to the deposition of a lower, sandstone-rich massive section of braided and meandering stream deposits. An overall transgressive sequence followed, leading to the deposition of an upper, shale-rich interval that contains lenticular sandstones of fluvial and subsequent littoral deposits (Fig. 4-2). Most of the hydrocarbons occur within the Lower Tuscaloosa fluvial and littoral sandstones in stratigraphic or combined structural/stratigraphic traps. The transgression culminated with the deposition of marine shales of the Middle Tuscaloosa; these sediments are the probable source rocks for the Lower Tuscaloosa. The contact of the Lower and Middle Tuscaloosa is characterized by a high-resistivity zone that is an excellent regional marker.

Structure and Stratigraphy

Figure 4-3 shows the structure of the Lower Tuscaloosa Q sand stratigraphic marker and the outline of the sandstone distribution. The structure is an elongate, unfaulted, north-south low-relief nose with maximum dips on the flanks of 1 to 2°. The feature trends gently south from Mallalieu Field and is about 14 miles (23 km) long and 4 to 6 miles (6.4–9.7 km) wide. Little Creek and Sweetwater fields are located near the southern end of this anticlinal nose. This gentle structure was recognized on 1950s-vintage seismic lines. Figure 4-4 is a more recent and much improved seismic line that illustrates the simple closure at the Lower Tuscaloosa objective horizon.

The approximately 100-foot (30.5-m) oil column at Little Creek is controlled by both structural and stratigraphic closure. The stratigraphic aspects of the trap result from the configuration of a sinuous belt of point-bar sandstones. These sandstones are encased in sealing delta-plain backswamp siltstones, mudstones, and claystones and often pinch out, on the channel margin, into channel abandonment

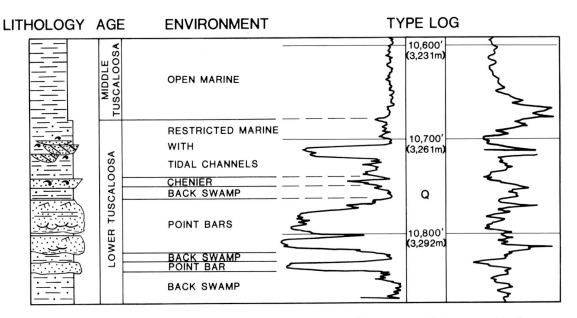

Fig. 4-2. Type log illustrating the lithology, depositional environment, and log response of the transgressive Lower to Middle Tuscaloosa interval. The point bars are the objectives at Little Creek Field.

deposits. The structural nose that formed later served to localize and increase closure.

Reservoir Characterization

Depositional Setting

Production is from the Lower Tuscaloosa Q and Q2 sandstones (Figs. 4-2, 4-5) at an average depth of 10,750 feet (3,277 m) or 10,350 feet (3,155 m) subsea. These sandstones were extensively cored during development drilling. The depositional sequence, lithologic data, and sandstone distribution patterns indicated the Q-Q2 sandstones to be a series of point bars deposited by highly sinuous meandering streams on an upper deltaic plain. Figure 4-6 is a schematic of the depositional model for the Little Creek and Sweetwater areas.

The depositional sequence in Little Creek Field consists of a transgressive sequence of sediments grading from braided and meandering fluvial deposits at the base, through the meandering Q-Q2 sandstone objectives, to overlying littoral sediments that are primarily fossiliferous, burrowed shales. Mudstones and claystones underlying and lateral to the Q-Q2 sandstones are mottled red and green and

are typical of floodbasin deposits; they serve as excellent lateral seals. Representative type logs for the various sequences penetrated in Little Creek Field are shown in Figure 4-5 and include wells with both the Q and Q2 sandstones, the Q sandstone only with abandonment facies, and total absence of the objective sandstones (laterally equivalent floodbasin sediments).

Major sedimentary features in the Q-Q2 sandstones and their lateral shale equivalents are described in Figures 4-7, 4-8, and 4-9. They illustrate features within the objective Q sandstone section that are typical of point-bar deposits. These include erosional bases with channel lags, large-scale cross-beds grading up through horizontal and small-scale ripple cross-laminae, clay drapes, local mud balls and intraclasts, micaceous and carbonaceous streaks, and some calcareous patches and discontinuous claystone layers.

In the abandoned channel fill, where lower energy deposition predominated, vertical accretion led to characteristic features that include interbedded very fine-grained sandstone, siltstone, and mudstone, flow structures formed during dewatering and soft sediment deformation, microfaulting, small-scale cross-stratification, clay intraclasts, and carbonaceous plant remains. The wells penetrating the

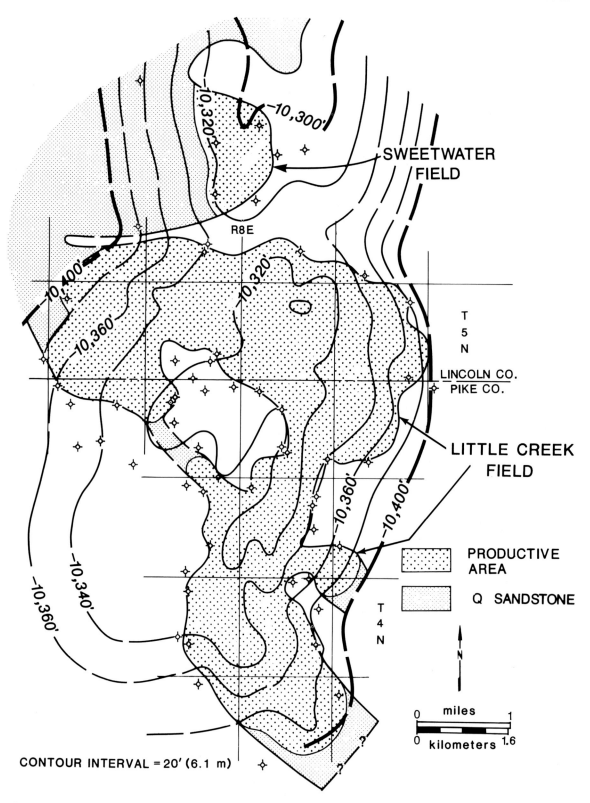

Fig. 4-3. Structure map on the Q marker horizon. Little Creek and Sweetwater fields are structurally (gentle anticlinal nose) and stratigraphically trapped. Sweetwater Field is believed to be part of the same fluvial system but is a separate reservoir. Only dry holes are shown.

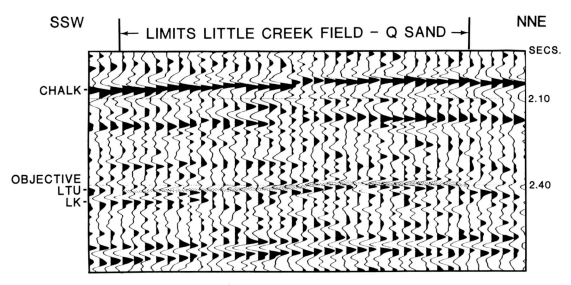

Fig. 4-4. Little Creek Field was discovered based on sub-surface mapping and on the seismic interpretation of minor simple closure at the Lower Tuscaloosa (LTU) horizon. This seismic line runs NNE-SSW through the field; the LTU objective is shown by the stippled pattern. Chalk is Upper Cretaceous Clayton chalk, and LK = Lower Cretaceous.

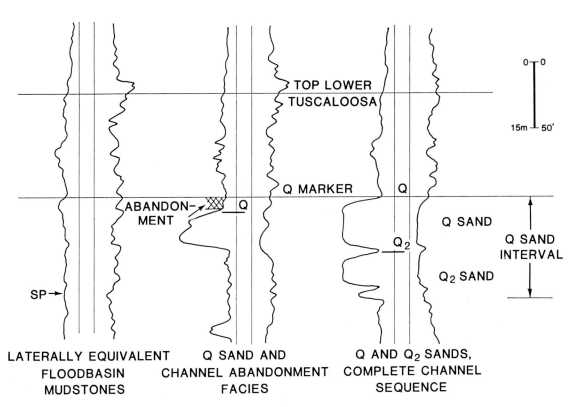

Fig. 4-5. The objective sandstones in Little Creek Field consist of the Q2 and Q. These type logs illustrate SP and resistivity responses of the three characteristic log types penetrated.

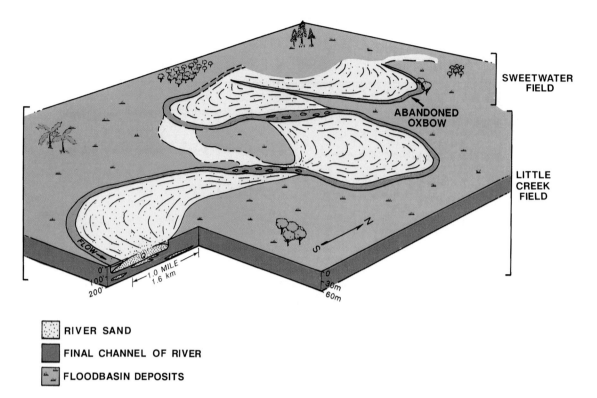

Fig. 4-6. Idealized depositional model of the large meander system that deposited the Q point-bar sands. The river is shown at its interpreted last position prior to abandonment.

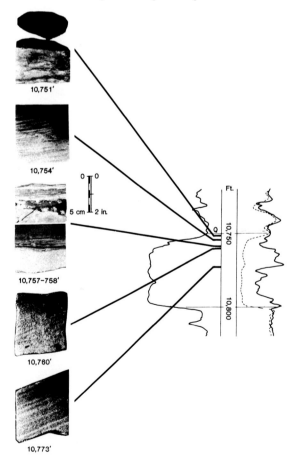

Fig. 4-7. The Q sandstone has characteristic point-bar features including large-scale cross-stratification, isolated shale laminae, small-scale ripple cross-stratification, and abandonment facies. Additional features are illustrated in Figures 4-8 and 4-9. The well shown here is the Skelly McCullough No. 1.

Fig. 4-8. Intraformational shale rip-up clasts above an erosional surface in the Q sandstone. The sequence grades through parallel-laminated and large- to small-scale cross-stratified sandstone. The Soloman Atkinson No. 1 well, 10,783.5-10,799.5 feet (3,286.8-3,291.7 m). The bottom of the cored interval is at the lower right. The scale bar is 1 foot (30.5 cm).

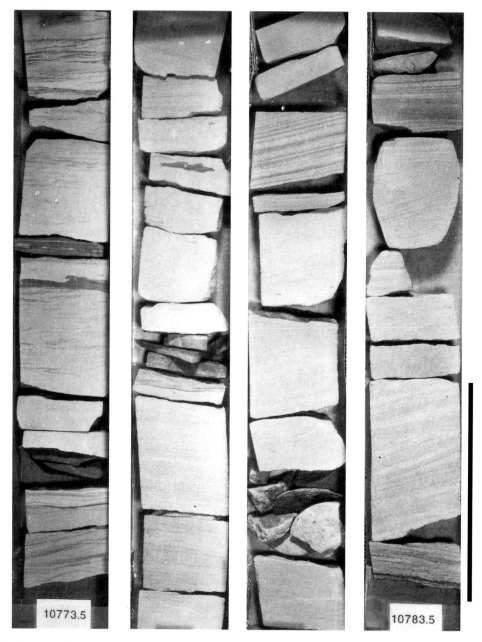

Fig. 4-9. Large-scale cross-stratified and parallel-laminated sandstone passing upward to ripple cross-laminated sandstone in the Q sandstone. The Soloman Atkinson No. 1 well, 10,770.5-10,783.5 feet (3,282.8-3,286.8 m). The bottom of the cored interval is at the lower right. The scale bar is 1 foot (30.5 cm).

stratigraphically equivalent zone without sandstones primarily contained mottled red, green, and brown floodbasin mudstone.

The distribution of the Q and Q2 sandstones is shown by the net pay isopach maps (Figs. 4-10, 4-11) and by the simplified fence diagram (Fig. 4-12). The sandstones have an average thickness of 40 net feet (12.2 m). The maximum sandstone thicknesses of the Q and Q2 are 55 net feet (16.8 m) and 30 net feet (9.1 m), respectively. The Q2 sandstone is separated from the overlying Q sandstone in part of the field by a 5- to 10-foot (1.5-3.1 m) thick mudstone bed, but elsewhere the Q channel sandstone has incised the Q2 sandstone, and a distinction between the two is difficult. These sandstones form a common reservoir based on pressure and production data.

Q2 Sandstone. The Q2 sandstone (Figs. 4-2, 4-5, 4-10) represents the oldest meander-belt system in Little Creek Field. The Q2 sandstone grades from a narrow belt 1,300 feet (400 m) wide in the north to a wider (approximately 2,500 feet; 760 m) belt in the south. The Q2 sandstone limits are not as well defined as those of the overlying Q sandstone, because channel widths are commonly less than the distance between the 40-acre (16.2-ha.) well spacing.

Q Sandstone. The Q sandstone is thick (40 ft, 12.1 m), laterally continuous (2,600 to 8,000 feet; 790 to 2,440 m), and widespread. Its limits are well defined by a number of dry holes which penetrated no sandstone (Figs. 4-3, 4-11, 4-12). The Q sandstone contained an estimated 87% of the in-place hydrocarbons.

The top of the Q sandstone shales out as the channel edge is approached (Fig. 4-13A). This map of the Q stratigraphic marker to the top of the actual Q sandstone clearly shows shale-out toward the edge, as does the cross section (Fig. 4-13B). The Q marker to top-of-Q sand top map is interpreted to be a "channel" map showing where the fine-grained siltstone and mudstone were deposited in the abandoned channel. Therefore, the map approximates the morphology of the top of the sandstone body (plan view of point bar as illustrated by the depositional model—Fig. 4-5) and the last position of the river prior to abandonment and transgression.

Using the above interpretation, the Q sandstone in Little Creek Field is made up of three large point-bar accretion loops formed by a river flowing generally from the northwest to the south. Sweetwater Field is another possible upstream, abandoned meander

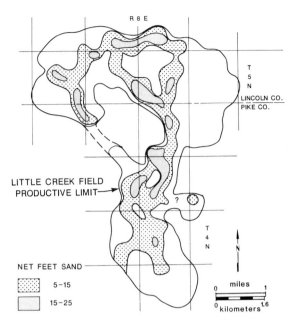

Fig. 4-10. Q2 sandstone net pay isopach map contoured in feet. The dimensions are much smaller (1,320 ft; 402 m) than those of the overlying Q sandstone.

loop of this river system and consists of point-bar and oxbow-lake deposits. This depositional interpretation suggests there should be few major discontinuities within the reservoir. An alternate, but not supported, interpretation is that the sand body was deposited as several smaller meander loops in a complex meander belt, rather than as three large, rather simple meanders. In this case, the depressed sand edge would be the envelope of many point bars along the edge of the meander belt. This seems unlikely because nowhere in the main fairway of sand has an abandoned channel plug or other major sandstone heterogeneity been encountered by a well. In addition, reservoir performance did not indicate the presence of any major discontinuities.

Petrography and Diagenetic History

The sandstones are very fine- to medium-grained, moderately to well-sorted, consolidated to semiconsolidated sublitharenites (Fig. 4-14). Quartz and igneous rock fragments are the most abundant framework minerals, but chert, metamorphic rock fragments, lignite, mica, and pyrite are also present. Minor cements, which are locally important to reservoir quality, include ferroan dolomite, quartz

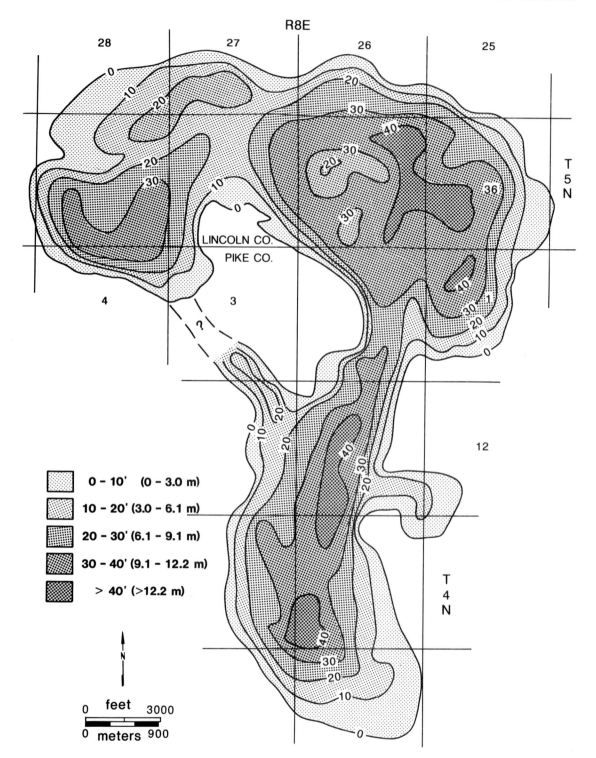

Fig. 4-11. Q sandstone net pay isopach map contoured in feet. The Q sandstone may be grossly divided into three main pay areas (three point bars) connected by narrower zones of channel sandstones.

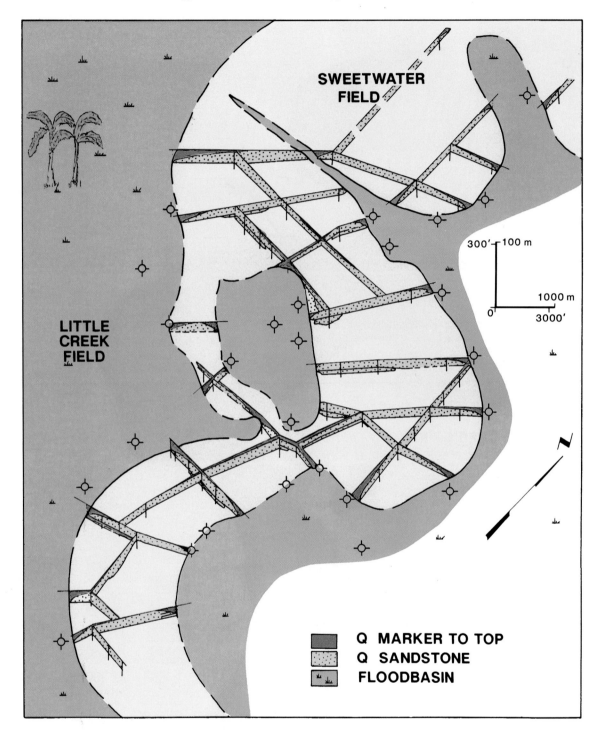

Fig. 4-12. Perspective view (fence diagram) of the Q2 and Q sandstone distribution in the Little Creek Field area. Note indications of channel abandonment and thicker central portions of the point bars.

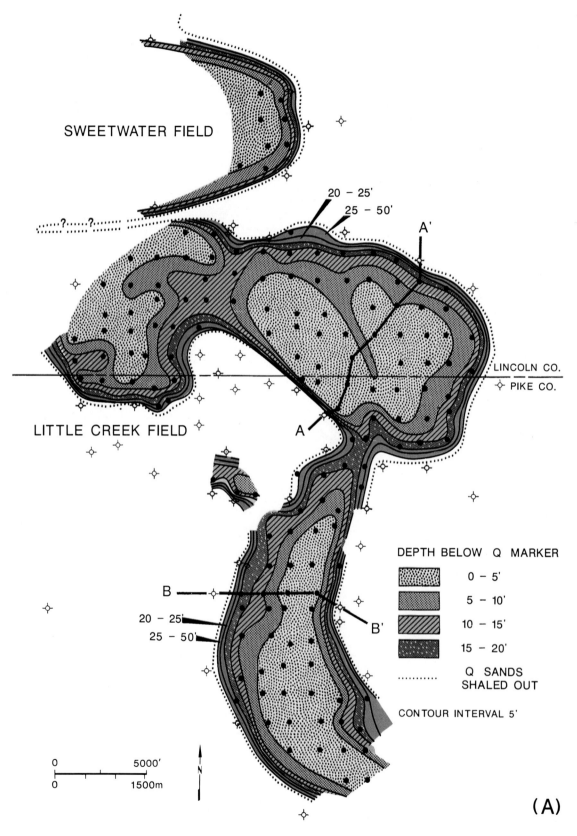

Fig. 4.13.

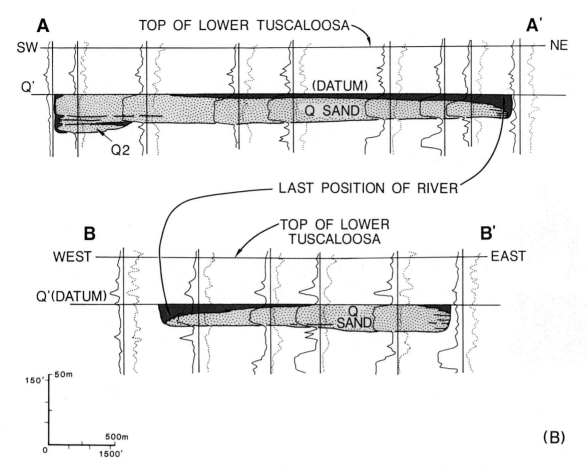

Fig. 4-13. (A) Q marker to top Q sandstone isopach map. This map was very useful in helping to predict the proximity of the channel edge. (B) Illustrates the loss of sand at the top of the section related to the abandonment. A to A' illustrates migration from SW to NE and B to B' illustrates channel migration from E to W.

◁

overgrowths, and siderite. Clays are present as detrital and authigenic types. Detrital clays occur as clay drapes and as matrix in the upper parts of upward-fining channel abandonment sequences. The authigenic clays include pore-lining chlorite and minor kaolinite.

In addition to depositional controls, diagenesis has also played a large role in determining the observed reservoir quality and petrophysical properties. A generalized sequence of the most important diagenetic events, many of which occurred relatively early, is shown in Figure 4-15.

Quartz overgrowths and the ferroan dolomite have greatly reduced or occluded porosity and permeability locally. Chlorite, which occurs throughout the reservoir, is considered most important in determining the reservoir quality and petrophysical properties (see next section). Chlorite may be important in inhibiting quartz cementation and in partially preventing compaction (Thomson, 1979). The chlorite occurs as a uniform "druse" of grain-coating hexagonal plates (Fig. 4-16) except in areas where abundant quartz overgrowths occur. Also present are partially and completely dissolved rock fragments that have "floating" chlorite rims and/or a matrix of chlorite. These degraded rock fragments are the probable source of iron for the formation of this early formed Fe-rich chlorite. Chlorite occurs throughout the Lower Tuscaloosa sandstones in southwestern Mississippi (Stancliffe and Adams, 1986; Shew, 1987) and in the downdip Tuscaloosa in south Louisiana (Thomson, 1979).

Fig. 4-14. The reservoir sandstones are predominantly moderately to well-sorted, fine- to medium-grained, quartz-rich sublitharenites. Most framework grains in this photo are quartz (q), but partially dissolved and relatively fresh volcanic rock fragments (rf) are present. Also shown is a small patch of pore-filling ferroan dolomite (fd), and most of the grains have a 5 to 10 micron coating of chlorite. This sample has good macroporosity (p) as well as abundant microporosity within the clays. Scale bar is 0.50 mm.

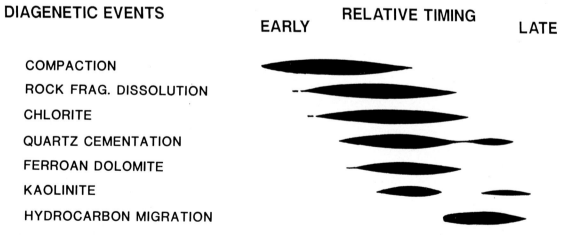

DIAGENETIC EVENTS   RELATIVE TIMING
                                 EARLY   LATE

COMPACTION
ROCK FRAG. DISSOLUTION
CHLORITE
QUARTZ CEMENTATION
FERROAN DOLOMITE
KAOLINITE
HYDROCARBON MIGRATION

Fig. 4-15. A generalized diagenetic sequence for the Lower Tuscaloosa Formation in Little Creek Field.

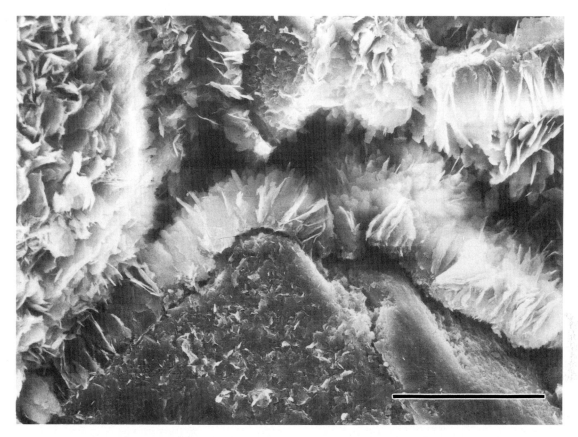

Fig. 4-16. Scanning electron photomicrograph of grain-coating chlorite. The chlorite is 5 to 10 microns in length and consists of thin hexagonal platelets which are oriented perpendicular to the grain surface. Chlorite is very impor-tant in controlling both reservoir quality and petrophysical properties. A large amount of bound water, leading to high S_w, occurs between the chlorite platelets. Scale bar is 0.03 mm.

Reservoir Quality and Petrophysics

Reservoir quality is generally good. The Q and Q2 sandstones have average air porosities and permea-bilities of 24% and 100 md, respectively. Figure 4-17 is a cross-plot of the porosity and permeability from a number of core measurements. This illus-trates some of the variability present in rock proper-ties within the point bar. Figure 4-18 also shows the porosities, permeabilities, saturations, and log data for two typical wells in Little Creek Field. Reservoir quality is controlled by the original depositional fabric with a diagenetic overprint. Shale drapes, beds of sandstone with increased detrital clay, and local cemented zones (primarily carbonate tight streaks) have led to reduced vertical permeabilities. The reduction ranges from slight in clean blocky sandstones to as much as 50% where tight streaks

and abandonment sequences occur. However, the shale beds and cemented zones have a very limited extent. The zones often occur in only one well or possibly several wells but do not cross the entire point-bar deposit. Lateral continuity of the reservoir is generally high and vertical communication, except at some individual wellbores, is also good. Detailed well-log correlations and reservoir per-formance (primary and waterflood operations) sup-port this interpretation.

Most of the porosity is primary intergranular but minor secondary dissolution porosity (from the degradation of the volcanic rock fragments) is also present. However, as much as 50% of the measured porosity may be microporosity associated with the pervasive grain-coating chlorite, thus reducing the effective porosity. Conversely, the chlorite is also important and beneficial in helping to maintain good

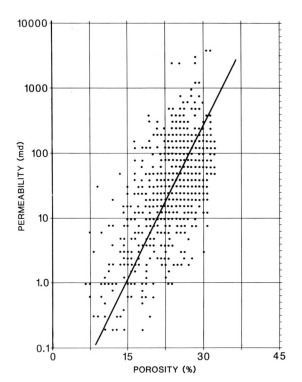

Fig. 4-17. Plot of porosity versus permeability for Lower Tuscaloosa sandstones. The large range of values (unstressed) is primarily related to the abundance of clays and to grain-size variations.

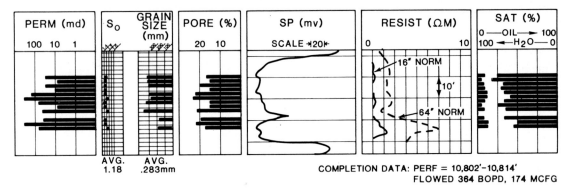

COMPLETION DATA: PERF = 10,802'–10,814'
FLOWED 364 BOPD, 174 MCFG

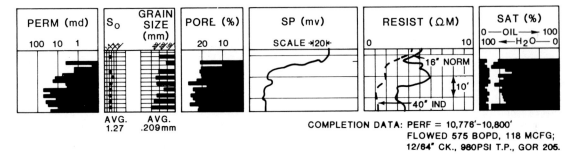

COMPLETION DATA: PERF = 10,776'–10,800'
FLOWED 575 BOPD, 118 MCFG;
12/84" CK., 980PSI T.P., GOR 205.

Fig. 4-18. Partial logs and rock properties of the Q sandstone in two wells. These illustrate the lithologic and petrophysical properties that are characteristic of the more blocky, middle part of the point-bar sequence. The Shell Atkinson well has a slightly reduced porosity and permeability at the top of the sandstone. Note the extremely high water saturation in these sandstones.

reservoir quality by inhibiting cementation and reducing compaction. The chlorite is also the most important compositional component in terms of its effects on the petrophysical properties (Shew et al., 1989). Extremely low resistivities (<1 ohm-m) and average high water saturations (>55%) are related to the bound water within the microporous chlorite. It is therefore possible to produce clean oil even with these extremely high water saturations.

A relatively long transition zone, averaging 30 feet (9.1 m), is present below a 90-foot (27.4-m) clean-oil producing zone. Figure 4-19 shows representative low-pressure capillary pressure curves for Little Creek Field and illustrates the importance of reservoir quality and position within the reservoir to clean-oil production. Based on capillary pressure curves, oil shows, initial production test data, and log analysis, it was determined that the free- and 100%-water levels were at 10,425 feet (3,178 m) and 10,420 feet (3,176 m) subsea, respectively. Water-free production was obtained above 10,390 feet (3,167 m) subsea. A large portion of the field was underlain by the oil-water transition zone, which was calculated to contain about 18% of the net acre feet and 13% of the OOIP.

Detailed mapping and rock property studies are important in understanding the occurrence and distribution of hydrocarbons in the Mid-Dip Tuscoloosa. For instance, Sweetwater and Little Creek fields (Fig. 4-3) are separate hydrocarbon accumulations, as indicated by different oil-water contacts, but appear to be part of the same fluvial depositional system based on detailed log and seismic correlations. The subsea depths at the crest of the trapped hydrocarbons and the characteristics of the Q sandstone capillary-pressure curve are the same in each field; however, Sweetwater Field has an 80-foot (24-m) higher average oil-water contact than does Little Creek Field. All Sweetwater Field wells produced both oil and water initially or shortly after production started, indicating they were entirely within the oil-water transition zone. Little Creek Field had an average 30-foot (9.1-m) transition zone leaving a 90-foot (27.4-m) clean-oil producing interval. As there are only 40 feet (12.2 m) of synclinal closure separating the fields, a shale plug or an area with relatively high capillary-displacement pressure must be present or sand must be absent to separate the fields. Also, the Sweetwater Field meander loop must have a spill point to the north, i.e., continuous open-ended sandstone.

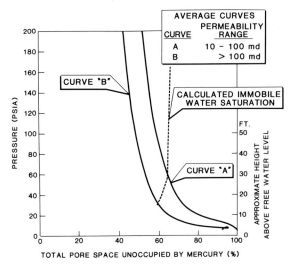

Fig. 4-19. Capillary pressure curves of typical rock types in Little Creek Field. A larger transition zone will be present in poorer quality rock samples. Higher-pressure capillary pressure curves would illustrate the bimodal nature of the pore size distribution, including the macroporosity shown here and the microporosity (within chlorite) apparent at higher injection pressures.

Rock/Log Correlation. In addition to the resistivity/chlorite association previously mentioned, other rock/log correlations are important in influencing reservoir quality. The amplitude of SP curve deflection generally increases with an increase in grain size. The distribution of typical SP shapes for the Q-Q2 sandstones is shown in Figure 4-20. In general, the SP curves tend to be blocky toward the center of the channel sandstone and bell-shaped toward the edge of the channel, as would be expected from the depositional model. The SP curve also tends to be more serrate near the channel edge; however, this was not consistent enough to be a reliable sand-edge indicator.

Permeability profiles vary from location to location, as might be expected from the log shapes shown in Figure 4-20: decreased permeability near the top of the sandstone in bell-shaped curves and fairly uniform permeabilities in blocky curves. No consistent layered sequence was observed, but the local barriers to vertical flow (clay drapes and cemented zones) as well as the grain size changes lead to a slightly higher horizontal versus vertical permeability. Porosity remains relatively uniform, except in the cemented or shaly zones, but the distri-

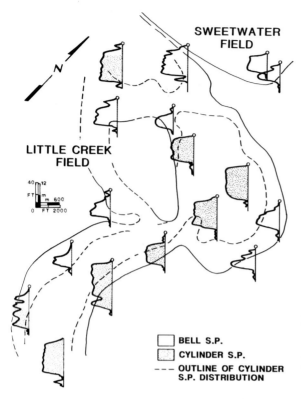

Fig. 4-20. Representative SP curve shapes (blocky in central areas and bell-shaped along the margins) are helpful but not unique in trying to delineate the sandstone boundary. However, when used in combination with the other maps they provide important information to define reservoir quality and distribution.

bution and amount of macroporosity and microporosity are significant to the true effective porosity.

Dipmeters. Model interpretations and the recognition in cores and logs that the sandstones shale out along the edge of the abandoning channel led geologists to use dipmeter logs to attempt to predict the strike of the channel margin during development drilling. Use was made of the shale drapes on the foreslope (lateral accretion beds) of the point bar to measure dips. This information was then used to determine the strike and direction to the edge of the sand body. The idea met with limited success in two wells. Improved high-resolution stratigraphic dipmeters have now been used extensively for channel axes determination, but few references are available concerning dipmeter interpretation of the channel margin.

Compaction Effects. A comparison was made of the interval thickness between two reliable resistivity

datums in two wells. One well contained no sandstone, and the other well contained both the Q and Q2 sandstones. The calculated average compaction factor for the shales was about 75%. Knowing the shale compaction value is important for the localization of sandstone highs and for the correct correlation of lithologic markers within the Tuscaloosa interval.

Production Characteristics

Little Creek Field was completely developed and delineated by Shell and competitors with 208 wells (162 completions and 46 dry holes). Development occurred rapidly on 40-acre (16.2-ha.) spacing, and it became apparent early that the primary producing mechanism would be liquid expansion and solution-gas drive with only limited water influx. Bottom-hole pressures steadily dropped from the initial reservoir pressure of 4,840 psi (3.3×10^4 kPa) to 3,000 psi (2.1×10^4 kPa). The bubble point pressure was 2,150 psi (1.5×10^4 kPa) at a subsea depth of 10,375 feet (3,162 m). Unitization efforts began in 1959 when it was recognized that pressure maintenance would be necessary for maximum recovery. Approximately 23 MMBO (3.7×10^6 m³) or 22.5% were recovered by primary production, and 25 MMBO (4.0×10^6 m³) or 24.5% of the initial 102 million barrels (1.6×10^7 m³) oil in place was recovered by the line-drive waterflood which occurred from 1962 to 1970. The results of the primary and waterflood production operation are shown in Figure 4-21. Thirteen percent of the oil in place was estimated to be present in the Q2 sandstone and the remainder in the Q sandstone. Although porosity and permeability variations are present in the reservoir, the flood results supported the interpretation that most of the observed heterogeneities within the wells are localized and not reservoir-wide barriers.

Even after producing 47% of the original oil in place, the large remaining oil volumes led Shell to consider alternative methods to produce part of the remaining oil. A CO_2 pilot study was conducted from 1974 to 1977 (Hansen, 1977). The small pilot area chosen was in the eastern part of the field and consisted of one CO_2 injector, three producers, and five back-up water injectors in an inverted nine-spot (Fig. 4-22). The results were encouraging and, after producing 125 MBO (2.0×10^4 m³), plans were formulated for a phased full-scale CO_2 flood for the field. The Phase I area to be flooded is shown in

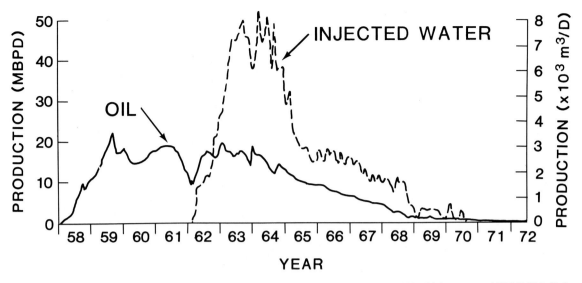

Fig. 4-21. Plot of primary and secondary (utilizing water injection) production to 1972 which recovered 23 MMBO (3.7 × 10⁶ m³) and 25 MMBO (4.0 × 10⁶ m³), respectively.

Figure 4-22. Injection of CO_2 in this area did not commence until December, 1985 because of further unitization efforts, construction of a pipeline from the carbon dioxide source area near Jackson, Mississippi, and construction of field facilities. Figure 4-23 shows the early project results. The estimated additional oil recovery of the CO_2 flood, based on results from the pilot study, is more than 20 MMBO (3.2 × 10⁶ m³).

Exploration and Production Strategy

Seismic Data

Little Creek Field was drilled and discovered based on subsurface mapping and on the seismic interpretation of minor structural closure at the Lower Tuscaloosa horizon. Most, if not all, of the larger structural or combined structural/stratigraphic traps have likely been discovered in the mature Mid-Dip trend. However, detailed seismic stratigraphic analysis, coupled with comprehensive geologic and petrophysical models, has been used to identify and map smaller stratigraphic traps (Shew, 1987). Seismic models have indicated that a "soft" seismic response results from porous sandstones being encased within "harder" floodbasin mudstones. Figures 4-4 and 4-24 illustrate the presence of the objective Q sandstone and its pinch-out into stra-

tigraphically equivalent mudstones and shales. Seismic stratigraphic analysis is probably the most useful tool for the discovery of these smaller fields and is extremely useful for delineating the reservoir.

Mapping Techniques

In addition to the typical maps generated for volumetric calculations, several other maps and cross sections were used during development in an effort to delineate more accurately the channel margins and to reduce the dry hole risk. Although no map proved to be a panacea, one map, the Q marker to Q sandstone (Fig. 4-13), proved to be a very useful, practical, and reliable tool for mapping reservoir boundaries. This map defines the abandonment sequence and delineates the last position of the river channel prior to abandonment. The technique was discovered from the results of a thin layer ("slice") map study which showed that the top of the sand was indeed shaling out along the edge as the point-bar geologic model would predict. Nowhere in the field was a successful well drilled on a 40-acre (16.2-ha.) location outside of and perpendicular to the 15-foot (4.6-m) contour of this "channel" map. Only in one area, on the inner part of the north-central meander loop, did an abrupt and complete shale-out occur with no deterioration of the upper part of the sand in the adjacent well. The layer map was extensively utilized in the selection of Shell's well locations.

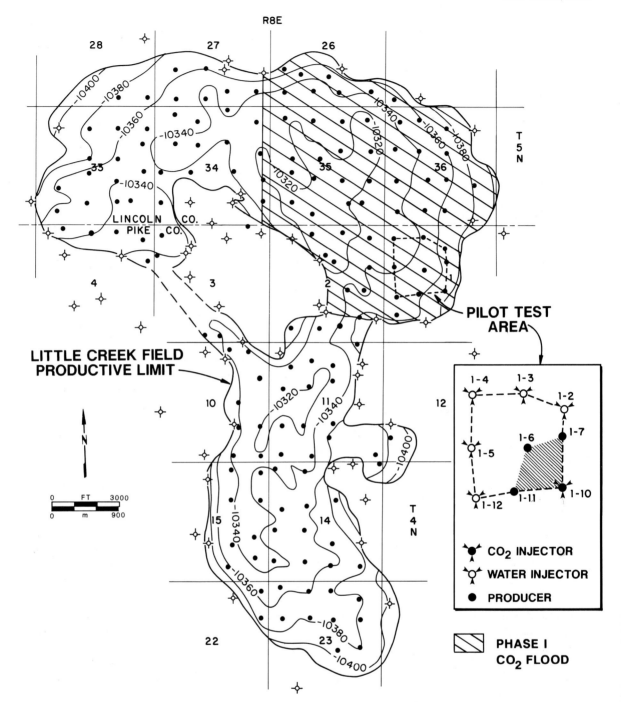

Fig. 4-22. Structure map on the top of Q sandstone showing the location of the pilot CO_2-injection project and the currently operating Phase I flood area.

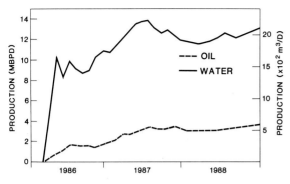

Fig. 4-23. Early project performance for the first phase of CO_2 injection at Little Creek Field.

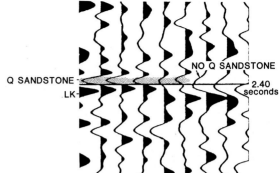

Fig. 4-24. Seismic expression of the "soft" porous sandstone (left-kicking stippled loop) and its termination into "hard" encasing mudstones.

When it was noted that the top of the sand was shaling out, more cautious outsteps were made, even if the total sandstone thickness remained high.

Further discoveries and proper development of fields in the mature Mid-Dip Tuscaloosa trend are dependent upon using an integrated geologic/geophysical approach for predicting the presence and distribution of smaller, stratigraphically trapped sandstones.

Conclusions

The Lower Tuscaloosa Q-Q2 sandstones are primarily point bars deposited in sinuous fluvial meanders on a deltaic plain. Reservoir quality is generally high and is controlled by both depositional and diagenetic processes. Although local reservoir heterogeneities are present, detailed correlations and reservoir performance during primary and waterflood production indicated good lateral continuity. Blocky channel sandstones in the heart of the field have relatively uniform porosities and permeabilities, whereas the channel margins often exhibit decreasing sandstone quality near the top of the Q sandstone, representing channel abandonment. Average porosities and permeabilities are 24% and 100 md, respectively. Clean-oil production from sandstones with high water saturations should not be over-looked because of the large amount of bound water associated with the presence of pervasive chlorite.

Recovery efficiencies are quite high in this reservoir. Of the 102 MMBO (1.6×10^7 m³) originally in place, 23 MMBO (3.7×10^6 m³) were recovered by primary and 25 MMBO (4.0×10^6 m³) by secondary waterflood production, leaving 54 MMBO (8.6×10^6 m³) as an enhanced oil recovery target. Using CO_2 in the enhanced recovery process, an additional 20 MMBO (3.2×10^6 m³) are estimated to be recoverable from the field. Meander-belt systems, like Little Creek Field, provide reservoirs that are inherently difficult to delineate but are generally high-quality sandstones with "built-in" top and side seals. Production from these reservoirs requires that close attention be paid to permeability stratification and potential shale baffles and barriers within the point-bar sands. Integrated geological, geophysical, and engineering studies are essential to the delineation and maximum hydrocarbon recovery from these types of reservoirs.

Acknowledgments. We thank Shell Oil Company for permission to publish this paper and express our appreciation to both the many former and current workers who provided a better understanding of the geological and engineering characteristics of Little Creek Field.

Reservoir Summary

Field: Little Creek
Location: Southwestern Mississippi
Operator: Shell Oil and many other operators
Discovery: 1958
Basin: Mississippi Salt basin; Gulf of Mexico
Tectonic/Regional Paleosetting: Subsiding passive margin
Geologic Structure: Anticlinal nose
Trap Type: Four-way closure and sandstone pinchout into shale
Reservoir Drive Mechanism: Solution gas; little water
  • **Original Reservoir Pressure:** 4,840 psi (3.3×10^4 kPa) at 10,340 feet (3,152 m) subsea
Reservoir Rocks:
  • **Age:** Late Cretaceous, Cenomanian
  • **Stratigraphic Unit:** Lower Tuscaloosa Formation
  • **Lithology:** Fine- to medium-grained sublitharenite
  • **Depositional Environment:** Fluvial meander belt
  • **Productive Facies:** Point bars
  • **Petrophysics**
    • **Porosity Types:** Total = 21 to 27%; 13 to 15% primary; 3 to 4% secondary dissolution; 5 to 8% microporosity
    • **ϕ:** Average 24%; range 10 to 35% (cores)
    • **k:** Average 100 md; range 0.1 to 1000 md (air, cores)
    • **S_w:** Average 55%; range 40 to 75% (logs)
    • **S_{or}:** NA
Reservoir Dimensions
  • **Depth:** 10,770 feet (3,283 m)
  • **Areal Dimensions:** 6 by 3 miles (9.7×4.8 km)
  • **Productive Area:** 6,200 acres (2,510 ha.)
  • **Number of Reservoirs:** 1 (composed of the Q and Q2 sandstone members)
  • **Hydrocarbon Column Height:** 100 ft (30.5 m)
  • **Fluid Contacts:** Oil-water at 10,390 ft (3,167 m) subsea
  • **Number of Pay Zones:** 1 (Q and Q2 sandstones)
  • **Gross Sandstone Thickness:** 40 ft (12.2 m); range 15 to 85 ft (4.6-25.9 m)
  • **Net Sandstone Thickness:** 30 ft (9.1 m)
  • **Net/Gross:** 0.75
Source Rocks
  • **Lithology and Stratigraphic Unit:** Marine shale, Middle Tuscaloosa
  • **Time of Hydrocarbon Maturation:** Tertiary
  • **Time of Trap Formation:** Late Cretaceous (stratigraphy); Tertiary (structural)
  • **Time of Migration:** Tertiary
Hydrocarbons
  • **Type:** Oil
  • **GOR:** 555:1
  • **API Gravity:** 39°
  • **FVF:** 1.32
  • **Viscosity:** 40 cP (4.0×10^4 Pa·s) at 200°F (93°C)
Volumetrics
  • **In-Place:** 102 MMBO (1.6×10^7 m³)
  • **Cumulative Production:** Total = 48 MMBO (7.6×10^6 m³); Primary = 23 MMBO (3.7×10^6 m³); Secondary = 25 MMBO (4.0×10^6 m³)
  • **Ultimate Recovery:** Est. 68 MMBO (1.1×10^7 m³)
  • **Recovery Efficiency:** 67% (primary, waterflood, CO_2)
Wells
  • **Spacing:** 40-acre (16.2-ha.); 1,320 ft (402 m)
  • **Total:** 208 (all operators)
  • **Dry Holes:** 46 (all operators)

Typical Well Production
- **Average Daily:** 300 BO (48 m³)
- **Cumulative:** 141 MBO (2.2 × 10⁴ m³) Primary

References

Bernard, H. A., and Major, C. F., Jr., 1963, Recent meander belt deposits of the Brazos River: An alluvial "sand" model [abst.]: American Association of Petroleum Geologists Bulletin, v. 47, p. 350.

Cronquist, C., 1968, Waterflooding by linear displacement in Little Creek Field, Mississippi: Journal of Petroleum Technology Transactions, v. 243, p. 525–533.

Hansen, P. W., 1977, A CO_2 Tertiary recovery pilot, Little Creek Field, Mississippi: Society of Petroleum Engineers, Paper 6747, 7 pp.

Shew, R. D., 1987, Seismic and geologic interpretation and reservoir properties of the Lower Tuscaloosa B Sandstone, Olive Field, southwestern Mississippi: Society of Exploration Geophysicists, Annual Meeting, New Orleans, p. 131–132.

Shew, R. D., Werren, E. G., Adams, E. R., and Stancliffe, R. J., 1989, Depositional, diagenetic, production, and seismic characteristics of a Mid-Dip Tuscaloosa point-bar complex, Little Creek Field, Mississippi [abst.]: American Association of Petroleum Geologists Bulletin, v. 73, p. 412.

Stancliffe, R. J., and Adams, E. R., 1986, Lower Tuscaloosa fluvial channel styles at Liberty Field, Amite County, Mississippi: Gulf Coast Association of Geological Societies Transactions, v. 36, p. 305–313.

Thomson, A., 1979, Preservation of porosity in the deep Woodbine/Tuscaloosa Trend, Louisiana: Gulf Coast Association of Geological Societies Transactions, v. 29, p. 396–403.

Key Words

Little Creek Field, Mississippi, Mississippi Salt basin, Lower Tuscaloosa Formation, Upper Cretaceous, Cenomanian, fluvial, meander belt, enhanced oil recovery, point-bar sandstones, chlorite, low resistivity pays, dipmeters, compaction, seismic stratigraphy.

5

Fluviodeltaic Reservoir, South Belridge Field, San Joaquin Valley, California

Donald D. Miller, John G. McPherson, and Thomas E. Covington

Mobil Exploration & Producing U.S., Inc., Denver, Colorado 80217; Mobil Research & Development Corporation, Dallas, Texas 75244; Mobil Exploration & Producing U.S., Inc., Denver, Colorado 80217

Introduction

A high percentage of the oil produced from sandstone reservoirs comes from fluviodeltaic sandstones. The abundance of high-quality reservoir sand, the ideal location relative to downdip source beds, and the high potential for stratigraphic trap development all make this depositional setting a favored target for explorationists. However, the sand-body architecture and reservoir quality of fluviodeltaic sandstones commonly display extreme variability over short vertical and lateral distances due to depositional controls. The fluvial-dominated delta system described in this chapter displays a wide spectrum of depositional settings from sandy braided fluvial to muddy delta front. This case study presents an opportunity to compare and contrast the reservoir characteristics of distal-bar, mouth-bar, meandering fluvial, and braided fluvial sands from the same depositional system in a single field, with well spacing close enough to make correlations relatively certain. Sand-body geometries range from sheets to shoestrings, and reservoir quality ranges from excellent to poor. The case study illustrates the importance of understanding a reservoir at all scales, from depositional setting to pore-throat geometry.

The Pleistocene Tulare Formation at South Belridge Field is a giant heavy-oil reservoir composed of unconsolidated sediments deposited in a fluviodeltaic setting. Small-scale reservoir geometries are defined and exploited by a high density of well penetrations. More than 6,000 closely spaced and shallow wells are the key to producing more than an estimated 1 billion barrels (1.6×10^8 m³) of oil from hundreds of layered and laterally discontinuous reservoir sands. Wells are typically spaced 200 to 500 feet apart (61–152 m) for optimal heavy-oil production in steamflood patterns and drilled to a total depth of 1,000 feet (305 m) within the 14 square mile (36 km²) producing area.

South Belridge Field's annual production ranks among the top five fields in the United States. It is one of several giant fields in the oil-rich San Joaquin Valley of California (Fig. 5-1). Although the field contains several producing formations, only the main producing reservoir, the Pleistocene Tulare Formation, is the subject of this chapter.

The discovery well was completed in 1911 at a depth of 782 feet (238 m) with initial production of 100 BOPD (16 m³/D) (California Division of Oil and Gas, 1950). The well was located next to an outcrop of dry oil sand in a creek bed. Historically, development was slow and sporadic, as dictated by fluctuating market demand for heavy asphaltic crude oil. Fieldwide development recently intensified, concurrent with the development of in situ combustion and steamflood technology. Continued development has resulted in 6,100 active wells (Fig. 5-2) with daily production of approximately 170 MBO (2.7×10^4 m³) (California Division of Oil and Gas, 1987) and annual production of 64 MMBO (1.0×10^7 m³) (Oil and Gas Journal, 1988). Cumulative field production has been approximately 700 MMBO (1.1×10^8 m³), and additional recoverable reserves are

Casebooks in Earth Science
Sandstone Petroleum Reservoirs
Eds.: Barwis/McPherson/Studlick

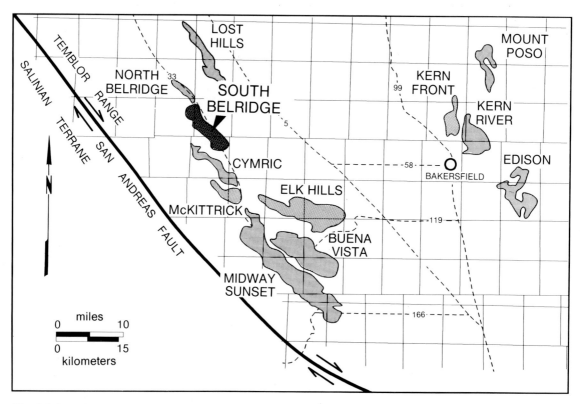

Fig. 5-1. Location map showing South Belridge Field, other giant heavy oil fields, and the San Andreas fault in the southern San Joaquin Valley, California.

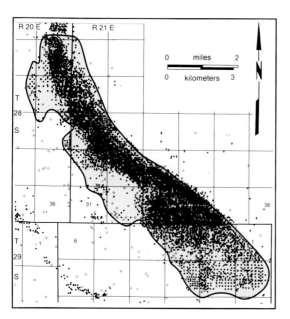

Fig. 5-2. Well location map of South Belridge Field showing locations of 6,100 active wells. (Permission to publish from Petroleum Information, 1988)

approximately 500 MMBO (8.0×10^7 m³) (California Division of Oil and Gas, 1987).

Regional Data

Location

South Belridge Field is located on the western side of the San Joaquin Valley in Kern County, California, 100 miles (160 km) north of Los Angeles and 40 miles (64 km) west of Bakersfield (Fig. 5-1). The most prominent local geologic feature is the San Andreas fault, which is 12 miles (19 km) southwest of the field. This fault system is the western transform margin of the North American plate and has been a major influence on San Joaquin basin geologic history.

Regional and Tectonic Setting

Regional tectonism strongly controls the character of central California sedimentary basins, as forearc basins were replaced in Tertiary time by localized borderland basins related to strike-slip faults (Graham, 1987). A number of detailed regional syntheses describe the dynamic basinal tectonic setting and the resulting framework for abrupt and complex facies changes in detail (Atwater, 1970; Nilsen and Clarke, 1975; Dickinson et al., 1979; Webb, 1981; Bartow, 1987; Namson and Davis, 1988). The elongate San Joaquin basin first developed as part of the Mesozoic Great Valley forearc basin (Ingersoll, 1979) and consisted chiefly of a marine shelf and slope largely open to the west (Clarke et al., 1975). It subsequently evolved into a late Cenozoic intermontane basin as strike-slip movement along the San Andreas fault and related tectonic events progressively closed off the basin. The change from convergent to transform margin imposed a new structural character and changed the style of sedimentation during the late Cenozoic.

Patterns of sediment accumulation in the western San Joaquin basin were directly influenced by the mountains which bordered the basin on the west (Lennon, 1976). Granitic highlands of the Salinian terrane west of the San Andreas fault and the adjacent Temblor uplift east of the San Andreas fault provided the provenance for the feldspar-rich sands of the Pleistocene Tulare Formation that constitute the Tulare reservoir at South Belridge Field. The final marine regression of the San Joaquin basin began in the late Miocene and continued through the Pliocene as progradation and aggradation of coarse clastic sediments from all sides of the basin outpaced subsidence (Lettis, 1982). This led to the final retreat of the sea near the end of the Pliocene (Bartow, 1987).

South Belridge Field lies on the crest of a large anticline along the western side of the San Joaquin basin. The anticline is one of many that are subparallel to the San Andreas fault system (Harding, 1976). Growth of the anticline before, during, and after deposition of the Tulare sands had an important influence on gross sand and reservoir distribution in the field area.

Stratigraphy

Two major Neogene erosional unconformities are present at South Belridge Field. A late Miocene unconformity is overlain by the Mio-Pliocene Etchegoin, Pliocene San Joaquin, and Pleistocene Tulare formations (Fig. 5-3). The Etchegoin Formation is predominantly marine and wedges out on the flanks

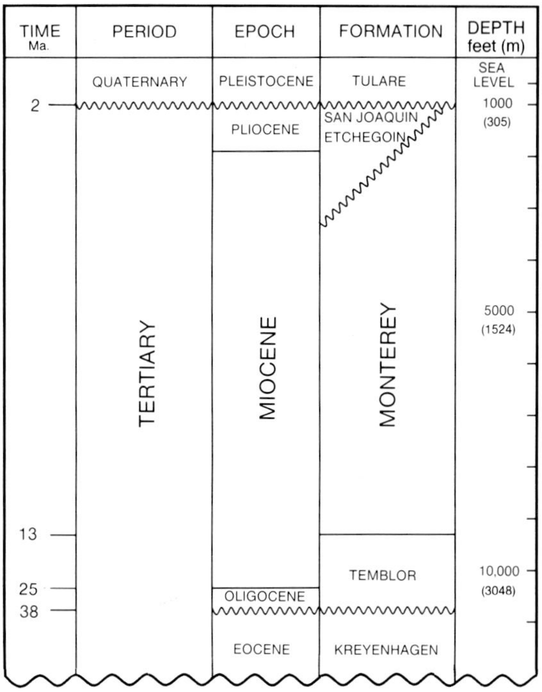

Fig. 5-3. Simplified stratigraphic section showing geologic ages and unconformable contacts between the Pleistocene Tulare Formation and the Pliocene Etchegoin and San Joaquin formations and the upper Miocene in South Belridge Field.

of the South Belridge anticline. The overlying San Joaquin Formation represents generally lacustrine and brackish-water environments (Barbat and Galloway, 1934). It also wedges out on the South Belridge anticline and overlaps onto the Miocene deposits.

The second unconformity occurs where the basal Tulare truncates the Etchegoin, San Joaquin, and Miocene Monterey formations on the South Belridge anticline. Structural deformation and erosion created a complex and irregular surface. The Tulare was then deposited unconformably on both the Pliocene and the Miocene cores of the structure, becoming conformable with upper Pliocene sediments toward the center of the basin to the east.

The Pleistocene Tulare Formation contains the youngest folded strata in the western San Joaquin basin (Fig. 5-3). This nonmarine sequence (Woodring et al., 1940; Stanton and Dodd, 1970) represents filling of the southern San Joaquin basin during Pleistocene time. Sediment was fed from nearby basin-margin highlands, via coarse-grained deltas, into the large brackish to freshwater Tulare Lake. Both fan-delta and braid-delta depositional systems

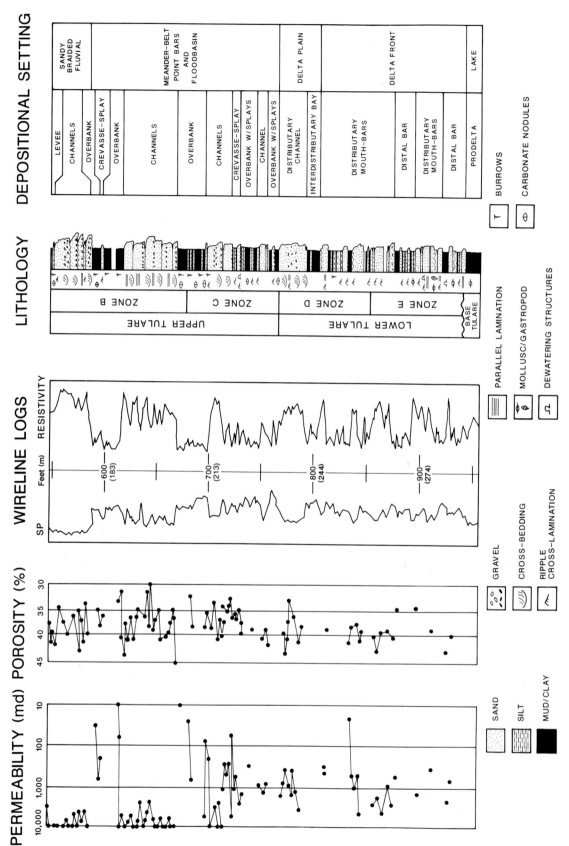

Fig. 5-4. Composite log showing reservoir zones, typical log responses, reservoir quality, and depositional lithofacies for lower Tulare deltaic sands and upper Tulare fluvial sands. The porosity values are from routine analyses at ambient conditions and permeability values (air) at 300 psi (2.1×10^3 kPa) confining pressure.

(terminology of McPherson et al., 1987) are represented in the Tulare Formation.

Tulare stratigraphy at South Belridge Field is locally subdivided into upper and lower divisions, each containing several reservoir zones (Fig. 5-4). The upper Tulare division includes zones A, B, and C. The lower Tulare division includes zones D, E, and in some places F and G. These divisions are based on general aspects of reservoir behavior and early mapping of oil-water contacts. They are not formally recognized elsewhere in the basin. This reservoir stratigraphy underlies the widespread lacustrine Tulare Corcoran Clay stratigraphic marker.

Reservoir Characterization

Geometry

Reservoir geometries in the Tulare occur on two major scales and show two influences: a large-scale structural configuration, and smaller-scale depositional controls. The South Belridge structure is a large southeast-plunging anticline, but within the structure are complex, small-scale, reservoir geometries created largely by the depositional setting. The geometry of individual sand bodies and the effective reservoir geometry of interconnected sand bodies show considerable vertical and lateral variations resulting from temporal and spatial changes in the depositional setting. The upper and lower Tulare reservoir sands consequently display quite different reservoir geometries as a result of their differing depositional settings. Syndepositional growth of the South Belridge anticline further complicated the gross sand-body distributions and resulting reservoir geometries.

Geometries have been interpreted from wireline-log correlations, core facies analysis, log character, producing characteristics, steam pathways, and comparison to nearby Tulare outcrops. Preserved cores from 21 wells and information from unpreserved, disaggregated cores from 221 older wells support log correlations of 1,700 wells in the southern area of the field, the primary area of this study. Previous fieldwide studies by Mobil Oil Corporation used more than 1,000 well logs in the adjacent northern area of the field.

The lower Tulare reservoir geometries are the product of distributary-channel, distributary-mouth-bar/inner-fringe, and distal-bar/outer-fringe depositional facies (Fig. 5-4). A typical lower reservoir unit comprises a 5 to 15-foot (1.5–4.6 m) thick sand

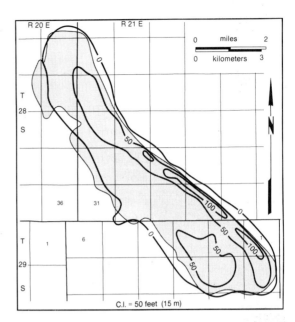

Fig. 5-5. Lower Tulare oil-sand thickness map indicating widespread distribution of sands on the flanks of the anticline in the southern part of South Belridge Field. Sands thicken basinward to the northeast and pinch out to the north and west, creating a stratigraphic trap at the northern end of the field.

body interbedded with clays of lacustrine and interdistributary-bay origin. Moderate reworking of the delta front created a shore-parallel (SE-NW) orientation with the exception of a few distributary channel sands. The lower Tulare sands are highly stratified and are distributed throughout the producing area of the field (Fig. 5-5) and broader region (Lennon, 1976). Reservoir sands and overlying clay strata typically have very good lateral continuity, as illustrated by each of the lower sand horizons displaying separate oil-water contacts. Nonproductive sands of greater thickness occur off structure toward the basin center and link the producing reservoir with an extensive, active aquifer (Fig. 5-6). Productive reservoir sands are thicker in the southern portion of the field. The total interval thickness of the lower Tulare ranges between 100 and 300 feet (30–91 m), and oil-producing reservoir sands total as much as 150 feet (46 m) thick.

The upper Tulare sand distribution is somewhat restricted over the field as a product of its fluvial deposition (Fig. 5-7). The sands thin and pinch out in the northwestern area of the field. The thickest producing sands occur on both flanks of the anticline and thin over the crest in the southern portion of the field

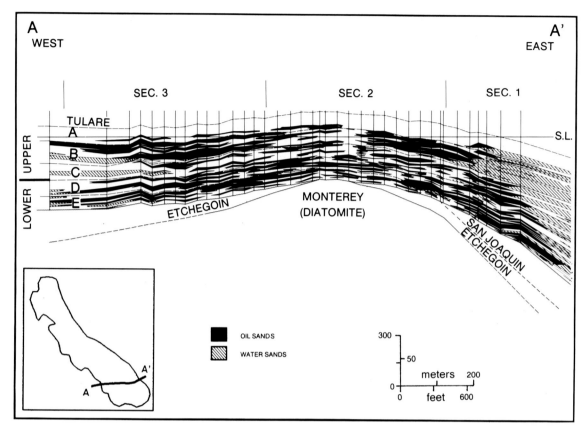

Fig. 5-6. Geologic cross section illustrating structural and stratigraphic relationships of Tulare reservoir sands. Well locations are represented by the vertical lines and the datum (S.L.) is sea level. Note the highly layered and laterally discontinuous nature of reservoirs sands, each horizon with its own oil/water contact.

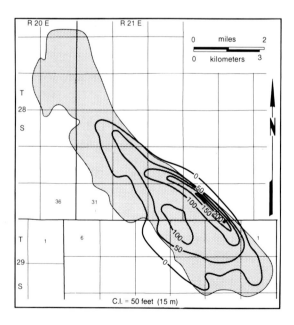

(Figs. 5-6, 5-7). This reflects a paleotopographic influence of the developing structure on gross sediment distribution. By contrast, the lower Tulare sands exhibit greater continuity over the structure (Fig. 5-6), because as deltaic sands they were probably less susceptible to paleotopographic control.

Regionally, the upper Tulare sands are thick to the south of the field and in the general direction of the sediment source. Nonproductive, water-bearing upper Tulare reservoir quality sands are thickest basinward to the east of the anticline, similar to the lower Tulare reservoir sands. West of the field these upper reservoir sands thin and pinch out. The total

Fig. 5-7. Upper Tulare oil-sand thickness map showing thickest sands on the flanks of the anticline in the southern part of South Belridge Field.

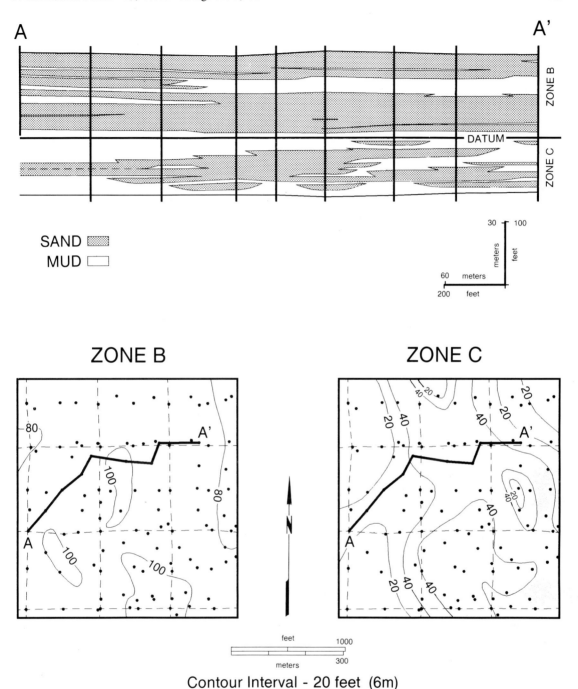

Fig. 5-8. Detailed upper Tulare stratigraphic cross section A-A′ with the corresponding net sand maps of reservoir zones B and C. Well locations on the cross section are represented by the vertical lines, and the stratigraphic datum is the contact between the two reservoir zones. The cross section illustrates the scale and variability of vertical and lateral reservoir continuity. Note the discontinuous and highly channelized character of the fine-grained meander-belt sands of zone C. By contrast, the overlying coarse-grained meander-belt sands (lower B zone) and braided fluvial sands (upper B zone) have a sheet-like character with high lateral continuity. The corresponding net sand maps clearly show these differences in sand distribution.

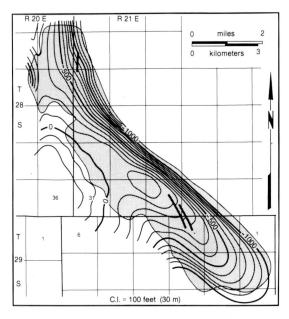

Fig. 5-9. Structural configuration of South Belridge Field reflecting the southeasterly plunging anticline in the southern area of the field and a broad saddle in the northern area of the field. (From and reprinted by permission of California Department of Oil and Gas, 1985.)

interval thickness of the upper Tulare reservoir within the producing area ranges from 300 to 450 feet (91–137 m), and the reservoir sands total as much as 230 feet (70 m) thick.

At a smaller scale, the upper Tulare reservoir geometries are the product of numerous fluvial channel sands. A reservoir unit typically consists of one or multiple stacked and amalgamated channel sands (Fig. 5-8). Reservoirs in the upper Tulare zones are commonly less continuous than for the lower Tulare zones due to the "shoestring" geometry of many individual channel sands. This is most evident in the zone A and zone C sands, which typically show less amalgamated stacking and are encased in overbank/interchannel muds. In contrast, the zone B sands show amalgamation and multilateral stacking which significantly increases the sand-body connectedness and creates a much larger effective reservoir geometry. The vertically stacked and laterally coalesced sand bodies produce a sheet-like reservoir flow unit (Fig. 5-8). Reservoir sand-body widths for the meander-belt sands in the upper Tulare zones have been calculated from paleohydrologic studies of the system (using the methodology of Ethridge and Schumm, 1978; Fielding and Crane, 1987) and average 500 feet (152 m), with a range from 150 to

1,500 feet (46–460 m). These calculated widths were confirmed by detailed sand-body mapping and coring in a selected 10-acre (4.1-ha.) pattern of the field. The sand-body widths have a direct bearing on well-spacing design.

Trapping Mechanism

Source Rock. The presumed source for the Tulare oil is the underlying Miocene Monterey Formation, which is an organic-rich unit in most areas of California. Total organic carbon (TOC) values greater than 5% (weight) are common in the San Joaquin basin and range from 0.40 to 9.16% (Graham and Williams, 1985). The heavy oil in the Tulare Formation at South Belridge is biodegraded and has characteristics of water washing as evidenced by oil fingerprint data.

Migration. The subcropping relationship of the Monterey Formation source rock to the Tulare Formation in the southern axial portion of the anticline provided the necessary pathway for hydrocarbon charge. The basal Tulare unconformity truncates Monterey rocks on the crest of the anticline, with the deeper Monterey Formation more sharply folded than the Tulare. Dips are commonly greater than 45°. Pre-Tulare and syndepositional hydrocarbon migration is suggested by tar mats in the sediments bounding the unconformity.

Seal. The Tulare hydrocarbon trap is controlled by a combination of structural and stratigraphic factors. Structurally, the overprint of the southeast-plunging Belridge anticline controls the reservoir's overall dimensions (Fig. 5-9). A basin-edge stratigraphic pinch-out provides the principal trapping mechanism in the northern portion of the field.

The greatest area of hydrocarbon entrapment occurs in the lowermost reservoir zones due to the history of continued folding and the stratigraphy of individual reservoir sand horizons encased in sealing clays and muds. The shallowest reservoir zones have minimal structural relief which results in a smaller area of hydrocarbon entrapment. The deepest horizons have the greatest structural relief and the largest area of hydrocarbon entrapment. The resulting "Christmas tree" shape of numerous stacked oil-water contacts in cross-sectional configuration is largely due to the structural arrangement (Fig 5-6).

Fluid-level traps (Foss, 1972) and tar-seal traps also provide local constraints at South Belridge Field. The fluid level, which is usually the ground-

Fig. 5-10. The scale bar is 2 inches (5 cm). (A) Pebble gravel of lithofacies G. The gravel is rich in diatomaceous shale clasts and is a channel-lag deposit. (B) The basal and erosional contact of a channel sand (oil-saturated lithofacies G and S_1) and underlying overbank mud (lithofacies M). Mud intraclasts eroded from the underlying unit can be seen as a lag gravel. (C) Cross-bedded and oil-saturated gravelly (pebbles and granules) sand of lithofacies S_1. (D) Parallel-laminated fine-grained sand of lithofacies S_2. (E) Dewatering structures in fine-grained, heavy-oil-saturated sand of lithofacies S_2. (F) Ripple cross-laminated very fine-grained sand of lithofacies S_3. (G) Carbonate-cemented and burrowed silt typical of lithofacies Si. (H) A mud unit with thin, interstratified silty lenses typical of lithofacies M. The bed has been extensively bioturbated.

water table, forms a trap when the oil cannot migrate upward beyond the plane of zero hydrostatic pressure. Capillary pressure is not a significant control in the highly porous and permeable Tulare reservoirs. Fluid-level traps are reported in a limited area of the field. Local tar-seal traps have been created where oil viscosity has dramatically increased to that of an immobile tar through exposure to either water washing or biodegradation processes involving oxidation and extensive bacterial activity (Connan and Coustau, 1987).

Lithofacies

Lithofacies and lithofacies associations are the key to determining depositional environments of the Tulare sands in South Belridge Field. Furthermore, reservoir quality and producibility of the Tulare are directly tied to lithofacies variability. Six principal lithofacies are recognized: (1) sandy pebble gravel; (2) cross-bedded, gravelly sand; (3) fine-grained sand; (4) ripple-laminated, very fine-grained sand; (5) silt; and (6) mud and clay. Sandy lithofacies are the principal reservoir facies, whereas insignificant reservoir volumes are associated with the gravel and silt lithofacies.

Sandy Pebble Gravel (Lithofacies G). Gravels of pebble and granule size are clast supported, mostly with a poorly sorted sand matrix (Figs. 5-10A, 5-10B). Horizontal stratification and large-scale cross-bedding are common. The clasts comprise both intrabasinal (mud "rip-ups") and extrabasinal types. The extrabasinal clasts are rich in diatomaceous shale eroded from the underlying Miocene Monterey Formation.

Cross-Bedded Gravelly Sand (Lithofacies S_1). These sands are mostly poorly sorted, fine- to coarse-grained with a gravel component that varies in size from granules to pebbles and in concentration from low to high (Fig. 5-10C). Bed thickness ranges from an inch to several feet (centimeters to meters), and normal size grading is common. Large-scale cross-bedding is common and is delineated by granule and small pebble layers and lenses. Slumped bedding and dewatering structures are present locally.

Fine-Grained Sand (Lithofacies S_2). Lithofacies S_2 consists of moderately to poorly sorted, fine-grained sands which are commonly parallel laminated or unstratified (Fig. 5-10D). Bed thickness ranges from inches to feet (centimeters to decimeters) with some

upward fining in grain size. Convolute bedding and dewatering features are evident in some of these sands (Fig. 5-10E).

Ripple-Laminated, Very Fine-Grained Sand (Lithofacies S_3). Lithofacies S_3 consists of moderately sorted, very fine-grained sands, commonly displaying ripple cross-lamination (Fig. 5-10F). Unstratified and parallel-laminated beds of this lithofacies are also common. Bed thickness ranges from inches to several feet (centimeters to meters). Thin-bedded sequences of this lithofacies are commonly interbedded with mud and clay beds. Bioturbation structures are abundant in these sands, especially where they are interbedded with muds.

Silt (Lithofacies Si). Silts cap many of the thicker sand beds in the Tulare Formation and are commonly overlain by mud beds. Bed thickness ranges from fractions of an inch to several feet (centimeters to meters). Ripple cross-lamination, bioturbation structures, and localized zones of carbonate cement are all common features of this lithofacies (Fig. 5-10G).

Mud and Clay (Lithofacies M). Greenish gray muds (including clays) occur both as thick beds separating the major sand intervals, by several tens of feet (meters), and as thin beds, a few inches (centimeters) thick, within sand units. The muds are of two types, unstratified and thinly laminated. Lenticular beds of very fine-grained sand and silt are common in some mud units (Fig. 5-10H). Bioturbation structures are very common, and the degree of bioturbation varies from slight to complete. Localized zones of carbonate cement, up to 1 foot thick (30 cm), are common in many of the thicker mud units.

Depositional Setting

Specific depositional subenvironments of a fluviodeltaic setting in the Tulare Formation are interpreted on the basis of distinctive and repeated lithofacies associations. These subenvironments include (1) channel and bar, (2) levee and crevasse splay, (3) overbank/interchannel, (4) distributary-channel-mouth bar, (5) distal bar, and (6) prodelta/lacustrine basin (Table 5-1).

Channel and Bar Deposits. The upper Tulare reservoir in the southern portion of the field consists of channel and bar sands mostly of point-bar origin. They occur in units of stacked and amalgamated beds (Fig. 5-4), with individual depositional sequences that average 10 feet (3.1 m) thick and range from 5 to

Table 5-1. Lithofacies associations and their interpretations for the Tulare Formation in South Belridge Field.

Lithofacies	Lithofacies associations and depositional settings					
	Channel/bar	Levee	Overbank	Mouth bar	Distal bar	Lake
Sandy pebble gravel (G)	o					
Cross-bedded gravelly sand (S$_1$)	●					
Fine-grained sand (S$_2$)	●	o		o		
Ripple-laminated vf sand (S$_3$)	o	●	o	●	o	
Silt (Si)		o	o		o	o
Mud and clay (M)	o	●	●	o	●	●

● dominant lithofacies
o associated lithofacies

25 feet (1.5–7.6 m). The depositional sequences typically fine upward and have a lower erosional surface overlain by a lag gravel (G) (Fig. 5-10B). They comprise cross-bedded, gravelly (pebbles and granules), fine- to medium-grained, poorly sorted sands (S$_1$ and S$_2$) at the base, with ripple cross-laminated, very fine-grained sands (S$_3$) at the top (Table 5-1). Levee and crevasse-splay deposits commonly directly overlie the point-bar sands.

Some channel and bar sands of the uppermost B zone are coarser grained and gravelly (pebbles and granules), more thinly bedded, and lack the well-developed upward fining cycles characteristic of the point bars of the underlying Tulare. These sequences are interpreted as braided fluvial deposits. Prominent erosional surfaces overlain by thin gravel beds delineate the base of each sequence. Large-scale cross-bedding and parallel lamination are common throughout most sequences. The high-angle cross-beds and relatively well-sorted texture suggest deposition in a downstream braided segment dominated by transverse bars.

Levee and Crevasse-Splay Deposits. Levee deposits are mostly interbedded sequences of very fine-grained sand (S$_3$) or silt (Si) and mud (M) (Table 5-1). Thickness varies from 2 to 10 feet (0.6–3.1 m). The sands are typically ripple cross-laminated, commonly with climbing ripples indicative of high aggradation rates. Both sands and muds are extensively bioturbated. Crevasse-splay sands are present as isolated beds of very fine- to fine-grained sand, usually ripple cross-laminated and bioturbated. Most splay sands are encased in overbank/interchannel mud and the lower contact is commonly erosional. Upper contacts are sometimes carbonate cemented and are interpreted as paleosol horizons.

Overbank/Interchannel Deposits. Overbank and interchannel deposits are a major component of the

Tulare Formation in the southern portions of the field. The deposits comprise thin (less than 1 foot (30 cm) muds and clays (M) and minor silts (Si) (Table 5-1). Wavy- and lenticular-bedded sands are present locally. These overbank deposits were emplaced by interchannel sedimentation from flood-stage channel overtopping and by channel crevassing. A high degree of bioturbation in many of the interchannel sequences points to a lower delta-plain lacustrine influence. Similar to the levee and crevasse-splay deposits, carbonate-cemented paleosol horizons indicate periods of nondeposition and soil weathering.

Distributary-Channel-Mouth-Bar Deposits. The distributary mouth-bar/deltaic inner-fringe sands are moderately well sorted and very fine- to fine-grained (S$_2$ and S$_3$ (Table 5-1). These deposits have sharp but nonerosive upper and lower contacts with mud beds of lithofacies M. Prominent sedimentary structures include ripple cross-lamination, wavy lamination, and parallel lamination, but unbedded units are also common. The sedimentary structures indicate reworking of these sands by wave and current action into more areal extensive fringe sands. Although the lacustrine energy was relatively low, the episodic nature of river mouth-bar deposition allowed time for the reworking. Mouth-bar sands that are sandwiched between distal-bar and prodelta sequences indicate lateral shifting of the distributary channel in the lower delta plain.

Distal-Bar Deposits. The distal-bar/deltaic outer-fringe deposits are composed of thin bedded, ripple-laminated sand (S$_3$) interbedded with laminated silt (Si) and mud (M) interbeds (Table 5-1). The sand content varies from less than 5 to 30%. Bioturbated beds are common. Distal-bar deposits are most abundant in the lower Tulare (zones D and E) associated with distributary-mouth-bar sediments.

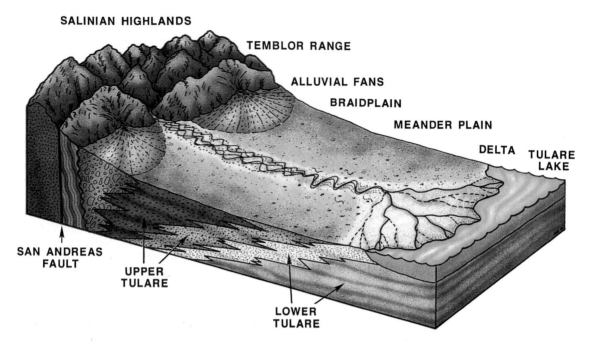

SALINIAN HIGHLANDS

TEMBLOR RANGE

ALLUVIAL FANS

BRAIDPLAIN

MEANDER PLAIN

DELTA TULARE LAKE

SAN ANDREAS FAULT

UPPER TULARE

LOWER TULARE

Fig. 5-11. Schematic representation of the depositional setting of the Tulare Formation at South Belridge Field.

Prodelta/Lacustrine Basin Deposits. The prodelta-lacustrine basin deposits are composed primarily of thick muds, silty muds, and clays (M) (Table 5-1). Very thin, lenticular beds of sandy silts in these deposits represent distal, delta-toe turbidite flows.

In summary, the overall depositional setting of the Tulare at South Belridge is that of a prograding fluviodeltaic system (Fig. 5-11). The lower Tulare reservoir sands (zones D and E) are mainly of delta-front and delta-plain origin, with distal-bar, distributary-mouth-bar, and a few distributary-channel sands (Fig. 5-4). The upper Tulare reservoir sands display a variety of fluvial styles. Somewhat more isolated (multistoried) fine-grained meander-belt sands (zone C) are overlain by more amalgamated (multilateral) coarse-grained meander-belt sands (lower zone B), in turn overlain by even less confined and braided fluvial sands (upper zone B). Laterally accreted point-bar deposits are common, with well-defined upward-fining sequences and associated levees and crevasse-splays. Zone A sands at the top of the reservoir sequence, although of lesser reservoir importance, establish a return to a fine-grained meander-belt system. Paleotopography created by the growing South Belridge anticline directly influenced the gross distribution of the fluvial system but had much less influence on the deltaic sands.

Petrography

The reservoir sands of the upper and lower Tulare zones are texturally and mineralogically immature (Fig. 5-12). The sands are mostly lithic feldsarenites (classification after Folk et al., 1970) comprising subequal amounts of feldspar and quartz (35–45%) and lesser amounts of rock fragments, mica, pyrite, carbonate, amorphous material, and clay minerals. Diatomaceous gravel and sand are particularly abundant in the fluvial channel sequences of the upper Tulare zones. The sands are unconsolidated and essentially "oil-cemented," although grain supported.

The clay mineral content in the reservoir facies is typically 5% and ranges from 1 to 8%. Clay minerals consist primarily of illite (60%) and smectite (40%), usually in mixed-layer morphology. These clays are only a part of the total formation fines in the reservoir which affect reservoir permeability. The total sediment fines (less than 44 microns) constitute 5 to 18% of the reservoir facies. In addition to clay minerals, they consist of clays and micas, clay- and silt-sized particles of quartz, feldspar, pyrite, calcite, and rock fragments (Fig. 5-12). Related production effects include migration of fines which plug pores and mineral dissolution and authigenesis in steam injection conditions. Higher plagioclase-to-quartz

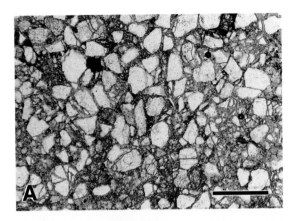

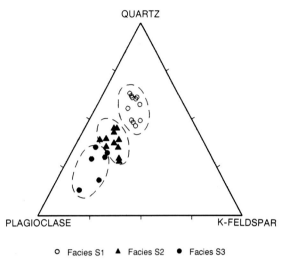

Fig. 5-13. Quartz, plagioclase, and potassium feldspar reservoir sand compositions plotted by lithofacies. S_1 is cross-bedded gravelly sand; S_2 is fine-grained sand; and S_3 is ripple-laminated, very fine-grained sand.

Fig. 5-12. A (plane polarized light) and B (cross polarizers): Photomicrographs of a poorly sorted, fine- to medium-grained lithic feldsarenite representative of lithofacies S_2. The sand is both texturally and compositionally immature and in places contains a matrix of silt-sized quartz and feldspar and minor (5%) clay. Porosity (blue areas in A) is mostly intergranular and well connected, but the matrix has abundant micropores that contribute to ineffective porosity. Scale bar is 0.50 mm.

ratios in the finer-grained lithofacies (S_2 and S_3, as compared to S_1) are a product of composition and size sorting (Fig. 5-13); feldspar is generally more abundant in the finer-grained lithofacies.

The reservoir sands display a wide range in grain size, sorting, and shape as a product of differing depositional processes and settings of the Tulare. Grain size, the most important textural element for reservoir considerations, ranges from gravel to very fine-grained sand (Fig. 5-14). Sorting ranges from moderate to poor. The poorest sorting is found in the coarser-grained, fluvial channel sands (lithofacies S_1), whereas the best sorting is found in the delta mouth-bar sands (lithofacies S_2 and S_3). In the

Tulare reservoir sands, sorting has much less influence on permeability, and therefore on oil saturation, than does grain size.

Diagenesis

The reservoir sands show almost no diagenetic modification due to their young age and very shallow burial. Compaction has been minimal because the maximum burial depths are generally less than 1,000 feet (305 m). The sands are uncemented, as are the silts and muds, with the exception of local zones of penecontemporaneous nodular carbonate.

Mineralogical changes associated with steam-rock reactions can lead to significant changes in porosity and permeability (cf. Sedimentology Research Group, 1981). Smectite-to-illite ratios in the Tulare reservoir sands show significant changes caused by thermal diagenesis resulting from steam injection. These changes result from chemical reactions involving quartz, carbon dioxide, and water:

$$\text{illite} + \text{calcite} + \text{quartz} \rightarrow \text{smectite} + \text{carbon dioxide} + \text{water}$$

Although the Tulare reservoir sands show an increased smectite-to-illite clay ratio in the steamed intervals, they do not exhibit significant reductions in porosity and permeability because of their overall low clay content. However, an increase in the smectite-to-illite ratio in the mud lithofacies greatly

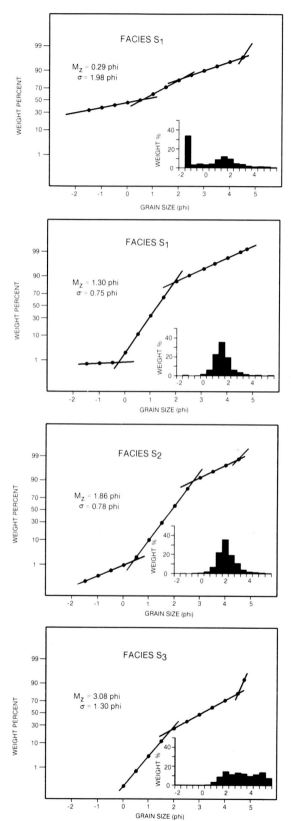

increases its effectiveness as a permeability barrier. In addition, the steam commonly precipitates secondary carbonate in the mud lithofacies. The combined effects of carbonate cementation and smectite growth create low-permeability zones or discontinuous baffles and impermeable barriers to fluid flow in the reservoir. These dynamic processes during production alter the geometry of the injected steam profile. This can occur even within thin (0.5 ft, 15 cm) mud units, as evident in cores of steamed reservoirs.

Petrophysics

Reservoir quality in the Tulare Formation is primarily controlled by lithofacies, which are the product of the depositional environment. This depositional lithofacies control of reservoir quality is clearly observed in the Tulare because of the absence of diagenesis, which overprints and complicates the depositional influences in most other sandstone reservoirs. The reservoir sands exhibit excellent reservoir quality, with an average effective porosity of 35% and 3,000 md permeability (Table 5-2) (Gates and Brewer, 1975).

Porosity, permeability, and fluid saturation measurements from routine core analysis and wireline logs have unique problems in the unconsolidated, heavy-oil sands of the Tulare Formation. Grain rearrangement during coring, wellsite, and laboratory procedures can alter significantly the physical properties of core samples (Elkins, 1972). Therefore, a correction factor is applied to ambient core porosity data in order to better calculate porosities at reservoir conditions. Core samples analyzed at ambient conditions average 2 to 3 porosity units (% absolute) greater than that measured at overburden pressure conditions. Permeability values of brine-saturated samples measured at reservoir stress conditions are approximately half of the routine permeability values measured in air. Permeability data from cores are also sensitive to alteration of the deli-

Fig. 5-14. Representative grain-size distributions of the reservoir sands plotted by lithofacies. Note that the uppermost lithofacies S_1 sample is from the base of a fluvial channel sequence and incorporates the channel-lag deposits. S_1 is cross-bedded gravelly sand; S_2 is fine-grained sand; and S_3 is ripple-laminated, very fine-grained sand.

Table 5-2. Typical values for grain size, porosity, permeability, and oil saturation for the principal lithofacies of the Tulare Formation.

Lithofacies	Mean grain size (Mz in phi)	Porosity* (ϕ)	Permeability** (md)		Oil saturation (So)
			(k_h)	(k_v)	
Sandy pebble gravel (G)	−2.5	34%	10,000	7,000	50%
Cross-bedded gravelly sand (S₁)	1.1	34%	10,000	8,000	85%
Fine-grained sand (S₂)	2.3	35%	6,000	5,000	75%
Ripple-laminated vf sand (S₃)	3.3	36%	700	300	60%
Silt (Si)	4.5	35%	50	10	15%
Mud and clay (M)	9.0	35%	0.1	0.1	0%

* Effective porosities corrected to reservoir conditions.
** Air permeabilities determined at 300 psi (2.1 × 10³ kPa).

cate clay fabric caused by fluid extraction and sample drying. Oil saturations from cores are relatively accurate in these heavy-oil sands (once corrected for reservoir porosities) because of reduced flushing during drilling and coring and use of large-diameter (5 in. (13 cm)) core barrels.

Total porosity in the Tulare ranges from 32 to 42% and correlates poorly with permeability and lithofacies type (Table 5-2; Fig. 5-15). However, porosity is not a limiting factor in this reservoir. Ineffective

porosity, up to 5% (absolute), is attributed to the clays and diatomaceous fragments in the finer-grained lithofacies. The diatomaceous fragments have been eroded from underlying Miocene Monterey strata.

Measured core permeability values (air at 300 psi (2.1 × 10³ kPa) confining pressure) range from less than 100 to 10,000 md. These permeabilities display a strong correlation with lithofacies (Fig. 5-15). This is primarily a function of grain size variability in the lithofacies; the coarser-grained lithofacies have the highest permeabilities (Table 5-2; Figs. 5-4, 5-16) (cf. Beard and Weyl, 1973).

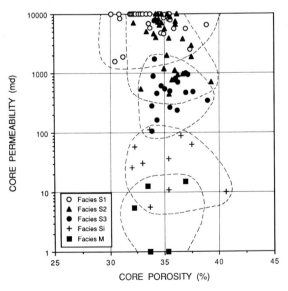

Fig. 5-15. Core porosity and permeability plotted by lithofacies. The porosities of the silt and mud lithofacies are low relative those of the sand lithofacies because of differences in compressibility. The porosity data have been corrected to reservoir conditions. Air permeability data were determined at 300 psi (2.1 × 10³ kPa). S₁ is cross-bedded gravelly sand; S₂ is fine-grained sand; S₃ is ripple-laminated, very fine-grained sand; Si is silt; and M is mud.

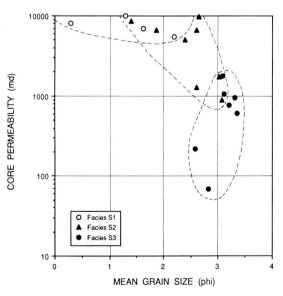

Fig. 5-16. Core permeability and mean grain size plotted by lithofacies. Permeability (air) is determined at 300 psi (2.1 × 10³ kPa) confining pressure. S₁ is cross-bedded gravelly sand; S₂ is fine-grained sand; and S₃ is ripple-laminated, very fine-grained sand.

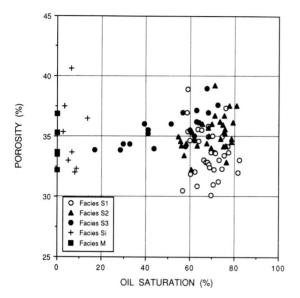

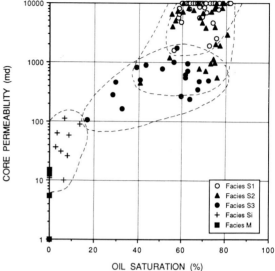

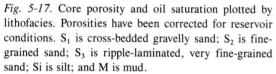

Fig. 5-17. Core porosity and oil saturation plotted by lithofacies. Porosities have been corrected for reservoir conditions. S_1 is cross-bedded gravelly sand; S_2 is fine-grained sand; S_3 is ripple-laminated, very fine-grained sand; Si is silt; and M is mud.

Fig. 5-18. Core permeability and oil saturation plotted by lithofacies. Permeability (air) is determined at 300 psi (2.1 $\times 10^3$ kPa) confining pressure. S_1 is cross-bedded gravelly sand; S_2 is fine-grained sand; S_3 is ripple-laminated, very fine-grained sand; Si is silt; and M is mud.

Original oil saturations are assumed to have averaged 76% in the Tulare reservoir (Gates and Brewer, 1975). Oil saturations range from a maximum of 85% in sands to less than 10% in silts. The saturations of the reservoir sands are unrelated to porosity (Fig. 5-17) but are strongly correlated to permeability, grain size, and lithofacies (Fig. 5-18; Table 5-2). Lower permeability sands have greater specific surface areas and higher irreducible water saturations; the higher oil saturations correspond to the more permeable lithofacies. Vertical and horizontal distribution of these lithofacies thus controls overall oil storage capacity and producibility of the reservoir.

Wireline-log analysis has its own set of problems in these unconsolidated heavy-oil sand reservoirs. Borehole washouts from drilling are common in these unconsolidated sediments, limiting the usefulness of some logs. In addition, resistivity logs, used to calculate oil saturations, are very sensitive to variations in formation water salinity induced by the close proximity to meteoric groundwater. This is indicated by decreasing baseline resistivities in clays, which can be used to calibrate and improve log calculations. On the positive side, porosity calculated from logs has an inherent accuracy advantage of in situ measurement, avoiding the grain rearrangement problem incurred during core acquisition.

Steam injection complicates wireline surveys because the multiphase porosity system contains formation water, oil, steam, and fresh water from condensed steam. Steamed reservoir horizons produce a suppressed or inverted SP log response, a heat-induced resistivity suppression, and a "gas effect" signature on neutron-density logs (Fig. 5-19). The steamed Tulare reservoirs also have a high gamma-ray signature due to mobilized uranium roll-front concentrates carried along with the steam front.

Heterogeneities

The Tulare is a highly layered and laterally discontinuous reservoir at South Belridge Field (Figs. 5-6, 5-8). Heterogeneities in the reservoir sands control the vertical and lateral continuity on several scales. The scale of reservoir heterogeneities, relative to well spacing, determines the control on reservoir flow characteristics. The variable reservoir layering is a product of the depositional stacking of sands and muds typically creating 1 to 4 separate layers in each of the 5 reservoir zones. Lateral flow restrictions are a function of the three-dimensional sand-body geometries. The overall reservoir sequence contains laterally continuous reservoir sands in the lower zones and both coalescing and isolated "shoestring"

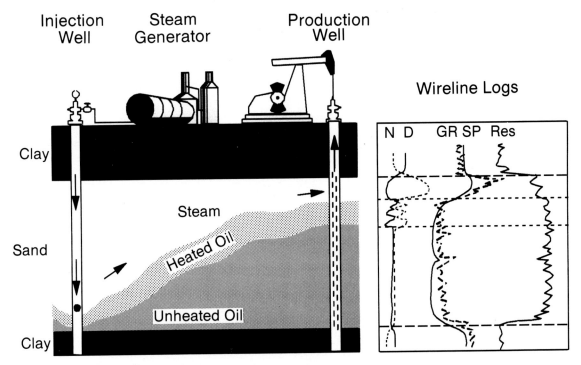

Fig. 5-19. Steamflood schematic showing an injection well and production well profile. Steam injected through a single (limited-entry) perforation near the base of a sand heats the oil which buoyantly rises to the top of the reservoir sand unit. Sand thickness typically ranges from 5 to 20 feet (1.5–6.1 m). Typical wireline log responses include a gas effect on neutron and density logs, a uranium roll-front on the gamma-ray log, and reversed SP signature in steamed intervals. (N = neutron porosity; D = density porosity; GR = gamma ray; SP = spontaneous potential; and Res = resistivity curve.)

geometries in the upper zones. These predictable characteristics, based on the depositional model, are verified by producing characteristics.

Generally the lower Tulare reservoir zones (D and E) are continuous and highly layered, because they represent delta-front sheet sands. Occasional thicker bedded mouth-bar sands link the thinner bedded and highly stratified distal-bar sands and serve as permeability conduits.

By contrast, the upper Tulare sands (zones A, B, and C) are channelized fluvial sequences that display considerable differences in flow-unit scale and form, reflecting the differences in character of the fluvial system from which they were deposited. The zone A and C sands are typical fine-grained meander-belt sands in that they are sinuous, elongate, and narrow (shoestring) bodies encased in mud (Fig. 5-8). Vertical and lateral communication between these sands is poor because of their limited interconnectedness. Accordingly, lateral flow-unit geometry is restricted. The zone B sands, representing coarse-grained meander-belt and braided fluvial

systems, are commonly stacked and amalgamated to the extent that they perform as one highly connected reservoir (Fig. 5-8). Stacked channel sequences containing more than 75% sand can act effectively as a single [flow] unit due to their connectedness (Allen, 1978). However, with frontal displacement of oil during enhanced recovery processes, the scale of geologic heterogeneities relative to well spacing distinguishes baffles from barriers. Thin muds separating these sands are discontinuous over distances of tens of feet (meters) and are not effective barriers to steam with the existing well density, although they serve to baffle vertical flow. However, mud horizons which split the reservoir into layers over an area of several wells require both injection and production on both sides of that heterogeneity.

The effective reservoir-scale permeabilities throughout the Tulare are lower than measured core values by more than two orders of magnitude due to reservoir heterogeneities; effective reservoir permeability values used to simulate reservoir production behavior are as low as 1% of the measured core

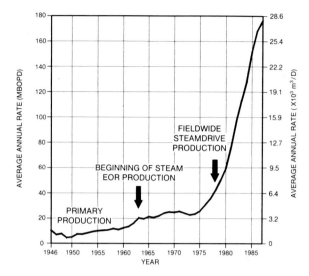

Fig. 5-20. Production history of the South Belridge Field. Up to 25% of this production is from non-Tulare reservoirs.

values (Dietrich, 1988). Both vertical and horizontal permeabilities are affected considerably by discontinuous clay interbeds which act as baffles or local barriers to vertical flow. As a consequence, effective reservoir-scale vertical permeability is calculated by the stochastic method of Begg and others (1985). Lateral permeability estimates are similarly decreased by the tortuosity created as a product of the amalgamation of sand bodies.

Production Characteristics

Well and Reservoir Performance

The Tulare reservoir at South Belridge Field has been produced by primary production, cyclic steaming, in situ combustion, and steamflood. Current production operations focus on steamflood development to produce the heavy (13–15° API gravity) oil (Fig. 5-20). Data presented here are mostly from the Mobil Oil Corporation-operated southern portion of the field, reflecting the authors' primary area of interest and data availability.

Primary field development wells were drilled on 10-acre (4.1-ha.) spacing, a distance between wells of 660 feet (201 m). These wells initially produced between 20 and 200 BOPD (3.2–32 m³/D) and generally experienced a rapid decline. Wells typically produced 50 to 150 MBO (0.8–2.4 × 10⁴ m³) cumulative with 90% of the production occurring

over a 20-year life. The primary producing mechanism was solution-gas drive with some assistance from gravity drainage and aquifer support on the flanks of the field. Initial reservoir pressure was probably hydrostatic at approximately 400 psi (2.8 × 10³ kPa).

The first South Belridge Field thermal recovery pilot was an in situ combustion thermal-recovery (fireflood) experiment conducted from 1955 to 1958 (Gates and Ramey, 1958; Gates et al., 1978). This pilot demonstrated that viscous oil could be readily moved and oil recovery increased to 40 to 60% compared with a primary recovery of approximately 10%.

Cyclic steaming ("huff-n-puff") was started in 1963 and a fieldwide continuous steam injection line-drive operation began in 1969 (Fig. 5-20). The steamflood project was converted to 10-acre (4.1-ha.), inverted nine-spot steam patterns in the mid-1970s. Steam injectors are located at the center of a pattern with eight surrounding producing wells. This pattern configuration contains one injection well and three net producing wells, when patterns are developed side by side. With this arrangement, a well typically produces more than 200 MBO cumulative (3.2 × 10⁴ m³) of oil.

Existing steamflood patterns at South Belridge Field range from 2.5 to 10 acres (1.0–4.1 ha.) in size and each contains four to eight producing wells. Limited-entry perforations (Small, 1986) control the vertical steam injection profile in each cased injection well. This method can minimize the effects of lithofacies-controlled permeability variations at the well. Producing wells have traditionally used open-hole gravel-flow-pack, slotted liners, or wire-wrapped liners to provide sand control. Inside-casing gravel-packs have also been used to avoid completion in intermediate water sands or to avoid steam-desaturated reservoir sands.

Fluid properties in the Tulare change dramatically with the introduction of steam to the reservoir during steamflood. Oil saturations at South Belridge Field attain residual saturations commonly less than 5% and as low as 0% in steam-swept reservoir intervals. This is similar to other reported steamflood projects (Ali, 1982; Traverse et al., 1983). Due to density contrasts (gravity override), the steam rises to the top of reservoir flow units, commonly resulting in the lowest residual oil in the upper portions of the reservoirs (Ali and Meldau, 1979) (Fig. 5-19). Oil saturations below these swept zones are typically greater than 40%, and essentially original saturations occur at the base. Vertical and lateral distribution of reservoir lithofacies, structural dip, gravity override of steam, and seal continuity and thickness are natural factors influencing steam pathways.

Lithofacies-Related Considerations

The reservoir model of the Tulare at South Belridge Field shows strong dependence on depositional controls. This is evident at both the sand-body geometry scale and the smaller pore-throat scale. From an understanding of the depositional system, the reservoir model allows the prediction of reservoir geometry and quality and oil trapping potential. Permeability and oil saturation in the Tulare are directly related to depositional lithofacies, whereas the reservoir flow-unit geometries are dictated by the depositional system. Understanding the reservoir's three-dimensional geometry and flow-path continuity between wells is important for selecting completion intervals and well spacing, which are paramount to a successful steamflood.

The major differences in the reservoirs of the upper and lower Tulare can be explained by lithofacies types and their distribution. The upper Tulare fluvial-channel sands display the highest permeabilities and oil saturations, primarily reflecting their coarser grain sizes (Fig. 5-4). These channel sands have appreciably higher permeabilities than their associated finer-grained crevasse-splay and levee sands. The lower Tulare deltaic sands generally have lower than average permeabilities due primarily to their finer grain sizes and a small matrix-clay content.

Exploration And Production Strategy

Opportunities to discover another South Belridge Field are very limited. There are probably few onshore structures the size of the South Belridge anticline that have not been tested. The days of discovering major fields by surface expressions of anticlines with oil seeps in creek beds, as was the case for South Belridge, are long gone. However, existing fields with similar economic development potential to South Belridge have undoubtedly been overlooked.

The fundamental criterion for developing another South Belridge Field is to realize the extraordinary development potential of shallow, heavy-oil reservoirs, even when the discovery well is a poor producer. Isolated reservoir sand bodies in the initial wells may provide only a hint of the development potential in a fluviodeltaic reservoir complex. A crit-

ical early step is to look carefully at the geologic scale of potential reservoir units. Isolated well performance provides little indication of infill potential due to the highly variable scale of reservoir flow-unit geometries. During project planning, primary field development, and subsequent EOR operations, aspects of well spacing, pattern size and geometry, injection schedule, and completion plans can be matched with sand continuity models to make optimal use of EOR techniques.

Summary

The unconsolidated reservoir sands of the Pleistocene Tulare Formation at South Belridge Field contain heavy oil that is produced by more than 6,000 wells from depths generally less than 1,000 feet (305 m). Numerous stratigraphically discontinuous reservoir sands were deposited in a prograding fluviodeltaic depositional environment. The overall reservoir sequence contains laterally continuous reservoir sands in the lower zones and both coalescing and isolated "shoestring" geometries in the upper zones. Syndepositional structural growth of the anticline altered fieldwide distribution of sands. The reservoir model can predict the range of reservoir geometries and connectivity of the various reservoir horizons between wells.

Depositional lithofacies directly control reservoir properties due to the absence of diagenetic modification. Permeability and oil saturation are directly related to lithofacies, whereas the reservoir sand-body and reservoir flow-unit geometries are dictated by the depositional system. Depositional geometries of sands and clays and the structural overprint of the South Belridge anticline control the numerous oil-water contacts.

Primary production, cyclic steaming, steamflood injection, and in situ combustion methods have been used during the life of the field. Current production comes primarily from steamflood development of the field. The high density of wells involved in steamflood production defines small-scale reservoir flow units which would be geologically isolated and nonproductive with more widely spaced wells.

Opportunities for economic development of similar reservoirs with poor primary production benefit from comprehensive reservoir characterization and applications of new technology. These qualities are vital to engineering studies necessary to exploit the full potential of the field.

Acknowledgments. The authors appreciate and acknowledge the history and producing information contributed by E. Forrester, Jr. and mineralogical work performed by J.D. Cocker. We are grateful to M.G. Acosta, D.S. Anderson, J.M. Eagan, I.A. Fischer, R.J. Pashuck, and B.J. Welton for discussion and manuscript review. The manuscript benefited from review by T.H. Nilsen, G.F. Canjar, and the editors of this volume. Special thanks are due H.H. Mottern and Mobil Oil Corporation for sponsoring this work. Thanks to Mobil Oil Corporation and Mobil Research and Development Corporation for permission to publish this work.

Reservoir Summary

Field: South Belridge
Location: San Joaquin Valley, central California
Operators: Shell, Mobil, Exxon, Santa Fe Energy, Unocal, Mission Resources, and others
Discovery: 1911
Basin: San Joaquin basin
Tectonic/Regional Paleosetting: Wrench-faulted basin margin
Geologic Structure: Plunging anticline.
Trap Type: Stratigraphic pinch-out on a plunging anticline
Reservoir Drive Mechanism: Solution-gas drive with limited gravity drainage and aquifer support (primary)
  • **Original Reservoir Pressure:** 400 psi (2.8×10^3 kPa) (assumed hydrostatic)
Reservoir Rocks:
  • **Age:** Pleistocene
  • **Stratigraphic Unit:** Tulare Formation
  • **Lithology:** Gravel to very fine-grained sand; lithic feldsarenite
  • **Depositional Environment:** Fluviodeltaic
  • **Productive Facies:** Braided/meandering fluvial, distributary-channel, and distributary-mouth-bar sands
  • **Petrophysics**
    • ϕ: Average 35%, range 32 to 42%; all primary intergranular (cores)
    • k: Average 3,000 md (air 300 psi [2.1×10^3 kPa]), range 100 to 10,000 md (cores)
    • S_w: Average 24%, range 15 to 90% (cores, logs)
    • S_o: 76% (cores)
Reservoir Dimensions
  • **Depth:** 500 feet (152 m)
  • **Areal Dimensions:** 2 by 10 miles (3.2×16 km)
  • **Productive Area:** 8,700 acres (3,525 ha.)
  • **Number of Reservoirs:** Multiple; 5 zones with 1 to 5 sands per zone
  • **Hydrocarbon Column Height:** 900 feet (275 m)
  • **Fluid Contacts:** Multiple oil-water contacts
  • **Number of Pay Zones:** 5 to 7
  • **Gross Sandstone Thickness:** 400 to 750 feet (122–229 m)
  • **Net Sandstone Thickness:** 50 to 275 feet (15–84 m)
  • **Net/Gross:** 0.3 to 0.7
Source Rocks
  • **Lithology & Stratigraphic Unit:** Siliceous shale, Miocene Monterey Formation
  • **Time of Hydrocarbon Maturation:** Late Miocene to Recent
  • **Time of Trap Formation:** Pleistocene to Recent
  • **Time of Migration:** Pleistocene to Recent
Hydrocarbons
  • **Type:** Oil
  • **GOR:** 45:1 SCF/bbl
  • **API Gravity:** 13 to 14°
  • **FVF:** 1.03
  • **Viscosity:** 1,800 cP at 90°F (32°C); 7.9 cP at 300°F (149°C)

Volumetrics
  • **In-Place:** N/A
  • **Cumulative Production:** 700 MMBO (1.1 × 10⁸ m³)
  • **Ultimate Recovery:** 1,200 MMBO (1.9 × 10⁸ m³) (estimated)
  • **Recovery Efficiency:** N/A

Wells
  • **Spacing:** 200 to 500 feet (61–152 m); 10 acres (4.1 ha.) primary, 2.5 to 10 acres (1.0–4.1 ha.) EOR patterns
  • **Total:** 6,100 (currently active)
  • **Dry Holes:** Less than 1%

Typical Well Production
  • **Average Daily:** Primary 40 BO (6.4 m³); Steam EOR 80 BO (12.7 m³)
  • **Cumulative:** Primary 50 to 100 MBO (0.8–1.6 × 10⁴ m³); Steam EOR 200 MBO (3.2 × 10⁴ m³)

References

Ali, F.A., 1982, Steam injection theories—A unified approach: Society of Petroleum Engineers, Paper 10746, p. 309–315.

Ali, F.A., and Meldau, R.F., 1979, Current steamflood technology: Journal of Petroleum Technology, October, p. 1332–1341.

Allen, J.R.L., 1978, Studies in fluviatile sedimentation: An exploratory quantitative model for the architecture of avulsion-controlled alluvial suites: Sedimentary Geology, v. 21, p. 129–147.

Atwater, T., 1970, Implications of plate tectonics for the Cenozoic tectonic evolution of western North America: Geological Society of America Bulletin, v. 97, p. 97–109.

Barbat, W.F., and Galloway, J., 1934, San Joaquin Clay, California: American Association of Petroleum Geologists Bulletin, v. 18, p. 476–499.

Bartow, J.A., 1987, The Cenozoic evolution of the San Joaquin valley, California: U.S. Geological Survey Open-File Report 87-581, 74 p.

Beard, D.C., and Weyl, P.K., 1973, Influence of texture on porosity and permeability of unconsolidated sand: American Association of Petroleum Geologists Bulletin, v. 57, p. 349–369.

Begg, S.H., Chang, D.M., and Haldorsen, H.H., 1985, A simple statistical technique for calculating the effective vertical permeability of a reservoir region containing discontinuous shales, in Proceedings, 60th Annual Meeting of the Society of Petroleum Engineers, Paper 14271, 15 p.

California Division of Oil and Gas, Department of Natural Resources, 1950, 36th Annual Report State Oil and Gas Supervisor 1950: v. 36, p. 18–25.

California Division of Oil and Gas, Department of Conservation, 1985, California Oil and Gas Fields, Central California, Vol. I.

California Division of Oil and Gas, Department of Conservation, 1987, 72nd Annual Report of the State Oil and Gas Supervisor 1986: v. 72, p. 57–58.

Clarke S.H., Jr., Howell, D.G., and Nilsen, T.H., 1975, Paleogene geography of California, in Weaver, D.W., Hornaday, G.R., and Tipton, A., eds., Paleogene Symposium and Selected Technical Papers: Long Beach, California, American Association of Petroleum Geologists—Society of Economic Paleontologists and Mineralogists—Society of Exploration Geophysicists, Pacific Sections, p. 121–154.

Connan, J., and Coustau, H., 1987, Influence of the geological and geochemical characteristics of heavy oil on their recovery, in Meyer, R. F., ed., Exploration for Heavy Crude Oil and Natural Bitumen: American Association of Petroleum Geologists Studies in Geology 25, p. 261–279.

Dickinson, W.R., Ingersoll, R.V., and Graham, S.A., 1979, Paleogene sediment dispersal and paleotectonics in northern California: Geological Society of America Bulletin, v. 90, Pt I, p. 898–899, Pt II, p. 1458–1528.

Dietrich, J., 1988, Steamflooding in a waterdrive reservoir: Upper Tulare sands, South Belridge field, in Proceedings, 1988 California Regional Meeting: Society of Petroleum Engineers, Paper 17453, p. 479–494.

Elkins, L.F., 1972, Uncertainty of oil in place in unconsolidated sand reservoirs—A case history: Journal of Petroleum Technology, November, p. 1315–1319.

Ethridge, F.G., and Schumm, S.A., 1978, Reconstructing paleochannel morphologies and flow characteristics: Methodology, limitations, and assessment, in Miall, A.D., ed., Fluvial Sedimentology: Canadian Society of Petroleum Geologists Memoir 5, p. 703–721.

Fielding, C.R., and Crane, R.C., 1987, An application of statistical modelling to the prediction of hydrocarbon recovery factors in fluvial reservoir sequences, in Ethridge, F. G., Flores, R. M., and Harvey, M. D., eds., Recent developments in fluvial sedimentology: Society of Economic Paleontologists and Mineralogists Special Publication 39, p. 321–327.

Folk, R.L., Andrews, P.B., and Lewis, D.W., 1970, Detrital sedimentary rock classification and nomenclature for use in New Zealand: New Zealand Journal of Geology and Geophysics, v. 13, p. 937–968.

Foss, C.D., 1972, A note on the fluid level traps in the San

Joaquin valley, *in* Rennie, E.W., Jr., ed., Geology and Oil Fields, West Side Central San Joaquin Valley: American Association of Petroleum Geologists Pacific Section, Guidebook, p. 15.

Gates, C.F., and Brewer, S.W., 1975, Steam injection into the D&E zones, Tulare Formation, South Belridge field, Kern County, California: Journal of Petroleum Technology, v. 27, p. 343–348.

Gates, C.F., and Ramey H.J., Jr., 1958, Field results of South Belridge thermal recovery experiment: American Institute of Mining, Metallurgical, and Petroleum Engineering Petroleum Transactions, v. 213, p. 236–244.

Gates, C.F., Jung, K.D., and Surface, R.A., 1978, In-situ combustion in the Tulare Formation, South Belridge Field, Kern County, California: Journal of Petroleum Technology, v. 30, p. 798–806.

Graham, S.A., 1987, Tectonic controls on petroleum occurrence in California, *in* Ingersoll, R.V. and Ernst, W.G., eds., The Geotectonic Development of California, Rubey Volume VI: Englewood Cliffs, New Jersey, Prentice-Hall, p. 47–63.

Graham, S.A., and Williams, L.A., 1985, Tectonic, depositional, and diagenetic history of Monterey Formation (Miocene), central San Joaquin basin, California: American Association of Petroleum Geologists Bulletin, v. 69, p. 385–411.

Harding, T.P., 1976, Tectonic significance and hydrocarbon trapping consequences of sequential folding synchronous with San Andreas faulting, San Joaquin valley, California: American Association of Petroleum Geologists Bulletin, v. 60, p. 356–378.

Ingersoll, R.V., 1979, Evolution of the Late Cretaceous forearc basin, northern and central California: Geological Society of America Bulletin, v. 90, p. 813–826.

Lennon, R.B., 1976, Geological factors in steam-soak projects on the west side of the San Joaquin basin: Journal of Petroleum Technology, v. 4, p. 741–748.

Lettis, W.R., 1982, Late Cenozoic stratigraphy and structure of the western margin of the central San Joaquin valley, California: U.S. Geological Survey Open-File Report 82-526, 26 p.

McPherson, J.G., Shanmugam, G., and Moiola, R.J., 1987, Fan-deltas and braid deltas: Varieties of coarsegrained deltas: Geological Society of America Bulletin, v. 99, p. 331–340.

Namson, J.S., and Davis, T.L., 1988, Seismically active fold and thrust belt in the San Joaquin valley, central California: Geological Society of America Bulletin, v. 100, p. 257–273.

Nilsen, T.H., and Clarke, S.H., Jr., 1975, Sedimentation and tectonics in the early Tertiary continental borderland of central California: U.S. Geological Survey Professional Paper 925, 64 p.

Oil and Gas Journal, 1988, Forecast and Review, v. 86, p. 60.

Sedimentology Research Group, 1981, The effects of in-situ steam injection on Cold Lake oil sands: Bulletin of Canadian Petroleum Geology, v. 29, p. 447–478.

Small, G.P., 1986, Steam-injection profile control using limited-entry perforations: Society of Petroleum Engineers Production Engineering, September, p. 388–394.

Stanton, R.J., and Dodd, J.R., 1970, Paleoecologic techniques—Comparison of faunal and geochemical analyses of Pliocene paleo-environments, Kettleman Hills, California: Journal of Paleontology, v. 44, p. 1092–1121.

Traverse, E.F., Deibert, A.D., and Sustek, A.J., 1983, San Ardo—A case history of a successful steamflood: Society of Petroleum Engineers, Paper 11737, 8 p.

Webb, G.W., 1981, Stevens and earlier Miocene turbidite sandstones, southern San Joaquin valley, California: American Association of Petroleum Geologists Bulletin, v. 65, p. 438–465.

Woodring, W.P., Steward, R., and Richards, R.W., 1940, Geology of the Kettleman Hills Oilfield, California: U.S. Geological Survey Professional Paper 195, 170 p.

Key Words

South Belridge Field, California, San Joaquin basin, Tulare Formation, Pleistocene, fluviodeltaic, distal bar, mouth bar, meandering fluvial, braided fluvial, heavy oil, steamflood, enhanced oil recovery, in situ combustion, San Andreas fault, petrophysics, reservoir heterogeneity.

Desert Environments

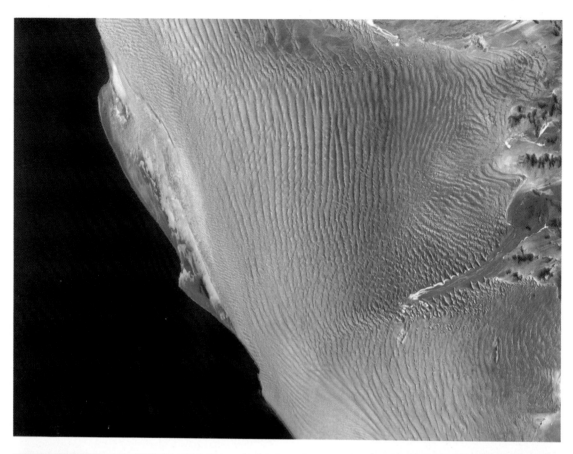

6

Reservoir Properties of the Desert Shattuck Member, Caprock Field, New Mexico

Ariel Malicse and Jim Mazzullo

Department of Geology, Texas A&M University, College Station, Texas 77843

Introduction

The deposits of deserts are often viewed as homogeneous sequences of eolian sand. However, studies of modern deserts show them to be the product of a complex interaction of wind- and water-driven processes. Eolian, fluvial, sabkha, and lacustrine settings are common in deserts, giving rise to a variety of sand-body geometries and a wide range of possible reservoir characteristics.

This study examines desert sedimentary deposits in the Shattuck Member, the principal pay zone in the Caprock Field of New Mexico (Fig. 6-1), and illustrates the importance of various desert sedimentary processes and environments to the formation of reservoir and nonreservoir rocks. The major reservoir facies of the Shattuck are gray sandstones deposited in fluvial sand-flat and eolian sand-sheet settings, and their porosity is a product of the dissolution of anhydrite or dolomite cements. By contrast, the reddish-brown sandstones of the Shattuck are impermeable and water-saturated trap rocks that accumulated in continental sabkhas.

Regional Data

The study area is the southern part of Caprock Field, which is an elongate northeast-to-southwest trending field in the southeastern corner of Chaves County, New Mexico. The field was discovered by Union Oil Company of California (UNOCAL) with completion of the Livermore No. 1-D State well in 1940 and was intensely drilled in the 1940s and 1950s by them. The southern part of the field (T15S, R31E) contains 116 wells for which wireline logs were made available for this study by UNOCAL (Fig. 6-2). In addition, seven cores of Shattuck and its bounding lithologies from this field were provided for detailed sedimentological analysis.

Tectonic Setting

Caprock Field is located on the Northwest shelf of the Permian basin, which covers about 115,000 square miles (3.0×10^5 km²) of Texas and New Mexico (Fig. 6-1). The basin was formed in the early Paleozoic as part of the Ouachita-Marathon fold-thrust belt. It evolved into its present configuration of two small deepwater basins (Delaware and Midland basins) separated by the Central Basin platform and flanked by the Northwestern, Eastern, and Southern shelves during the Late Pennsylvanian to Early Permian Marathon-Ouachita orogeny. By Late Permian time, the Permian basin was tectonically stable, the Midland basin was filled with sediment, and the deepwater Delaware basin was bordered by the Capitan and Goat Seep reefs. The Shattuck was deposited on the Northwest shelf, which during this time was a relatively flat, featureless, and slowly subsiding platform that extended from the shelf-margin reefs to a series of low-lying foothills in northern New Mexico and Oklahoma (Hills, 1972).

Casebooks in Earth Science
Sandstone Petroleum Reservoirs
Eds.: Barwis/McPherson/Studlick
© 1990 by Springer-Verlag New York, Inc.

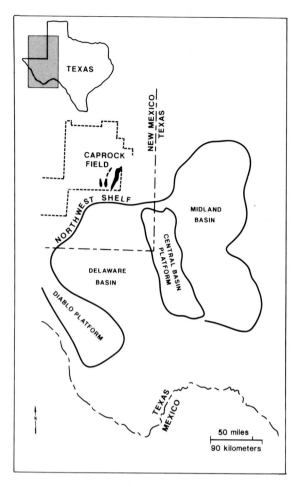

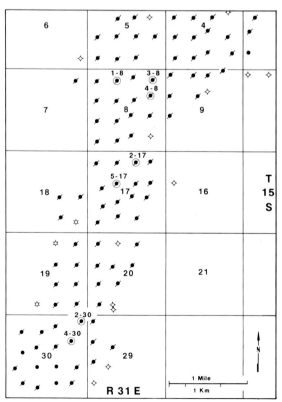

Fig. 6-1. Permian basin of Texas-New Mexico, showing the major physiographic features and the location of the Caprock Field (modified after Silver and Todd, 1969 and reprinted by permission of American Association of Petroleum Geologists).

Stratigraphy

During Late Permian (Guadalupian) time, deposition on the back-reef shelves of the Permian basin was controlled by glacio-eustatic fluctuations in sea level (Crowell, 1982; Veevers and Powell, 1987). The resultant back-reef deposits are composed of the complexly interbedded and interfingering carbonates, evaporites, and siliciclastics of the Artesia Group (King, 1942; Tait et al., 1962), which were deposited in lagoonal and continental environments during alternating periods of high and low stands of sea level (Meissner, 1972). The Artesia Group consists of five formations (from base to top): Grayburg, Queen, Seven Rivers, Yates, and Tansill (Fig. 6-3). The Shattuck is the uppermost member of the Queen Formation (Newell et al., 1953) and is a major reservoir across the Northwest shelf region.

Reservoir Characterization

Geometry and Structure

The Shattuck Member in the subsurface of Caprock Field has a slightly wedge-planar geometry that is superimposed by a north-to-northeast trending shoestring sand body (Figs. 6-4, 6-5). This shoestring sand body has an average thickness of 16 to 22 feet (4.9–6.7 m), and it is thicker along its central axis than on its flanks (Fig. 6-4). Most oil production from the Shattuck occurs along the trend of this sand body, which is at a high angle to the orientation of the shelf-margin reef (Figs. 6-1, 6-2, 6-4).

The structure on top of the Shattuck is homoclinal and gently dips to the east and southeast at a gradient of 0.9° (80 ft/mi or 15 m/km) (Fig. 6-6). There is no structural closure in the hydrocarbon entrapment of the Shattuck reservoir sandstones. Rather, the trap is stratigraphic and is caused by the enclosing impermeable facies.

Fig. 6-2. Reference map for a portion of Caprock Field; the cored wells are circled and numbered.

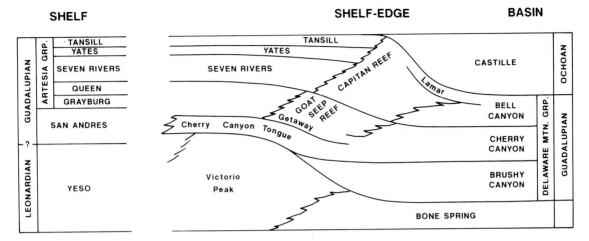

Fig. 6-3. Stratigraphy of Late Permian sedimentary rocks on the Northwest shelf and the adjacent shelf margin and basin. The Shattuck Member marks the top of the Queen Formation in this region (modified after Ward et al., 1986 and reprinted by permission of American Association of Petroleum Geologists).

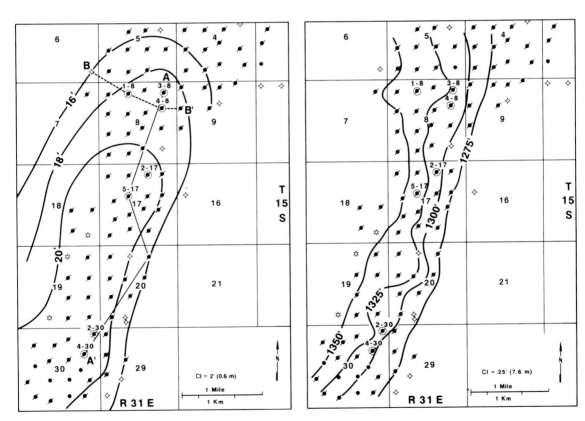

Fig. 6-4. Isopach map for the Shattuck Member in a portion of Caprock Field. Note the linear trend of the shoestring sandstone body.

Fig. 6-5. Structure contour map at the top of the Shattuck Member showing monoclinal westward dip.

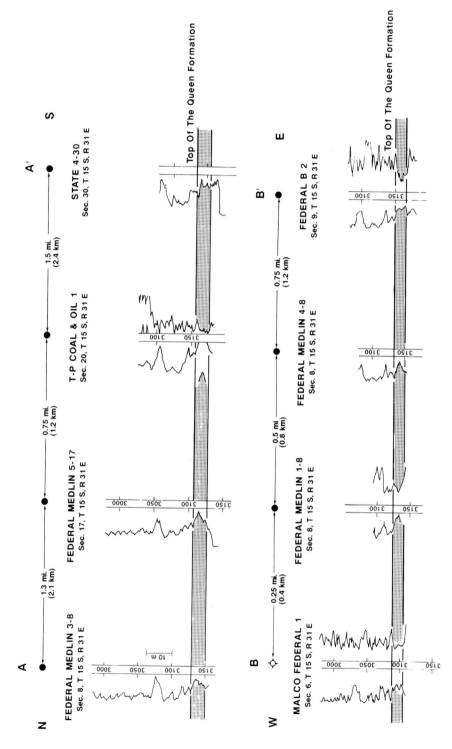

Fig. 6-6. Cross sections A-A' and B-B' through the Caprock Field. The Shattuck Member is indicated by the dotted pattern. The Shattuck Member is bounded at the bottom by anhydrite and at the top by dolomite. See Figure 6-4 for locations of cross sections.

In cores from Caprock Field, the Shattuck is underlain and overlain by interbedded dolomite and anhydrite, and its contact with these bounding lithologies is typically sharp. The dolomite is light brown and pinkish gray, very fine- to medium-grained, and sparsely fossiliferous. It is generally thinly wavy-laminated and contains cryptalgal laminae and tepee structures. The anhydrite is light purple and thickly wavy-laminated. Both lithologies were deposited in subtidal and intertidal parts of a shallow and periodically hypersaline lagoon that existed on this part of the Northwest shelf during high stands of sea level (Kendall, 1969).

Sandstone Facies and Depositional History

Three desert sedimentary environments are recognized for the Shattuck of Caprock Field (Fig. 6-7): fluvial sand flat, fluvial-dominated sabkha, and dry eolian sand sheet.

Fluvial Sand Flats. Fluvial sand flats are broad, low-relief, and unvegetated fluvial plains that are common to sandy deserts. They are flat and featureless and traversed by a few small, shallow channels. The channels become engorged and overflow during the flash floods that periodically occur in deserts (Glennie, 1970). Thus, the major depositional agents on a fluvial sand flat are nonchannelized overbank or sheetflood waters (Handford, 1982; Stear, 1983; Turnbridge, 1984).

Fluvial sand-flat deposits constitute the bulk of the Shattuck Member in Caprock Field. They are composed primarily of fluvial sheetflood deposits, which consist of:

1. Ripple cross-laminated very fine-grained sandstones (Figs. 6-7A, 6-7B, 6-7C);
2. Wavy-laminated silty sandstones with silty mudstone drapes which were formed by the migration of low-amplitude sinusoidal ripples with poorly developed slipfaces under low-flow regime conditions (Jopling and Walker, 1968); and
3. Planar-laminated siltstones separated by silty mudstone laminae, which were deposited from suspension under low-flow regime conditions (Fig. 6-7D).

Channel deposits are rarely present in the Shattuck and comprise thin (less than 1 ft; 0.3 m) medium- to very fine-grained cross-bedded and planar-bedded sandstones that overlie scour surfaces (Figs. 6-7E, 6-7F).

Fluvial-Dominated Sabkhas. Fluvial-dominated sabkhas are poorly vegetated salt flats wherein fluvial-deposited siliciclastic sediment is modified by the evaporation of surface or subsurface brine waters and resultant precipitation of displacive evaporite nodules and surficial salt crusts (Kendall, 1984; Warren, 1989). Sabkha deposits are common to the lower Shattuck in the southernmost part of Caprock Field, but not in the northern part. The siliciclastic component in the sabkha deposits of the Shattuck is similar in texture to the sheetflood deposits of fluvial sand flats and contains the following sedimentary structures:

1. Wavy ripple laminae, with associated silty mudstone drapes and laminae, which were deposited by sinusoidal ripples but were not later deformed by evaporites; and
2. Haloturbation structures, which consist of wavy sinusoidal ripple laminae that were deformed and disrupted by the displacive growth of evaporite nodules (Smith, 1971; Kendall, 1984) (Figs. 6-7G through 6-7J). The nodules are always composed of anhydrite, and they are generally small (0.12–2.0 in. or 3.0 mm-5.0 cm), white to gray in color, ovate, and discrete (Fig. 6-7K).

Dry Eolian Sand Sheets. Dry eolian sand sheets are broad, low-relief and poorly vegetated flats that are dominated by wind-ripple migration, eolian grainfall, and dune migration (Hunter, 1977; Kocurek, 1981). Dry eolian sand-sheet deposits are not very abundant in the Shattuck in Caprock Field and generally form thin and laterally discontinuous lenses within fluvial sand-flat or sabkha deposits. They are typically composed of fine- to very fine-grained, well-sorted sandstone that is mud-free and are recognized by the following sedimentary structures:

1. Planar laminae, which are thin (less than 0.08 in. or 2.0 mm) and dip at angles between 0 and 15 degrees, and which were deposited by wind ripples and eolian grainfall processes (Hunter, 1977) (Fig. 6-7N); and
2. Dune cross-beds, which are parallel, evenly spaced, and steeply dipping (maximum of 30°) (Fig. 6-7O).

A typical core of the Shattuck in Caprock Field is the State of Texas No. 2-30 (Fig. 6-8), which contains 22 feet (6.7 m) of the Shattuck. The basal unit (Unit 1) consists of 7 feet (2.1 m) of sabkha deposits

Fig. 6-7

B

Fig. 6-7. Sedimentary structures in the Shattuck Member and associated lithologies. Scale bar is 1 inch (2.5 cm). (A) Oil-stained, ripple cross-laminated sandstone of the fluvial sand flat. (B) and (C) Ripple cross-laminated and wavy-laminated sandstone of a fluvial sand flat. (D) Planar and thinly laminated siltstone of the fluvial sheet-flood environment. (E) and (F) Cross-bedded channel sandstones (with a basal scour in E) of a fluvial sand flat. (G) and (H) Wavy and discontinuously laminated (haloturbated) sandstones of a continental sabkha. (I) Wavy, discontinuously, laminated sandstone (below) of the sabkha environment overlain by planar-laminated sandstone (oil-stained color) of an eolian sand-sheet environment. (J) Photomicrograph of boxed area in I, showing the sharp (and presumably erosive) contact between sabkha and eolian sand-sheet deposits. (K) Massive sandstone in sabkha deposits with small displacive anhydrite nodules (in white). (L) Thinly laminated dolomite with tepee structure. (M) Thinly laminated and interbedded dolomite and anhydrite overlain by the Shattuck Member (white) and separated by a thin cryptalgal lamina (al). (N) Planar-laminated sandstone, formed by the migration of wind ripples in an eolian sand sheet. (O) Dune cross-bedded sandstone of the eolian sand sheet.

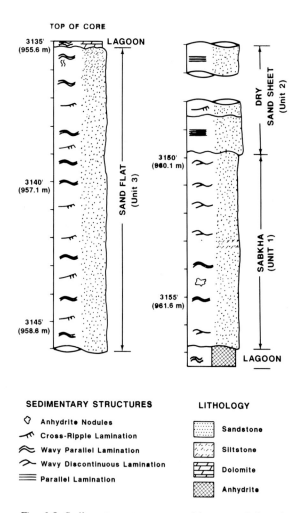

SEDIMENTARY STRUCTURES

◇ Anhydrite Nodules
⤙ Cross-Ripple Lamination
≈ Wavy Parallel Lamination
⌇ Wavy Discontinuous Lamination
≡ Parallel Lamination

LITHOLOGY

▫ Sandstone
Siltstone
Dolomite
Anhydrite

Fig. 6-8. Sedimentary structures and interpreted depositional environments for the Shattuck Member and its bounding lithologies in the State No. 2-30 core.

and is separated from underlying lagoonal evaporites by a sharp, wavy surface. This is overlain by 4 feet (1.2 m) of an eolian sand-sheet deposit (Unit 2), which also contains a thin interval of fluvial-reworked sediment, which is then overlain by 11 feet (3.4 m) of fluvial sand-flat deposits (Unit 3). The Shattuck is overlain by 0.2 feet (6.0 cm) of thinly wavy-laminated lagoonal dolomite, and the contact is abrupt.

Throughout the field, the fluvial sand-flat deposits can be traced along the elongate north-to-northeast axis of the shoestring sandstone. On the other hand, the sabkha deposits form a lenticular body that pinches out updip and toward the axis of this shoestring body. The eolian sand-sheet deposits form

thin lenses within these sand-flat and sabkha deposits (Fig. 6-9A).

This stratigraphic sequence suggests that the Shattuck of the Caprock Field was deposited during a relative low stand of sea level, which exposed the back-reef shelf and thereby allowed northerly coastal-plain environments to prograde southward toward the shelf-margin reef (Fig. 6-9B). The deposition of the Shattuck was finally terminated by a transgression that caused these coastal-plain environments to retreat landward (northward) and returned lagoonal conditions to the area. This transgression was evidently rapid, because the upper Shattuck is immediately overlain by lagoonal sediments and shows little evidence of reworking by marine processes and biota.

Petrography and Diagenesis

The Shattuck is generally a well-sorted, very fine- to fine-grained sandstone with varying amounts of siltstone, silty mudstone, and medium- to coarse-grained sandstone (Fig. 6-10). Generally, the fluvial sand-flat deposits are the most variable in size and have mean grain sizes that range from 45 microns (in the rippled sheetflood deposits) to 363 microns (in the cross-bedded channel deposits). The sabkha deposits are the finest and muddiest deposits, with mean grain sizes that vary from 40 to 121 microns. Lastly, eolian sand-sheet deposits are the least texturally variable and best-sorted deposits, with mean grain sizes between 80 and 176 microns.

The Shattuck sandstones are typically feldsarenites (arkosic) or subfeldsarenites (subarkosic) (Figs. 6-11, 6-12). The principal detrital grains are monocrystalline quartz (70–85% of the total detrital-grain composition) and feldspar (15–30%; commonly K-feldspar and rarely plagioclase). Minor detrital species include biotite and muscovite micas, sedimentary and metamorphic rock fragments, and altered iron-rich grains.

The principal cements in the Shattuck are dolomite and anhydrite (Figs. 6-12, 6-13), which are generally present in all types of deposits. Dolomite constitutes between 4 and 30% of the total-rock composition. It occurs as a fine- to medium-grained, subhedral to anhedral pore filler and grain replacer (Figs. 6-13A, 6-13B) and is most abundant in finer-grained fluvial sand-flat deposits. Anhydrite constitutes 6 to 25% of the total-rock composition. It occurs as a pore-filling and grain-replacing poikilotopic cement (Figs.

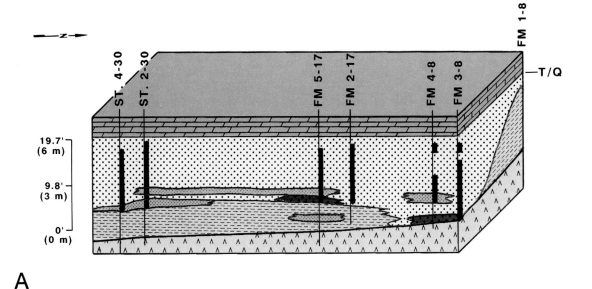

A

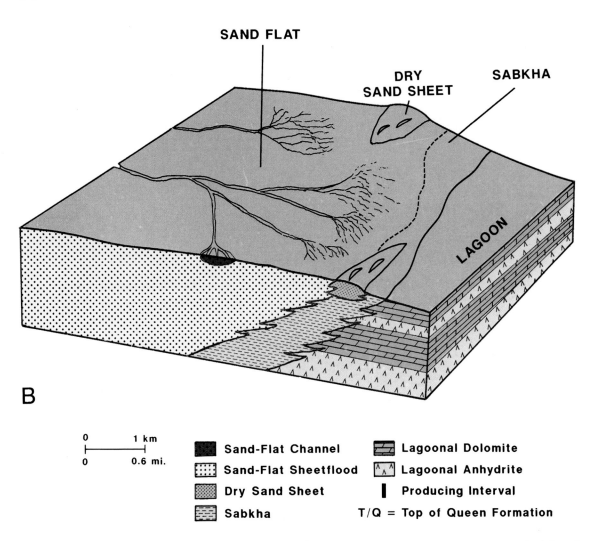

B

■ Sand-Flat Channel	▤ Lagoonal Dolomite
⠿ Sand-Flat Sheetflood	⋀⋀ Lagoonal Anhydrite
▨ Dry Sand Sheet	▮ Producing Interval
▥ Sabkha	T/Q = Top of Queen Formation

0   1 km
├──┼──┼──┤
0   0.6 mi.

Fig. 6-9. (A) Block diagram through the Caprock Field showing the north-south and east-west geometry of fluvial, eolian, sabkha, and lagoonal deposits. See Figure 6-2 for location of wells. (B) Three-dimensional figure show-ing the progradation of siliciclastic sabkha, fluvial, and eolian sediments of the Shattuck Member over the back-reef lagoon evaporites. Note greatly exaggerated vertical scale.

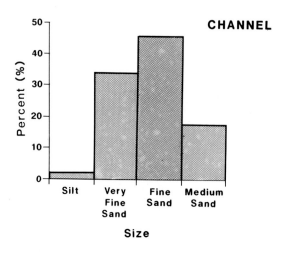

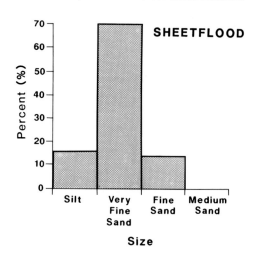

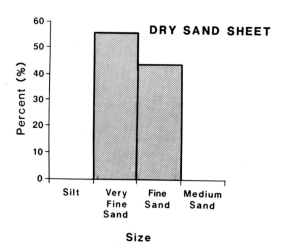

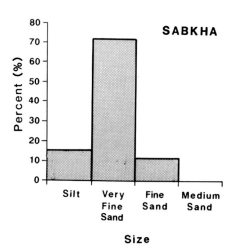

Fig. 6-10. Textural characteristics of the clastic sediments in the Shattuck showing the different textures by environment. Generally the reservoir sandstones (channels and dry sand sheets) are coarser grained than the nonreservoir sandstones.

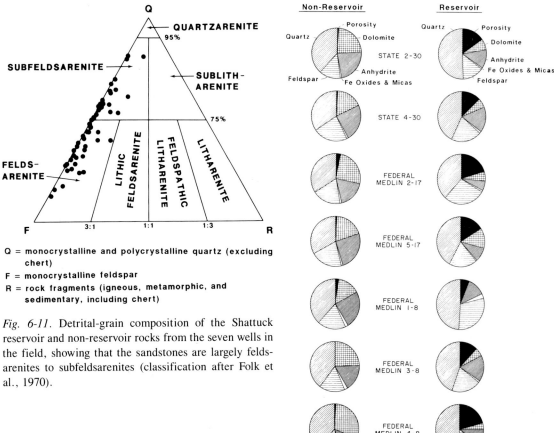

Q = monocrystalline and polycrystalline quartz (excluding chert)

F = monocrystalline feldspar

R = rock fragments (igneous, metamorphic, and sedimentary, including chert)

Fig. 6-11. Detrital-grain composition of the Shattuck reservoir and non-reservoir rocks from the seven wells in the field, showing that the sandstones are largely feldsarenites to subfeldsarenites (classification after Folk et al., 1970).

Fig. 6-12. Total-rock composition (including porosity) of Shattuck reservoir and nonreservoir rocks showing the effect of a varying cement content on porosity development. Also, note the increased quartz content in the coarser-grained reservoir rocks.

6-13A, 6-13C) and is most abundant in the sabkha, eolian sand-sheet, and coarse-grained fluvial sand-flat deposits (Figs. 6-12, 6-13).

The precipitation of dolomite and anhydrite cement was preceded by the mechanical infiltration of hematitic detrital illite-smectite clay and the precipitation of small amounts of authigenic feldspar (Fig. 6-13D) and hematite (Figs. 6-13E, 6-13F, 6-13G). The hematite and hematitic illite-smectite are most abundant within the nonreservoir horizons of the Shattuck and are responsible for their reddish-brown color. However, they are rarely present in the reservoir horizons, which instead contain well-crystallized, grain-lining corrensite (chlorite-smectite) (Figs. 6-13H, 6-14D). This corrensite may have been formed by the reduction of the hematitic illite-smectite by the organic acid solvents during late-stage diagenesis (Glennie et al., 1978; Dixon et al., 1989) (Fig. 6-15).

The porosity of the reservoir horizons in the Shattuck is secondary in origin and was formed by the dissolution of dolomite and anhydrite cements and labile grains within the Shattuck reservoirs, probably by organic acid solvents that formed as a by-product of kerogen maturation (Schmidt and

McDonald, 1979; Surdam et al., 1984; Crossey et al., 1986) (Fig. 6-14). The leaching of the pore-filling cements produced about 95% of the porosity in the reservoirs, which is equivalent to removal of about 34 to 60% of their original cement.

Petrophysics and Heterogeneities

The State of Texas No. 2-30 well was tested at an interval of 1,298 to 1,311 feet (395.6–399.6 m) above sea level and produced 34° API oil (at 90°F or 32°C) at an initial production rate of 296 BOPD (47 m³/D). The well has an average residual oil saturation of about 14% and an average water saturation of about 50% (Fig. 6-16).

Fig. 6-13. Cements in the Shattuck Member. Scale bar is 0.05 mm. (A) Pore-filling anhydrite (a) and dolomite (d). (B) Pore-filling dolomite (d) and metamorphic rock fragment (indicated by arrow). (C) Pore-filling anhydrite (a) surrounding quartz (q) grains. (D) Feldspar (microcline) with overgrowth (indicated by arrow). (E) and (F) Mechanically infiltrated hematitic clay on detrital sand grains. (G) A hematized sand grain (g) and surrounding aureole of hematite that was deformed by compaction during burial of the sandstone. (H) Authigenic corrensite (mixed-layer smectite and chlorite), found as a grain liner in the Shattuck reservoir sandstones.

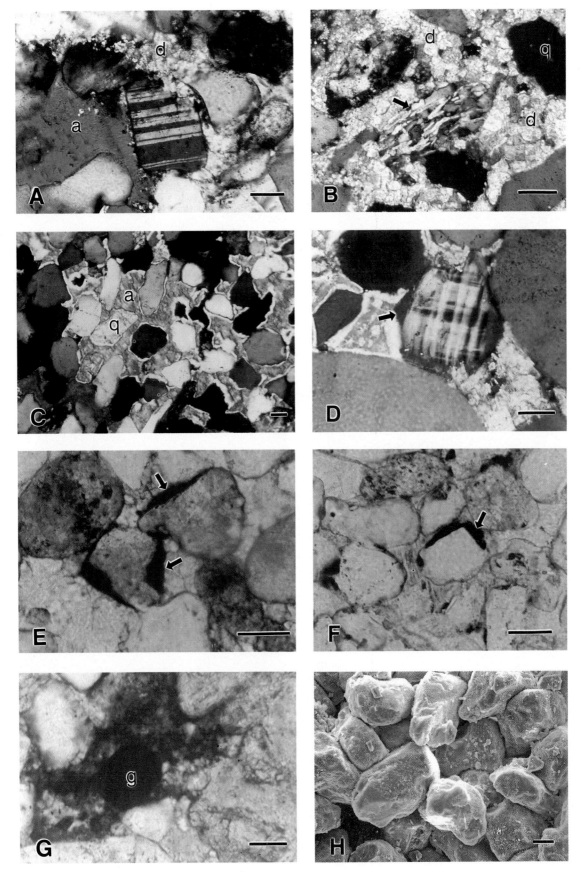

145

Fig. 6-14. (A) and (B) Intergranular porosity (p) in the reservoir horizons of the Shattuck Member. (C) SEM photomicrograph of intragranular porosity (p) which was formed by the dissolution of a labile detrital grain. Note the grain-coating corrensite (indicated by arrow). (D) SEM photomicrograph of K-feldspar rhombs (k) devel-oped over a grain surface coated with authigenic corrensite in the Shattuck reservoir sandstone. Note bald areas are points of grain contact. (E) Thin-section photomicrograph of partly dissolved anhydrite (a) cement and the resultant intergranular porosity (p). Scale bar is 0.10 mm.

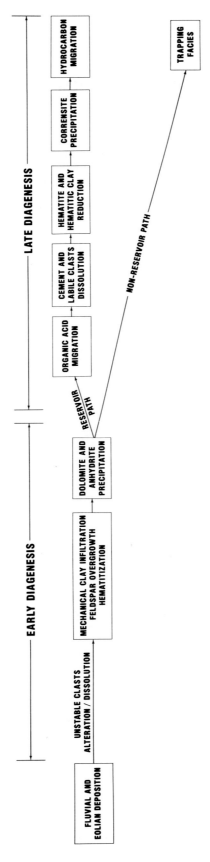

Fig. 6-15. The diagenetic sequence of the Shattuck reservoir and nonreservoir sandstones. Note the change in diagenetic path of the reservoir and nonreservoir sandstones at the start of late diagenesis. The nonreservoir rocks become the trapping facies.

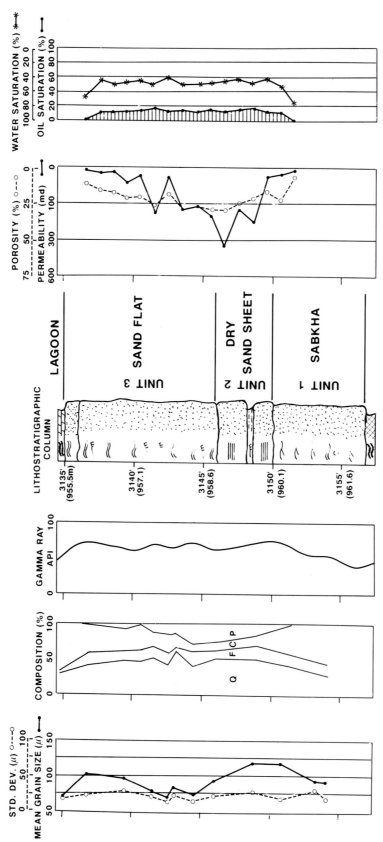

Fig. 6-16. Summary logs for the Shattuck Member and its bounding lithologies in the State No. 2-30 well, illustrating the lithology, porosity, permeability, and water and oil saturation of the sandstone and the gamma-ray log for the entire core. In the Composition column, Q = quartz, F = feldspar, C = cement, and P = porosity.

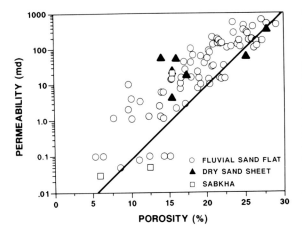

Fig. 6-17. Porosity versus permeability plot of fluvial sand-flat (open circles), dry sand-sheet (black triangles), and sabkha (open squares) reservoir rocks in Caprock Field. Note the sabkha sandstones have few data points because they are generally nonreservoir and were not tested.

Figure 6-17 shows the relationship between porosity and permeability of the fluvial, eolian, and sabkha deposits in the cores at Caprock Field. This relationship is exponential and shows a drop of permeability to 0.1 md at about 5% porosity. The fluvial sand flat exhibits the widest range of porosity and permeability values: the coarsest-grained deposits have porosities that range from 5 to 30% (mean = 18%) and permeabilities that range from 5 to 630 md (mean = 100 md), whereas the finest-grained deposits have porosities of about 0 to 15% (mean = 5%) and permeabilities from 0 to 5 md.

The eolian sand-sheet deposits also have consistently high porosities of 14 to 28% (mean = 18%) and permeabilities of 21 to 341 md (mean = 102 md). On the other hand, the sabkha deposits are "tight sands" with porosities of less than 5% on the average and permeabilities of 0 to 10 md (mean = less than 0.1 md). Consequently, the hydrocarbon production is principally from the eolian sand-sheet and coarsest-grained fluvial sand-flat deposits, whereas sabkha and finer-grained sand-flat deposits are generally nonreservoirs. The poor reservoir quality of sabkha and sand-flat sequences is attributed to their high content of siltstone and silty mudstone, which greatly reduced their permeability and prevented the infiltration of the cement-dissolving, porosity-forming subsurface fluids.

Production History

Since its discovery in 1940, about 740 wells have been drilled in Caprock Field. Production from 1943 to 1986 yielded about 72.6 MMBO (1.2 × 10⁷ m³) from 665 wells at an average rate of 5 to 40 BOPD (0.8–6.4 m³/D) per well. More recent production (1973 to present) has declined to an average rate of 2 BOPD (0.3 m³/D) per well (Fig. 6-18). Between 1979 and 1982, about 76% of the wells were shut in because they watered out (produced very high proportions of water versus oil).

The normal completion practice for Shattuck reservoirs is to set pipe on the top of the pay zones and then continue drilling with cable tools. After the drilling, neutron and gamma-ray logs are run, the

Fig. 6-18. Union Oil of California's production data and number of producing wells for 1973–1988 for the Caprock Field. The sudden decrease in production was due to shutting in of wells due to high water production. Increased production resulted from better reservoir management by water injection.

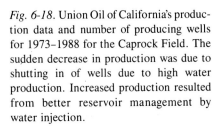
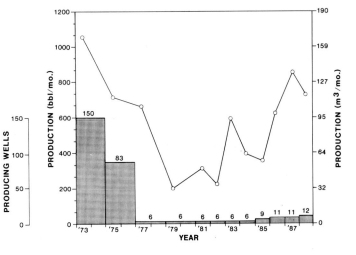

reservoirs are perforated and fractured, and pumps are installed in order to control the paraffin. The wells have initial pressures between 850 and 965 psi (5.9–6.7 × 10³ kPa), and they are driven by dissolved gas (high nitrogen and low BTU). The producing-formation zone water is generally saline with a resistivity of 0.44 ohm-meter at 75°F (24°C). Production stimulation by water injection has been attempted. Any future attempts at water injection should consider the presence of smectite-rich clays and the possibility of formation damage.

Exploration Strategy

The study of the reservoir characteristics of desert sedimentary deposits in the Shattuck emphasizes and reiterates the importance of secondary porosity to reservoir development and the influence of depositional textures on diagenetic porosity enhancement. Shattuck reservoirs offer good examples of isolated porous sandstone bodies encased within impermeable silty sandstone deposits. The successful exploration for similar shoestring bodies may be difficult, since they are rarely indicated by structural deformations in overlying formations (Levorsen, 1967). Successful exploration requires:

1. The ability to recognize fluvial sand-flat and eolian sand-sheet deposits from cores and wireline logs, using the gamma-ray, SP, and short-spacing resistivity curves;

2. Careful mapping of fluvial sand-flat and eolian sand-sheet deposits and their adjacent areas; and

3. The drilling of test wells to establish the relative positions of the pools with respect to the sandstone body geometry. In the case of Caprock Field, most hydrocarbons are situated on the elongate axis of the shoestring sandstone body.

Conclusion

The Shattuck Member of the Queen Formation in Caprock Field, New Mexico has a slightly wedge-planar geometry that is superimposed by a north-to-northeast trending shoestring sandstone body. The Shattuck was deposited in fluvial sand-flat, dry eolian sand-sheet, and sabkha environments on the Northwest shelf of the Permian basin during a low stand of sea level. The Shattuck reservoirs are composed of elongate fluvial channel and sand-flat deposits and interbedded eolian sand-sheet deposits. These reservoir rocks are trapped by relatively impermeable silty sabkha and sand-flat deposits. The reservoir porosity is secondary in its origin and was created by the dissolution of anhydrite and dolomite cement and labile grains. The reservoir quality in the Shattuck is controlled by the relative abundances of siltstone and silty mudstone within the sandstone, which controlled their permeability to the leaching fluids that created secondary porosity during late-stage diagenesis.

Reservoir Summary

Field: Caprock
Location: Chavez County, New Mexico
Operators: UNOCAL, Texas Pacfic, Sinclair, O'Neill, and others
Discovery: 1940
Basin: Northwest shelf of the Permian basin
Tectonic/Regional Paleosetting: Slowly subsiding continental margin
Geologic Structure: Homoclinal dip
Trap Type: Stratigraphic
Reservoir Drive Mechanism: Water and dissolved gas
  • **Original Reservoir Pressure:** 850–965 psi (5.9–6.7 × 10³ kPa) at 1,350 feet (410 m) above sea level
Reservoir Rocks
  • **Age:** Permian, Guadalupian
  • **Stratigraphic Unit:** Shattuck Member of the Queen Formation
  • **Lithology:** Very fine-grained feldsarenite
  • **Depositional Environments:** Eolian, desert fluvial, and sabkha
  • **Productive Facies:** Coarser-grained fluvial sand flat and eolian sand sheet

- **Petrophysics:** All data from cores
  - **ϕ:** Average 18%; range 15 to 30%; secondary reduced intergranular and intragranular (logs)
  - **k:** Average 250 md; range 30 to 650 md (cores)
  - **S_w:** Average 35%; range 30 to 85% (logs)
  - **S_o:** Average 14% (logs)

Reservoir Dimensions
- **Depth:** 3,100 feet (945 m)
- **Areal Dimensions:** 1.7 by 23 miles (2.7×37 km)
- **Productive Area:** 40 square miles (104 km²)
- **Number of Reservoirs:** 1
- **Hydrocarbon Column Height:** 10 feet (3.0 m)
- **Number of Pay Zones:** 1
- **Gross Sandstone Thickness:** 20 feet (6.1 m)
- **Net Sandstone Thickness:** 10 feet (3.0 m)
- **Net/Gross:** 0.50

Source Rocks
- **Lithology:** Unknown
- **Time of Hydrocarbon Maturation:** Unknown
- **Time of Trap Formation:** Permian (Guadalupian)
- **Time of Migration:** Unknown

Hydrocarbons
- **Type:** Oil
- **GOR:** 215:1
- **API Gravity:** 34° to 38°
- **FVF:** 1.125 at 1,000 psi
- **Viscosity:** 6.8 to 7.0 cP (6.8–7.0×10^3 Pa·s) at 90°F (32°C)

Volumetrics
- **In-Place:** 290 MMBO (4.6×10^7 m³)
- **Cumulative Production:** 72.6 MMBO (1.15×10^7 m³)
- **Ultimate Recovery:** 75.5 MMBO (1.2×10^7 m³)
- **Recovery Efficiency:** 26%

Wells
- **Spacing:** 1,320 feet (402 m), nominal 40 acres (16.2 ha.)
- **Total:** 740
- **Dry Holes:** 75

Typical Well Production:
- **Average Daily:** 5 to 40 BO (0.8–6.4 m³)
- **Cumulative:** 100 MBO (1.6×10^4 m³)

References

Crossey, L.J., Surdam, R.C., and Lehann, 1986, Application of organic-inorganic diagenesis to porosity prediction, *in* Gautier, L., ed., Roles of Organic Matter in Sedimentary Diagenesis: Society of Exploration Paleontologists and Mineralogists Special Publication 38, p. 147–155.

Crowell, J.C., 1982, Continental glaciation through geologic time, *in* Climate in Earth History, Studies in Geophysics: Washington, D.C., National Academy Press, p. 77–82.

Dixon, S. A., Summers, D. M., and Surdam, R. C., 1989, Diagenesis and preservation of porosity in Norphlet Formation (Upper Jurassic), Southern Alabama: American Association of Petroleum Geologists Bulletin, v. 73, p. 707–728.

Folk, R.L., Andrews, P.B., and Lewis, D.W., 1970, Detrital sedimentary rock classification and nomenclature for use in New Zealand: New Zealand Journal of Geology and Geophysics, v. 13, p. 937–968.

Glennie, K.W., 1970, Desert sedimentary environments: Developments in sedimentology, v. 14, Amsterdam, Elsevier, 222 p.

Glennie, K.W., Mudd, G.C., and Nagtegaal, P.J.C., 1978, Depositional environments and diagenesis of Permian Rotliegendes sandstones in Leman Bank and Sole Pit areas of the U.K. southern North Sea: Journal of Geological Society of London, v. 135, p. 25–34.

Handford, C., 1982, Sedimentology and evaporite diagenesis in a Holocene continental sabkha: Bristol Dry Lake, California: Sedimentology, v. 29, p. 239–253.

Hills, J., 1972, Late Paleozoic sedimentation in West Texas Permian Basin: American Association of Petroleum Geologists Bulletin, v. 56, p. 2302–2322.

Hunter, R.E., 1977, Basic types of stratification in small eolian dunes: Sedimentology, v. 24, p. 361–387.

Jopling, A.V., and Walker, R.G., 1968, Morphology and origin of ripple drift cross-lamination with examples from the Pleistocene of Massachusetts: Journal of Sedimentary Petrology, v. 38, p. 971–984.

Kendall, C., 1969, An environmental reinterpretation of the Permian evaporite-carbonate shelf sediments of the Guadalupe Mountains: Geological Society of America Bulletin, v. 80, p. 2503–2526.

Kendall, A.C., 1984, Evaporites, in R.G. Walker, ed., Facies Models (Second Edition): Geoscience Canada Reprint Series 1, p. 259–296.

King, P.B., 1942, Permian of west Texas and southeastern New Mexico: American Association of Petroleum Geologists Bulletin, v. 26, p. 535–763.

Kocurek, G., 1981, Erg reconstruction: The Entrada Sandstone (Jurassic) of north Utah and Colorado: Paleogeography, Paleoclimatology, Paleoecology, v. 36, p. 125–153.

Levorsen, A.I., 1967, Geology of Petroleum: San Francisco, W. H. Freeman and Co., 724 p.

Meissner, F.J., 1972, Cyclic sedimentation in Middle Permian strata of the Permian Basin, west Texas and New Mexico, in Elam, J. and Chuber, S., eds., Cyclic sedimentation in the Permian Basin: West Texas Geological Society, p. 203–231.

Newell, N.D., Rigby, J.K., Fisher, A.G., Whiteman, A.J., Hickox, I.E., and Bradley, J.S., 1953, The Permian reef complex of the Guadalupe Mountain region, Texas and New Mexico: San Francisco, W.H. Freeman and Co., 236 p.

Schmidt, V., and McDonald, D.A., 1979, Role of secondary porosity in the course of sandstone diagenesis, in Scholle, P.A. and Schluger, P.R., eds., Aspects of Dia-genesis: Society of Exploration Paleontologists and Mineralogists Special Publication 26, p. 175–207.

Silver, B.A., and Todd, R.G., 1969, Permian cyclic strata, northern Midland and Delaware basins, West Texas and southeastern New Mexico: American Association of Petroleum Geologists Bulletin, v. 53, p. 2223–2251.

Smith, D.B., 1971, Possible displacive halite in the Permian Upper Evaporite Group of Northeast Yorkshire: Sedimentology, v. 17, p. 221–232.

Stear, W.M., 1983, Morphological characteristics of ephemeral stream channel and overbank splay sandstone bodies in Permian Lower Beau-Group, Karoo Basin, South Africa, in Collinson, J.D. and Lewin, J., eds., Modern and Ancient Fluvial Systems: International Association of Sedimentologists Special Publication 6, p. 405–420.

Surdam, R.C., Boese, S.W., and Crossey, L.J., 1984, The chemistry of secondary porosity, in McDonald, D. and Surdam, R.C., eds., Clastic Diagenesis: American Association of Petroleum Geologists Memoir 37, p. 127–149.

Tait, D.B., Ahlen, J.L., Gordon, A., Scott, G.L., Motts, W.S., and Spitler, M.E., 1962, Artesia Group of New Mexico and west Texas: American Association of Petroleum Geologists Bulletin, v. 46, p. 504–517.

Turnbridge, I.P., 1984, Facies model for a sandy ephemeral stream and clay playa complex; the Middle Devonian Trentishoe Formation of North Devon, U.K.: Sedimentology, v. 31, p. 697–715.

Veevers, J., and Powell, C., 1987, Late Paleozoic glacial episodes in Gondwanaland reflected in transgressive-regressive depositional sequences in Euroamerica: Geological Society of America Bulletin, v. 98, p. 475–487.

Ward, R.F., Kendall, G.St.C., and Harris, P.M., 1986, Upper Permian (Guadalupian) facies and their association with hydrocarbons—Permian Basin, west Texas and New Mexico: American Association of Petroleum Geologists Bulletin, v. 70, p. 239–262.

Warren, J.K., 1989, Evaporite Sedimentology: Englewood Cliffs, New Jersey, Prentice Hall, 285 p.

Key Words

Caprock Field, New Mexico, Permian basin, Queen Formation, Permian Guadalupian, eolian, desert fluvial, sabkha, Shattuck Member, secondary porosity, stratigraphic trap, fluvial sand flat, fluvial-dominated sabkha, shoestring sandstone body.

7

Desert Environments and Petroleum Geology of the Norphlet Formation, Hatter's Pond Field, Alabama

Ernest A. Mancini, Robert M. Mink, Bennett L. Bearden, Steven D. Mann, and David E. Bolin

Geological Survey of Alabama, Tuscaloosa, Alabama 35486

Introduction

Reservoirs of the Norphlet Formation provide good examples of the varied influences of depositional and diagenetic controls on reservoir distribution and quality. Norphlet sandstones are representative of a variety of desert and near-desert depositional settings, and show a considerable range in reservoir quality. The degree to which deep burial of the Norphlet has overprinted and modified the depositional controls of reservoir quality is varied, as seen in this chapter and the two which follow. The Norphlet reservoir in the Hatter's Pond Field provides an opportunity to compare and contrast the reservoir quality of several desert and near-desert sequences, including wadi, playa dune, and nearshore-marine lithofacies. Reservoir quality varies greatly, and the most productive lithofacies are dune and shoreface sandstones.

The Hatter's Pond Field located in Mobile County in southwestern Alabama was discovered in 1974 with the drilling of the Getty Peter Klein 3-14 No. 1 well (Fig. 7-1). This field is unique in that production includes gas condensate at depths of more than 18,000 feet (5,490 m) from Jurassic marine and eolian Norphlet sandstones and marine Smackover Formation dolostones that are in communication across the formational contact (Fig. 7-2). The Klein 3-14 No. 1 well tested 2.2 MBCPD (350 m³/D) and 6.2 MMCFGPD (1.8 × 10⁵ m³/D) from Smackover carbonates at a depth of 18,042 to 18,060 feet (5,499–5,505 m) and 1.4 MBCPD (225 m³/D) and 4.4 MMCFGPD (1.3 × 10⁵ m³/D) from Norphlet sandstones at a depth of 18,180 to 18,200 feet (5,541–5,547 m) (Fig. 7-2).

Twenty-two Norphlet fields have been discovered in the onshore and offshore Alabama area. Norphlet fields in Alabama have cumulative production of more than 309 BCFG (8.7×10^9 m³), 3.1 MMBO (4.9×10^5 m³), and 55 MM BGC and BLG (8.7×10^6 m³), with the Hatter's Pond Field being the most productive of these fields. To date, 40.4 MM BGC and BLG (6.4×10^6 m³) and 128 BCFG (3.6×10^9 m³) have been produced from the Hatter's Pond Field. Enhanced recovery operations recently initiated at the field are expected to increase ultimate recovery by 15 MM BGC and BLG (2.4×10^6 m³) and 49 BCFG (1.4×10^9 m³) over the next 22 years. Recent oil discoveries in onshore Alabama and significant natural gas discoveries in offshore Alabama have aroused considerable interest in the drilling of additional Norphlet prospects. An understanding of the structural and stratigraphic relationships and reservoir heterogeneity at the Hatter's Pond Field will assist in the formulation of exploration and development strategies and the design of efficient enhanced recovery operations not only in onshore and offshore Alabama but in deep eolian sandstones elsewhere.

Regional Setting

Jurassic sedimentation in Alabama was affected by rifted continental margin tectonics associated with the opening of the Gulf of Mexico basin (Woods and Addington, 1973; Wood and Walper, 1974; Salvador, 1987). The established plate-tectonic overprint influenced the accumulation of Norphlet

Casebooks in Earth Science
Sandstone Petroleum Reservoirs
Eds.: Barwis/McPherson/Studlick
© 1990 by Springer-Verlag New York, Inc.

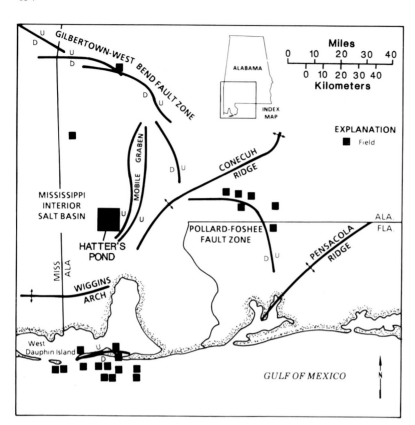

Fig. 7-1. Major structural features and Norphlet fields in southwestern Alabama.

sediments during the Jurassic. Norphlet deposition was controlled, in part, by a combination of differential basement subsidence and the presence of pre-Jurassic paleohighs and Jurassic highs (Wilson, 1975; Sigsby, 1976; Mancini and Benson, 1980). Differential subsidence in the established rift graben and various basinal areas, such as the Mississippi interior salt basin, resulted in thick accumulations of Norphlet sandstone (Fig. 7-1). Structurally, the Norphlet was affected by postdepositional movement of the underlying Louann Salt. Louann Salt movement in updip areas accentuated movement along the regional peripheral fault trend (Gilbertown, West Bend, and Pollard-Foshee Fault systems) near the updip limit of Norphlet deposition (Fig. 7-3). Regionally, the Norphlet dips 80 feet per mile (15 m/km) to the southwest.

Several pre-Jurassic paleohighs and Jurassic highs are present in the Alabama area (Fig. 7-1). Many of these are unnamed; however, two of the more pronounced features are the Conecuh and the Pensacola ridges (Fig. 7-1). Another positive feature is the Wiggins arch. The Wiggins arch, which extends westward into Mississippi, may represent a continental block that foundered during rifting or possibly a southwestward extension of the Appalachian structural front. The radiometric date of the metamorphic rocks recovered from the Betty Joe Anderson No. 1 well drilled in Mobile County, Alabama, on the Wiggins arch indicates a Mississippian age (Forest Oil Corporation, personal communication, 1982). Several of these positive features, like the Conecuh ridge and the Wiggins arch, served as source areas for Norphlet sediments. The Norphlet is absent on portions of several of these features, including the Conecuh ridge and the Wiggins arch (Fig. 7-4).

In Alabama, the Norphlet Formation is overlain by the Smackover Formation (Fig. 7-5). The contact between the Norphlet and the Smackover can be gradational or abrupt. At Hatter's Pond Field, the contact is conformable, grading upward from dolomitic or calcitic sandstone to silty dolostone or limestone. The Norphlet can overlie the Pine Hill Anhydrite Member, the Louann Salt, the Werner Formation, the Eagle Mills Formation, Mesozoic volcanic rocks, or Paleozoic strata (Fig. 7-5).

The accumulation of thick Jurassic salt deposits, anhydrite, red beds, and widespread eolian sandstones indicates that arid climatic conditions were

Fig. 7-2. Type log for the productive intervals in Hatter's Pond Field.

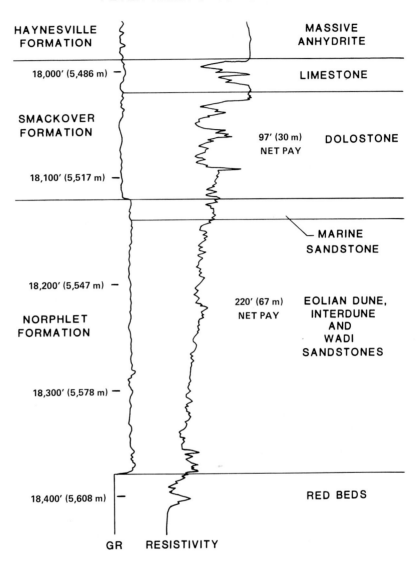

**PERMIT NO. 1978
GETTY OIL COMPANY
PETER KLEIN 3-14 NO.1**

HAYNESVILLE FORMATION	MASSIVE ANHYDRITE
18,000' (5,486 m)	LIMESTONE
SMACKOVER FORMATION	97' (30 m) NET PAY DOLOSTONE
18,100' (5,517 m)	
	MARINE SANDSTONE
18,200' (5,547 m)	220' (67 m) NET PAY EOLIAN DUNE, INTERDUNE AND WADI SANDSTONES
NORPHLET FORMATION	
18,300' (5,578 m)	
18,400' (5,608 m)	RED BEDS

GR   RESISTIVITY

prevalent during Norphlet deposition, as documented by studies of both modern and ancient desert environments (Walker, 1967; Glennie, 1970).

· Norphlet deposits are part of two Jurassic depositional sequences (Mancini et al., 1988) (Fig. 7-6). The lower Norphlet continental sandstone is the progradational regressive highstand deposits of a Type 2 depositional sequence that is comprised of Louann evaporite (transgressive), Pine Hill anhydrite and shale (condensed section), and Norphlet sandstone (progradational, regressive highstand). The upper Norphlet marine sandstone is part of an overlying Type 2 depositional sequence consisting of

Norphlet sandstone and Smackover carbonate (transgressive), Smackover carbonate (condensed), and Smackover and Haynesville carbonate and anhydrite (progradational, regressive highstand) (Fig. 7-6).

The sequence of deposition of the Norphlet and associated lithofacies began with the accumulation of intertidal mud in isolated lagoons or bays along an emerging shoreline (Fig. 7-7). Subsequent Norphlet sand accumulation was initiated with the erosion of the Appalachian Mountains. Gravelly sands were deposited in alluvial fans, and red bed sand, silt, and mud accumulated in distal-fan and alluvial-plain

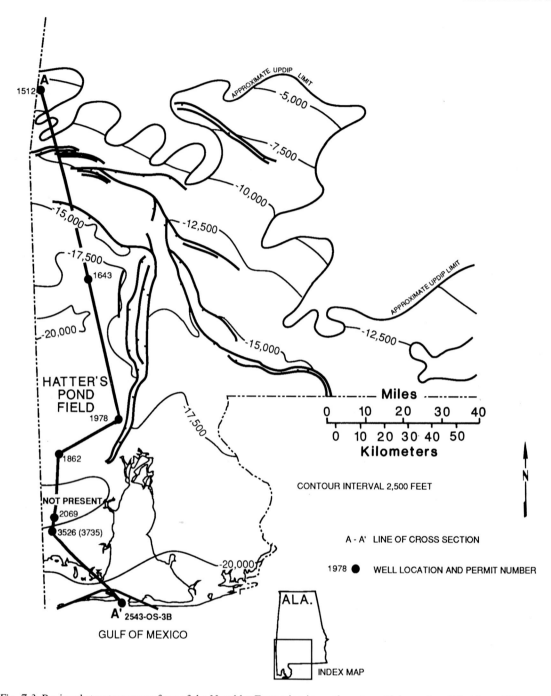

Fig. 7-3. Regional structure map of top of the Norphlet Formation in southwestern Alabama (modified from Wilson and Tew, 1985). Contour interval is 2,500 feet (762 m). Shown is cross section A-A' of Figure 7-4.

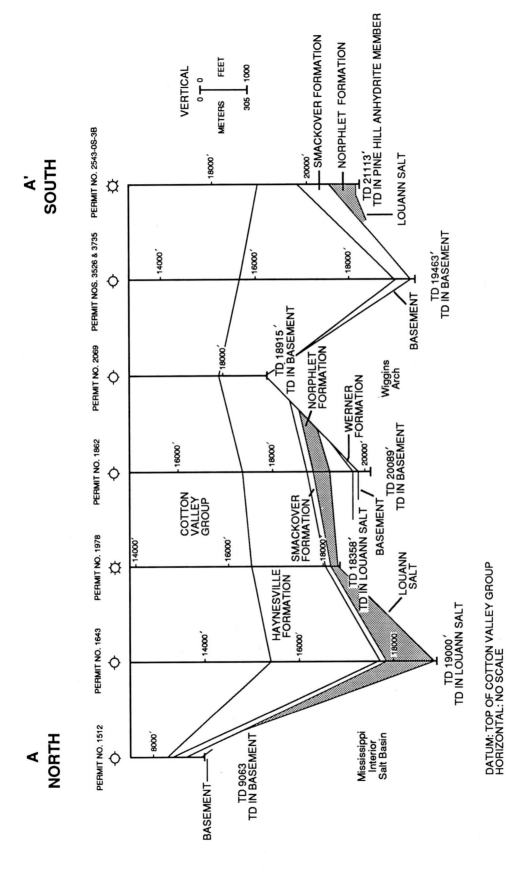

Fig. 7-4. Regional stratigraphic cross section A-A' (modified from Mink et al., 1985). (See Figure 7-3 for location.)

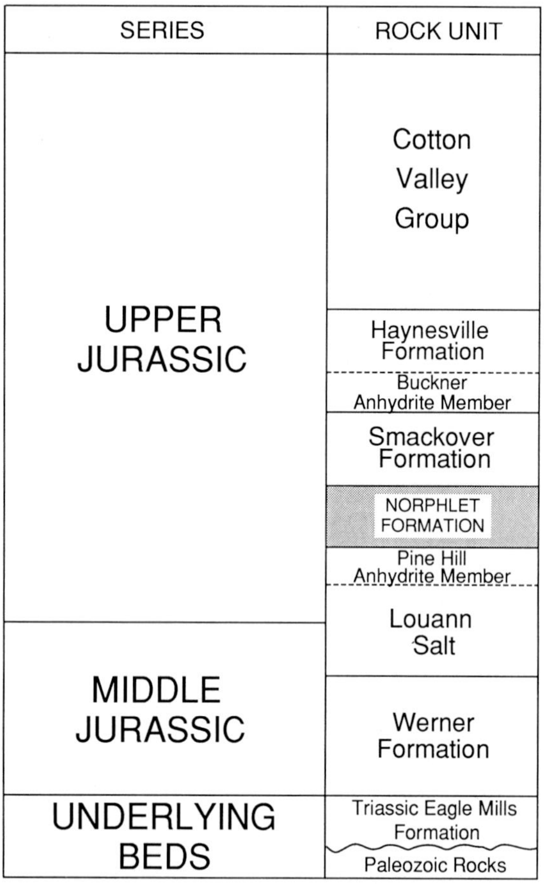

SERIES	ROCK UNIT
UPPER JURASSIC	Cotton Valley Group
	Haynesville Formation
	Buckner Anhydrite Member
	Smackover Formation
	NORPHLET FORMATION
	Pine Hill Anhydrite Member
	Louann Salt
MIDDLE JURASSIC	Werner Formation
UNDERLYING BEDS	Triassic Eagle Mills Formation
	Paleozoic Rocks

Fig. 7-5. Generalized Jurassic stratigraphy for southwestern Alabama.
◁

Fig. 7-6. Depositional sequences in Jurassic strata in the area of the Mississippi interior salt basin.   ▽

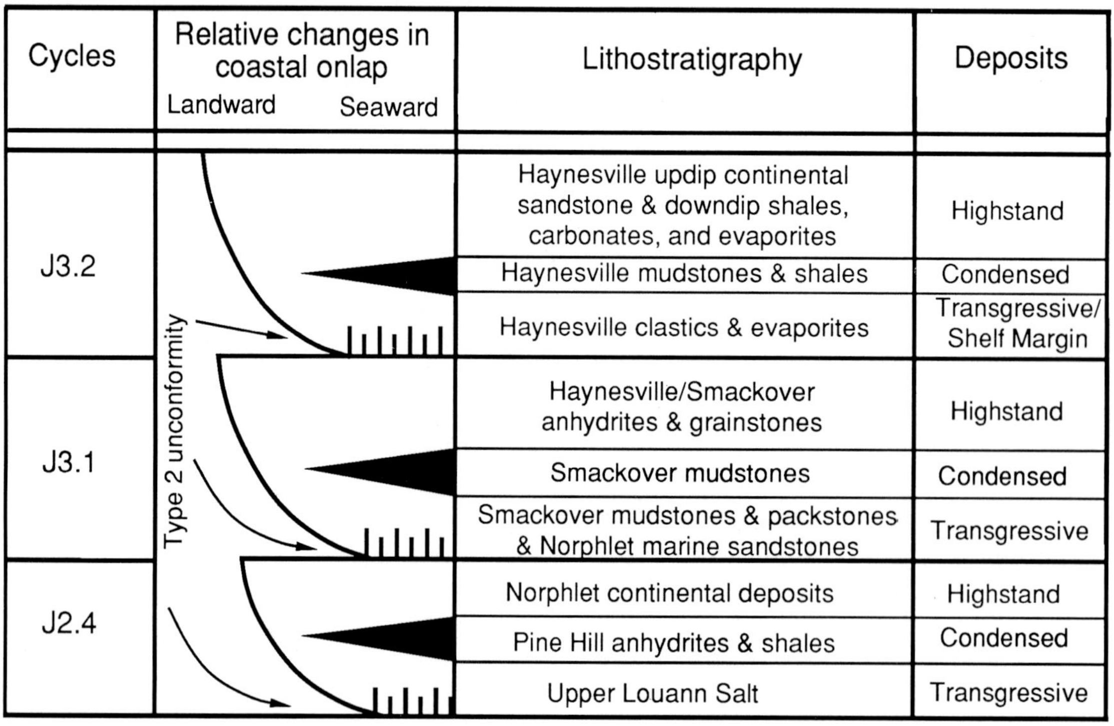

Cycles	Relative changes in coastal onlap		Lithostratigraphy	Deposits
	Landward	Seaward		
J3.2			Haynesville updip continental sandstone & downdip shales, carbonates, and evaporites	Highstand
			Haynesville mudstones & shales	Condensed
			Haynesville clastics & evaporites	Transgressive/ Shelf Margin
J3.1	Type 2 unconformity		Haynesville/Smackover anhydrites & grainstones	Highstand
			Smackover mudstones	Condensed
			Smackover mudstones & packstones & Norphlet marine sandstones	Transgressive
J2.4			Norphlet continental deposits	Highstand
			Pine Hill anhydrites & shales	Condensed
			Upper Louann Salt	Transgressive

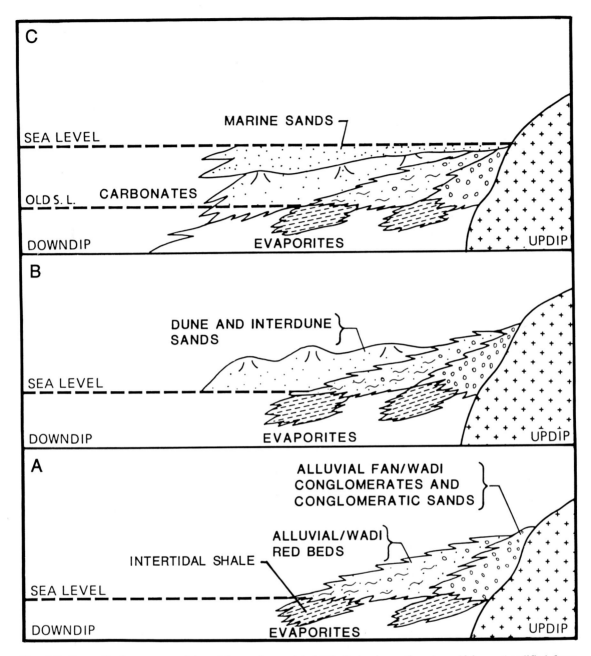

Fig. 7-7. Generalized sequence of deposition and associated lithofacies in southwestern Alabama (modified from Wilkerson, 1981; Mancini et al., 1985 and reprinted by permission of American Association of Petroleum Geologists).

environments. Sand from these environments was transported toward the coast across a desert plain. This sand was reworked into the dune and interdune Norphlet lithofacies. The marine transgression which was initiated in late Norphlet time and continued into Smackover time resulted in the reworking of the underlying sediments.

Reservoir Characterization

Seismic Attributes

Seismic response in the Hatter's Pond Field area (Fig. 7-8) is characterized by high amplitude and lateral continuity at the Smackover-Norphlet level

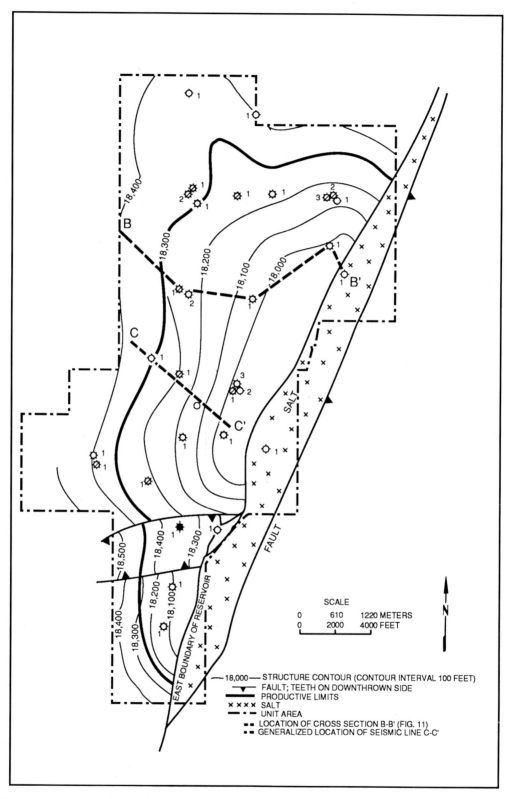

Fig. 7-8. Structure, top of the Norphlet in Hatter's Pond Field contoured in 100-foot (30.5-m) intervals. Modified from Exhibit 12 as presented by Getty Oil Company at the July 1984 meeting of the State Oil and Gas Board of Alabama, Docket 4-11-841 and reprinted by permission.

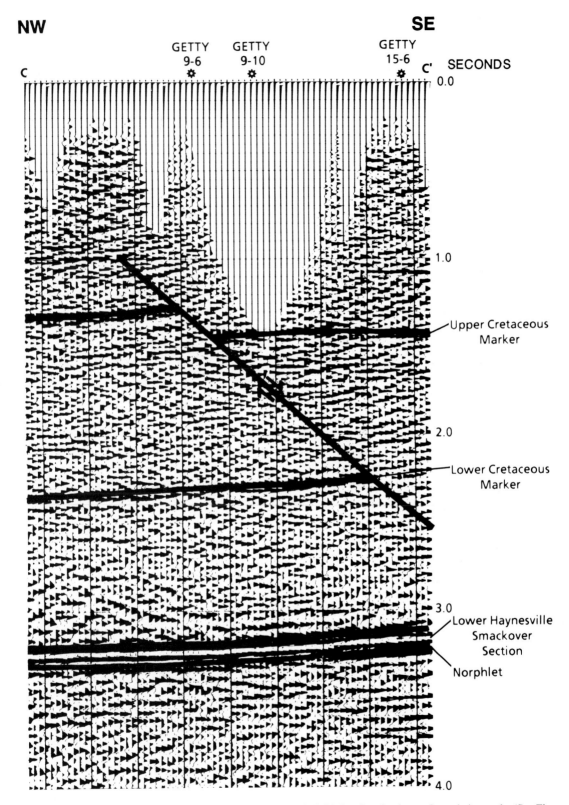

Fig. 7-9. Northwest-southeast seismic line across Hatter's Pond Field showing the deeper Jurassic intervals. (See Figure 7-8 for orientation of seismic line.)

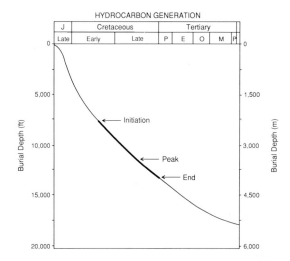

Fig. 7-10. Lopatin diagram illustrating hydrocarbon generation potential for Smackover source rocks in the Hatter's Pond Field area.

(Fig. 7-9). On both synthetic seismograms and observed seismic data, the sequence in the Jurassic is exemplified by a strong positive reflection coefficient (peak) at the high-velocity Haynesville-Smackover level, while the low-velocity Norphlet sandstone beneath the high-velocity Haynesville-Smackover section appears as a negative reflection-coefficient (trough) response. Good correlations can be drawn between synthetic seismograms in the area and the observed sequence at the Jurassic level.

Seismic line C-C' (Fig. 7-9) illustrates the western flank of the faulted salt anticline which forms the petroleum-trapping mechanism in the Hatter's Pond Field. The west-to-northwest dip in the field is apparent on the line. The fault depicted on Figure 7-9 truncates the Jurassic level further to the southeast along the eastern terminus of the field (Fig. 7-8). The structural style in the Hatter's Pond Field is due to the influence of early salt tectonism along the western margin of the Mobile graben.

Trapping Mechanism

The hydrocarbon trap at Hatter's Pond Field involves salt movement along the west side of the Mobile graben fault system that has resulted in a northeast-southwest trending, salt-pierced, faulted anticline (Fig. 7-8). The trap is believed to have formed during Late Jurassic to Early Cretaceous time. Hydrocarbon generation and primary migration (Fig. 7-10) were initiated during the Early Cretaceous and continued through the Late Cretaceous (Honda and McBride, 1981; McBride et al., 1987). The feature encompasses an area of approximately 9,000 acres (3,650 ha.) and has more than 400 feet (122 m) of structural closure at the top of the Norphlet level. An east-west cross section through the field (Fig. 7-11) indicates that maximum Norphlet reservoir development is along the eastern side of the field adjacent to

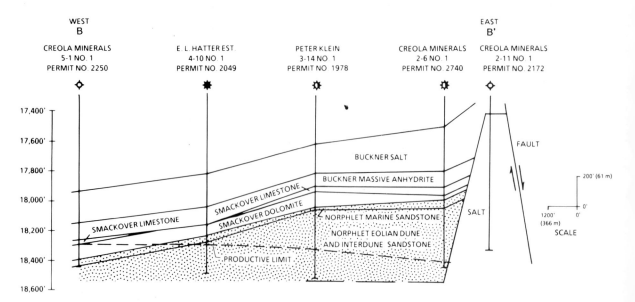

Fig. 7-11. East-west structural cross section of Hatter's Pond Field showing west dip. Modified from Exhibit 7 as presented by Getty Oil Company at the August 1982 meeting of the State Oil and Gas Board of Alabama, Docket 7-19-821 and reprinted by permission. (See Figure 7-8 for location.)

the salt piercement and the fault. The reservoir seal for the field consists of the overlying Smackover limestone and Buckner massive anhydrite.

Smackover carbonate mudstones were probably the source of the Norphlet hydrocarbons, as these carbonates are locally rich in algal and amorphous kerogen (Mancini and Benson, 1980). Analyses of samples from the black shale associated with the Norphlet indicate that it is barren of any significant organic matter at the Hatter's Pond Field. Geochemically, gas condensate recovered from the Norphlet is identical to the gas condensate produced from the Smackover, indicating vertical and lateral communication in the field. The condensate in the field has an API gravity of 54.0°, and the C_{15+} composition consists of 35.7% asphaltenes, 39.8% saturates, 4.7% aromatics, and 19.8% nitrogen-sulfur-oxygen compounds. The carbon isotope (ΔC^{13}) value based on PDB standard for the saturate fraction is $-23.8‰$, and the aromatic fraction is $-22.1‰$. The sulfur isotope (ΔS^{34}) value is 3.1‰ (Claypool and Mancini, 1989).

Depositional Setting

The environment of deposition of the Norphlet Formation in southwestern Alabama has been interpreted as eolian by Wilkerson (1981), Honda and McBride (1981), Pepper (1982), Mancini and others (1985), Marzano and others (1988), and Dixon and others (1989). The Norphlet desert plain of southwestern Alabama was broad and extended westward into eastern and central Mississippi, where the upper Norphlet has been determined as eolian by Hartman (1968), Badon (1975), and McBride (1981). The Norphlet dunes have been interpreted as barchan or transverse ridges with the general direction of wind transport being north to south (Marzano et al., 1988; Dixon et al., 1989). Principal sources of the sandstone deposited in the area of the Hatter's Pond Field were updip alluvial fan, alluvial plain, and braided stream deposits and wadi systems (Fig. 7-12). The sandstone in the Hatter's Pond Field area was, for the most part, transported by wind into and across the sand-sheet area. Water-deposited sediments accumulated in wadis and playa lakes in interdune areas. Deposition in this area of Alabama was dominated by eolian processes, with a sand erg developing in the Hatter's Pond Field area (Fig. 7-13).

Lithofacies and Petrography

At the Hatter's Pond Field, the upper part of the Norphlet reservoir is usually unstratified (Figs. 7-14, 7-15A); however, wavy, horizontal laminae are sometimes apparent (Table 7-1). The upper Norphlet has been interpreted as marine by Sigsby (1976), Wilkerson (1981), Pepper (1982), and Mancini and others (1985). These upper Norphlet beds usually consist of moderately to well-sorted, subrounded to well-rounded, very fine- to medium-grained subfeldsarenites-feldsarenites (subarkosic to arkosic) (Fig. 7-16A). Sandstone in the marine lithofacies is texturally mature. Illite and to a lesser extent chlorite are authigenic components. The upper unstratified Norphlet sandstone is finer grained than the underlying eolian sandstone and usually contains a higher percent of calcium carbonate cement. The upper Norphlet has been interpreted to have originated as a shoreface deposit and probably reflects coastal reworking of the underlying eolian sediment by sea-level rise which was initiated in late Norphlet time and continued into Smackover time (Wilkerson, 1981).

Norphlet eolian sandstones are usually texturally mature subfeldsarenites to feldsarenites (subarkosic to arkosic). Grains are well-sorted, mostly very fine- to medium-grained, angular to well-rounded quartz and feldspar (Fig. 7-16E). Angular and rounded grains of similar size and composition occur together and are indicative of multiple source areas. The presence of a significant percentage of feldspar in the Norphlet is attributed to the proximity and nature of the eroding Wiggins arch and to the arid climate during transportation and deposition of the Norphlet. Feldspar in the sand fraction is predominantly orthoclase, with albite, the dominant type of plagioclase, much less common (Vaughan, 1985). A majority of the quartz grains are the unstrained, monocrystalline plutonic variety (Wilkerson, 1981). Rock fragments comprise only about 1% of the framework grains (Honda and McBride, 1981).

Norphlet eolian sequences in the Hatter's Pond Field are composed primarily of dune and interdune deposits, with sand sheets forming a minor component (Table 7-1). Dune deposits are the most common part (75%) of the reservoir and consist mainly of grainflow, grainfall, and climbing wind-ripple deposits. They form sandstone bodies which are usually 30 to 100 feet (9.1–30.5 m) thick and have a tabular shape. Interdune deposits (10% of the reservoir) comprise wadi, playa, grainfall, climbing wind ripples, and deflation lag sediments. These sequences are usually much thinner than the dune deposits, most often 2 to 10 feet (0.6–3.0 m) thick, and they generally are not continuous across the field.

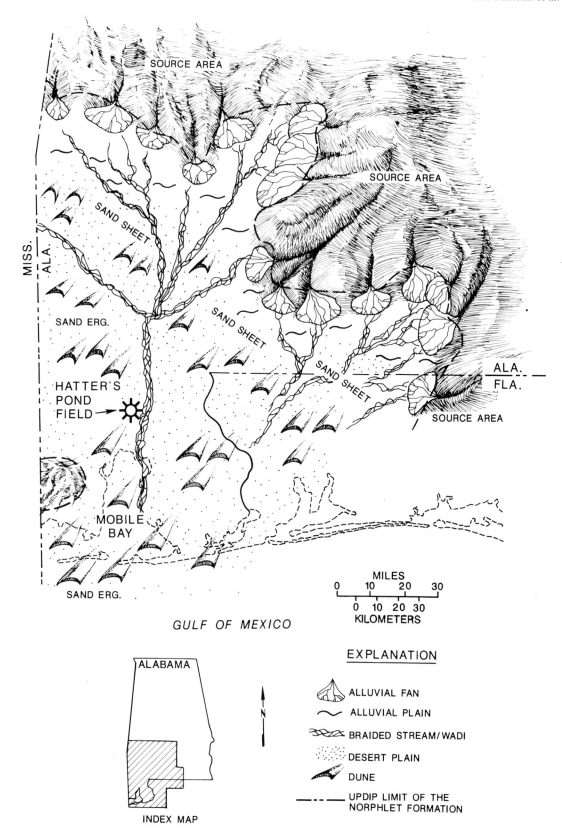

Fig. 7-12. Generalized Norphlet Formation paleogeography for southwestern Alabama.

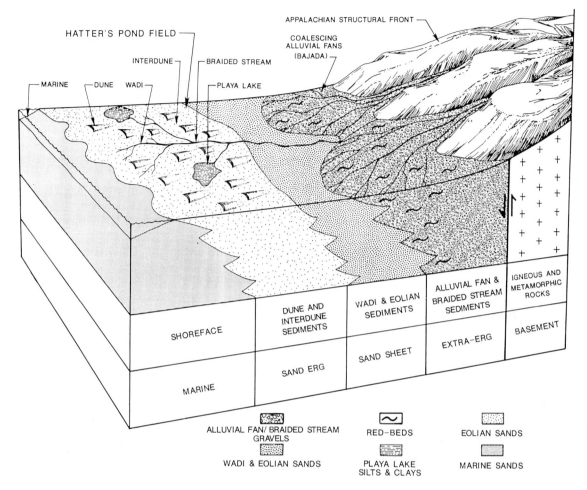

Fig. 7-13. Block diagram illustrating Norphlet Formation depositional systems and lateral lithofacies associations in southwestern Alabama.

The high-angle (up to 30°) cross-bedded sandstone of the Norphlet was deposited as eolian dunes (Fig. 7-15B) that accumulated as a result of oversteepening of dune lee slopes and subsequent avalanching down the slipface as mostly grainflow (sandflow) deposits (Table 7-1). The direction of dip in beds produced by grainflow is variable (Fig. 7-15B). In recent eolian sediments, these beds dip in the direction of dune migration (Bagnold, 1954; Reineck and Singh, 1975; Kocurek and Dott, 1981). Grainflow sediment is well sorted and ranges from very fine to medium sand. Bed contacts are sharp and often normally or reverse graded. Occasionally, beds wedge out into sandflow toes.

The horizontal to low-angle laminated sandstone of the Norphlet is also eolian (Fig. 7-15C) and accumulated as a result of flow separation on the leeward slopes as grainfall deposits (Table 7-1). Lami-

nae produced by grainfall are often inversely graded and rarely normally graded. They are typically parallel and continuous and have sharp bed contacts.

Climbing wind-ripple cross laminae (Table 7-1) are common in the Norphlet and are formed by migrating ripples that are accompanied by net deposition (Hunter, 1981). Climbing translatent (Fig. 7-15D) and rippleform strata are the components of climbing wind ripples and both occur in the Norphlet. These deposits consist of well-sorted, very fine- to medium-grained sand and are typically inversely graded.

The thick cross-bedded dune sandstone of the Norphlet is interbedded with thinner lithologies which originated as wind- and water-deposited interdune sediment. The water-deposited sediments consist primarily of wadi (Fig. 7-15F) and playa lake deposits (Fig. 7-15E) and are usually not as well

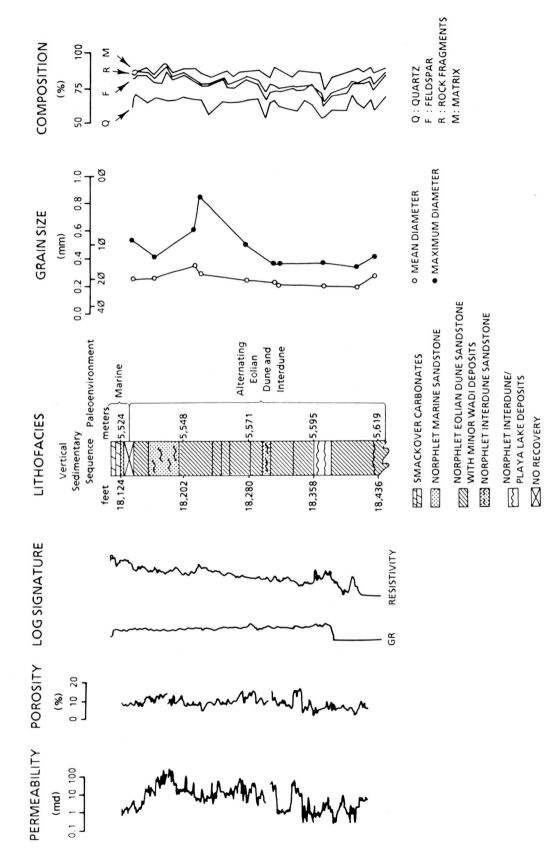

Fig. 7-14. Typical vertical sequence of paleoenvironments, wireline-log signatures, porosity, permeability, textural, and compositional changes observed in the Norphlet Formation, Getty-Peter Klein 3-14 No. 1 well, Mobile County, Alabama.

Fig. 7-15. Photographs of core slabs illustrating paleoenvironments and sedimentary structures of the Norphlet Formation in Hatter's Pond Field. Scale bar is 1 inch (2.5 cm). (A) Marine shoreface sandstone depicting massive nature, Getty-Creola Minerals 33-10 No. 1 well, Mobile County, depth 18,440 feet (5,621 m). (B) Eolian dune sandstone depicting bimodal high-angle foreset laminae, Getty-Peter Klein 3-14 No. 1 well, Mobile County, depth 18,400 feet (5,608 m). (C) Eolian sandstone depicting horizontal to low-angle laminae, Getty-Peter Klein 3-14 No. 1, Mobile County, depth 18,307 feet (5,580 m). (D) Eolian sandstone depicting ripple laminae, Getty-Creola Minerals 35-11 No. 2 well, Mobile County, depth 18,448 feet (5,623 m). (E) Playa lake deposits in interdune area, Getty-Peter Klein 3-14 No. 1 well, Mobile County, depth 18,422 to 18,423 feet (5,615.0–5,615.3 m). (F) Wadi sandstone in interdune area, Getty-Peter Klein 3-14 No. 1 well, Mobile County, depth 18,214 feet (5,552 m).

Table 7-1. Summary of lithofacies in the Norphlet reservoir at Hatter's Pond Field.

Facies	Percent of reservoir	ϕ range (%)	k range (md)	Types of deposits	Characteristics
Marine	10	4.2–15.9	0.03–739	Marine reworked	Structureless. Wavy horizontal laminae.
Dune	75	3.7–20	0.14–891	Grainflow	20 to 30° dip; 23° mean dip; variable dip direction; well-sorted, fine to medium sand size; 0.2 to 2.0 inch (0.5–5.0 cm) bed thickness; continuous and usually parallel bedding; normal, reverse, and ungraded beds; sharp bed contacts; beds wedge out down slipface (sand-flow toes); soft-sediment faulted and folded.
				Grainfall	0 to 20° dip; <0.4 inch (<1.0 cm) bed thickness; inversely graded laminae; rarely normally graded; continuous and parallel beds; sharp contact beds; well-sorted, very fine to fine sand; soft-sediment faulted and folded.
				Climbing wind ripples	Inversely graded; usually <15° dip; sharp bed contacts; 0.08 to 0.24 inch (2 to 6 mm) bed thickness; rare rippleform strata; well-sorted, very fine to medium sand; soft-sediment faulted and folded.
Interdune	10	2.9–9.4	0.04–9.5	Wadi	Usually poorly sorted; indistinct bedding; sand ripples; granule ripples; inverse grading; normal grading; silt and clay to coarse sand granules.
				Playa	Poorly bedded; poorly sorted; adhesion ripples; anhydrite nodules.
				Grainfall	As in dune except no soft-sediment deformation.
				Climbing wind ripples	As in dune except no soft-sediment deformation.
				Deflation lags	Granule-rich horizons; granule ripples; bimodal and poor sorting; inversely graded ripples.
Sand sheet	5	3.9–17.7	0.24–188	Grainfall	As in dune except no soft-sediment deformation.
				Climbing wind ripples	As in dune except no soft-sediment deformation.
				Wadi	As in interdune.

sorted as the dune sandstone (Fig. 7-16E). In modern desert settings, the interdune areas are depressions in which temporary wadi or playa lake deposits commonly develop (Glennie, 1970; Reineck and Singh, 1975; Ahlbrandt and Fryberger, 1981; Fryberger et al., 1983).

Wadi deposits are a minor component of the Norphlet at Hatter's Pond Field (Table 7-1). They are composed mainly of sandy conglomerate to conglomeratic sandstone (Figs. 7-15F, 7-17B) in association with low- to moderate-angle cross-bedded sandstone. The cross-bedded sandstone probably represents small dunes deposited within wadi channels during dry periods. The wadi deposits are

generally feldsarenites to subfeldsarenites (arkosic to subarkosic) and are fine- to medium- grained. The quartz grains in this lithofacies are angular to subrounded. Sorting within this lithofacies is usually poorer than in the dune deposits (Fig. 7-17C), and this can adversely affect primary depositional porosity. The low initial porosity and permeability are factors influencing diagenesis, since relatively nonporous rocks are less susceptible to the effects of migrating fluids. As a result, secondary porosity development in the eolian interdune deposits is limited to sporadic occurrences of intragranular microporosity in feldspars (Fig. 7-17E) and rare decementation. Where wind-

Fig. 7-16. Photomicrographs illustrating sandstone compositional and textural characteristics and porosity types of Norphlet Formation in Hatter's Pond Field. Scale bar is 0.2 mm. (A) Marine sandstone depicting composition and texture, Getty-Creola Minerals 2-6 No. 1 well, Mobile County, depth 18,055 feet (5,503 m). (B) Intragranular microporosity in feldspar grain, Getty-Creola Minerals 2-6 No. 1 well, Mobile County, depth 18,055 feet (5,503 m). (C) Advanced stage of feldspar dissolution, Getty-Creola Minerals 2-6 No. 1 well, Mobile County, depth 18,055 feet (5,503 m). (D) Framework grain completely dissolved. Former grain boundary outlined by illite clay coating, Getty-Creola Minerals 5-1 No. 1 well, Mobile County, depth 18,573 feet (5,661 m). (E) Eolian dune sandstone depicting composition, Getty-Peter Klein 3-14 No. 1 well, Mobile County, depth 18,403 feet (5,609 m). Note feldspar grain at f. (F) Eolian dune sandstone secondary dissolution porosity at p, Getty-Peter Klein 3-14 No. 1 well, Mobile County, depth 18,403 feet (5,609 m).

Fig. 7-17. Photomicrographs illustrating sandstone compositional and textural characteristics and porosity types of Norphlet Formation in Hatter's Pond Field. Scale bar is 0.2 mm. (A) Eolian dune sandstone secondary dissolution creating intragranular microporosity in feldspar, Getty-Creola Minerals 5-1 No. 1 well, Mobile County, depth 18,588 feet (5,666 m). (B) Wadi interdune deposit, Getty-Creola Minerals 2-6 No. 1 well, Mobile County, depth 18,270 feet (5,569 m). (C) Eolian interdune sandstone depicting composition and poor sorting, Getty-Creola Minerals 2-6 No. 1 well, Mobile County, depth 18,088 feet (5,513 m). (D) Eolian interdune sandstone depicting composition and good sorting, Getty-Creola Minerals 5-1 No. 1 well, Mobile County, depth 18,593 feet (5,667 m). (E) Eolian interdune porosity consisting of intragranular microporosity in a feldspar grain, Getty-Creola Minerals 2-6 No. 1 well, Mobile County, depth 18,088 (5,513 m). (F) Playa lake interdune deposits, Getty-Peter Klein 3-14 No. 1 well, Mobile County, depth 18,359 feet (5,596 m). Note matrix at m.

Fig. 7-18. Paragenetic sequence of major diagenetic events in the Norphlet sandstone in northern Mobile County, Alabama (modified from Vaughan and Benson, 1988 and reprinted by permission of the Gulf Coast Association of Geological Societies).

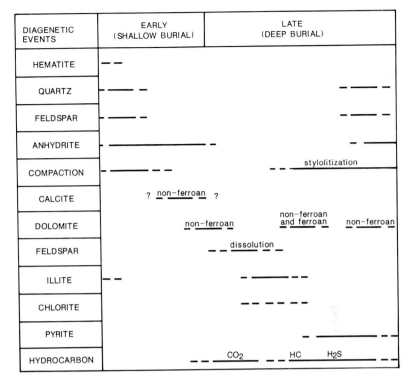

Diagenesis

Diagenetic events have played a major role in producing Norphlet sandstone porosity and permeability in the Hatter's Pond Field (Fig. 7-18). As much as 46% of the intergranular and intragranular porosity

deposited, interdune sediment sorting is better, and because primary depositional porosity is higher, secondary porosity is more common.

Playa lake deposits (Fig. 7-15E, Table 7-1) consist of finer grained and more poorly sorted sediment containing a significant amount of clay matrix. This lithofacies is texturally the least mature sandstone in the Norphlet eolian depositional system (Fig. 7-17F). Often, these interdune playa lake sediments are characterized by an irregular wavy-bedded (adhesion ripples) fabric formed when water is pulled to the sediment-air interface by capillary action due to evaporation, and windblown sand then adheres to the damp surface (Glennie, 1970). Deposition of fines is common in these wet interdune areas because silt and clay are much less likely to be resuspended after adhering to a moist sediment surface. Core analyses indicate that playa lake deposits in the Hatter's Pond Field area are of limited lateral extent.

occurring in the reservoir rocks has been determined to be secondary (McBride et al., 1987). Early-stage (shallow burial) diagenetic events included hematite and illite grain coatings, which were followed by quartz and feldspar overgrowths and anhydrite and carbonate cementation (Honda and McBride, 1981; McBride et al., 1987; Marzano et al., 1988; Vaughan and Benson, 1988). The amount of quartz overgrowth cementation was influenced by the amount of illite clay coatings (McBride et al., 1987). Early anhydrite and carbonate cementation was generally patchy in its distribution. Mechanical compaction due to packing readjustments significantly reduced porosity (McBride et al., 1987).

Decementation of early carbonate and sulfate cement was followed by illite and/or chlorite precipitation in association with, but prior to, hydrocarbon migration (Vaughan, 1985). Dissolution of framework grains, mainly feldspars but also rock fragments and quartz grains, also took place at this time (Honda and McBride, 1981; Vaughan, 1985). Late deep-burial cementation by carbonate and anhydrite further reduced porosity, although some of these cements occur as replacements of feldspar and rock fragments (McBride et al., 1987). Anhydrite cementation and dolomitization of earlier calcite cement may be the result of downward perco-

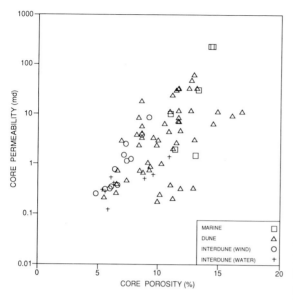

Fig. 7-19. Porosity-permeability cross-plot of the various sandstone lithofacies at Hatter's Pond Field.

Petrophysics

At Hatter's Pond Field, the reservoir rocks are Smackover marine dolostone and Norphlet marine and eolian sandstones (Fig. 7-14). Although reservoir geometry is principally structurally controlled, reservoir porosity is an important factor in the productivity of the field. Reservoir porosity in the Smackover is late-stage vuggy and/or moldic porosity and appears to be lithofacies selective and preferential to the coarsely crystalline dolomite lithology (Benson and Mancini, 1982). High-energy oolitic grainstone, which accumulated as a series of oolitic bars, is probably the precursor of the coarsely crystalline dolomite. Dolomite porosity is secondary and averages 12.2% with an average permeability of 2.1 md. Norphlet sandstone porosity ranges from 2.9 to 20% and averages 10.4% with an average permeability of 0.5 md. Reservoir quality is best in

the marine and dune sandstones (Figs. 7-16B, 7-16F, 7-19) and poorest in the wadi and playa interdune sandstones (Figs. 7-17F). Water saturations average 31% for the Norphlet reservoir.

Porosity in the marine lithofacies is predominantly intergranular (Fig. 7-16B) with a significant amount of intragranular porosity. The intergranular porosity is the result of both primary depositional pore space and diagenetic secondary dissolution of feldspar and quartz overgrowths and dissolution of calcium carbonate and other cements. Evidence of quartz dissolution is observed in etched and embayed grains (Fig. 7-16A). Intragranular porosity results principally from dissolution within feldspar grains and rarely in rock fragments. Intragranular porosity in feldspars begins with the formation of micron-sized micropores (Fig. 7-16B). In more advanced stages, the micropores join to form pores that appear to follow cleavage traces (Fig. 7-16C). This process can result in the complete dissolution of grains (Fig. 7-16D). The observation that pore spaces are surrounded by isopachous illite (Vaughan, 1985) supports this hypothesis.

Porosity in the dune lithofacies is similar to that of the marine lithofacies (Fig. 7-16F). Porosity is predominantly intergranular with a significant amount of intragranular porosity (Fig. 7-17A). Etching and embaying of feldspar overgrowths are common. Calcium carbonate cements are patchy in distribution and volumetrically less important than in the marine lithofacies. Intragranular porosity consists predominantly of microporosity in feldspars (Fig. 7-17A) to complete dissolution of grains.

Porosity in the interdune lithofacies is primarily intergranular with minor intragranular porosity. Intergranular porosity is both primary and secondary, with secondary porosity resulting from dissolution of cement. Intragranular porosity, as in the dune and marine lithofacies, is the result of feldspar dissolution. The overlap in the porosity-permeability fields of the dune and interdune lithofacies in Figure 7-19 is due in part to the interbedding of small eolian dunes occurring within water-laid wadi deposits. In these interbedded wind- and water-laid deposits, the small dune deposits may have provided local avenues of greater porosity and permeability enabling greater access of diagenetic solvents to certain interdune deposits. On the other hand, the extent of early cementation in the small dune deposits was variable. Dune sediment that was more thoroughly cemented provided a greater obstacle for diagenetic solvents and, therefore, porosity and permeability remained

lation of hypersaline brines from the overlying Smackover Formation (Vaughan, 1985). Pyrite may have formed from the reaction of sour gas with hematite coatings on framework grains (McBride et al., 1987). Pressure solution (stylolitization) of framework grains, including feldspar, resulted in porosity reduction and may have provided components for the formation of late-stage illite (Vaughan, 1985).

relatively low. Porosity in the wadi lithofacies is usually low, and the playa lake lithofacies is generally nonreservoir.

In summary, reservoir quality in the Norphlet was initially controlled by depositional texture, but burial conditions have greatly modified this. Factors that have maintained and/or enhanced primary reservoir porosity are cement and grain dissolution and, to some degree, clay coatings on quartz grains. Factors that have acted to reduce reservoir porosity are mechanical compaction, cementation, pressure solution, and late-stage formation of authigenic clay.

Heterogeneities

Although the sandstone reservoir lithofacies at the Hatter's Pond Field are heterogeneous, no substantial intrareservoir barriers to hydrocarbon flow are evident within the field. The interdune beds of the Norphlet may locally act as barriers to vertical flow, but because they are discontinuous, they have a minimal effect on reservoir drainage.

The Smackover reservoir is also a heterogeneous body. However, this portion of the reservoir consists of high reservoir-quality zones that are well connected. Furthermore, the Smackover-Norphlet constitutes a single reservoir with good communication between the Smackover and the Norphlet formations. Cores taken from wells in the field indicate that the contact of the Smackover and the Norphlet is gradational or interbedded and contains effective porosity (Benson and Mancini, 1982).

Although the nature of the reservoir flow path in the Hatter's Pond Field is complex, the geologic evidence of a single fieldwide reservoir is supported by engineering data. Twelve years of production data combined with more than 140 bottom-hole pressure tests indicate communication among wells in the reservoir. The results of a detailed computer reservoir-model study of the Hatter's Pond Field by Getty Oil Company (O'Dell, 1982) provide support for a single reservoir. A good history match was obtained between the results of the model study and the actual performance of the wells in the field indicating communication among the wells.

Production Characteristics

Individual well and overall reservoir performance in the Hatter's Pond Field are influenced by a complex combination of factors including reservoir geometry and heterogeneity, fluid properties, and operational constraints. The reservoir boundaries include a salt intrusion along a major north-south fault on the east side of the field and a gas-water contact and/or a loss of porosity and permeability down-structure on the north, west, and south sides of the field (Fig. 7-20). The Smackover average reservoir porosity increases up-structure (to the east) reflecting ooid-bar development and favorable diagenetic events, while the Norphlet average reservoir porosity increases down-structure (to the west) because only the upper part of the Norphlet (marine and dune lithofacies) comprises the sandstone portion of the reservoir due to structural position (Fig. 7-11). The permeability of the Norphlet reservoir is heterogeneous, ranging from 0.03 to 891 md with an average of 0.5 md.

At original reservoir conditions of 9,170 psi (6.3 $\times$ 10^4 kPa) and 324°F (162°C), the hydrocarbons exist as a single-phase gas with a dew point of 3,030 psi (2.1 $\times$ 10^4 kPa). A typical reservoir fluid compositional analysis is shown in Table 7-2. Due to decreases in hydrocarbon pressure and temperature during the production process, a rich gas condensate is formed. Total hydrocarbon liquid yield is 255 BGC (41 m^3) and 48 BLG (7.6 m^3) per million cubic feet (2.8 $\times$ 10^4 m^3) of gas. Production for the field has been maintained at a relatively constant level since 1981, as indicated in Figure 7-21A, with 128 BCFG (3.6 $\times$ 10^9 m^3) and 40.4 MM BGC and BLG (6.4 $\times$ 10^6 m^3) produced from an average of 13 wells. Constant production levels have been attributed primarily to gas sales-contract limitations, processing-plant capacity, downhole mechanical problems, and other operational constraints. Individual well performance generally has been influenced to a greater extent by downhole mechanical problems than by reservoir depletion. One of the most important mechanical problems has been the collapse of casing caused by massive salt movement in the overlying Haynesville Formation, resulting in decreased production and, eventually, extensive workovers or drilling replacement wells. Prior to the fieldwide unit being established in May, 1985, each producing well, spaced on 640-acre (259-ha.) units, had generally been replaced at least once during the life of the field to maintain production on that unit. The combination of producing interval depth, temperature, pressure, and wellbore mechanical problems requires that particular attention be given to wellbore completion and construction.

As illustrated in Figure 7-21B, declines in production result from mechanical problems because of the

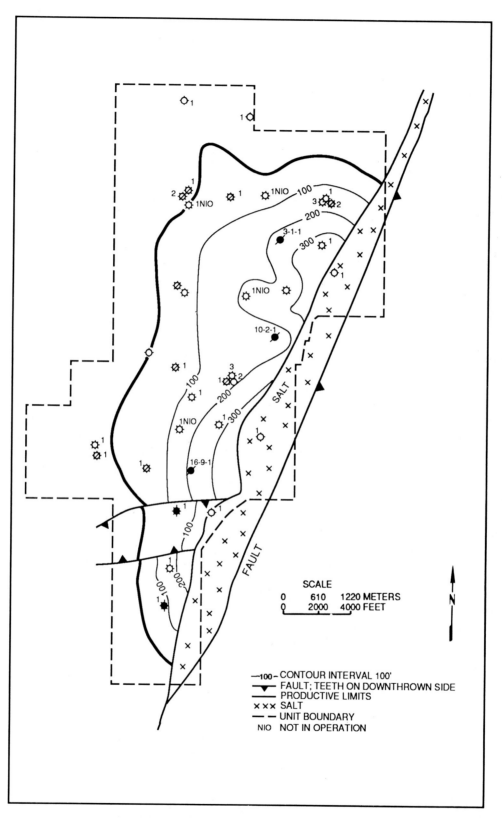

Fig. 7-20. Norphlet net pay isopach map at Hatter's Pond Field contoured in 100-foot (30.5-m) intervals. Modified from Exhibit 11 as presented by Getty Oil Company at the March 1988 meeting of the State Oil and Gas Board of Alabama, Docket 3-10-884 and reprinted by permission.

Table 7-2. Reservoir fluid composition, Peter Klein 3-14 No. 1, Hatter's Pond Field.

Component	Full well stream (mole percent)
Hydrogen sulfide	NIL
Carbon dioxide	6.42
Nitrogen	3.16
Methane	50.26
Ethane	9.68
Propane	6.20
iso-Butane	2.19
n-Butane	3.73
iso-Pentane	1.88
n-Pentane	1.85
Hexanes	3.04
Heptanes plus	11.59
	100.00

downtime associated with the drilling of replacement wells. Production decline rates are variable among the wells and are influenced by the heterogeneity of the reservoir. Wells completed in zones of low porosity and/or reduced permeability contain less recoverable hydrocarbons. Declines in production for such wells are naturally more rapid than the declines of wells completed in more productive areas of the reservoir. Average daily production from each well is 2.8 MMCFGPD (7.9 × 10^4 m³/D) and 730 BCPD (116 m³/D). Cumulative production from wells that are initially acidized is typically 2 MMBGC (3.2 × 10^5 m³) and 7 BCFG (2.0 × 10^8 m³).

Original in-place hydrocarbons for the Hatter's Pond Field are 754 BCFG (2.1 × 10^{10} m³) and 228 MM BGC and BLG (3.6 × 10^7 m³). Ultimate primary recovery by gas depletion has been estimated at 193 BCFG (5.5 × 10^9 m³) and 58 MM BGC and BLG (9.2 × 10^6 m³). Prior to being depleted, the field was unitized in 1985, and an enhanced recovery project consisting of pressure maintenance by gas cycling was initiated. Residue gas injection began with one well in 1985 and averaged 150 MMCFG (4.2 × 10^6 m³) per month through 1985. The average monthly gas-injection volume was 122 MMCFG (3.5 × 10^6 m³) in 1986. Injection operations were suspended in January, 1987 and were not resumed until April, 1988. Monthly injection volumes have been 380 MMCFG (1.1 × 10^7 m³) since that time. Cumulative injection volume is 3.4 BCFG (9.6 × 10^7 m³). This enhanced recovery project is expected to increase the ultimate recovery of the field by 25% (increase recovery of 6% of original-in-place hydrocarbons).

Exploration And Production Strategy

Structure and stratigraphy are critical factors in exploration for Norphlet reservoirs in southwestern Alabama, as exemplified by these relationships in the Hatter's Pond Field. The Hatter's Pond Field is typical of Norphlet fields in this area inasmuch as the trap is structural as the result of salt tectonism, whereas stratigraphy determines the productivity of the reservoir. Structural features in southwestern Alabama that are underlain by salt, such as the regional peripheral fault trend (Mancini et al., 1985), the Mobile graben, and the Lower Mobile Bay fault trend (Bearden, 1987), are excellent locations to explore for potential Norphlet traps and productive reservoirs.

The stratigraphic relationships between the Norphlet and the overlying Smackover carbonate are important factors in the formation and productivity of Norphlet reservoirs. The Norphlet Formation is barren of in situ organic carbon and does not serve as a petroleum source rock in southwestern Alabama. Therefore, a stratigraphic relationship with Smackover source rocks which facilitates oil and gas migration is a crucial factor for ensuring a productive Norphlet reservoir. Ideally, hydrocarbons generated from Smackover carbonate mudstone in basinal areas migrate updip into Norphlet reservoir lithofacies located on basin margins. The hydrocarbon seal is either nonporous upper Smackover or lower Haynesville evaporite, as present at Hatter's Pond Field, or nonporous upper Norphlet or lower Smackover lithofacies, which is the case in much of southwestern Alabama. The potential seal rock varies because the Smackover section can be porous and because the lower Haynesville evaporite is not present in all areas.

A thick eolian section is essential for a Norphlet reservoir as productive as that at Hatter's Pond Field. In the Mississippi interior salt basin and in portions of offshore Alabama, Norphlet eolian sandstones 400 to 600 feet (122–183 m) thick are common. Porosity in the onshore Norphlet fields, including Hatter's Pond Field, has been interpreted chiefly as secondary intergranular and intragranular (Honda, 1981; McBride, 1981; Honda and McBride, 1981; Laird, 1985; Mancini et al., 1985;

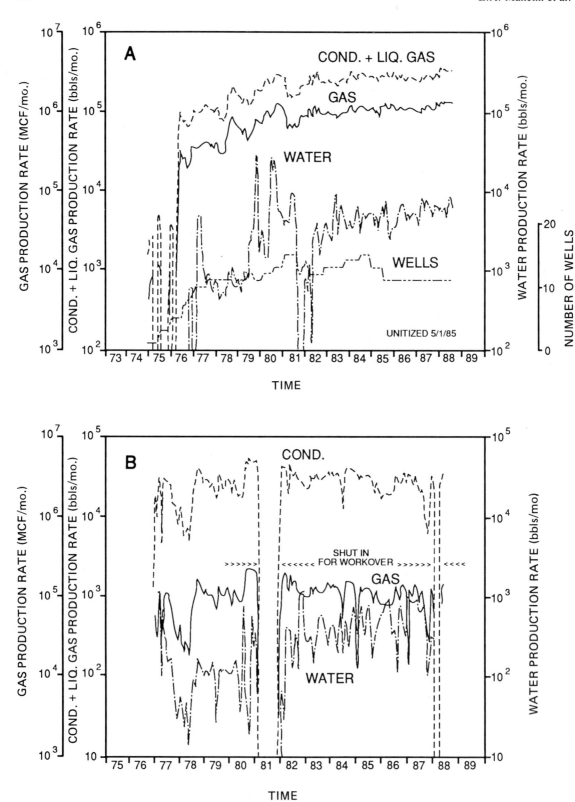

Fig. 7-21. (A) Reservoir performance graph and number of wells for Hatter's Pond Field. (B) Typical well performance graph for Hatter's Pond Field.

Vaughan, 1985; McBride et al., 1987; Marzano et al., 1988; Vaughan and Benson, 1988); however, porosity in the offshore Norphlet fields has been interpreted principally as preserved primary depositional (Dixon et al., 1989). Broad low-relief structural closures, such as the Lower Mobile Bay-Mary Ann Field, or high-relief structures with hundreds of feet (more than 60 m) of elevation, such as the Hatter's Pond Field, are important components to the formation of productive Norphlet reservoirs.

Prospective areas for Norphlet reservoirs similar to Hatter's Pond Field include the thick eolian sand seas or ergs of the Mississippi interior salt basin, offshore Alabama, and offshore Florida in the Destin Dome Area (Marzano et al., 1988). Salt features associated with these sand ergs serve as the petroleum traps (Mink et al., 1985). In order for these traps to be productive, they must occur in association with adequate Norphlet, Smackover, or Haynesville seals. In addition, traps need favorable stratigraphic and structural relationships with overlying organically rich, thermally mature Smackover source rocks to ensure timely hydrocarbon migration and trapping.

Most Norphlet reservoirs occur below 15,000 feet (4,570 m) in hostile subsurface conditions of high temperatures and pressures associated with nonhydrocarbon gases such as H_2S, CO_2 and N_2, which adversely affect production methods and economics. Special consideration must be given to wellbore completions to address these subsurface conditions adequately, as well as to problems associated with reservoir heterogeneity. Determination of the most efficient recovery procedure is dependent upon reservoir hydrocarbon type (oil, gas condensate, or dry gas) and initial reservoir drive mechanism. Operational constraints, such as Haynesville salt flowage causing casing collapse, processing-plant size, and gas sales contracts, can affect the economics and vary from field to field. Early unitization of Norphlet reservoirs providing for strategic well placement and enhanced recovery is one of the best means available to optimize the recovery of hydrocarbons.

Conclusions

The hydrocarbon trap at Hatter's Pond Field involves salt movement along the west side of the Mobile graben fault system that has resulted in a northeast-southwest trending, salt-pierced, faulted anticline. The trap formed during Late Jurassic to Early Cretaceous time. The Norphlet reservoir at the Hatter's Pond Field is a large sandstone wedge containing gas-condensate hydrocarbons. The geometry of the reservoir is controlled by a northeast-southwest trending faulted anticline with more than 350 feet (107 m) of net pay on the east side of the field near the fault and decreasing net pay to the west toward the gas-water contact. The primary controls on reservoir quality are Norphlet lithofacies and diagenesis. The marine and eolian dune lithofacies are the best reservoir rocks, whereas interdune lithofacies have less potential as reservoir rocks because of higher heterogeneity and lower porosity and permeability. Reservoir porosity includes both primary (depositional) and secondary (diagenetic) types, the latter as a result of decementation and grain dissolution.

The large volume of hydrocarbons found in the Hatter's Pond Field probably was generated from Smackover algal carbonate mudstone downdip in the Mississippi interior salt basin. The rich gas condensate in the field is a result of the deeply buried (greater than 18,000 ft; 5,490 m) source and reservoir rocks. Hydrocarbon generation and primary migration were initiated during the Early Cretaceous and continued through the Late Cretaceous.

The depth of the reservoir at Hatter's Pond Field results in subsurface conditions hostile to the production of these hydrocarbons. Well losses due to these conditions have had a major impact on individual well producibility in the field. Particular attention must be given to well design and completion techniques. Development requires close coordination between reservoir and production engineers in order to achieve an acceptable balance among a complex set of factors. Fieldwide reservoir unitization for the purposes of drilling wells at strategic locations and for the initiation of enhanced recovery operations appears to provide an ideal plan for optimizing production from reservoirs such as the Hatter's Pond Field.

Reservoir Summary

Field: Hatter's Pond
Location: Mobile County, Alabama
Operator: Getty Oil Company
Discovery: 1974
Basin: Mississippi Interior Salt Basin; Gulf of Mexico
Tectonic/Regional Paleosetting: Rifted continental margin
Geologic Structure: Salt-pierced, faulted anticline
Trap Type: Anticlinal and fault closure
Reservoir Drive Mechanism: Pressure maintenance by gas cycling
 • **Original Reservoir Pressure:** 9,170 psi (6.3 × 10^4 kPa) at 18,023 feet (5,493 m) subsea
Reservoir Rocks
 • **Age:** Late Jurassic, Oxfordian
 • **Stratigraphic Units:** Norphlet (sandstone) and Smackover (carbonate) Formations
 • **Lithology:** Fine- to medium-grained subfeldsarenite
 • **Depositional Environments:** Eolian, marine, fluvial (wadi), playa
 • **Productive Facies:** Eolian dune, interdune, and marine sandstones
 • **Petrophysics**
   • **Porosity Types:** ϕ Total = 10.4% with primary possibly as much as 5.6% and secondary intergranular and
                intragranular at least 4.8%
   • **ϕ:** Average = 10.4%, range = 2.9 to 20.0%, cutoff = 6% (unstressed conventional cores)
   • **k:** Average = 0.5 md, range = 0.03 to 891 md, cutoff = 0.1 md (unstressed conventional cores)
   • **S_w:** Average = 30.5% (unstressed conventional cores and log data)
Reservoir Dimensions
 • **Depth:** 18,100 feet (5,520 m)
 • **Areal Dimensions:** 6.5 by 3 miles (10.5 × 4.8 km)
 • **Productive Area:** 9,000 acres (3,650 ha.)
 • **Number of Reservoirs:** Smackover (carbonate) and Norphlet (sandstone) in communication producing as one
                reservoir
 • **Hydrocarbon Column Height:** 349 feet (106 m) in Creola Minerals 2-6 No. 1 well
 • **Fluid Contacts:** Gas-water contact varies between 18,300 and 18,437 feet (5,578–5,620 m)
 • **Number of Pay Zones:** 1
 • **Gross Sandstone Thickness:** > 400 feet (122 m)
 • **Net Sandstone Thickness:** > 400 feet (122 m)
 • **Net/Gross:** 1.0
Source Rocks
 • **Lithology & Stratigraphic Unit:** Algal carbonate mudstone, Smackover Formation, Late Jurassic
 • **Time of Hydrocarbon Maturation:** Early Cretaceous to Late Cretaceous
 • **Time of Trap Formation:** Late Jurassic to Early Cretaceous
 • **Time of Migration:** Early Cretaceous to Late Cretaceous
Hydrocarbons
 • **Type:** Gas, condensate, and natural gas liquids
 • **GOR:** NA
 • **API Gravity:** 54°
 • **FVF:** Bg = 0.0035
 • **Viscosity** NA
Volumetrics
 • **In-Place:** 754 BCFG (2.1 × 10^1) m^3) and 228 MM BGC and BLG (3.6 × 10^7 m^3)
 • **Cumulative Production:** 128 BCFG (3.6 × 10^9 m^3) and 40.4 MM BGC and BLG (6.4 × 10^6 m^3)
 • **Ultimate Recovery:** 242 BCFG (6.9 × 10^9 m^3) and 73 MM BGC and BLG (1.2 × 10^7 m^3)
 • **Recovery Efficiency:** gas = 32%; liquids = 32%
Wells
 • **Spacing:** Under primary operations, 5,280 feet (1,609 m) between wells and 640-acre (259-ha.) drilling units;
                currently the field is unitized for the purposes of enhanced recovery and optimum well spacing.

- **Total:** 40; most producing wells in the field are either sidetracks or replacement wells due to operational problems
- **Dry Holes:** 16

Typical Well Production:
- **Average Daily:** 2.8 MMCFG (7.9×10^4 m³) and 730 BC (116 m³)
- **Cumulative:** 7.0 BCFG (2.0×10^8 m³) and 2.0 MMBGC (3.2×10^5 m³)

Other:
- **Reservoir Temperature:** 324°F (162°C) at 18,150 feet (5,532 m) measured depth
- **Reservoir Dew Point Pressure:** 3,030 psi (2.1×10^4 kPa)

References

Ahlbrandt, T.S., and Fryberger, S.G., 1981, Sedimentary features and significance of interdune deposits, *in* Ethridge, F.G. and Flores, R.M., eds., Recent and ancient nonmarine depositional environments: Models for exploration: Society of Economic Paleontologists and Mineralogists Special Publication 31, p. 293–314.

Badon, C.L., 1975, Stratigraphy and petrology of Jurassic Norphlet Formation, Clarke County, Mississippi: American Association of Petroleum Geologists Bulletin, v. 59, p. 377–392.

Bagnold, R.A., 1954, The physics of blown sand and desert dunes: London, Methuen, 265 p.

Bearden, B.L., 1987, Seismic expression of structural style and hydrocarbon traps in the Norphlet Formation offshore Alabama: State Oil and Gas Board of Alabama, Oil and Gas Report 14, 29 p.

Benson, D.J., and Mancini, E.A., 1982, Petrology and reservoir characteristics of the Smackover Formation, Hatter's Pond Field: Implications for Smackover exploration in southwestern Alabama: Gulf Coast Association of Geological Societies Transactions, v. 32, p. 67–75.

Claypool, G.E., and Mancini, E.A., 1989, Geochemical relationships of petroleum in Mesozoic reservoirs to carbonate source rocks of Jurassic Smackover Formation, southwestern Alabama: American Association of Petroleum Geologists Bulletin, v. 73, p. 904–924.

Dixon, S.A., Summers, D.M., and Surdam, R.C., 1989, Diagenesis and preservation of porosity in Norphlet Formation (Upper Jurassic) southern Alabama: American Association of Petroleum Geologists Bulletin, v. 73, p. 707–728.

Fryberger, S.G., Al-Sari, A.M., and Clisham, T.J., 1983, Eolian dune, interdune, sand sheet, and siliciclastic sabkha sediments of an offshore prograding sand sea, Dhahran Area, Saudi Arabia: American Association of Petroleum Geologists Bulletin, v. 67, p. 280–312.

Glennie, K.W., 1970, Desert sedimentary environments: Developments in Sedimentology, v. 14, Amsterdam, Elsevier, 222 p.

Hartman, J.A., 1968, The Norphlet Sandstone, Pelahatchie Field, Rankin County, Mississippi: Gulf Coast Association of Geological Societies Transactions, v. 18, p. 2–11.

Honda, H., 1981, Diagenesis and reservoir quality of the Norphlet sandstone (Upper Jurassic), the Hatter's Pond area, Mobile County, Alabama [M.S. thesis]: Austin, Texas, University Texas, 213 p.

Honda, H., and McBride, E.F., 1981, Diagenesis and pore types of the Norphlet Sandstone (Upper Jurassic), Hatter's Pond area, Mobile County, Alabama: Gulf Coast Association of Geological Societies Transactions, v. 31, p. 315–322.

Hunter, R.E., 1981, Stratification styles in eolian sandstones: Some Pennsylvanian to Jurassic examples from the western interior, U.S.A., *in* Ethridge, F.G. and Flores, R.M., eds., Recent and ancient nonmarine depositional environments: Models for exploration: Society of Economic Paleontologists and Mineralogists Special Publication 31, p. 315–329.

Kocurek, G., and Dott, R.H., Jr., 1981, Distinctions and uses of stratification types in the interpretation of eolian sand: Journal of Sedimentary Petrology, v. 51, p. 579–595.

Laird, J.W., 1985, Diagenetic controls on reservoir characteristics and development in the Jurassic Norphlet Formation, Escambia County, Alabama [M.S. thesis]: Tuscaloosa, Alabama, University Alabama, 114 p.

Mancini, E.A., and Benson, D.J., 1980, Regional stratigraphy of Upper Jurassic Smackover carbonates of southwest Alabama: Gulf Coast Association of Geological Societies Transactions, v. 30, p. 151–165.

Mancini, E.A., Mink, R.M., Bearden, B.L., and Wilkerson, R.P., 1985, Norphlet Formation (Upper Jurassic) of southwestern and offshore Alabama: Environments of deposition and petroleum geology: American Association of Petroleum Geologists Bulletin, v. 69, p. 881–898.

Mancini, E.A., Mink, R.M., and Tew, B.H., 1988, Jurassic sequence stratigraphy in the Mississippi interior salt basin: An aid to petroleum exploration in the eastern Gulf of Mexico area: American Association of Petroleum Geologists Bulletin, v. 72, p. 217.

Marzano, M.S., Pense, G.M., and Andronaco, P., 1988, A comparison of the Jurassic Norphlet Formation in Mary Ann Field, Mobile Bay, Alabama to onshore regional Norphlet trends: Gulf Coast Association of Geological Societies Transactions, v. 38, p. 85–100.

McBride, E.F., 1981, Diagenetic history of Norphlet Formation (Upper Jurassic), Rankin County, Mississippi: Gulf Coast Association of Geological Societies Transactions, v. 31, p. 347–351.

McBride, E.F., Land, L.S., and Mack, L.E., 1987, Diagenesis of eolian and fluvial feldspathic sandstones, Norphlet Formation (Upper Jurassic), Rankin County, Mississippi, and Mobile County, Alabama: American Association of Petroleum Geologists Bulletin, v. 71, p. 1019–1034.

Mink, R.M., Bearden, B.L., and Mancini, E.A., 1985, Regional Jurassic geologic framework of Alabama coastal waters area and adjacent federal waters area: State Oil and Gas Board of Alabama, Oil and Gas Report 12, 58 p.

O'Dell, H.G., 1982, Engineering and economic analysis of primary depletion and gas cycling operations for proposed unit, in Hatter's Pond Field, Mobile County, Alabama: Alabama State Oil and Gas Board Docket No. 8-19-821 (Getty Oil Company), Exhibit 13, 43 p., plus plates and tables.

Pepper, C.F., 1982, Depositional environments of the Norphlet Formation (Jurassic) for southwestern Alabama: Gulf Coast Association of Geological Societies Transactions, v. 32, p. 17–22.

Reineck, H.E., and Singh, I.B., 1975, Depositional sedimentary environments, with reference to terrigenous clastics: New York, Springer-Verlag, 439 p.

Salvador, A., 1987, Lower Triassic-Jurassic paleogeography and origin of the Gulf of Mexico: American Association of Petroleum Geologists Bulletin, v. 71, p. 419–451.

Sigsby, R.J., 1976, Paleoenvironmental analysis of the Big Escambia Creek-Jay-Blackjack Creek Field area: Gulf Coast Association of Geological Societies Transactions, v. 26, p. 258–278.

Vaughan, R.L., Jr., 1985, Diagenetic effects on reservoir development in the Upper Jurassic Norphlet Formation, Mobile and Baldwin Counties and offshore Alabama [M.S. thesis]: Tuscaloosa, Alabama, University Alabama, 143 p.

Vaughan, R.L., Jr., and Benson, D.J., 1988, Diagenesis of the Upper Jurassic Norphlet Formation, Mobile and Baldwin Counties and offshore Alabama: Gulf Coast Association of Geological Societies Transactions, v. 38, p. 543–551.

Walker, T.R., 1967, Formation of red beds in modern and ancient deserts: Geological Society of America Bulletin, v. 78, p. 353–368.

Wilkerson, R.P., 1981, Environments of deposition of the Norphlet Formation (Jurassic) in south Alabama [M.S. thesis]: Tuscaloosa, Alabama, University Alabama, 141 p.

Wilson, G.V., 1975, Early differential subsidence and configuration of the northern Gulf Coast basin in southwest Alabama and northeast Florida: Gulf Coast Association of Geological Societies Transactions, v. 25, p. 196–206.

Wilson, G.V., and Tew, B.H., 1985, Geothermal data for southwest Alabama: Correlations to geology and potential uses: State Oil and Gas Board of Alabama, Oil and Gas Report 10, 125 p.

Wood, M.L., and Walper, J.L., 1974, The evolution of the interior Mesozoic basin and the Gulf of Mexico: Gulf Coast Association of Geological Societies Transactions, v. 24, p. 31–41.

Woods, R.D., and Addington, J.W., 1973, Pre-Jurassic geologic framework northern Gulf basin: Gulf Coast Association of Geological Societies Transactions, v. 23, p. 92–108.

Key Words

Hatter's Pond Field, Alabama, Mississippi Interior Salt basin, Norphlet Formation, Late Jurassic, Oxfordian, eolian, desert wadi, playa, dune, marine sandstones, erg, diagenesis, enhanced recovery.

8

A Giant Carbon Dioxide Accumulation in the Norphlet Formation, Pisgah Anticline, Mississippi

Joseph R.J. Studlick, Roger D. Shew, George L. Basye, and Johnny R. Ray

Shell Oil Company, Houston, Texas 77002; Shell Development Company, Houston, Texas 77001; Shell Western E&P Inc., Bakersfield, California 93389; Pecten International Company, Houston, Texas 77001

Introduction

General

In the search for oil and gas during the past century, other gases (helium, nitrogen, carbon dioxide) have been encountered. These uncommon instances were traditionally classified as failures, i.e., the gases had little or no economic value, and areas known to contain them were commonly avoided during further drilling. However, two factors have made carbon dioxide (CO_2) an attractive resource target in some areas. First, research has shown that the injection of CO_2 into some "watered-out" reservoirs can substantially increase oil recovery. Second, the increase in the price of oil in the 1970s has made some of these enhanced recovery operations economically viable.

This chapter discusses the geologic characteristics of a large CO_2 accumulation. In addition, the important factors, some serendipitous, in development of this type of resource are discussed. Some of these factors are large accumulation size, high purity of the CO_2 gas, excellent recovery efficiency and producibility (requiring fewer source wells), nearby fields amenable to CO_2 floods that had existing infrastructures (requiring the drilling of fewer drainage wells), and the development of technology that allowed the safe use of corrosive CO_2 in oil-field operations. Although recent oil-price declines have led to increased caution and project delays, several fields are undergoing successful CO_2 floods. The CO_2 accumulation described in this chapter is the largest of the CO_2 fields in the area, which could play a major role in recovering additional oil from fields in the Gulf of Mexico area. Thus, CO_2 accumulations in the right place and at the right time may become exploration targets in the future.

High-purity CO_2 occurs in central Mississippi, in an area north and east of the intrusive Jackson igneous dome (Studlick et al., 1987). Several salt-cored anticlinal and domal structures contain large volumes of CO_2 within the Jurassic (Kimmeridgian/Oxfordian) Buckner, Smackover, and Norphlet formations. The largest of these structures, the Pisgah anticline, has estimated reserves of more than 4 TCFG (1.1×10^{11} m³). The Norphlet Formation contains about one-third of these reserves. Carbon dioxide for tertiary CO_2 floods is being transported from the Pisgah anticline to several mid-dip Cretaceous Tuscaloosa fields in southwestern Mississippi and to a Miocene field in South Louisiana (Fig. 8-1).

Development History

Beginning in 1936, prolific Smackover oil fields were discovered in southern Arkansas and northern Louisiana. The industry used improved seismic methods in the early 1950s to drill along trend into Mississippi in the search for hydrocarbons. Previously, drilling was limited to depths of 15,000 feet (4,570 m), as seismic data were poor below a

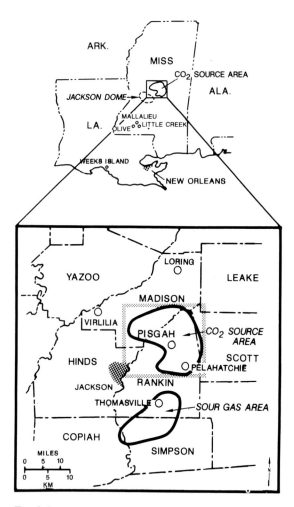

Fig. 8-1. Upper location map shows the CO_2 source area and Jackson Dome in central Mississippi. The four fields (open circles) in southwestern Mississippi and southern Louisiana are undergoing enhanced oil recovery operations using CO_2 from the CO_2 source area. Lower map shows the CO_2 source area and the Pisgah anticline in Rankin County, Mississippi. The five fields (open circles) were discovered in the 1950s and 1960s and delineated the area of high-purity CO_2.

regional Cretaceous reflector (Rodessa Formation). Drilling in the area north of the Jackson Dome in central Mississippi found mainly high-purity CO_2 or variable gas mixtures with high concentrations of CO_2 and some hydrogen sulfide (H_2S). Only sporadic exploration efforts continued in the area in the absence of a CO_2 market.

Further drilling for hydrocarbons in the 1950s and 1960s delineated an area with large volumes of high-

purity CO_2 (Fig. 8-1). It occurs within Jurassic sandstone and carbonate reservoirs, updip (north and east) of the Jackson igneous dome (i.e., Northeast Jackson Dome). These discoveries included Virlilia (1950; 94.0% CO_2), Loring (1953; 75.0% CO_2), Pisgah (1967; 98.0% CO_2), and Pelahatchie (1968; 65.0% CO_2) fields. Thomasville Field (see chapter by Shew and Garner, this volume) which was discovered in 1969, south of Pelahatchie and the Northeast Jackson Dome (CO_2 source area), has much lower CO_2 concentrations (9.0% CO_2, 35.0% H_2S).

Interest in CO_2 development remained low until pilot studies, conducted by Shell Oil Company at Little Creek Field in southwestern Mississippi in the early 1970s, indicated that the injection of CO_2 could lead to the successful recovery of additional oil reserves even after waterflood operations (see chapter by Werren et al., this volume). It was realized that large volumes of CO_2 would be required for potential enhanced recovery operations in the onshore and offshore U.S. Gulf Coast. Interest in the Northeast Jackson Dome increased because it was the closest and best source of CO_2 in the area.

The Pisgah anticline (now referred to as South Pisgah and Goshen Springs fields) was discovered in 1967 based on the seismic mapping of four-way closure at the top of the Jurassic section (Fig. 8-2). The discovery well, the Chevron Cox No. 1, encountered high-purity CO_2 (98.0–99.2%) in geopressured Jurassic (Kimmeridgian) Buckner sandstones, (Oxfordian) Smackover sandstones and dolomites, and (Oxfordian) Norphlet sandstones at depths of 15,000 to 16,000 feet (4,570–4,880 m). Production tests in the Buckner, Smackover, and Norphlet ranged from 2.8 to 10.6 MMCFGPD (0.8–3.0 × 10^5 m^3/D). Shell began to acquire leases in 1973 and drilled the first well specifically for CO_2 in 1978, the Hauberg No. 1 (Fig. 8-2).

Numerous structures in the area have been defined as CO_2 targets (Fig. 8-2). Ten of these structures have been tested by a total of 25 wells (22 successful and 3 dry holes). Only one structure, Langford, failed to encounter a CO_2 accumulation when nonreservoir quality rock was penetrated. Proven CO_2 reserves in the area are estimated to exceed 6 TCFG (1.7 × 10^{11} m^3). The CO_2 is more than 98% pure but contains variable amounts of contaminants. The Buckner and Norphlet formations contain "sweet" CO_2 (<10 ppm H_2S), whereas the Smackover contains H_2S in amounts that require treatment before transporting or injecting into oil reservoirs.

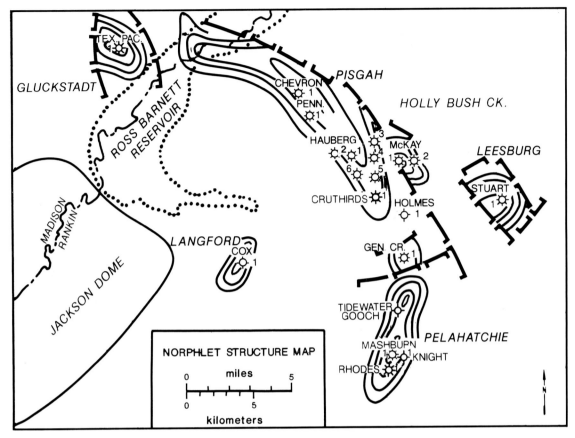

Fig. 8-2. Map of the southern two-thirds of the CO$_2$ source area showing Jurassic structures (mapped at the top of Norphlet Formation) and wells that penetrate the Jurassic. All wells were drilled by Shell except Texas Pacific (Sun/Oryx) Yandell No. 1 (Gluckstadt Field), Chevron Cox No. 1 and Pennzoil Board of Supervisors No. 1 (Pisgah anticline), Cox Busick No. 1 (Langford area), Tidewater Gooch No. 1 (Pelahatchie), and General Crude Spann No. 1 (south of Pisgah).

Neither field on the anticline has been fully developed. Wells have been drilled on 640-acre (259-ha.) spacing to develop Buckner, Smackover, and Norphlet reservoirs. Governmental units of 1,280 acres (518 ha.) have subsequently been approved for development of the Buckner and Norphlet formations.

Although tests have proven the wells are capable of producing at rates of more than 20 MMCFGPD (5.7 × 10^5 m^3/D), production has been limited by the injection needs of the EOR fields. Shell is producing 70 MMCFGPD (2.0 × 10^6 m^3/D) from six wells in the Northeast Jackson Dome area, which are completed in the Buckner, and is currently completing three additional wells in the Norphlet. Major lease holders on the anticline are Shell, Chevron, and Pennzoil. Chevron and Pennzoil each have one

shut-in well. Texas Pacific (Sun/Oryx) operates one well at nearby Gluckstadt Field (Fig. 8-2).

Regional Data

Overview

The Northeast Jackson Dome area is located in central Mississippi, approximately 100 miles (161 km) north of New Orleans, Louisiana (Fig. 8-1). It occurs within the Mississippi Interior Salt Basin which is bounded on the north by the Pickens-Gilbertown fault system, the approximate updip limit of the Jurassic Louann Salt, and on the south by basement highs of the Wiggins, South Mississippi, and Lasalle uplifts (Fig. 8-3).

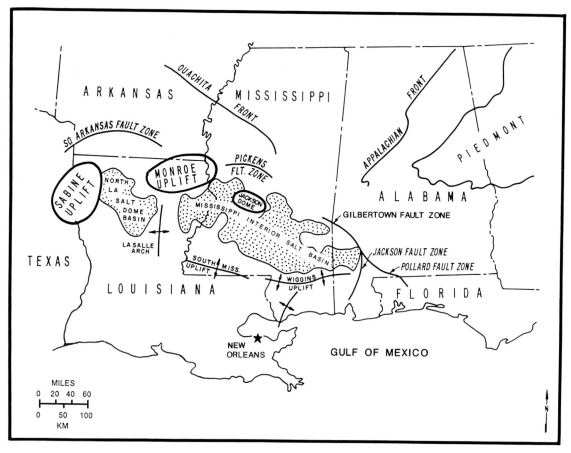

Fig. 8-3. Regional map of the central U.S. Gulf Coast showing the Mississippi Interior Salt basin and bounding structural features.

The Middle to Upper Jurassic (Callovian) Louann Salt is a thick, widespread salt deposit (Fig. 8-4) and, as a mobile substrate, is involved in the formation of all deep structures in the area. Salt is the economic basement for all exploration (Oxley et al., 1967), and no subsalt objectives have been drilled. The Louann Salt was deposited as a wedge which thins updip to the north against an eroded Paleozoic basement. These extensive salt deposits resulted from the intermittent flooding and evaporation of marine waters in rifted basins of the ancestral Gulf of Mexico (Salvador, 1980, 1982). Salt flowage from sediment loading began soon after deposition, and the salt is now typically 2,000 to 3,000 feet (610–915 m) thick in the cores of structures in comparison to 200 to 500 feet (61–152 m) in areas of salt withdrawal. These thicknesses contrast markedly to more basinward salt domes, which can be more than 10,000 feet (3,050 m) thick. These smaller salt-thick

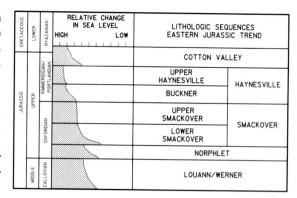

Fig. 8-4. Stratigraphic nomenclature used in central Mississippi and associated sea-level curves. (Modified from Salvadore, 1987, and Sassen and Moore, 1988; reprinted by permission of the American Association of Petroleum Geologists.)

areas are described as salt roller features (Parker, 1974), and, as domes, ridges, and pillows, they form traps for hydrocarbons and CO$_2$. Hughes (1968) described the types of salt structures in extreme eastern Mississippi which explain structural orientation and shapes in the Northeast Jackson Dome area. Near the basin rim where salt is thinner, salt-cored ridges, pillows, and anticlines are formed.

Early salt movement (Oxfordian) with associated faulting occurred along the Pisgah anticline and other nearby structures. Four-way and fault closures, along with the thickened salt in the Pisgah anticline, are evident in the seismic section shown in Figure 8-5. The Norphlet thickens in the vicinity of the Chevron Cox No. 1 and along the southwest side of the northwest-southeast-trending fault (Figs. 8-5, 8-6, 8-7). This thickening is related to faulting during deposition of the Norphlet and/or localization of the Norphlet dune complex. The top of the Louann Salt has local displacements of more than 600 feet (183 m) along the major faults which separate the Pisgah anticline from other structures.

The upper Smackover and Buckner sections expand on the downthrown sides of these faults, indicating syndepositional movement. On the downthrown sides of these faults, the Smackover-Buckner sections exhibit rollover into the faults, a structural style which is typical of "growth faults." Structural closure dies out rapidly up-section and is relatively unimportant at the Cotton Valley and shallower levels (Fig. 8-5). Both four-way and fault closures are the focus of exploration activities.

Early wildcats in the area were drilled based on seismic interpretation of four-way closure at the Norphlet level. Fault closures were avoided as it was believed that the late (Buckner to post-Buckner age) growth of structures could result in erosion of the seal and reservoir. Also, the large throw along these faults could place permeable Cotton Valley sandstones against Jurassic reservoirs; it was theorized that hydrocarbons or CO$_2$ would then escape up regional dip. However, the Shell Stuart No. 1, drilled in 1980 in the Leesburg area (Fig. 8-2), encountered a CO$_2$-productive reservoir in the faulted Smackover Formation. The 1,500 feet (460 m) of throw on that fault juxtaposes Smackover reservoir rocks against permeable Cotton Valley sandstones.

The area's dominant structural feature is the Jackson Dome, an igneous intrusion which probably occurred during the Late Cretaceous (Fig. 8-8). It intrudes rocks up to the Cretaceous Austin Chalk.

The occurrence of hydrocarbon accumulations in the Austin Chalk interval, and the absence of CO$_2$ accumulations in that interval, probably indicate that magma did not intrude the chalk. Numerous igneous intrusions of probable similar age are present throughout the U.S. Gulf Coast.

The geopressured, high-purity CO$_2$ in the area originated by thermal metamorphism of Jurassic carbonates by the Jackson Dome igneous intrusion (Fig. 8-8). The CO$_2$ laterally charged and inflated (geopressured) adjacent and updip structures (Parker, 1974). Although numerous modes of CO$_2$ generation are known, all known high-purity CO$_2$ accumulations are associated with igneous activity and/or subsequent thermal metamorphism of intruded carbonate rocks (Farmer, 1965). An estimate of 10 to 20 TCFG (2.8–5.6 $\times$ 10^{11} m^3) is suggested as the amount of CO$_2$ available in central Mississippi. Conservatively, we assume that only a small amount was trapped, indicating a large generation of CO$_2$ by igneous intrusions. Such large accumulations of CO$_2$ in the United States are rare (Irwin and Barnes, 1982).

The igneous intrusion had little or no effect on the structures formed during Norphlet-Smackover deposition. Hydrocarbons were probably present in the Norphlet-Smackover structures before or were concurrently migrating into the structures during emplacement of the Jackson Dome and carbon dioxide generation. Local occurrences of solid hydrocarbons indicate the presence of former oil pools in some reservoirs. At the time of the intrusion, the Smackover and Norphlet formations were overlain by 10,000 to 12,000 feet (3,050–3,660 m) of sediments, and were within the oil-generating window (Sassen and Moore, 1988). The CO$_2$ migrated through these reservoirs and displaced most hydrocarbons contained there. The CO$_2$ present in the Upper Buckner sandstones, above the Buckner carbonate seal, may have leaked from the underlying geopressured formations through faults and/or fractures.

Small amounts of H$_2$S contaminants are present in the CO$_2$ and resulted from an earlier phase of thermal "cracking" in the primary hydrocarbon pools. Hydrogen sulfide may also have been generated by thermal sulfate reduction, i.e., reduction of reservoir sulfate and anhydrite at slightly elevated temperatures. However, this is a less likely possibility, as CO$_2$ formation and migration were relatively early. The oils are from a carbonate source rock, the microlaminated zone of the Smackover Formation.

186   J.R.J. Studlick et al.

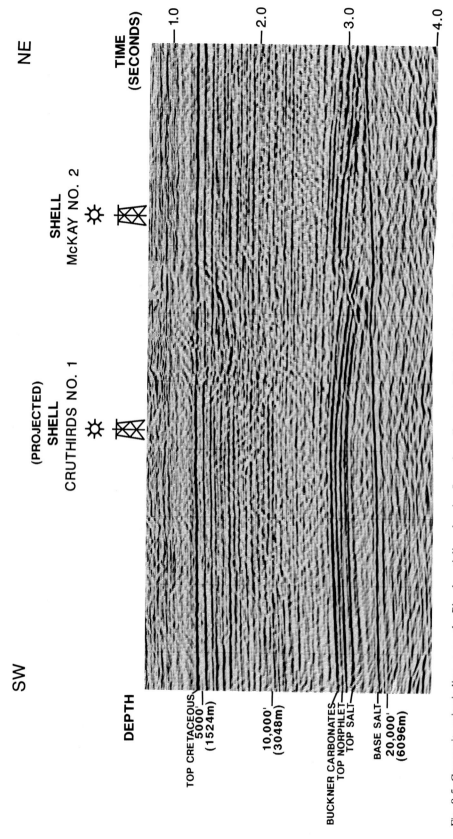

Fig. 8-5. Composite seismic line across the Pisgah anticline showing Jurassic sediments uplifted by thickened Louann Salt. The location of this seismic line is shown in Figure 8-7.

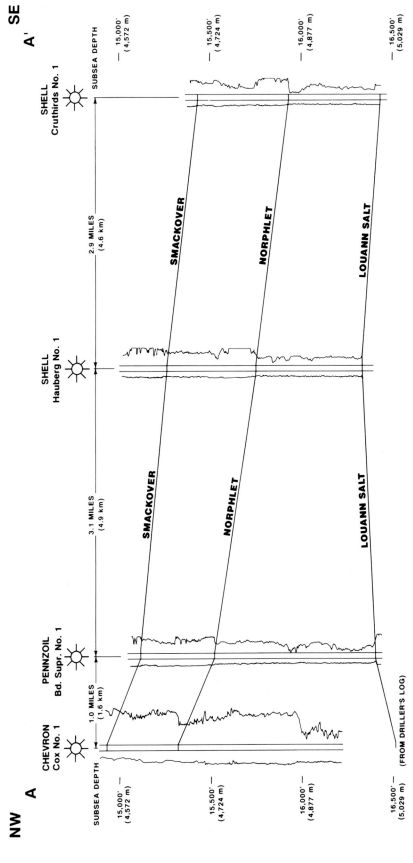

Fig. 8-6. Structural cross-section along the axis of the Pisgah anticline. Note the Norphlet thickening near the Chevron Cox No. 1 well. The line is shown in Figure 8-7.

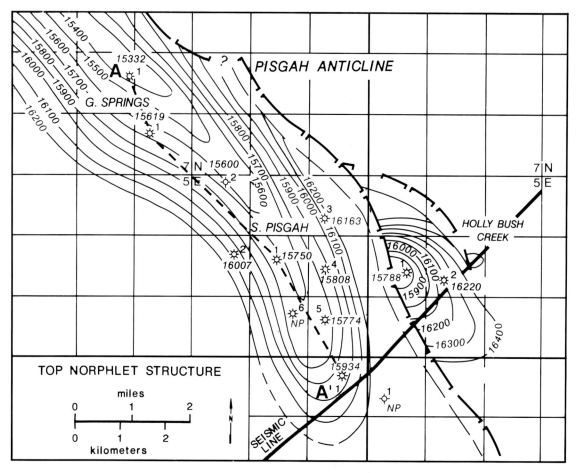

Fig. 8.7. Norphlet structure map of the southern portion of the Pisgah anticline which includes Goshen Springs Field (north) and South Pisgah Field (south). The depths are subsea. To the east is Holly Bush Creek Field, which is separated from the Pisgah anticline by a major fault. The locations of a seismic line (Fig. 8-5) and cross section A-A' (Fig 8-6) are also shown.

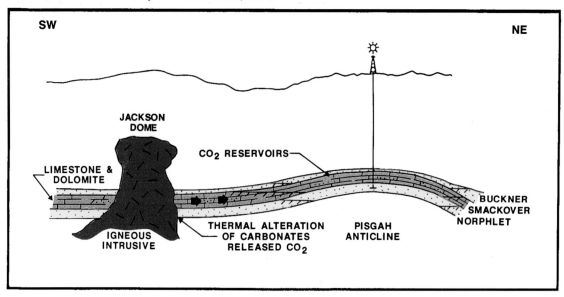

Fig. 8-8. Schematic illustration of the Northeast Jackson Dome area showing generation of CO_2 by the igneous intrusion and CO_2 migration updip and trapping in three separate reservoirs at the Pisgah anticline (modified from and reprinted by permission of the American Association of Petroleum Geologists, 1983).

Stratigraphy

Regional stratigraphic nomenclature of Dickinson (1968) is used in this chapter (Fig. 8-4). Three formations of Jurassic age, Buckner, Smackover, and Norphlet, have been proven to be CO_2-bearing in the Northeast Jackson Dome area. These formations contain five reservoirs, which include from the base the Lower Norphlet, the Upper Norphlet, the Smackover, the Main Buckner, and the Upper Buckner. These reservoirs occur at depths of 14,000 to 17,000 feet (4,270–5,180 m). The Norphlet Formation is commonly more than 500 feet (152 m) thick, unconformably overlies the Louann Salt, and is overlain by a dense carbonate of the lower Smackover (Fig. 8-9). The Norphlet shows the fewest facies and thickness variations of the Jurassic formations in the area. Predominant Norphlet facies changes are thinning (onlap) to the east, illite formation in the lower Norphlet, and salt plugging of reservoir rock to the south. The Smackover consists of shallow-water carbonate and sandstone, capped by anhydrite of the Buckner Formation. An anomalous CO_2-productive sandstone, 50 to 100 feet (15–30 m) thick, is present in the Main Buckner interval at Northeast Jackson Dome. Sandstones are uncommon in this portion of the Buckner in central Mississippi. Typical Buckner carbonates and evaporites underlie the sandstone. The top of the Upper Buckner is defined as the base of the Bossier Shale, an impermeable subregional marker more than 100 feet (30 m) thick. The Cotton Valley Formation and the Upper Buckner Member are a fluvial and deltaic section of sandstone and shale, 2,300 to 3,500 feet (700–1,070 m) thick.

Reservoir Characterization

Seismic Interpretation

The Pisgah anticline is a large structure, 18 miles (29 km) long by 5 miles (8 km) wide with a more than 1,000-foot (305-m) CO_2 column in the Jurassic. Seismic and well data clearly delineate the four-way closure at Jurassic levels (Figs. 8-2, 8-5). With the exception of field-bounding faults, faults within the field are either nonexistent or too small to resolve on seismic data or with the current large well spacing.

Seismic data are essential for the identification of four-way and fault closures at the Jurassic level. The quality of available data is sufficient to map structure (Fig. 8-5) but insufficient to evaluate seismic

stratigraphy. This is because the seismic data are mostly older vintage (1970s) and because relatively similar acoustic properties throughout the Norphlet prevent resolution of internal facies architecture. Only pronounced density-velocity contrasts (e.g., as tight carbonate versus porous sandstone) can be mapped, and these occur only at the Cotton Valley sandstone-Buckner carbonate and Smackover dense carbonate-Norphlet sandstone contacts.

Depositional Environment and Facies

The Norphlet is widespread throughout the Gulf of Mexico basin and consists primarily of eolian sandstones intercalated with variable amounts of alluvial fan and fluvial deposits (Mancini et al., this volume). Terrigenous sediments were derived from the Appalachians and Ouachitas and transported by rivers southward into a desert setting, possibly near the coast of a shallow saline sea (McBride et al., 1987; Marzano et al., 1988; Mancini et al. and Thomson and Stancliffe, this volume). Central Mississippi contains some of the thicker accumulations of Norphlet sediments. At the Pisgah anticline, the formation is mostly eolian and is 500 to 1,200 feet (152–365 m) thick. White to gray dune, sheet, and some interdune sandstones predominate, but hematite-stained eolian sandstones are locally present. These are locally underlain by red fluvial (braided) sandstones, which are a very minor part of the Norphlet Formation at the Pisgah anticline.

Eolian sandstones are the only productive Norphlet sediments at the Pisgah anticline. Figure 8-10 shows a block diagram of the interpreted Norphlet dune deposition, and Figures 8-10 and 8-11 show the rock types in core photographs. The sandstones may be generally divided into dominant dune and very minor interdune deposits. The dune facies is characterized by high-angle planar, massive, and laminated sandstones formed by grain-avalanche or grain-flow processes. Horizontal to low-angle cross-stratification is also present and indicates probable saltation (wind ripple) and lesser grain fall deposition. Wind-ripple cross-strata often grade upward into dune (grain-flow) cross-strata (similar to that described by Lindquist, 1988). Interdune deposits may be very thin (first-order bounding surface) or absent (second- and third-order surfaces) to separate sets of cross-strata. The dune sets are variable in thickness, ranging from only a few feet (1 m) to sets or cosets tens to a hundred feet (3–30 m) thick. Interdune deposits may be thin or up to 10 feet (3.0 m) thick and are transitional with the dune sand-

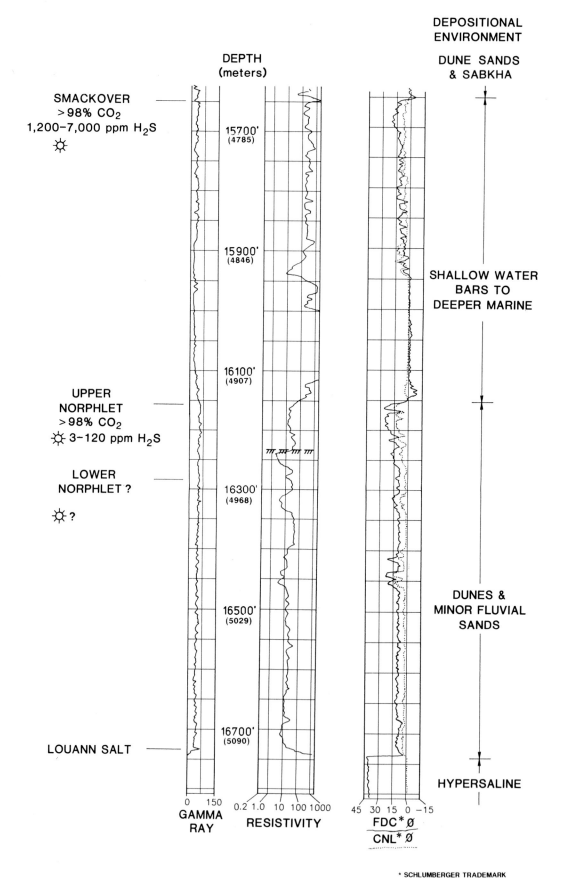

Fig. 8-9. Type log of the lower portion of the Jurassic productive interval at the Pisgah anticline; example from the Shell Hauberg No. 1, South Pisgah Field. The interpreted depositional environments are noted on the right.

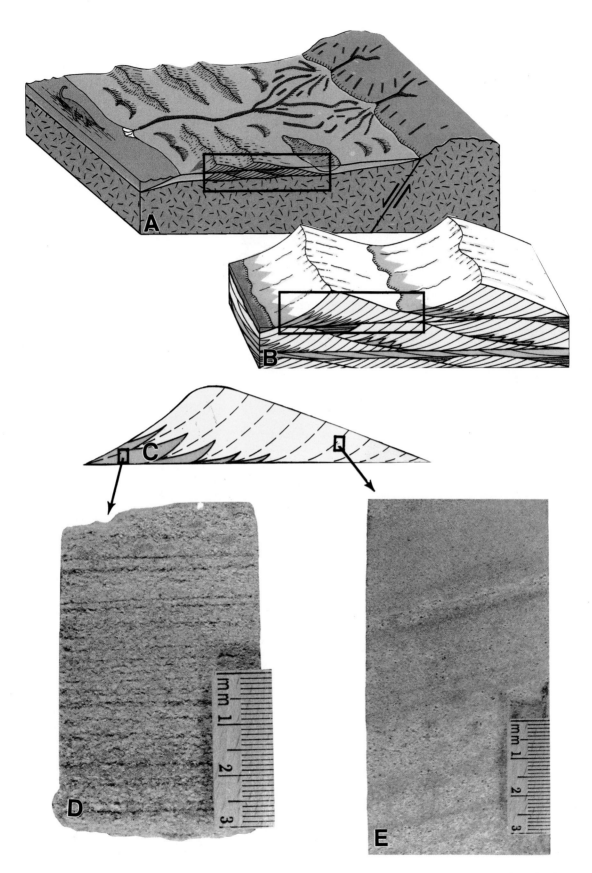

Fig. 8-10. Block diagrams of the interpreted eolian depositional setting and associated facies for the Norphlet Formation at the Pisgah anticline. Panels B and C are successive enlargements from panel A. Core D shows typical grain-flow strata, whereas core E illustrates typical grain-fall deposits.

191

16,369 ft

rx

c

c

c

hx

hx

hx

hx

A

16,381 ft

Fig. 8-11. Core photographs from the Shell Cruthirds No. 1, South Pisgah Field, showing the typical better reservoir quality of the upper (A) versus lower (B) Norphlet. Drilling mud has locally invaded the better porosity and permeability zones. Dune deposits are much more abundant than are interdune deposits and are characterized by high-angle,

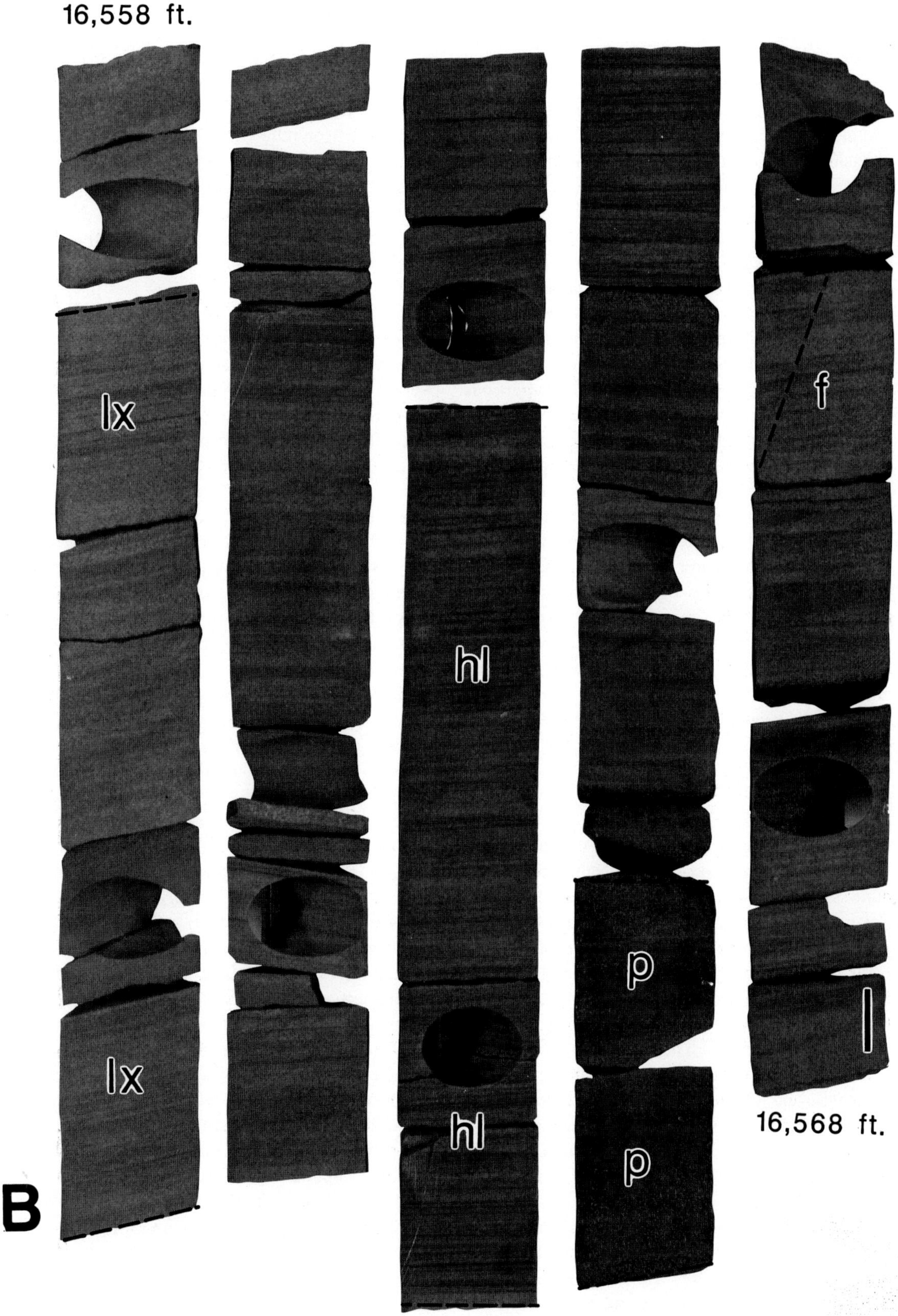

16,558 ft.

lx

lx

hl

hl

p

p

f

l

16,568 ft.

B

low-angle, and planar cross-strata. Note the patchy anhydrite and quartz cement (light colored spots). Scale bar is 1 inch (2.5 cm). (hx = high-angle cross-bedding; lx = low-angle cross-lamination; hl = horizontal laminae; rx = wind ripples; c = patchy cement; p = more porous zone; f = minor offset)

stones. They are composed of nearly horizontal
wind-ripple cross-strata and horizontal laminae.
Contorted bedding and microfaults occur in these
and in the dune deposits. These are all dry interdune
deposits based on the criteria of Ahlbrandt and
Fryberger (1981). However, no root tubules, shale,
dry lake deposits, or fossils are present. Additional
features of the eolian deposits are hematite coatings
on some quartz grains and the absence of detrital
matrix, mica, marine trace fossils, and shale beds.
Normal and inverse grading are common within
laminae, but the individual laminae are usually well
sorted (Sackheim, 1985; McBride et al., 1987).

Although Norphlet facies are not homogeneous
throughout, displaying significant local differences
in texture, porosity, and permeability, no major heterogeneities have been shown by production tests or
limited production. However, the usual limited
duration of production tests, the large well spacing,
and the common practice of perforating the entire
interval above the water level are inadequate to
establish heterogeneities and discontinuities within
the reservoir at this time. The major control on
reservoir quality is the diagenetic overprint,
although vertical textural variations (grain size
changes associated with laminae) substantially
decrease the vertical relative to the horizontal
permeability.

Diagenesis

The Norphlet is informally divided into upper and
lower members based on diagenetic alteration of the
dune sandstones. These members may or may not be
separated by a thin tight sandstone (Fig. 8-9). The
upper and lower members would be included in the
Denkman Member (Tyrrell, 1972). The upper
Norphlet is 125 to 250 feet (38–76 m) thick, and the
lower Norphlet ranges from 375 to 1,000 feet
(115–305 m) thick. The primary differences
between the upper and lower members are the
diagenetic and productive properties, as both members are composed of dune, sand-sheet, and very
minor amounts of interdune deposits. This
diagenetic overprint has dramatically changed the
original or depositional reservoir quality of both
members, which was generally high although variable throughout, similar to that described by Chandler and others (1989) in the Jurassic Page
Sandstone in Arizona. Porosity and permeability are
highest in the grain-flow deposits, lower in the wind-
ripple deposits, and lowest in the interdune deposits.
The reservoir quality of all these deposits is greatly

reduced by compaction, cements (anhydrite, halite,
quartz, carbonate), and authigenic illite clays. The
original differences in reservoir quality between
facies have been reduced and the difference between
lower and upper members increased.

Conventional cores from six wells at the Pisgah
anticline (401 feet (122 m) in the upper Norphlet,
1,457 feet (444 m) in the lower Norphlet) show the
vast predominance of dune versus interdune stratification. The interdune deposits are less porous and
permeable and may act as local barriers to flow.
However, these minor deposits cannot be traced
from well to well and are probably discontinuous
over distances of hundreds to thousands of feet
(hundreds of meters). Therefore, they may not significantly affect well performance but may decrease
vertical permeability within a wellbore. The Norphlet is 100% sandstone at the Pisgah anticline. Pay
quality is limited to those sandstones with porosity
greater than 5.0% and permeability greater than 0.1
md. Differentiation of rock quality into upper and
lower members, resulting from the strong diagenetic
overprint (discussed below), clearly overwhelms any
reservoir rock differences related to variable eolian
deposition.

The lower Norphlet has an order of magnitude
lower permeability and a somewhat reduced
porosity compared with the upper member (Fig.
8-12). This extreme permeability reduction is
caused by abundant fibrous illite and by occasional
halite cementation. The lower member also appears
to be slightly more compacted and cemented. This
may be related to increased pressure solution
associated with abundant grain-coating clays
described by Thomson (1959) and Weyl (1959).

A similar vertical zonation in diagenetically
induced facies characteristics may not be present in
the northern portion of the Pisgah anticline (Goshen
Springs Field). Core data from the Chevron Cox No.
1 well shows the absence of illite. The sandstones
are, as a result, more permeable and may be productive from the upper and lower members. However,
measured porosity and permeability may also have
been artificially increased due to removal of salt during coring operations. Salt is not pervasive but is
locally abundant in cores from South Pisgah Field.
Drilling there was conducted with oil-base muds.

Petrophysics

Norphlet sandstones are fine- to medium-grained,
moderately to well-sorted subfeldsarenites (classification of Folk et al., 1970). Textural properties are

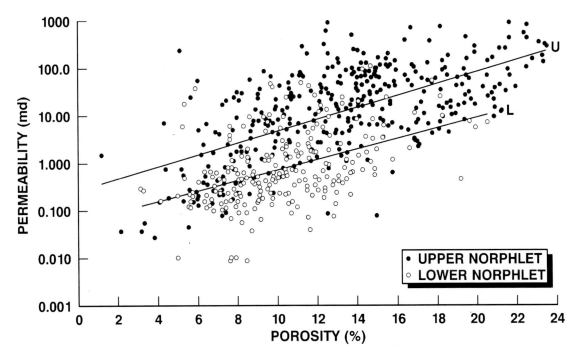

Fig. 8-12. Plot of porosity versus permeability of the Norphlet Formation. Note that the permeability of the upper Norphlet (best-fit regression line labeled U) is one order of magnitude higher than the lower Norphlet due to the absence of pervasive illite. Pervasive illite greatly reduces permeability but not porosity. All data are from unstressed conventional cores from six wells at South Pisgah Field.

similar throughout the interval except for changes associated with dune and interdune beds and laminations. Compositions are very uniform, with normalized major framework components of quartz (77%), feldspar, mainly orthoclase (16%), and rock fragments (7%) (Fig. 8-13). The major differences occur in the relative amounts of cement, including quartz overgrowths, halite, anhydrite, pyrite, calcite, and dolomite (Hartman, 1968; McBride, 1981; McBride et al., 1987). Illite is the dominant authigenic clay.

Norphlet porosities range from 1 to 24%, and permeabilities vary from 0.05 to 1,200 md (Fig. 8-12). The sediments, particularly the dune sandstones, had excellent estimated original primary porosity (40–45%) and permeability (a few thousand md). Average porosity is now 12%, and porosity reduction is primarily related to compaction as indicated by both numerous tangential and sutured grain contacts and by the precipitation of various cements. Cements range from 5 to 15% (bulk volume) so minus-cement porosity ranges from 17 to 27%. The cements may completely occlude porosity within individual laminae or within an entire interval of core. However, patchy cements are most common, giving cores a spotted appearance (Fig. 8-11). Both altered primary and secondary porosities are present.

The interpretation of altered primary pores is related to a reduction in pore size, but there is no petrographic evidence such as corroded cements and/or grains to indicate that they are secondary. Secondary porosity is 25% of the total porosity but is quite variable and is associated with feldspar, rock-fragment, and cement dissolution.

Permeability is generally low, averaging 1.0 md for the lower member and 10 md for the upper member. In addition to vertical permeability reduction associated with alternating tight and porous laminae, a majority of the reduction is related to pervasive fibrous illite that completely fills some pores and pore throats (Fig. 8-14). Compaction and the other cements occlude pores and in particular reduce vertical permeability. However, the order of magnitude difference in permeability between the upper and lower members is primarily the result of increased illite in the lower member. Petrographic evidence indicates that this is late-stage illite, but no age dates are available. McBride and others (1987) cited radiometric age dates for Norphlet illite of 55 Ma (Paleocene) from a well in Mobile County, Alabama.

Figure 8-12 shows porosity-permeability relationships for the upper and lower members determined

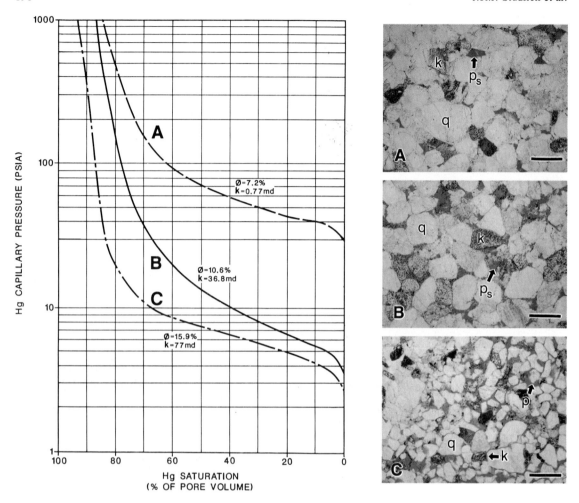

Fig. 8-13. Thin sections and associated capillary pressure curves of typical Norphlet sandstones in South Pisgah Field. Lower to higher entry pressures correspond with good to poor quality sandstones. Scale bar is 0.50 mm. (q = quartz; k = K-feldspar; p = primary porosity; p_s = secondary porosity)

from core data from six wells at South Pisgah Field. The wide data scatter represents the effects of variable depositional, but primarily diagenetic controls. Typical capillary pressure curves for the upper and lower Norphlet sandstones that have been diagenetically altered to different degrees at South Pisgah Field are shown in Figure 8-13. These curves demonstrate that sandstones with good porosity and permeability have low entry pressure and higher macroporosity. Sandstones with increased cement and illite have higher entry pressures and reduced macroporosity.

Permeability trends in the Norphlet cannot be predicted regionally from wireline logs. Permeability reduction is not a simple function of depth, as the Norphlet is permeable and produces at high rates at greater depths in offshore Alabama (Honda and McBride, 1981; Mancini et al., 1985; McBride et al., 1987; Vaughan and Benson, 1988; Dixon et al., 1989). However, the Norphlet offshore has chlorite grain-coating clays and has maintained a large volume of primary porosity. The Norphlet at Mobile Bay is separated from the underlying Louann Salt by an anhydrite (Pine Hill), which may have prevented halite cementation by limiting brine migration from below.

The central reason for vertical zonation in reservoir quality is the presence of locally pervasive illite (Fig. 8-14). It is possible that this illite resulted from the dissolution of K-feldspars and rock fragments. The origin of illite through the conversion of smectite is not possible here, as the shale needed

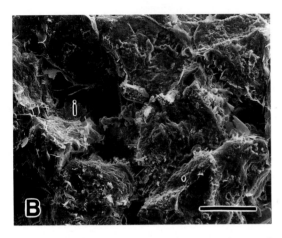

Fig. 8-14. Thin-section and SEM photomicrographs showing pervasive grain-coating and pore-filling illite common to the lower Norphlet at South Pisgah Field. (A) Note dark rims around quartz grains indicative of illite (scale bar is 0.25 mm); (B) SEM photomicrograph of grains coated with illite (i) (scale bar is 0.10 mm); and (C) enlargement shows pervasive illite filling pore throats (scale bar is 0.01 mm).

to supply cations is absent. A second possibility is that the clays are related to the types and distribution of brines. The Norphlet in central Mississippi directly overlies the Louann Salt, which may have provided potassium-rich brines to the reservoir. In addition, Lee and others (1989) described an anomalous increase in the occurrence of illite associated with a large fluid flow along a fault. Perhaps the bounding faults at the Pisgah anticline acted as conduits in this manner. More faults would be expected near the major fault separating South Pisgah Field from Holly Bush Creek Field (Fig. 8-7). A third possibility is that illite formed in the structurally lower part of the Pisgah anticline, perhaps below the original oil-water contact, suggesting that the Goshen Springs Field may have been

entirely hydrocarbon bearing. This is conjectural because CO_2 has removed all evidence of oil-water contacts.

From the previous discussion, it is obvious that coring is essential for reservoir evaluation, permeability determination, and salt identification. Particular care must be exercised in coring operations as an oil-base mud must be used with a salt-saturated water phase.

Production Characteristics

A gross sandstone isopach map of the Norphlet is shown in Figure 8-15. The CO_2-bearing interval with reservoir quality (net pay) sandstone is 125 to 250 feet (38–76 m) thick and is evaluated as produc-

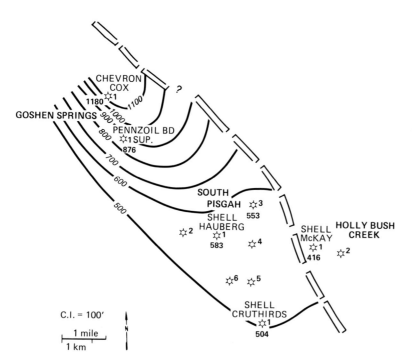

Fig. 8-15. Gross-sandstone isopach map of the Norphlet Formation for the southern two-thirds of the Pisgah anticline. Values are shown for wells that penetrate the entire Norphlet interval. Contour interval is 100 feet (30.5 m).

tive in an area 11 by 2 miles (18 × 3.2 km). Closure is estimated at 13,000 acres (5,270 ha.).

The lower Norphlet is evaluated as noncommercial based on production tests and core data in the southern part of the anticline. Well stimulation methods have been evaluated to improve production but with limited success. For example, an acid treatment in the lower member in the Shell Cruthirds No. 1 led to a short-lived increase in production rate, but the rate as well as pressure fell rapidly. The upper Norphlet has higher porosity and permeability and has been tested long-term (28 days) at an average rate of 30 MMCFGPD (8.5×10^5 m³/D).

Wells drilled in the area have tested hydrocarbons, CO_2, and water with abnormally high pressure gradients (0.65 psi/ft; 14.7 kPa/m). Sediments from the surface to the Buckner carbonate are normally pressured or slightly geopressured. Abnormal pressures occur mainly in the Smackover and Norphlet formations but also occur to a lesser extent in the Buckner. These geopressures are different from those commonly present in the U.S. Gulf Coast Tertiary, as they underlie nonshale, evaporite, and carbonate seals with no transition zones, thus giving no warning to drillers. The geopressures are not a result of "undercompaction" as in the typical Gulf Coast normal geopressuring mechanism. Geopressures cannot be predicted from compaction trends, since the

geopressured Jurassic sandstones are highly compacted. Geopressure prediction can only be approximated in the area by using seismic data to resolve the pressure seal (Buckner carbonates) and recognizing the first porosity below the seal. Therefore, casing is set in the carbonates above the objective Jurassic section before drilling ahead.

High purity (greater than 90%) CO_2 is limited to the area updip (north and east) of the Jackson Dome (Fig. 8-16). Gases in most of the area in the Buckner, Smackover, and Norphlet formations are 98 to 99% CO_2 with small amounts of methane, nitrogen, and hydrogen sulfide. Hydrogen sulfide concentrations vary from 1 to 7,000 ppm. CO_2 containing more than 10 ppm must be treated for H_2S removal or blended with low H_2S-content CO_2 to meet pipeline specifications due to H_2S corrosion. Away from this area, gas accumulations increase in hydrocarbon and H_2S content, including heavier hydrocarbons and condensate.

Stratigraphy plays an important role in CO_2 and H_2S content. Within the 90% CO_2 contour (Fig. 8-16), gas compositions remain relatively constant within individual formations. The Smackover gas is characterized as "sour" CO_2 throughout the area (H_2S concentrations of 1,600–7,000 ppm), whereas the Upper Buckner and Norphlet formations produce "sweet" CO_2 (< 10 ppm H_2S). These variations

Fig. 8-16. Map of the carbon dioxide trend (contoured in % CO₂) in central Mississippi derived from well data for gases from the Jurassic Buckner, Smackover, and Norphlet formations.

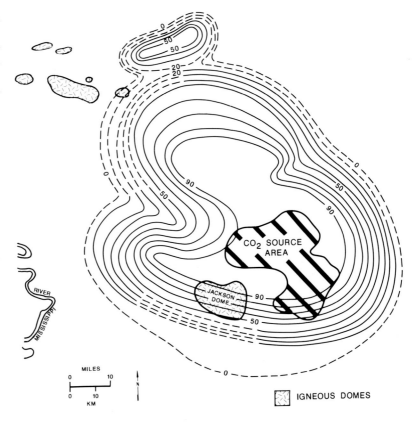

MILES
0   10
0   10
KM

N

IGNEOUS DOMES

in H₂S content may result from several processes. The Smackover, Norphlet, and possibly Buckner formations most likely had precursor oil accumulations. A high-sulfur source rock and/or maturation of a high-sulfur oil may have led to the formation of small amounts of H₂S. Although H₂S could also have formed through the reaction of hydrocarbons and sulfate (thermal sulfate reduction), the probable short time and lower temperature that were available before CO₂ displaced the hydrocarbons make this a less likely possibility. The lower H₂S in the Norphlet and Buckner may be related to H₂S removal by the formation of pyrite. Iron-bearing minerals such as rock fragments and clays are generally present in the Buckner and Norphlet but not in the Smackover.

Norphlet gas contains no hydrocarbons at the Pisgah anticline, because either the CO₂ swept all reservoir hydrocarbons or the CO₂ was emplaced early (before hydrocarbon migration in the area). The local occurrence of solid hydrocarbons as well as mixed gas compositions outside of the high-purity CO₂ trend indicates oil was probably present. The gas is high-purity CO₂ with small amounts of trace contaminants (Table 8-1). Of particular importance

Table 8-1. Typical composition of Norphlet gases at the Pisgah anticline showing the high CO₂ purity and low H₂S content. Data from the Shell Hauberg No. 1, South Pisgah Field.

Gases	%	ppm
CO₂	99.111	
H₂S		5
N₂	0.475	
CH₄	0.414	

is the consistently low H₂S content (1–5 ppm), which does not require treatment for removal. The wells also produce minimal amounts of formation water (1 bbl/MMCFG; 5.6×10^{-6} m³/m³).

The reservoir drive is assumed to be a depletion drive, although a small water influx is possible. Recovery efficiency is assumed to be 65%, primarily because the Norphlet is a regionally continuous, homogeneous sandstone unit. The limited production tests have revealed no major intrareservoir heterogeneities, so the entire Norphlet above the field water level is perforated.

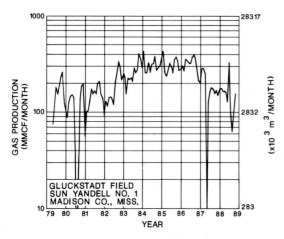

Fig. 8-17. Plot of typical CO_2 production from the Norphlet in the Northeast Jackson Dome area showing limited decline in production. Norphlet production has been limited at the Pisgah anticline as EOR requirements are met by production from the Buckner. The only well in the area which has been produced long-term from the Norphlet is shown. The rate is determined by supply needs; downward spikes result from well shutdowns. The well is located about 10 miles (16 km) west of the Pisgah anticline at Gluckstadt Field (see Fig. 8-2 for location).

Production at the Pisgah anticline has been curtailed as CO_2 injection demands have been met at the EOR fields. Only one Norphlet well from another field, the Texas Pacific (Sun/Oryx) Yandell No. 1 at Gluckstadt, has been on production for a number of years and shows no or little production decline (Figs. 8-2, 8-17).

Exploration and Production Strategy

Eolian sandstones of the Norphlet Formation, although they thicken and thin locally, are regionally widespread. Dune sandstones are the highest quality reservoirs and are more abundant than the very minor, generally poorer quality, interdune sandstones present at the Pisgah anticline. However, despite these obvious lithofacies differences, reservoir quality cannot be predicted ahead of the drill bit. Therefore, the questions to be addressed in prospect exploration and development are (1) CO_2 charge, (2) trap integrity, (3) localization of variable gas compositions, and (4) degree of diagenetic overprint on the original lithofacies. Charge limitation is not a problem in this area, as structures with seals

are filled to the structural spill point. The other concerns are discussed below.

Structural mapping at untested structures is based solely on reflection seismic data. Early prospecting in these deep Jurassic intervals was only done where the seismic data indicated four-way closure. However, drilling has confirmed that faults can be sealing and provide traps which support thick column heights. Thus, four-way and fault closures are now both prospective targets where there is an indication that the top seal is present.

The most important variable controlling reservoir quality and productivity is diagenetic alteration of sandstone fabrics. In general, porosity and permeability in the entire Norphlet have been reduced by compaction and cementation. However, we have also divided the Norphlet informally, particularly in the southern part of the Pisgah anticline, into upper and lower members because of rock quality and resulting productivity differences. An increased amount of fibrous illite and a slight increase in compaction and quartz cementation in the lower Norphlet make it noncommercial at South Pisgah Field. Reserve estimates have been adjusted downward to account for this interval that has porosity sufficient to produce CO_2, but a permeability that is too low to sustain flow. The occurrence and distribution of illite may be predicted locally but not regionally nor ahead of the drill bit for separate structures. The Norphlet in the Northeast Jackson Dome area is likely to exhibit reduced permeability associated with compaction, illite formation, and halite cementation.

Gas composition is variable both laterally and vertically within the Jurassic section. Lateral variation is related to the proximity and direction away from Jackson Dome. High-concentration, high-purity CO_2 accumulations occur updip (northeast) of the dome. An additional complication, even within the high-purity CO_2 belt, is the presence of 1,600 to 7,000 ppm H_2S within the Smackover Formation. Drilling and production have focused on the Buckner and Norphlet formations, which have H_2S concentrations less than 10 ppm.

The three U. S. Gulf Coast studies in this volume are focused on the Norphlet in a relatively small area, and their reservoirs were probably deposited in similar environments. Their sand sources are slightly different but the average framework compositions are not dramatically different. However, diagenesis (abundance of authigenic clay) has resulted in large differences in reservoir properties, especially in vertical permeability and porosity distributions. There is no consensus on the cause of these diagenetic differences, especially controls on illite-chlorite formation. Several areas that require additional study to address these differences include brine analyses, framework mineralogy, age dates of clays, and possible relations to hydrocarbon zones.

Conclusions

Numerous salt-generated structures of four-way and complex fault closures in central Mississippi are productive of high-purity CO_2. The Pisgah anticline is the largest of these structures and contains estimated reserves of more than 4 TCFG (1.1×10^{11} m³) in Jurassic sandstones and carbonates.

The Norphlet Formation is the deepest Jurassic reservoir at the Pisgah anticline, occurring at depths of 16,000 to 17,000 feet (4,880–5,180 m) and consisting of a 500- to 1,200-foot (152–365-m) thick

eolian sandstone. A variable diagenetic overprint of generally high depositional reservoir quality within the Norphlet led to a division at South Pisgah Field into two members. Core interpretations indicate that these members consist primarily of dune and very minor amounts of interdune deposits. The upper Norphlet is thinner but is capable of being produced at high rates. The thicker lower member is CO_2-bearing but noncommercial due to illite and halite cements reducing permeability as well as to slightly increased compaction. The CO_2 is currently being produced for use in EOR floods in Mississippi and Louisiana.

Acknowledgments. We thank Shell Oil Company for permission to publish and the many early Shell workers who contributed to our understanding of the area. C.A. Parker and J.A. Hartman provided excellent suggestions for improvement in the review process.

Reservoir Summary

Field: Pisgah Anticline (Goshen Springs, South Pisgah Fields)
Location: Central Mississippi, U.S.A.
Operators: Shell, Chevron, Pennzoil
Discovery: 1967
Basin: Mississippi Interior Salt basin; Gulf of Mexico
Tectonic/Regional Paleosetting: Salt tectonics along a rifted continental margin
Geologic Structure: Faulted anticline
Trap Type: Four-way closure
Reservoir Drive Mechanism: Depletion
• **Original Reservoir Pressure:** 10,600 psi (7.3×10^4 kPa) at 15,750 feet (4,800 m) subsea
• **Recovery Factor:** 1.29 MMCFG/NAF
Reservoir Rocks
• **Age:** Late Jurassic, Oxfordian
• **Stratigraphic Unit:** Norphlet Formation
• **Lithology:** Medium- to fine-grained subfeldsarenites
• **Depositional Environment:** Eolian
• **Productive Facies:** Dune sandstone
• **Petrophysics**
  • **Porosity Type:** Total (average) = 12% (Primary = 6.0%; Secondary (dissolution of feldspar, rock fragments, and cements) = 6.0%)
  • **φ:** Average 12%, range 1 to 24%, cutoff 5% (unstressed cores)
  • **k:** Average 1.0 md (lower Norphlet), 10 md (upper Norphlet); range 0.05 to 1,200 md, cutoff 0.10 md (air, unstressed cores)
  • **S_w:** Average 20%, range 10 to 35%, cutoff 50% (logs)
Reservoir Geometry
• **Depth:** 16,000 to 17,000 feet (4,880–5,180 m)
• **Areal Dimensions:** 11 by 2 miles (18 × 3.2 km)
• **Productive Area:** 13,000 acres (5,270 ha.)
• **Number of Reservoirs:** 1

- **CO_2-Column Height:** 505 feet (154 m)
- **Fluid Contact:** CO_2-water contact at 15,837 feet (4,827 m) subsea
- **Number of Pay Zones:** 2 (Upper and lower members of Norphlet)
- **Gross Sandstone Thickness:** 500 to >1,200 feet (152 - >365m) (Note that "greater than" symbology has been used as the assumed thickest part of the reservoir has not been penetrated.)
- **Net Sandstone Thickness:** 495 to >1,190 feet (151 - >362 m)
- **Net/Gross:** 1.0

Hydrocarbon Source, Migration
- **Lithology and Stratigraphic Unit:** Microlaminated carbonate, Jurassic Smackover
- **Time of CO_2 Formation:** Late Cretaceous/Early Tertiary
- **Time of Trap Formation:** Early Mesozoic
- **Time of CO_2 Migration:** Late Cretaceous/Early Tertiary

Gases
- **Type:** CO_2
- **Specific Gravity:** 0.78 g/cc
- **FVF:** 435 SCFG/RCF (12.2 m³/Rm³)

Volumetrics (Norphlet only)
- **In-Place:** 2.0 TCFG (5.7×10^{10}) m³
- **Ultimate Recovery:** 1.3 TCFG (3.7×10^{10}) m³
- **Recovery Efficiency:** 65%

Wells
- **Spacing:** 5,280 feet (1,609 m), 640 acres (259 ha.); 10,560 feet (3,219 m), 1,280 acres (518 ha.)
- **Total:** 9
- **Dry Holes:** 0

Typical Well Production
- **Average Daily:** 10 to 30 MMCFG (2.8–8.5×10^5 m³)
- **Anticipated Ultimate Recovery:** 50 to 100 BCFG (1.4–2.8×10^9 m³)

Other: CO_2 is transported via pipeline for injection in oil fields in Mississippi and Louisiana to increase recoveries.

References

Ahlbrandt, T.S., and Fryberger, S.G., 1981, Sedimentary features and significance of interdune deposits, *in* Ethridge, F.G. and Flores, R.M., eds., Recent and ancient nonmarine depositional environments: Models for exploration: Society of Economic Paleontologists and Mineralogists Special Publication 31, p. 293–314.

American Association of Petroleum Geologists, 1983, CO_2 projects to test recovery theories: American Association of Petroleum Geologists Explorer, June, p. 1, 20.

Chandler, M.A., Kocurek, G., Goggin, D.J., and Lake, L.W., 1989, Effects of stratigraphic heterogeneity on permeability in eolian sandstone sequence, Page Sandstone, northern Arizona: American Association of Petroleum Geologists Bulletin, v. 73, p. 658–668.

Dickinson, K.A., 1968, Upper Jurassic stratigraphy of some adjacent parts of Texas, Louisiana, and Arkansas: U.S. Geological Survey Professional Paper 594-E, 25 p.

Dixon, S.A., Summers, D.M., and Surdam, R.C., 1989, Diagenesis and preservation of porosity in Norphlet Formation (Upper Jurassic) southern Alabama: American Association of Petroleum Geologists Bulletin, v. 73, p. 707–728.

Farmer, R.E., 1965, Genesis of subsurface carbon dioxide:

American Association of Petroleum Geologists Memoir 4, p. 378–385.

Folk, R.L., Andrews, P.B., and Lewis, D.W., 1970, Detrital sedimentary rock classification and nomenclature for use in New Zealand: New Zealand Journal of Geology and Geophysics, v. 13, p. 937–968.

Hartman, J.A., 1968, The Norphlet sandstone, Pelahatchie Field, Rankin County, Mississippi: Gulf Coast Association of Geological Societies Transactions, v. 18, p. 2–11.

Honda, H. and McBride, E.F., 1981, Diagenesis and pore types of the Norphlet sandstone (Upper Jurassic), Hatter's Pond area, Mobile County, Alabama: Gulf Coast Association of Geological Societies Transactions, v. 31, p. 315–322.

Hughes, D.J., 1968, Salt tectonics as related to several Smackover fields along the Northeast rim of the Gulf of Mexico basin: Gulf Coast Association of Geological Societies Transactions, v. 18, p. 320–330.

Irwin, W.P. and Barnes, I., 1982, Map showing relation of carbon dioxide-rich springs and gas wells to the tectonic framework of the conterminous United States: U.S. Geological Survey Miscellaneous Investigations Series I-1301, 9 p.

Lee, M., Aronson, J.L., and Savin, S.M., 1989, Timing and conditions of Permian Rotliegende sandstone diagenesis, southern North Sea: K/Ar and oxygen isotopic data: American Association of Petroleum Geologists Bulletin, v. 73, p. 195–215.

Lindquist, S.J., 1988, Practical characterization of eolian reservoirs for development: Nuggett Sandstone, Utah-Wyoming thrust belt, in Kocurek, G., ed., Late Paleozoic and Mesozoic eolian deposits of the western interior of the United States: Sedimentary Geology, v. 56, p. 315–339.

Mancini, E.A., Mink, R.M., Bearden, B.L., and Wilkerson, R.P., 1985, Norphlet Formation (Upper Jurassic) of southwestern and offshore Alabama: Environments of deposition and petroleum geology: American Association of Petroleum Geologists Bulletin, v. 69, p. 881–898.

Marzano, M.S., Pense, G.M., and Andronaco, P., 1988, A comparison of the Jurassic Norphlet Formation in Mary Ann Field, Mobile Bay, Alabama to onshore regional Norphlet trends: Gulf Coast Association of Geological Societies Transactions, v. 38, p. 85–100.

McBride, E.F., 1981, Diagenetic history of the Norphlet Formation (Upper Jurassic), Rankin County, Mississippi: Gulf Coast Association of Geological Societies Transactions, v. 31, p. 347–351.

McBride, E.F., Land, L.S., and Mack, L.E., 1987, Diagenesis of eolian and fluvial feldspathic sandstones, Norphlet Formation (Upper Jurassic), Rankin County, Mississippi, and Mobile County, Alabama: American Association of Petroleum Geologists Bulletin, v. 71, p. 1019–1034.

Oxley, M.L., Minihan, E., and Ridgeway, J.M., 1967, A study of Jurassic sediments in portions of Mississippi and Alabama: Gulf Coast Association of Geological Societies Transactions., v. 17, p. 24–48.

Parker, C.A., 1974, Geopressures and secondary porosity in the deep Jurassic of Mississippi: Gulf Coast Association of Geological Societies Transactions, v. 24, p. 69–80.

Sackheim, M.J., 1985, Regional stratigraphy of the Louark Group (Upper Jurassic), Norphlet, Smackover, and Buckner Formations in Rankin and northern Simpson Counties, Mississippi [M.S. thesis]: New Orleans, Louisiana, Tulane University, 165 p.

Salvador, A., 1980, Late Triassic-Jurassic paleogeography and the origin of the Gulf of Mexico, in Pilger, R.H., ed., Origin of the Gulf of Mexico and the early opening of the central north Atlantic Ocean: Symposium Proceedings, Baton Rouge, Louisiana State University, p. 101.

Salvador, A., 1982, The Jurassic of the Gulf of Mexico basin—A regional synthesis: Gulf Coast Section Society of Economic Paleontologists and Mineralogists, Jurassic of the Gulf Rim, Third Annual Research Conference, Baton Rouge, Louisiana, November 28–December 1, p. 93.

Salvador, A., 1987, Late Triassic-Jurassic paleogeography and origin of the Gulf of Mexico Basin: American Association of Petroleum Geologists Bulletin, v. 71, p. 419–451.

Sassen, R., and Moore, C.H., 1988, Framework of hydrocarbon generation and destruction in eastern Smackover trend: American Association of Petroleum Geologists Bulletin, v. 72, p. 649–663.

Studlick, J.R.J., Shew, R.D., Basye, G.L., and Ray, J.R., 1987, Carbon dioxide source development, Northeast Jackson Dome, Mississippi [abst.]: American Association of Petroleum Geologists Bulletin, v. 71, p. 619.

Thomson, A., 1959, Pressure solution and porosity: Silica in sediments symposium: Society of Economic Paleontologists and Mineralogists Special Publication 7, p. 92–110.

Tyrrell, W.W., 1972, Denkman sandstone member—An important Jurassic reservoir in Mississippi, Alabama, and Florida [abst.]: Gulf Coast Association of Geological Societies Transactions, v. 22, p. 32.

Vaughan, R.L., Jr., and Benson, D.J., 1988, Diagenesis of the Upper Jurassic Norphlet Formation, Mobile and Baldwin counties and offshore Alabama: Gulf Coast Association of Geological Societies Transactions, v. 38, p. 543–551.

Weyl, P.K., 1959, Pressure solution and the force of crystallization—A phenomenological theory: Journal of Geophysical Research, v. 64, p. 2001–2025.

Key Words

Goshen Springs Field, South Pisgah Field, Mississippi, Mississippi Interior Salt basin, Norphlet Formation, Smackover Formation, Late Jurassic, Oxfordian, eolian, Pisgah anticline, dune sandstone, carbon dioxide, illite, diagenesis, igneous intrusion, enhanced oil recovery, erg, halite.

9
Diagenetic Controls on Reservoir Quality, Eolian Norphlet Formation, South State Line Field, Mississippi

Alan Thomson and Richard J. Stancliffe

University of New Orleans, New Orleans, Louisiana 70148; Shell Offshore Inc., New Orleans, Louisiana 70161

Introduction

The Norphlet Formation (Upper Jurassic) at South State Line Field, Greene County, Mississippi, has poor reservoir quality in its upper part and good reservoir quality in its lower part. This chapter shows that these properties are a consequence of the formation of diagenetic clay minerals.

The Norphlet Formation has produced gas and condensate from normally pressured eolian sandstones at depths of more than 18,000 feet (5,500 m). The field is in the center of the Norphlet producing trend in the Mississippi Interior Salt Basin of the eastern Gulf Coast region (Fig. 9-1). Interest in the Norphlet took a dramatic upswing in 1980 with the discovery of prolific gas reserves in offshore Alabama. In this area as well as in the offshore of the Florida panhandle, Norphlet sandstones have excellent reservoir properties. This portion of the Norphlet productive trend is the "eastern productive trend." Norphlet sandstones in the "western productive trend" of Mississippi, Arkansas, Louisiana, and Texas have poorer reservoir properties. These differences are interpreted to be the result of different diagenetic histories. Specifically, authigenic chlorite is common in the east, and authigenic illite is common in the west. Authigenic chlorite is believed to be responsible for preserving porosity, whereas authigenic illite is thought to be instrumental in reducing porosity, mainly by pressure solution and related processes.

The Norphlet at South State Line Field is of particular interest because it contains chlorite in the lower, more porous and permeable part of the formation and illite in the upper, less porous and permeable part. Can the presence of chlorite be correlated with better reservoir quality? If so, why did chlorite develop in the lower part, whereas illite developed in the upper part? Is this a consequence of diagenesis (fluids, etc.) or does it reflect the provenance of the Norphlet sediments (primary mineralogic composition) or a combination of both? South State Line Field, on the Mississippi-Alabama border, is in an ideal location to record contributions from two major sources (if such existed), since it is at the boundary between the western and eastern productive trends.

Stratigraphy and Structure

The Norphlet Formation at South State Line Field is underlain by the Louann Salt and overlain by carbonates of the Smackover Formation, both of Jurassic age (Fig. 9-2). By late Middle Jurassic (Callovian) time, salt deposits had begun to form in two areas in the present-day Gulf of Mexico (Salvador, 1987). The saltwater influx was from the Pacific side, and salt was deposited in shallow lakes, thick accumulations of salt being the result of almost continuous subsidence (Salvador, 1987).

In early Late Jurassic (Oxfordian) time, sands of the Norphlet Formation were transported southward over Louann Salt deposits along the northern rim of the Gulf of Mexico Basin. Norphlet deposits are primarily of eolian facies; however, fluvial and alluvial-fan facies occur along the northern extremity of Norphlet deposition (Mancini and Benson, 1980).

Casebooks in Earth Science
Sandstone Petroleum Reservoirs
Eds.: Barwis/McPherson/Studlick
© 1990 by Springer-Verlag New York, Inc.

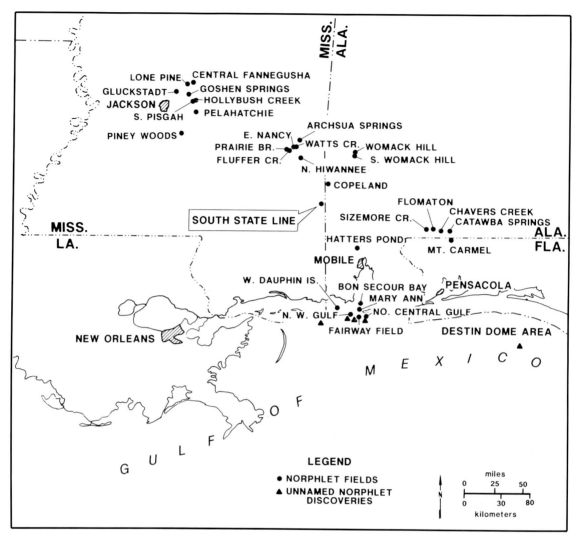

Fig. 9-1. Index map of South State Line Field and other fields in the Norphlet producing trend.

The remainder of the Oxfordian sequence consists mainly of carbonates of the Smackover Formation. Low-energy, deepwater lime mudstones abruptly overlying eolian deposits of the Norphlet attest to a rapid widespread transgression in the Gulf Coast region. Whether or not this transgression was due to a eustatic rise in sea level is uncertain (Salvador, 1987). The resulting stratigraphic sequence places sandstones of the Norphlet Formation, often with excellent reservoir properties, between seals of the underlying Louann Salt and the overlying Smackover Formation.

Rifted continental-margin tectonics exerted a strong control on Jurassic sedimentation in the eastern Gulf Coast region (Mancini et al., 1985). A combination of differential basement subsidence and the presence of pre-Jurassic paleohighs greatly influenced local depositional patterns. In established rift grabens (e.g., Mississippi Interior Salt Basin), thick accumulations of Jurassic sediments were deposited over the Louann Salt.

The structure at the Norphlet level at South State Line Field is a faulted, salt-cored, northeast-trending dome with more than 1,000 feet (305 m) of relief and 350 feet (107 m) of hydrocarbon column (Fig. 9-3). Faulting has separated the field into several crestal fault blocks (Fig. 9-3). Seismic isopach mapping as well as facies distribution patterns in the overlying Smackover and Buckner carbonates indicates early structural growth related to early movement of the underlying Louann Salt.

Reservoir Characterization

Seismic Attributes

The Norphlet top is clearly seen on conventional seismic lines due to its high density and velocity contrast with the overlying Smackover carbonates (Fig. 9-4). The contact of the Norphlet with the underlying Louann Salt is much less apparent due to the similar acoustic impedance properties of both formations.

Depositional Facies and Geometry

Regional stratigraphic control, coupled with two complete well penetrations (Shell Lucas No. 1 and Shell Glenpool No. 2), indicates that the Norphlet Formation occurs as a broad, continuous sand sheet 600 to 700 feet (183–213 m) thick. All of the 341 feet (104 m) of Norphlet examined in core is interpreted to be the result of eolian processes, being composed entirely of dune and interdune deposits as described by Fryberger and others (1983). The dune facies consists of thick, low- to high-angle cross-bed sets that range up to 25 feet (8 m) in thickness (Fig. 9-5A). Preserved dune foresets are composed of both avalanche deposits 1 to 2 inches (2.5–5.1 cm) thick and inversely graded wind-ripple deposits a fraction of an inch (<1 cm) thick, each representing approximately 50% of the preserved foresets (Hunter, 1981; Kocurek and Dott, 1981). This relatively high percentage of incorporated wind ripples is suggestive of dune-associated apron or plinth deposits typical of oblique and longitudinal dunes (Kocurek, 1986) but may also occur in transverse systems. Interdune deposits are characterized by roughly horizontal, wind-ripple deposits suggestive of a dry depositional setting (Fig. 9-5B). Locally, interdune stratification appears associated with damp or wet conditions (Glennie, 1970).

The Norphlet can be characterized in cores as comprising alternating large-scale cross-beds (dune) and thin horizontal deposits (interdune). Four complete "cycles" and portions of two others are observed in the Lucas No. 1 cores (Fig. 9-6). The horizontal interdune deposits range between 6 and 15 feet (1.8–4.6 m) thick, whereas dune deposits represented by cosets of cross-strata are 25 to 120 feet (8–37 m) thick. Unfortunately, high-resolution stratigraphic dipmeter data are not available at South State Line Field. Such data would be useful in distinguishing different eolian facies, establishing dune-migration directions, and possibly substantiating dune type and inferred directional permeability.

The cyclical pattern of dune and interdune deposits observed in the Norphlet cores suggests net

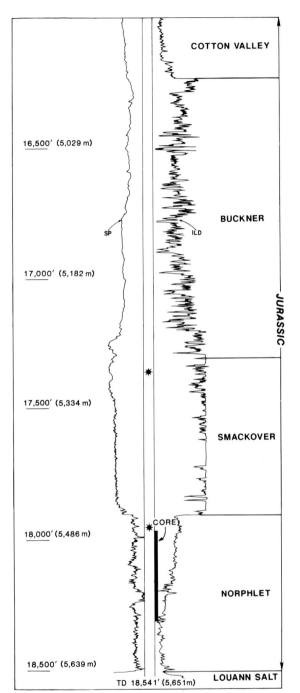

Fig. 9-2. Type log for South State Line Field, showing productive formations and cored interval. Shell Lucas No. 1.

lateral migration of large dunes and associated interdune areas. Kocurek (1981) documented a similar facies pattern for the Jurassic Entrada Sandstone of northern Utah and Colorado. These deposits were interpreted to have resulted from the lateral migra-

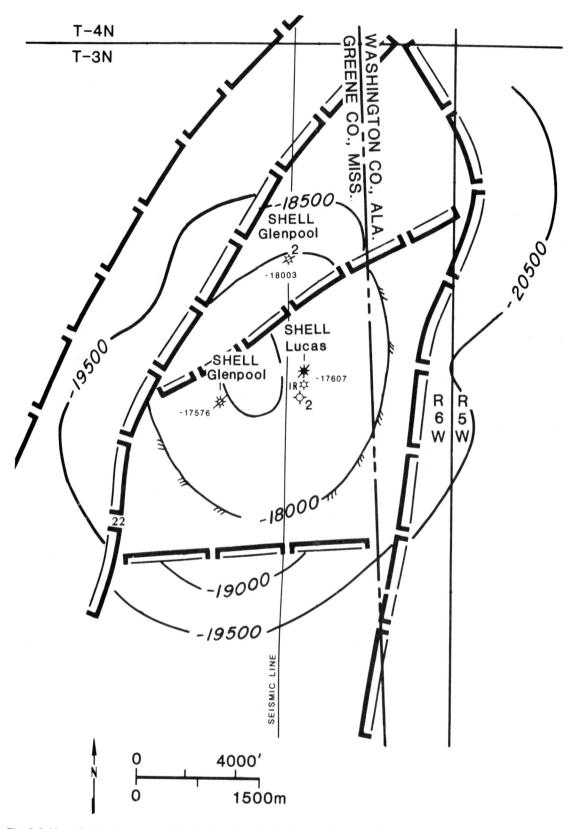

Fig. 9-3. Norphlet structure map of South State Line Field, showing location of wells and hydrocarbon-water contact (contours in feet).

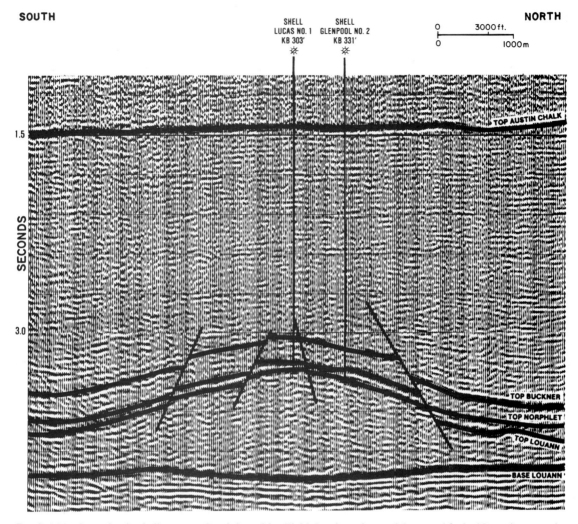

Fig. 9-4. North-south seismic line across South State Line Field showing salt-cored dome and faulted Jurassic reservoirs. See Figure 9-3 for line location.

Entrada erg." The Norphlet of South State Line Field is similarly located within an erg; however,the abundant rippled apron (plinth) deposits suggest the dunes were not purely transverse but were more oblique in character (Kocurek, 1986). Figure 9-7 is a hypothetical model of South State Line Field's Norphlet eolian system.

The Norphlet Formation is approximately 600 feet (183 m) thick across the South State Line structure, according to well penetration and seismic mapping (Fig. 9-8). Because of the field's limited well data base, a net sandstone isopach map was not prepared. Of greater significance is the net reservoir thickness (Fig. 9-8).

Trapping Mechanism, Hydrocarbon Expulsion and Migration, Seal

The development of the structural trap at South State Line Field was nearly contemporaneous with deposition of Norphlet sediments. Seismic mapping as well as lithofacies mapping in overlying Smackover carbonates indicates early salt movement and structural growth. Smackover carbonates immediately over the dome reflect deposition in shallower water than do equivalent deposits on the flank of the dome. The resulting salt-cored and fault-crested dome was formed long before hydrocarbons were generated to fill it.

Fig. 9-5. Representative core photos of the Norphlet sandstone at South State Line Field. Base of cores at lower right and scale bar is 2.4 inches (6 cm). Shell Lucas No. 1: (A) High-angle ripple and avalanche stratified dune cross-bedding, 18,144 to 18,152 feet (5,530–5,533 m). (B) Roughly horizontal, discontinuous ripple-stratified laminae of interdune, 18,098 to 18,105 feet (5,516–5,518 m).

The source of most Norphlet hydrocarbons is believed to have been the immediately overlying, lipid-rich lime mudstones of the Smackover Formation (Oehler, 1980; Mancini et al., 1985). Temperatures were probably not sufficiently elevated for peak hydrocarbon generation until the source rocks were buried on the order of 13,000 to 14,000 feet (3,960–4,270 m) (Honda, 1981; McBride et al., 1987). By assuming (1) a temperature of expulsion of oil at 250°F (121°C); (2) a temperature gradient of 1.3°F/100 feet (2.3°C/100 m); (3) that a suitable migration path between source and reservoir existed; and (4) that the time of oil arrival at the reservoir was essentially identical to the time of expulsion from the source, the migration probably occurred in mid-Cretaceous time.

The overlying lime mudstone source rocks of the Smackover Formation are also the seal for the Norphlet reservoir. However, a later section (Controls on Reservoir Quality) will show that the hydrocarbon distribution is controlled by more than just simple structural configuration.

Petrography

Norphlet sandstones are composed of very fine- to medium-grained sand. Grains are typically well-rounded to subangular. Bulk samples of sandstones from both dune and interdune facies are moderately to well-sorted, although individual laminae in dune facies sandstones may be very well-sorted.

Norphlet sandstones consist of nearly 50% framework quartz (Table 9-1). Other major framework constituents include potassium feldspar (9–16%), plagioclase feldspar (3–9%), and lithic fragments of various types (7–15%). The sandstones are classified as lithic feldsarenites (Fig. 9-9), with framework components in the Shell Lucas No. 1 core averaging Q = 61%, F = 25%, and R = 14%.

Authigenic constituents include quartz, potassium feldspar, ferroan dolomite, and clay minerals (Table 9-1). It should be noted that, whereas total clay measured in thin section averages about 2%, it averages 10% of bulk volume by X-ray diffraction analysis (Table 9-2). This discrepancy is due in part to an

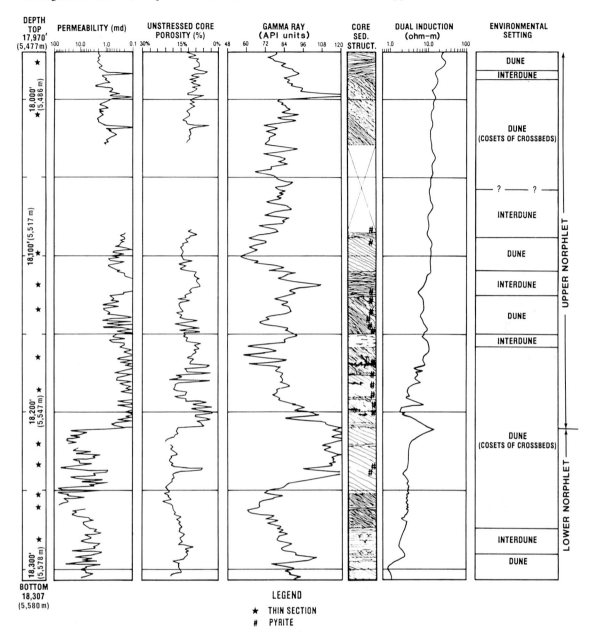

Fig. 9-6. Norphlet core section (core depth corrected to log) showing wireline logs, porosities, permeabilities, and environmental settings. Shell Lucas No. 1.

underestimation of total clay in thin-section analysis, the inclusion in X-ray diffraction analysis of clay incorporated in lithic fragments, and the fact that X-ray diffraction data are at best semiquantitative.

Diagenesis

In terms of reservoir quality, the most important authigenic constituents in the Norphlet sandstones are the

clay minerals illite and chlorite. Illite occurs as standalone or pore-bridging fibers in the upper several hundred feet of the Norphlet (Figs. 9-6, 9-10; Tables 9-1, 9-2). Because of its high surface area and pore throat-clogging nature, illite has a detrimental effect on reservoir quality. It often completely blocks pore spaces and may severely reduce porosity and permeability.

Chlorite occurs in amounts visible under the microscope only in the lower Norphlet sandstones (Table

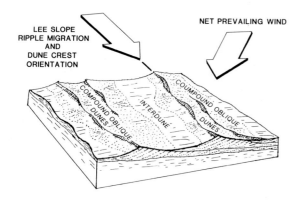

LEE SLOPE
RIPPLE MIGRATION
AND
DUNE CREST
ORIENTATION

NET PREVAILING WIND

Fig. 9-7. Model for the eolian Norphlet sandstones at South State Line Field, showing compound oblique dunes climbing and laterally migrating over interdune deposits. ◁

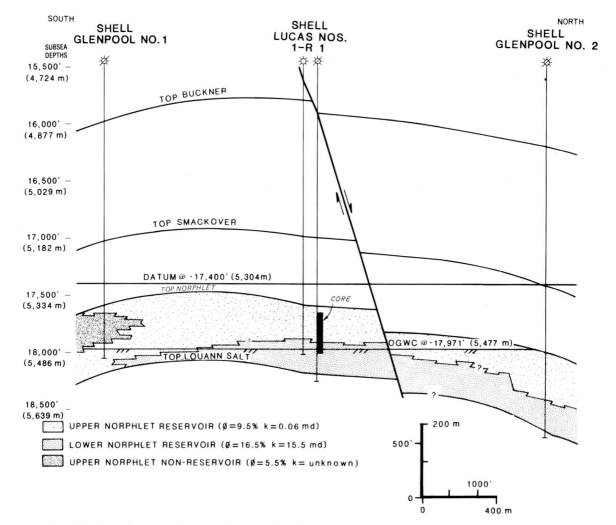

Fig. 9-8. North-south structural cross section showing distribution of reservoir types in the Norphlet sandstone. See Figure 9-3 for well locations.

Table 9-1. Petrographic point-count data. Shell Lucas No. 1, South State Line Field.

Depth (ft)	Framework components (%)								Authigenic components (%)								
	Feldspar			Rock fragments					Cement				Clay		Porosity		
	Quartz	Plagio	K-spar	SRF[1]	VRF[2]	MRF[3]	PRF[4]	Misc.[5]	Quartz	K-spar	Dol.[6]	Anhyd.	Illite	Chlorite	Primary	Secondary	
17,975.7	Not Point Counted																
18,010.5	49.3	6.8	9.1	0.7	2.6	5.9	2.3	2.0	9.3	3.3	0.7	0.7	2.0	—	5.0	2.0	
18,080.3	Not Point Counted																Upper SS
18,102.5	44.8	4.9	15.5	0.3	1.9	5.5	2.1	2.8	6.9	4.4	5.0	0.9	2.5	—	1.9	1.6	
18,117.1	50.1	6.0	13.6	0.3	1.9	6.3	2.5	3.5	2.6	3.2	—	—	1.3	—	6.9	3.8	
18,149.2	49.1	4.7	12.7	0.6	3.2	6.3	0.3	1.6	10.0	2.3	0.3	1.0	2.6	—	1.0	2.3	
18,169.7	45.3	3.9	11.2	1.6	2.5	7.3	0.3	4.9	9.3	1.0	—	0.3	2.3	—	3.7	4.3	
18,205.0	43.9	3.5	14.6	—	4.8	3.5	1.0	1.4	2.9	6.2	—	0.7	—	1.3	12.1	4.6	Lower SS
18,220.3	36.7	5.5	15.0	0.4	4.9	7.5	2.0	1.7	4.6	6.6	—	—	0.3	1.0	12.8	2.3	
18,239.5	45.3	8.4	12.4	—	3.1	3.4	1.2	—	6.0	3.2	—	—	—	1.6	12.7	3.5	
18,246.3	50.5	5.0	11.0	1.0	3.3	3.3	1.5	—	3.3	4.9	—	—	—	3.0	7.4	3.9	
18,271.0	43.7	7.1	15.9	1.0	3.2	3.2	—	1.6	2.3	7.0	—	1.3	—	0.7	9.3	3.0	

[1]Sedimentary rock fragment
[2]Volcanic rock fragment
[3]Metamorphic rock fragment
[4]Plutonic rock fragment
[5]Includes pyrite, bitumen, heavy minerals, etc.
[6]Ferroan dolomite

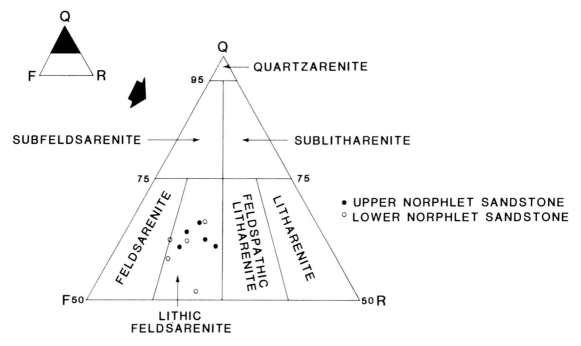

Fig. 9-9. Q-F-R compositional diagram (classification after Folk et al., 1970) for Norphlet sandstone samples studied. 300 point counts recorded per thin section.

9-1), as coatings on detrital grains. The coatings consist of 5–10μ wide pseudohexagonal plates, aligned perpendicular to grain surfaces (Fig. 9-11). Significantly, the chlorite coatings are the thickness of one crystal only and do not totally occlude pore space.

The presence of illite in upper Norphlet sandstones has been accompanied by significant compaction-related porosity loss as a result of pressure solution. Sutured contacts and tight packing of grains are common (Fig. 9-12A). Thin, closely spaced stylolite seams occur in zones rich in pyrite (Fig. 9-12B).

Chlorite, on the other hand, has the reverse effect. Lower Norphlet sandstones with chlorite coatings on grains display less evidence of pressure solution: grain contacts are not sutured and packing of grains is looser than in upper Norphlet sandstones (Figs. 9-13A, 9-13B).

Illite appears to promote pressure solution, thereby reducing porosity (by chemical or mechanical means, or both). Material lost in the dissolution process may be reprecipitated as cement and, in this regard, upper Norphlet sandstones have about twice as much authi-

Table 9-2. Semiquantitative X-ray diffraction data. Shell Lucas No. 1, South State Line Field.

Depth (ft)	Environment	Crystalline components (%)					Clay (<2 μ) (%)		
		Quartz	Feldspar	Clay	Dolomite	Hematite	Illite	Chlorite	
17,975.7	Dune	50	40	10			93	7	
18,010.5	Dune	64	26	10			95	5	
18,080.3	Dune	50	39	11			94	6	
18,102.5	Interdune	43	41	6	3	7	93	7	
18,117.1	Dune	69	26	5			93	7	
18,149.2	Dune	65	24	12			100	0	
18,169.7	Dune	44	37	19			99	1	Upper SS
18,205.0	Dune	71	22	7			84	16	Lower SS
18,220.3	Dune	52	32	16			77	23	
18,239.5	Dune	71	23	6			72	28	
18,246.3	Dune	62	28	10			73	27	
18,271.0	Interdune	56	37	7			76	24	

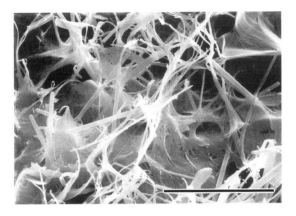

Fig. 9-10. Scanning electron microscope photo of fibrous illite in upper Norphlet sandstone. Shell Lucas No. 1, 18,010 feet (5,489 m). $\phi = 9.6\%$; k = 0.61 md. Scale bar is 0.01 mm.

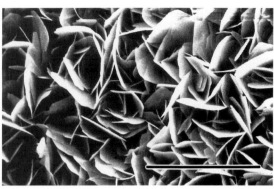

Fig. 9-11. Scanning electron microscope photo of chlorite grain coatings in lower Norphlet sandstone. Shell Lucas No. 1, 18,426 feet (5,616 m). $\phi = 19.9\%$; k = 78.0 md. Scale bar is 0.01 mm.

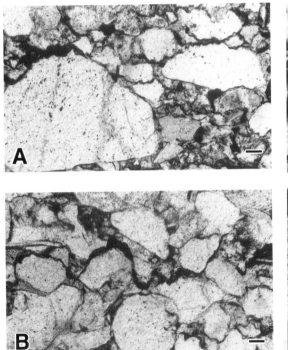

Fig. 9-12. Photomicrographs of upper Norphlet sandstone, Shell Lucas No. 1. Scale bar is 0.05 mm. (A) Pressure solution and tight packing of framework grains. 18,102 feet (5,517 m). $\phi = 17.1\%$; k = 0.19 md. (B) Detail of stylolite. 18,080 feet (5,511 m). $\phi = 11.1\%$; k = 0.65 md.

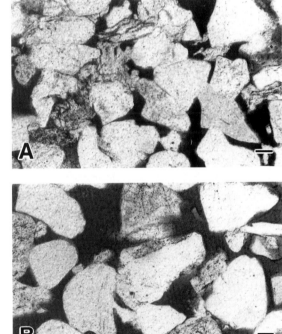

Fig. 9-13. Photomicrographs of lower Norphlet sandstone, Shell Lucas No. 1. Scale bar is 0.05 mm. (A) Typical view showing well-developed porosity. 18,220 feet (5,553 m). $\phi = 18.5\%$; k = 36.0 md. (B) Same as above. 18,246 feet (5,561 m). $\phi = 19.9\%$; k = 78.0 md.

Table 9-3. Intergranular volume data. Shell Lucas No. 1, South State Line Field.

Depth (ft)	Cement (%)	Primary porosity (%)	IV[1] (%)	
18,010.5	16.0	5.0	21.0	
18,102.5	19.7	1.9	21.6	
18,117.1	7.1	6.7	13.8	
18,149.2	16.2	1.0	17.2	
18,169.7	12.9	3.7	16.6	Upper SS
18,205.0	11.1	12.1	23.2	Lower SS
18,220.3	12.5	12.8	25.3	
18,239.5	10.8	12.7	23.5	
18,246.3	11.2	7.4	18.6	
18,270.0	11.3	9.3	20.6	

[1]Intergranular volume

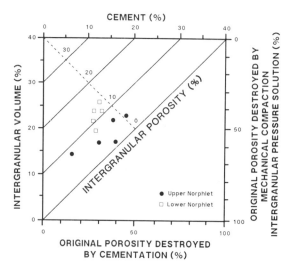

Fig. 9-14. Plot of intergranular volume vs. cement for Norphlet sandstones studied, indicating that compaction has been more important than cementation in reducing porosity. Lower Norphlet sandstones have lost less porosity by compaction than have upper Norphlet sandstones.

genic quartz as do lower Norphlet sandstones (Table 9-1). Conversely, chlorite appears to reduce or even halt chemical compaction. Furthermore, well-developed coatings of chlorite on detrital grains hinder the development of secondary overgrowths by effectively removing these grains as nucleation sites (Pittman and Lumsden, 1968; Thomson, 1979).

The possibility that a soluble cement was dissolved late in the burial history of the lower Norphlet must be considered. Halite, calcite, and anhydrite are likely candidates. Several lines of reasoning suggest that this process did not occur. First, halite dissolution would have to have been absolutely complete, since no trace of halite remains. Admittedly, halite is a prime suspect, inasmuch as the lower Norphlet immediately overlies the Louann Salt. In this regard, the authors have observed high-porosity sandstones containing chlorite in settings with no readily available evaporite source. These include the mid-dip and downdip Cretaceous Tuscaloosa Formation of Louisiana, the Eocene Wilcox Formation of Gulf coastal Texas and Louisiana, and various Miocene sandstones of offshore Texas and Louisiana. One of these, the downdip Tuscaloosa, has porosities of more than 25% at depths of more than 20,000 feet (6,100 m) in the presence of pristine calcareous fossils (Thomson, 1979). Second, the small amounts of carbonate and anhydrite present occur as euhedral crystals showing no evidence of optical continuity. Finally, pores, although large, are not "oversized" and except for occasional partially dissolved lithic grains, none of the standard criteria (Schmidt and McDonald, 1979) for recognition of dissolution (elongate pores, moldic pores, corroded grains) are present.

A quantitative expression of porosity lost by compaction is given by a measurement of "intergranular volume" (Houseknecht, 1987)—the "minus-cement porosity" of Heald (1956). The data show higher values for this parameter in the lower chlorite-coated Norphlet sandstones (Table 9-3). A plot of intergranular volume versus percent cement (from Table 9-3), following the scheme of Houseknecht (1987), shows that compaction has been more important than cementation in reducing porosity in all but one sample (Fig. 9-14). Furthermore, lower Norphlet sandstones containing chlorite appear to have lost less porosity by compaction than have upper Norphlet sandstones containing illite.

Cements in the Norphlet are predominantly quartz and potassium feldspar, both of which occur as clear overgrowths on their respective detrital counterparts. The distribution and amount of each are directly related to the occurrence of diagenetic illite and chlorite. The upper Norphlet reservoir contains on the average 8% quartz and 3% potassium feldspar cement, whereas the lower Norphlet contains an average of 4% quartz and 6% potassium feldspar cement (Table 9-1). Other local cements include carbonate (up to 5%) and anhydrite (up to 1%).

Petrophysical Properties

Porosities measured from core plugs in lower Norphlet samples studied ranged from 13% to 20% and averaged 18% (Fig. 9-6). Similar measurements in upper Norphlet samples ranged from 6% to 15% and averaged 10%. Thin-section measured porosities in lower Norphlet samples studied ranged from 11% to 17% and averaged 14%. Similar measurements in upper Norphlet samples ranged from 3% to 11% and averaged 7%. The reason for the discrepancy between core plug measurements and thin-section measurements may be attributed to the authigenic illite and chlorite, both of which contain considerable "microporosity."

Porosity types may be classified as "macro" (relict primary and secondary) and "micro." Relict primary porosity is the most common type in thin section. Secondary porosity results from dissolution of feldspar and certain labile rock fragments and is of least importance in the best reservoir rocks (Table 9-1). The absolute contribution of microporosity to total porosity is unknown but must be related to water saturations.

Measured air permeabilities (unstressed) in lower Norphlet samples studied ranged from 9 md to 78 md and averaged 47 md. Similar measurements in upper Norphlet samples ranged from 0.2 md to 4 md and averaged 1 md.

Routine core analysis of all samples, in conjunction with petrophysical interpretation and wireline-log correlation, allows subdivision of the Norphlet Formation at South State Line Field into two reservoir types (Figs. 9-6, 9-8). The lower Norphlet reservoir averages 16.5% porosity and 15.5 md permeability; the upper Norphlet reservoir averages 9.5% porosity and 0.6 md permeability. In addition, a third rock type encountered in one well (Shell Glenpool No. 1) has porosities which average 5.5% and unknown but presumably very low permeabilities. This rock type is referred to as "upper Norphlet nonreservoir" (Fig. 9-8).

Capillary-pressure data are generally not useful. These were systematically biased in the upper Norphlet reservoir containing illite. This was due to the "matting" of illite crystals as the core dried (Cocker, 1984), resulting in an underestimation of microporosity. Because water is retained in this extremely fine pore network, the capillary-pressure estimates of water saturation were too low when compared to in situ values. Capillary-pressure data in the lower Norphlet reservoir containing chlorite were closer to what was expected; however, estimates of water saturation were also unsuccessful. This is because the data fell in the transitional portion of the capillary-pressure curve, where estimates of water saturation are inherently less accurate.

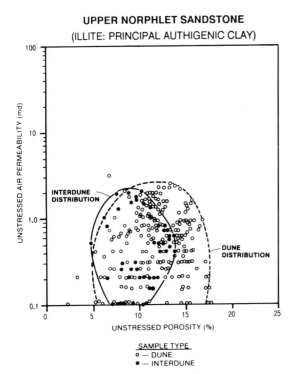

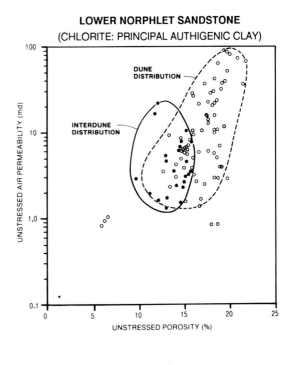

Fig. 9-15. Porosity-permeability relationships for upper and lower Norphlet sandstones, showing the control of depositional facies on these parameters, particularly for the lower Norphlet sandstone.

Table 9-4. Students t-test data. Shell Lucas No. 1, South State Line Field.

	v (degrees of freedom)	t	Critical values	Reject null hypothesis?
Upper Norphlet permeability	176	1.71	1.96 & −1.96	No
Upper Norphlet porosity	176	−3.45	1.96 & −1.96	Yes
Lower Norphlet permeability	89	3.34	1.99 & −1.99	Yes
Lower Norphlet porosity	89	9.77	1.99 & −1.99	Yes

The control by depositional environment on porosity and permeability can be seen in Figure 9-15. This is particularly evident in data from the lower Norphlet reservoir. Although data overlap exists, the dune facies samples tend to plot with higher porosities and permeabilities than do interdune samples.

Student's t-tests of significance were performed on the data in Figure 9-15 (Table 9-4). The t-tests support the interpretation that porosity and permeability are significantly different for dune versus interdune facies in the Lower Norphlet. In addition, porosity is significantly different between the two facies in the Upper Norphlet. The "null hypothesis" states that the means of the two samples are equal, the inference being that both samples came from the same population. If the null hypothesis is rejected ("YES"), then the alternate hypothesis must be accepted, which is that the two samples ARE derived from separate populations. The critical values correspond to a significance level of 95%.

Controls on Reservoir Quality

The major control on reservoir quality in the Norphlet Formation in South State Line Field was the introduction of authigenic chlorite into the lower reservoir and illite into the upper reservoir. The distribution of these two reservoir types is directly related to the distribution of these authigenic clay minerals. Neither the present structural configuration nor the original lithofacies architecture appears to have controlled their distribution (see following section). Although reservoir properties of dune sandstones are superior to those of interdune sandstones, it is significant that interdune reservoirs of the lower, chlorite-bearing Norphlet are of better quality than are dune reservoirs of the upper Norphlet. Prediction of similar Norphlet reservoirs here and in less intensely explored areas to the east therefore depends to some degree on understanding the distribution and origin of illite and chlorite.

From a regional study of the Norphlet Formation, Ryan (1986) concluded that sediments in the east (Florida and Alabama) were derived from provinces on the east side of the southern Appalachians, the Piedmont, Talladega Slate Belt, Valley and Ridge provinces, and Triassic-Jurassic volcanics. These rocks are mainly acid plutonic, gneissic, silicic and basic volcanics, and metasediments (Ryan, 1986). On the other hand, Norphlet sediments in the west (Mississippi, Arkansas, Louisiana, and Texas) were derived from the craton interior.

Our initial interpretation, based on the foregoing evidence, postulated an Appalachian source for the lower Norphlet and a craton interior source for the upper Norphlet, with the former introducing magnesium to form chlorite and the latter potassium to form illite. A Q_m-F-L_t (monocrystalline quartz-feldspar-total rock fragments) diagram (Dickinson, 1985), however, indicated a transitional-continental source for all samples and no evidence of a craton interior source (Fig. 9-16). Petrographic data showed a higher content of volcanic rock fragments in the lower Norphlet than in the upper Norphlet (3.9% vs. 2.4%) and a higher content of metamorphic rock fragments in the upper Norphlet than in the lower Norphlet (6.3% vs. 4.2%). Figure 9-17 is a vertical variation diagram of plutonic, volcanic,

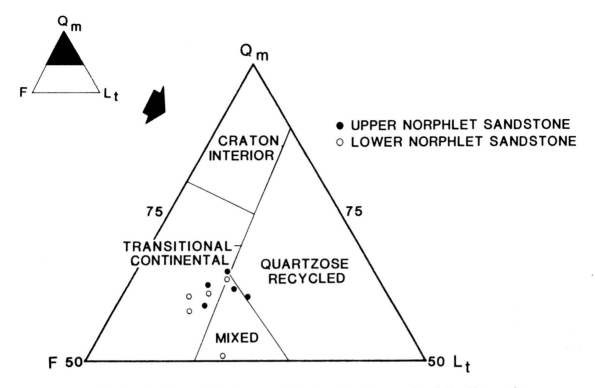

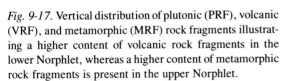

Fig. 9-16. Q_m-F-L_t diagram indicating a transitional-continental source for all Norphlet samples.

and metamorphic rock fragments illustrating this variation. These variations could indicate a subtle shift in contribution of compositional terranes on the east side of the southern Appalachians, with a volcanic terrane initially contributing the magnesium necessary for the formation of chlorite.

An alternate source of the diagenetic constituents necessary for the precipitation of authigenic chlorite is the Louann Salt, which immediately underlies the lower Norphlet sandstones. Salvador (1987) has suggested that the nonuniform thickness of Louann Salt deposits (perhaps as much as 13,000 feet (3,960 m) in some areas and absent in others) indicates differential subsidence. Different subbasins within the Louann may have exhibited differences in brine chemistry as a result of variations in evaporation

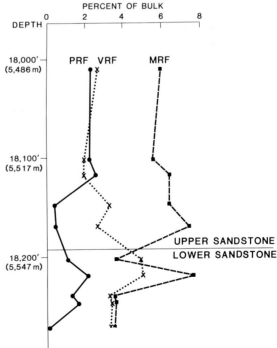

Fig. 9-17. Vertical distribution of plutonic (PRF), volcanic (VRF), and metamorphic (MRF) rock fragments illustrating a higher content of volcanic rock fragments in the lower Norphlet, whereas a higher content of metamorphic rock fragments is present in the upper Norphlet.

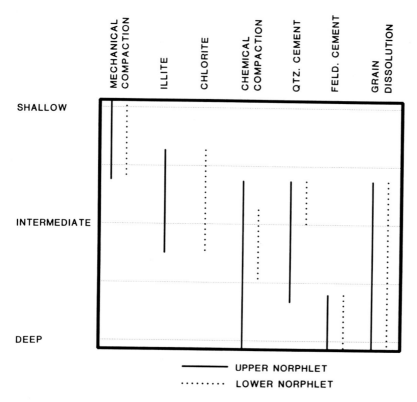

Fig. 9-18. Suggested relative timing of major diagenetic events in the Norphlet of South State Line Field.

rates (Carpenter, 1978). The Norphlet itself may not have contained sufficient magnesium-bearing detrital grains to account for the amount of authigenic chlorite present (McHugh, 1987). Furthermore, few shales are available to supply magnesium through clay diagenesis.

The suggested relative timing of major diagenetic events is shown in Figure 9-18. Iron for authigenic chlorite could have been originally present as detrital hematite grain coatings. Hematite coatings are no longer present in the Norphlet Formation of South State Line Field. The absence of hematite coatings in the lower Norphlet sandstones may be the result of early reaction with magnesium-bearing solutions derived from the underlying Louann Salt. The product of this reaction would have been chlorite. In the upper Norphlet sandstones, the present-day iron mineral is pyrite, suggesting that hydrogen sulfide (H_2S) in sour gas reacted with the hematite grain coats (McBride, 1981). The source for the H_2S may have been the basal Smackover Formation (Moore and Druckman, 1981). The underlying Louann Salt is the most likely source for the illite in the upper Norphlet and the potassium feldspar overgrowths in the lower Norphlet.

Heterogeneities

Dune sandstones comprise the highest quality reservoirs within the eolian system. Conversely, intimately related interdune deposits, especially those associated with wet depositional conditions, typically contain finer grained and more poorly sorted sediment and thus are poorer reservoirs. Interdune deposits often form permeability barriers to vertical flow (partial or complete) and, when cyclically stacked between multiple dune packages, can compartmentalize a reservoir (Lindquist, 1983). This dune-interdune depositional heterogeneity is often preserved despite complex structural and diagenetic overprints. At South State Line Field, dune and interdune deposits in the lower Norphlet can be crudely discriminated by porosity and permeability values despite a major diagenetic overprint (Fig. 9-15).

The principal internal heterogeneity in the Norphlet Formation at South State Line Field is the contact between the chlorite-bearing lower Norphlet reservoir and the illite-bearing upper Norphlet reservoir (Figs. 9-6, 9-8). The hydrocarbon accumulation may be classified as a "combination structural-diagenetic trap." No shale breaks were observed in cores. Solid

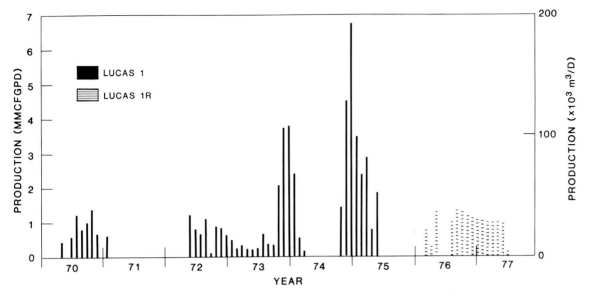

Fig. 9-19. Plot of gas production from the two Norphlet wells in South State Line Field. Daily rates are averaged on a monthly basis. Intermittent and fluctuating production of the Lucas No. 1 was due to mechanical problems and reworking of the well. The Lucas No 1-R was drilled after the No. 1 was incapable of producing due to wellbore problems.

zones of authigenic pyrite, of undetermined lateral extent, were seen in certain upper Norphlet sandstones (Fig. 9-6).

Production Characteristics

Well and Reservoir Performance

South State Line Field was discovered in 1970 when the Shell Lucas No. 1 was tested in the normally pressured Norphlet Formation at an initial rate of 11.4 MMCFGPD (3.2×10^5 m³/D). A total of six wells have been drilled including one sidetrack and one replacement well. The wells produce by depletion drive. One of three original 640-acre (259.2-ha.) gas units, the Shell Lucas No. 2, is still operative and producing 4.5 MMCFGPD (1.3×10^5 m³/D) and 90 BCPD (14 m³/D) from the Smackover. Cumulative production from the lower Norphlet is 2.7 BCFG (7.6×10^7 m³) and 157 MBC (2.5×10^4 m³). The production history of the wells is shown in Figure 9-19.

Special Problems

All wells are normally pressured down to the Buckner Formation, and wells can be drilled to the Buckner with few significant problems. Potential problems,

however, include hole-angle control, rough drilling, hole sloughing, and saltwater flows. The Buckner Formation is an abnormally pressured section containing low concentrations of sour gas. Mud weights required to control the pressures ran as high as 19.0 ppg (0.99 psi/ft; 22.4 kPa/m). Normal pressures are encountered in the underlying Smackover and Norphlet formations.

Mechanical problems, caused by severe corrosion of tubulars, were a major concern. In that regard, the H_2S content in the produced Norphlet gas from the Lucas No. 1 well rose from 2.7% to 19.0% in four years because of probable communication behind pipe with the overlying Smackover. Improved wellbore tubulars used in a sidetrack well eliminated this communication, reduced the H_2S content to a few percent, and increased the well life by two years.

Delineation of the tight, upper Norphlet and the permeable, lower Norphlet is crucial in determining the net feet of pay. In the existing wells, it could be reliably determined only from the available core and test data. The high versus low reservoir quality rocks were indistinguishable on the basis of the log data alone as neither had any unique log responses. More traditional methods of establishing a porosity-permeability trend and applying a porosity cutoff were unsuccessful since high-porosity rock is contained in both intervals.

Exploration and Production Strategy

A regional exploration strategy should concentrate on the "eastern productive trend" where reservoir quality is better than those in the west. Traps require structural closure, so the best areas are those where salt-cored structures are known to occur. These features should appear prominently on conventional seismic lines.

Drilling to and producing from these deep Norphlet reservoirs will require care to overcome unexpected problems with overpressured zones or gas with contaminants, such as H_2S, which will require special tubulars as well as gas treatment after producing. Communication with the overlying Smackover should be avoided in order to eliminate problems with high H_2S concentrations.

Norphlet eolian reservoirs like those at South State Line Field may be expected to be compartmentalized into zones of high and low permeability due both to vertical facies changes and to the presence of diagenetic clay minerals.

Conclusions

South State Line Field's unique location provided an opportunity to study regionally important diagenetic-clay mineralization in rocks of a single environmental facies. The primary control on Norphlet sandstone reservoir quality is the distribution of authigenic chlorite and illite. The lower Norphlet reservoir contains chlorite, which serves to preserve porosity by inhibiting cementation and pressure solution; the upper Norphlet reservoir contains illite, which tends to destroy porosity by promoting pressure solution. As a consequence, porosity and permeability values are higher in the lower Norphlet reservoir. The fact that both reservoirs are composed of eolian dune and interdune facies, the former having better reservoir properties than the latter, is believed to have had less effect on reservoir quality than did the influence of the clay minerals.

Petrographic data expressed as a Q_m-F-L_t diagram indicate the source of Norphlet sediment was the southern Appalachians, with little or no contribution from the craton interior. Vertical variations in detrital composition may be attributed to shifting contributions from different terranes within a single overall source. The magnesium and potassium necessary for the growth of diagenetic chlorite and illite may have been derived from detritus originating in these terranes or may have been introduced into the Norphlet via brines moving upward out of the underlying Louann Salt, or a combination of both.

The Norphlet Formation in the eastern Gulf Coast region contains thick accumulations of sands, principally of eolian origin. These sands are texturally mature and were very porous and permeable when deposited. Subsequent modification of these petrophysical properties was due to diagenetic processes, principal among which were the growth of authigenic chlorite and illite.

Acknowledgments. In preparing this chapter, data were drawn from the files of Shell Western E&P Inc., operator of South State Line Field. The authors are grateful to Shell for releasing this material for publication. Thanks are also extended to K.D. Karapasha, who helped with SEM and EDX analysis. The authors profited from critical reviews by E.D. Pittman, J.G. McPherson, and an anonymous reviewer. Many of their suggestions were incorporated in a revised version of the manuscript. Portions of this study were presented in poster form at the Annual AAPG and SEG Conventions of 1987.

Reservoir Summary

Field: South State Line
Location: Greene County, Mississippi
Operator: Shell Oil Company
Discovery: 1970
Basin: Mississippi Interior Salt basin; Gulf of Mexico
Tectonic/Regional Paleosetting: Rifted continental margin
Geologic Structure: Salt-cored faulted dome
Trap Type: Fault-bounded structural closure

Reservoir Drive Mechanism: Depletion
   • **Original Reservoir Pressure:** 8,600 psi (5.9 × 10⁴ kPa) at 17,600 feet (5,360 m) subsea
Reservoir Rocks
   • **Age:** Late Jurassic (Oxfordian)
   • **Stratigraphic Unit:** Norphlet Formation
   • **Lithology:** Fine-grained sandstone, lithic feldsarenite
   • **Depositional Environment:** Eolian
   • **Productive Facies:** Dune and interdune sandstones
   • **Petrophysics**
      • **Porosity Type:** Upper—total 6.5%, primary 3.7%, secondary 2.8%; Lower—total 14.4%, primary 10.9%, secondary 2.8% (thin-section data)
      • **ϕ:** Average—upper 9.5%, lower 16.5%; range—upper 1 to 19%, lower 6 to 21%; cutoff—upper 10% estimated, lower 8% (unstressed cores)
      • **k:** Average—upper 0.6 md, lower 15.5 md; range—upper 0.1 to 3 md, lower 1 to 84 md (unstressed cores)
      • **S_w:** Average—upper 40.4%, lower 46.9%; range—upper 25 to 90%, lower 39 to 54% (logs)
Reservoir Geometry
   • **Depth:** 17,900 feet (5,460 m)
   • **Areal Dimensions:** 1.9 by 1.3 miles (3.1 × 2.1 km)
   • **Productive Area:** 750 acres (300 ha.)
   • **Number of Reservoirs:** 1
   • **Hydrocarbon Column Height:** 366 ft (112 m)
   • **Fluid Contact:** Gas-water contact at 17,971 ft (5,478 m) subsea
   • **Number of Pay Zones:** 2 (upper, lower)
   • **Gross Sandstone Thickness:** 595 ft (181 m)
   • **Net Sandstone Thickness:** 595 ft (181 m)
   • **Net/Gross:** 1.0
Hydrocarbon Source, Migration
   • **Lithology and Stratigraphic Unit:** Basinal shaly limestone, Smackover Formation, Late Jurassic (Oxfordian)
   • **Time of Hydrocarbon Maturation:** Estimated at Mid-Cretaceous
   • **Time of Trap Formation:** Initiated during Late Jurassic (Oxfordian)
   • **Time of Migration:** Estimated at Mid-Cretaceous
Hydrocarbons
   • **Types:** Gas and condensate
   • **GOR:** 70 BC/MMCFG (0.4 m³/thousand m³)
Volumetrics
   • **In Place:** NA
   • **Cumulative Production:** 2.7 BCFG (7.7 × 10⁷ m³)
   • **Ultimate Recovery:** NA
   • **Recovery Efficiency:** NA
Wells
   • **Spacing:** 3,900 ft (1,190 m); 640 acres (259.2 ha.)
   • **Total:** 4
   • **Dry Holes:** 2
Typical Well Production
   • **Average Daily:** 5 MMCFG (1.4 × 10⁵ m³)
   • **Cumulative:** 1.4 BCFG (4.0 × 10⁷ m³)

References

Carpenter, A.B., 1978, Origin and chemical evolution of brines in sedimentary basins: Oklahoma Geological Survey Circular 79, p. 60–77.

Cocker, J.D., 1984, Critical point drying of illite-bearing sandstones: Morphology and permeability effects [abst.]: Clay Minerals Society Annual Meeting Program, Baton Rouge, Louisiana, p. 38.

Dickinson, W.R., 1985, Interpreting provenance relationships from detrital modes of sandstone, in Zuffa, G.G., ed., Provenance of Arenites: Boston, Reidel Publishing Company, p. 333–361.

Folk, R.L., Andrews, P.B., and Lewis, D.W., 1970, Detrital sedimentary rock classification and nomenclature for use in New Zealand: New Zealand Journal of Geology and Geophysics, v. 13, p. 937–968.

Fryberger, S.G., Al-Sari, A.M., and Clisham, T.J., 1983, Eolian dune, interdune, sand sheet, and siliciclastic sabkha sediments of an offshore prograding sand sea, Dhahran Area, Saudi Arabia: American Association of Petroleum Geologists Bulletin, v. 67, p. 280–312.

Glennie, K.W., 1970, Desert sedimentary environments, in Developments in Sedimentology 14: Amsterdam, Elsevier Scientific Publishing Company, 222 p.

Heald, M.T., 1956, Cementation of Simpson and St. Peter sandstones in parts of Oklahoma, Arkansas, and Missouri: Journal of Geology, v. 64, p. 754–761.

Honda, H., 1981, Diagenesis and reservoir quality of the Norphlet Sandstone (Upper Jurassic), Hatter's Pond Area, Mobile County, Alabama [M.S. thesis]: Austin, Texas, University of Texas, 213 p.

Houseknecht, D.W., 1987, Assessing the relative importance of compaction processes and cementation to reduction of porosity in sandstones: American Association of Petroleum Geologists Bulletin, v. 71, p. 633–642.

Hunter, R.E., 1981, Stratification styles in eolian sandstones: Some Pennsylvanian to Jurassic examples from the Western Interior, U.S.A., in Ethridge, F.G. and Flores, R.M., eds., Recent and ancient nonmarine depositional environments: Models for exploration: Society of Economic Paleontologists and Mineralogists Special Publication 31, p. 315–329.

Kocurek, G., 1981, Erg reconstruction: The Entrada Sandstone (Jurassic) of northern Utah and Colorado: Paleogeography, Paleoclimatology, Paleoecology, v. 36, p. 125–153.

Kocurek, G., 1986, Origins of low-angle stratification in aeolian deposits, in Nickling, W.G., ed., Aeolian Geomorphology, 17th Annual Binghamton Geomorphology Symposium: New York, Allen and Unwin Publishers, p. 177–193.

Kocurek, G., and Dott, R.H., Jr., 1981, Distinction and uses of stratification types in the interpretation of eolian sand: Journal of Sedimentary Petrology, v. 51, p. 579–595.

Lindquist, S.J., 1983, Nugget Formation reservoir characteristics affecting production in the Overthrust Belt of southwestern Wyoming: Journal of Petroleum Technology, v. 35, p. 1355–1365.

Mancini, E.A., and Benson, D.J., 1980, Regional stratigraphy of Upper Jurassic Smackover carbonates of southwest Alabama: Gulf Coast Association of Geological

Societies Transactions, v. 30, p. 151–165.

Mancini, E.A, Mink, R.M., Bearden, B.L., and Wilkerson, R.P., 1985, Norphlet Formation (Upper Jurassic) of southwestern and offshore Alabama: Environments of deposition and petroleum geology: American Association of Petroleum Geologists Bulletin, v. 69, p. 881–898.

McBride, E.F., 1981, Diagenetic history of the Norphlet Formation (Upper Jurassic), Rankin County, Mississippi: Gulf Coast Association of Geological Societies Transactions, v. 31, p. 347–351.

McBride, E.F., Land, L.S., and Mack, L.E., 1987, Diagenesis of eolian and fluvial feldspathic sandstones, Norphlet Formation (Upper Jurassic), Rankin County, Mississippi, and Mobile County, Alabama: American Association of Petroleum Geologists Bulletin, v. 71, p. 1019–1034.

McHugh, A., 1987, Styles of diagenesis in Norphlet Sandstone (Upper Jurassic), onshore and offshore Alabama [M.S. thesis]: New Orleans, Louisiana, University of New Orleans, 164 p.

Moore, C.H., and Druckman, Y., 1981, Burial diagenesis and porosity evolution, Upper Jurassic Smackover, Arkansas and Louisiana: American Association of Petroleum Geologists Bulletin, v. 65, p. 597–628.

Oehler, J.H., 1980, Carbonate source rocks in the Jurassic Smackover trend of Mississippi, Alabama, and Florida [abst.]: 93rd Annual Meeting of Geological Society of America, Abstracts with Programs, p. 494.

Pittman, E.D., and Lumsden, D.N., 1968, Relationship between chlorite coatings on quartz grains and porosity, Spiro Sand, Oklahoma: Journal of Sedimentary Petrology, v. 38, p. 668–670.

Ryan, W.P., Jr., 1986, Provenance of the Norphlet sandstone, northern Gulf Coast, Mississippi [M.S. thesis]: New Orleans, Louisiana, University of New Orleans, 136 p.

Salvador, A., 1987, Late Triassic-Jurassic paleogeography and origin of Gulf of Mexico Basin: American Association of Petroleum Geologists Bulletin, v. 71, p. 419–451.

Schmidt, V., and McDonald, D.A., 1979, Texture and recognition of secondary porosity in sandstones, in Scholle, P.A. and Schluger, P.R., eds., Aspects of Diagenesis: Society of Economic Paleontologists and Mineralogists Special Publication 26, p. 209–226.

Thomson, A., 1979, Preservation of porosity in the deep Woodbine/Tuscaloosa trend, Louisiana: Gulf Coast Association of Geological Societies Transactions, v. 29, p. 396–403.

Key Words

South State Line Field, Mississippi, Mississippi Interior Salt basin, Norphlet Formation, Smackover Formation, Late Jurassic, Oxfordian, eolian, chlorite, illite, diagenesis, petrography, sand sheet, erg, pressure solution, minus-cement porosity, Louann Salt, dune deposits, interdune deposits.

Deltaic Environments

10

Wave-Dominated Deltaic Reservoirs of the Brent Group, Northwest Hutton Field, North Sea

Iain C. Scotchman and L. Hywel Johnes

Amoco (UK) Exploration Co., London, England

Introduction

Wave-dominated deltas consist of sheet sands built by the regressive strand plains or barrier islands that flank the main distributary mouths. The meander belts and distributary channels which feed these shorelines comprise dip-oriented sand bodies which are coarser grained than the shoreface sands which they cut. In many basins, the best reservoirs in ancient wave-dominated deltas are these channel sequences, even though they are volumetrically less significant. This chapter describes the differences in reservoir characteristics between these two types of sandstone, using as an example wave-dominated deltas in a Mesozoic rift basin.

The Northwest Hutton Field is one of the most complex fields in the Brent Province of the U.K. North Sea. The field comprises a series of tilted fault blocks that are productive from Middle Jurassic (Bajocian-Bathonian) Brent Group sandstones. This structural complexity is combined with a reservoir that has a highly variable sand-body geometry and a strong depth-related diagenetic overprint on heterogeneous reservoir properties. The combination of these structural, stratigraphic, and diagenetic features has resulted in a field that is compartmentalized both laterally and vertically.

The field was discovered in 1975 by the Amoco-operated group comprising Enterprise Oil, Mobil North Sea, Texas Eastern North Sea, and Amerada Hess with the drilling of the 211/27-3 well which encountered a 396-foot (121-m) reservoir section at a depth of 11,100 feet (3,385 m) subsea. Development of the field began in 1979, and eight appraisal wells were drilled in the period up to 1983. The first seven development wells were drilled in 1979 to 1981 through a 20-slot seabed template prior to installation of a single steel jacket platform in 1981. The field came on stream in 1983, and production peaked at 86,680 BOPD (1.4×10^4 m³/D) when the last of the predrilled wells was put on production. Secondary recovery operations commenced in 1984, initially with water injection on a line-drive pattern and subsequently with gas lift beginning in late 1984. Currently the field has 26 producing wells and 9 injection wells.

Regional Data

Regional Setting and Structure

The Northwest Hutton Field is located within the Brent Province of the northern North Sea in UKCS Block 211/27 about 80 miles (130 km) northeast of the Shetland Islands (Fig. 10-1). The water depth is 475 feet (145 m).

Northwest Hutton is part of a series of major Mid-Late Jurassic, westerly dipping, tilted fault blocks which occupy the East Shetland basin, a broad relatively shallow shoulder on the western flank of the Northern Viking graben. The East Shetland basin is dominated by two major normal fault systems: a northwest-southeast trending system initiated in the

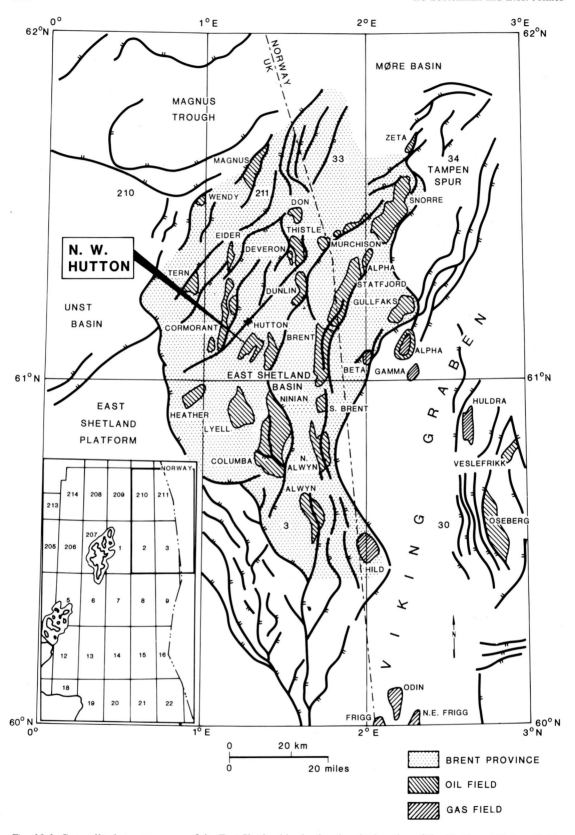

Fig. 10-1. Generalized structure map of the East Shetland basin showing the location of the Northwest Hutton Field.

Permo-Triassic, and an older Caledonide northeast-southwest trending fault system (Challinor and Outlaw, 1981; Eynon, 1981; Threlfall, 1981). North-south faults are also apparent and are related to the earliest phase of rifting in the Permian to Early Triassic (Badley et al., 1988). Rifting of the North Sea in the Permo-Triassic to Early Jurassic led to the formation of the Viking graben and the reactivation of the two main fault systems (northeast-southwest and northwest- southeast) during the Jurassic and Cretaceous. Interaction of the two systems resulted in the formation of complex tilted, fault block structures and movement of the faults acted as a major control on sedimentation during much of the Jurassic and Early Cretaceous. Normal faulting was dominant but an element of wrench movement is evident on the northeast-southwest and northwest-southeast trending faults.

Further tectonic activity occurred during the Late Jurassic to Early Cretaceous, as a second phase of rifting characterized by listric subsidence. Synsedimentary fault movement which began during the Middle Jurassic (Brown et al., 1987) is indicated by stratigraphic expansion across faults and resulted in broad thickness variations in the individual formations. Further movement on these faults took place during the Late Cretaceous, the Paleocene, and the Miocene and was accompanied by regional subsidence which has continued to the present.

Stratigraphy

The oldest rocks penetrated in the Northwest Hutton Field are Triassic red beds of the Cormorant Formation (Deegan and Scull, 1977; Morton et al., 1987) (Fig. 10-2). The sequence comprises continental alluvial fan or braided stream conglomerates and sandstones, interbedded with mudstones of lacustrine origin. The overlying lower Jurassic Statfjord Formation is fluvial and was deposited in a wetter climate, producing thick braided and meandering stream sandstones. Fluvial deposition was terminated by a regional marine transgression which deposited mudstones of the Dunlin Group unconformably on the Statfjord Formation. These mudstones contain minor sandstones and siltstones (Vollset and Doré, 1984) which may represent tidal and storm deposits in a shallow, mud-dominated, inner-shelf environment. Deeper water deposition occurred in the eastern part of the East Shetland basin and Viking graben where the shale thickens to more than 500 feet (152 m).

Regional basinal subsidence coupled with uplift of the platform area to the southwest ended deposition of the marine mudstones. High sediment-supply rates resulted in the formation of a large-scale deltaic system of the Middle Jurassic Brent Group, which prograded northeastwards into relatively deep water from the East Shetland platform (Richards et al., 1988). Deposition was controlled by syndepositional faulting which appears to have influenced the direction of stream flow across the delta plain.

The Brent Group is subdivided into five formations in the Northwest Hutton Field (Deegan and Scull, 1977) (Fig. 10-3). Two depositional systems are recognized: a lower system represented by the basal coarse sandstones of the Broom Formation deposited by an easterly prograding coarse-grained delta complex (Brown et al., 1987), and an upper system composed of a regressive to transgressive clastic wedge comprising the Rannoch to Tarbert formations.

In the present-day structural lows, thick shales of Bathonian or Callovian to Oxfordian age are conformable with the top of the Brent Group sandstones. In other areas, a sharp discordance delineates the Middle Jurassic unconformity. Upper Jurassic sedimentation in the Northwest Hutton Field is represented by basinal shales of the Oxfordian Heather Formation. Pelagic deposition with basinwide subsidence occurred in shallow to moderate water depths (Badley et al., 1988). At this time, considerable tectonic activity in the form of local fault-block displacements produced a series of uplifts and en-echelon depocenters. Some horst blocks may have been emergent, forming islands. To the west, the East Shetland platform was exposed in part, and sand deposition occurred on the downside of the active-margin fault system. Basinward, pelagic deposition with gradual subsidence and an overall rise in sea level continued throughout the Kimmeridgian and Volgian into the Ryazanian, with the deposition of anoxic mudstones across the area (Deegan and Scull, 1977). Local sandstones and thin limestones occur in these otherwise monotonous shale sequences (Doré et al., 1985). Several large-scale, sand-rich, submarine fan systems were deposited in association with areas of uplift and extreme erosion, especially on the northern basin margin (DeAth and Schuyleman, 1981; Brown, 1984).

The end of Jurassic sedimentation was marked by an extensive regional unconformity, the Base Cretaceous Unconformity. This late Cimmerian tectonic event affected much of the older underlying

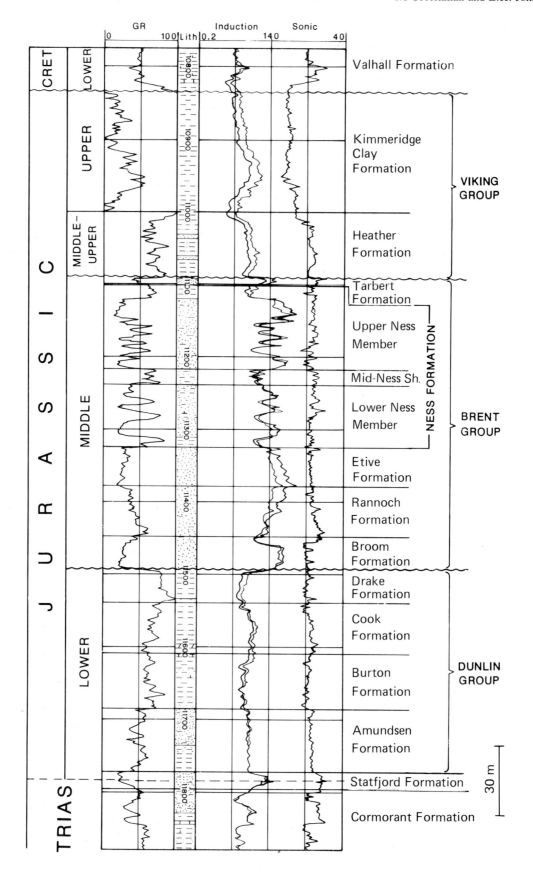

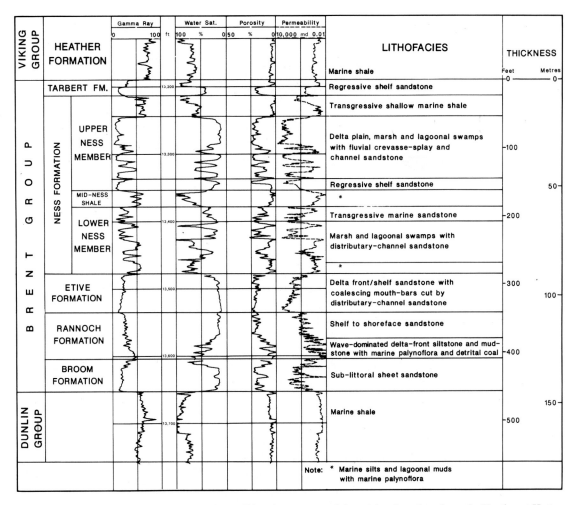

Fig. 10-3. Brent Group sandstone statigraphy, well log signatures, and depositional settings from the Northwest Hutton Field.

Fig. 10-2. Jurassic stratigraphic column from the Northwest Hutton Field, with well log signatures from crestal well 211/27-3.

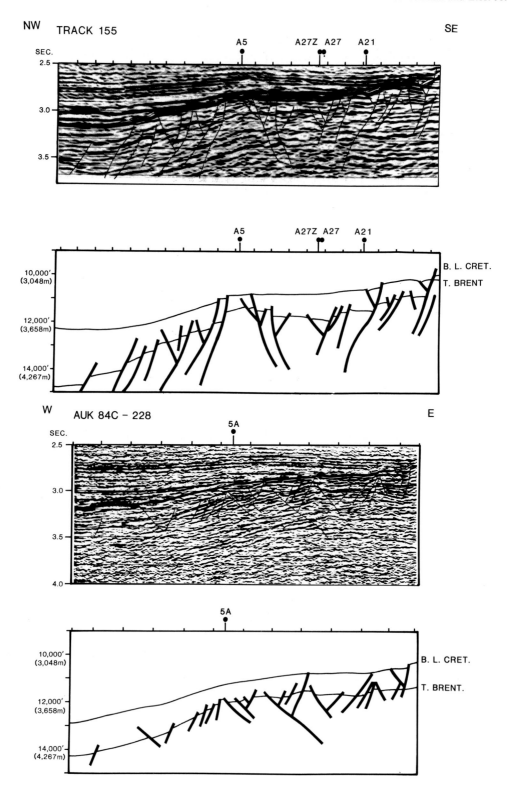

Fig. 10-4. Seismic data panels showing comparison between the 1979 3-D survey (reprocessed in 1984) (top two panels–TRACK 155) shot with air guns and the 1984 3-D survey (bottom two panels–AUK 84C-228) shot with water guns. Note the greatly enhanced resolution of faulting and Brent horizon picking in the 1984 survey data. The section lines are shown in Figure 10-5.

sequences due to accompanying regional uplift and fault-block rotation.

The Lower Cretaceous in the Northwest Hutton Field is marked by further deepwater marine deposition. Shales and limestones are preserved in the more synclinal areas of the present-day structural basin. Early Cretaceous erosion later removed much of this section over the palaeohigh regions. As a result, Upper Cretaceous shale was deposited unconformably on the Jurassic sediments. In the center of the basin, the Lower and Upper Cretaceous shales are conformable.

Tertiary sedimentation was dominated by claystone deposition. An extensive submarine fan developed in the southwestern part of the East Shetland basin in the Paleocene, but generally the Paleocene and Eocene were characterized by shallow to deep marine-shelf silt and clay deposition.

Reservoir Characterization

Seismic Mapping

Seismic data coverage of the field is provided by two overlapping 3-D seismic surveys: a northern survey covering the platform area acquired in 1979 (and reprocessed in 1984), and a southern survey acquired in 1984. Comparison of data quality between the surveys is illustrated in Figure 10-4. In addition, normal incidence vertical seismic profiles (VSPs) have been acquired in most of the recent wells to provide a tie between well and seismic data.

The top of the Brent Group is not a strong or easily identified seismic reflector. Well data are essential for mapping and have provided the basis for defining the Top Brent seismic reflector. Seismic interpretation away from well control requires study of the entire Brent Group sequence. Crestal erosion of the Brent can result in variations of the Top Brent reflector. Numerous faults make identification and correlation of the Top Brent difficult.

The seismic data were used to map the fault patterns and construct structure maps. The Top Brent structure map for the field is shown in Figure 10-5. It has not been possible to undertake seismic facies analyses as data are of insufficient quality to allow resolution of the individual Brent Group formations. The conversion from seismic time to depth involved the generation of an average velocity map that tied the wells to the seismic whose contours were shaped by the stacking velocity data.

Geometry

The Brent Group sandstones in the Northwest Hutton Field are between 320 and 500 feet (98–152 m) thick (Fig. 10-6). Thickness distribution shows a general lobate geometry, with the lobes orientated in a northeast-southwest trend (Fig. 10-7). Net-sand isopach maps for the Broom (Fig. 10-8), Rannoch (Fig. 10-9), Etive (Fig. 10-10), Ness (Figs. 10-11, 10-12), and Tarbert (Fig. 10-13) formations are presented. The Ness Formation is subdivided into upper and lower members in the Northwest Hutton Field by a laterally continuous marine shale, the Mid-Ness Shale. Net-sand distributions in the Broom, Rannoch, and Ness formations display a ridge-and-trough geometry in a generally northeast-southwest trend. The Etive Formation has a similar northeast-southwest pattern of "thicks" and "thins" in the net-sand trend, but a thickening into an east-west orientated body in the southern part of the field is also evident (Fig. 10-10). The Tarbert Formation net-sand isopach map exhibits a different distribution with a northwest-southeasterly trending thick covering the central area of the field, thinning northward due to post-Brent erosion. The net sand increases southwestward to more than 30 feet (9 m) in the vicinity of the 211/27-11 well (Fig. 10-13).

Hydrocarbon Trapping and Source

The Northwest Hutton Field comprises a series of complex-tilted fault blocks (Figs. 10-5, 10-14) resulting from the interaction of the northwest-southeast and the northeast-southwest fault trends. The productive area of the field is bounded by three structural elements, which are (1) a major northeast-southwest trending fault on the eastern side which has downthrown the Brent Group sandstones against Lower Jurassic shales, (2) a major northwest-southeast trending down-to-the-north fault which juxtaposes the Brent Group against Upper Jurassic shales on the northern side, and (3) structural dip to the south and west.

The field is further subdivided into four productive fault blocks by northeast-southwest trending faults parallel to the eastern field boundary fault (Fig. 10-5). The trap is sealed vertically by shale of the Heather and Kimmeridge Clay Formations of Mid to Late Jurassic age, and the whole structure is draped by thick Cretaceous shale which creates a regional seal.

Hydrocarbons were sourced from the organic-rich shale of Upper Jurassic Kimmeridge Clay For-

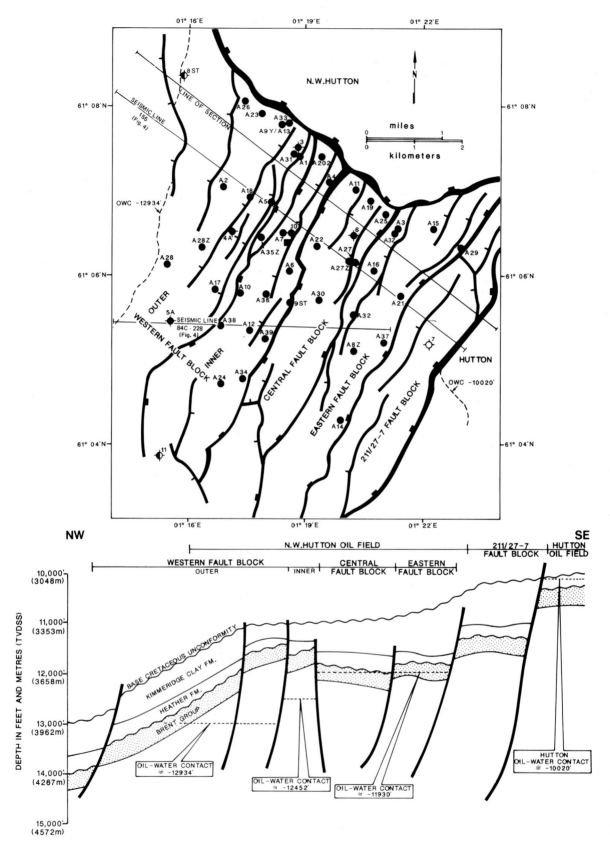

Fig. 10-5. Simplified top Brent Group sandstone structure map of the Northwest Hutton Field and diagrammatic northwest-southeast cross section showing the structural subdivision of the field into the four productive fault blocks.

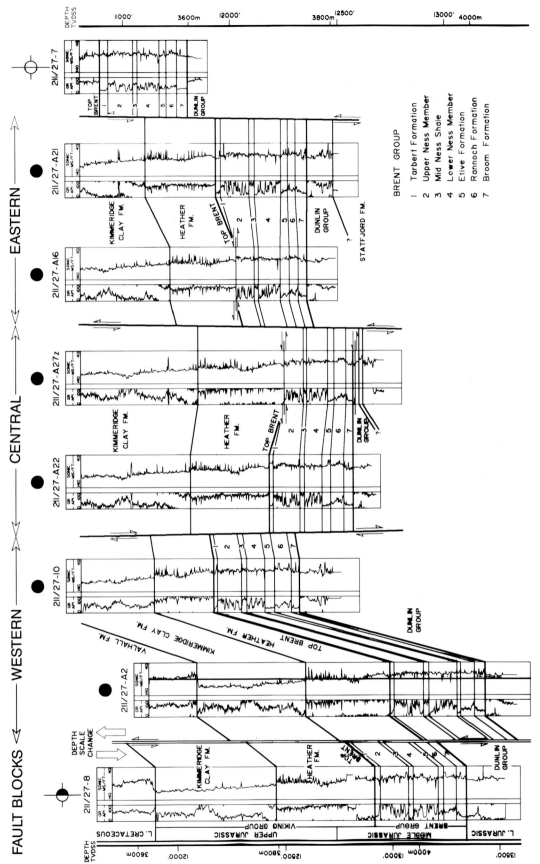

Fig. 10-6. Northwest-southeast structural well correlation across the Northwest Hutton Field. The section is generally parallel to that in Figure 10-5.

235

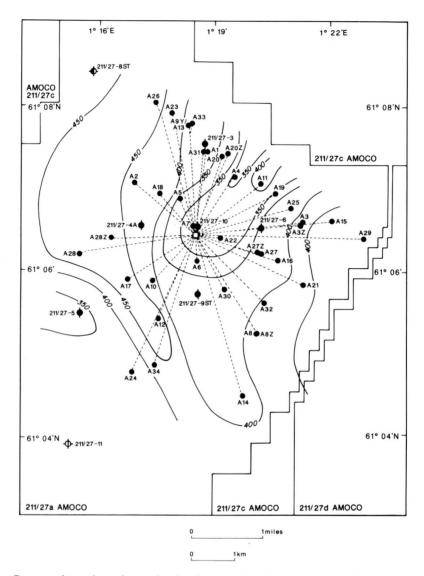

Fig. 10-7. Brent Group sandstone isopach map showing the general northeast-southwest oriented lobate geometry of the sandstone. Contour interval is 50 feet (15.2 m).

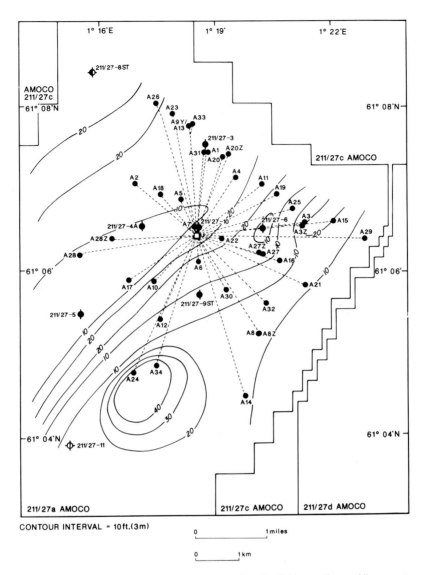

Fig. 10-8. Broom Formation net-sandstone isopach map showing the "ridge and trough" geometry with northeast-southwest orientation.

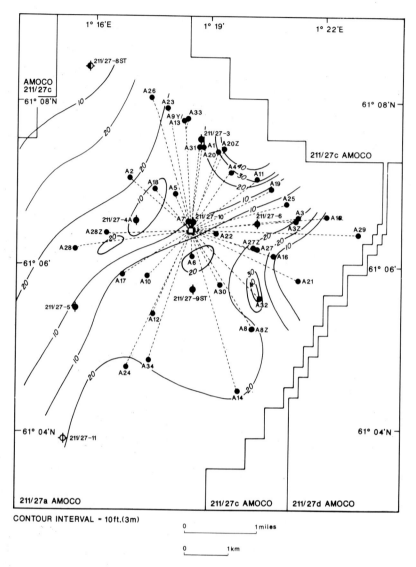

Fig. 10-9. Rannoch Formation net-sandstone isopach map with "ridge and trough" geometry similar to the underlying Broom Formation.

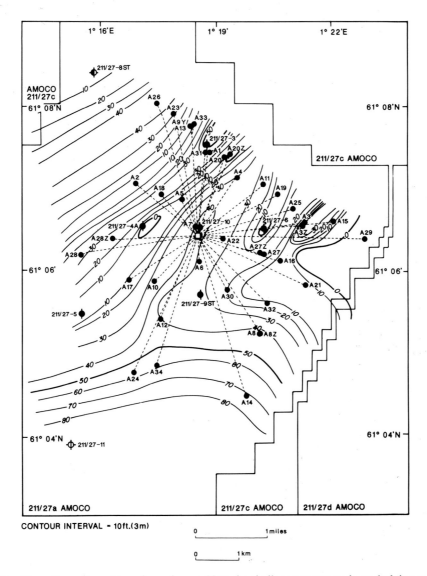

CONTOUR INTERVAL = 10ft.(3m)

Fig. 10-10. Etive Formation net-sandstone isopach map. Note the similar geometry to the underlying units but with an east-west oriented thickening of the sandstone in the southern part of the field.

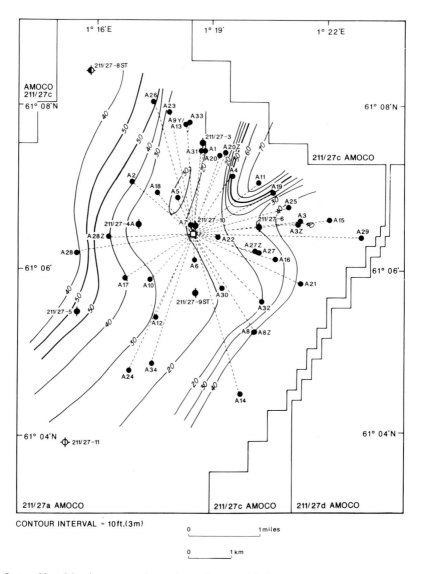

Fig. 10-11. Lower Ness Member net-sandstone isopach map with the "ridge and trough" sand body geometry.

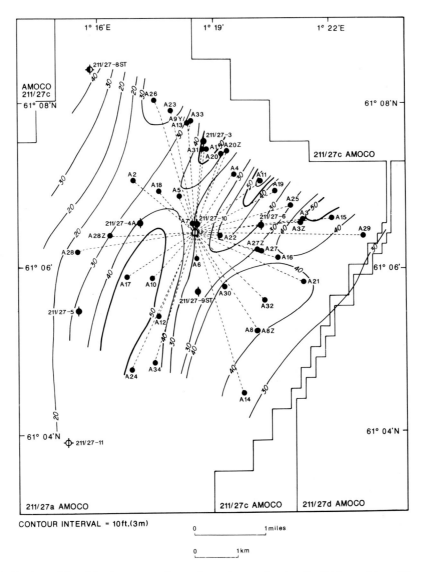

Fig. 10-12. Upper Ness Member net-sandstone isopach map showing the "ridge and trough" geometry.

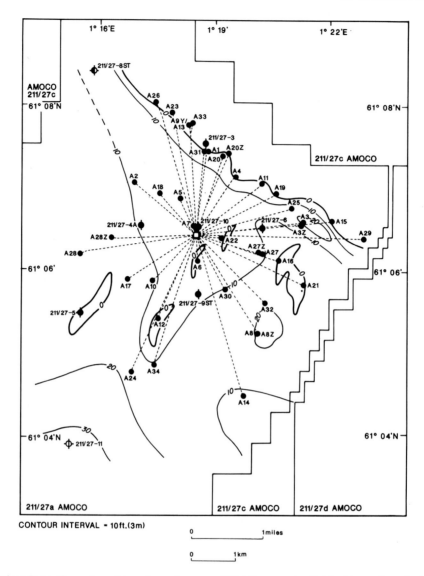

CONTOUR INTERVAL = 10ft.(3m)

0 _____ 1miles

0 _____ 1km

Fig. 10-13. Tarbert Formation net-sandstone isopach map showing a more sheet-like distribution than that of the underlying units. There is a northwest-southeast oriented "thick" across the field with northwards thinning due to post-Brent erosion of crestal parts of the field.

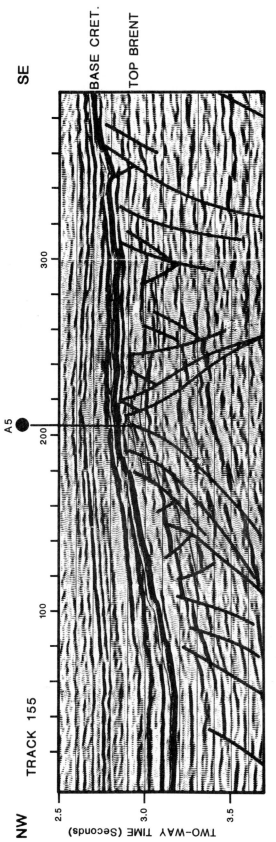

Fig. 10-14. Northwest-southeast-trending seismic section across the Northwest Hutton Field. Section line is shown on Figure 10-5.

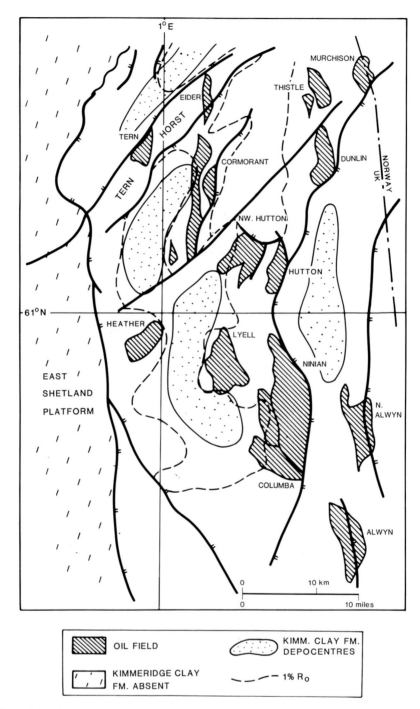

Fig. 10-15. Kimmeridge Clay Formation isopach map showing the juxtaposition of the Northwest Hutton Field to the mature hydrocarbon-generating basin to the southwest.

mation in the basinal depocenters to the southwest of the field (Fig. 10-15). Maturity modeling indicates that generation commenced in the Eocene (Goff, 1983) with migration updip and along faults from the Kimmeridge Clay Formation into the stratigraphically lower Brent Group sandstone reservoirs, successfully filling the fault blocks in the structure.

Depositional Setting

Two major depositional systems are recognized within the Brent Group in the Northwest Hutton Field (Brown et al., 1987). The lower system is represented by the Broom Formation, while the upper system is represented by five main depositional facies:

1. A marine offshore to shoreface transition, represented by bioturbated mudstone and thin sandstone;
2. A lower shoreface environment with hummocky cross-stratified sandstone deposited under storm-wave influence (Richards and Brown, 1986);
3. An upper regressive shoreface to coastal-plain transitional environment developed as either upward-shoaling or a channelized barrier-island complex with thick cross-bedded sandstones and thin coals;
4. A coastal-plain environment represented by interbedded sandstones, shales, and coals. Both lagoonal and fluvial influences are evident in the upward-coarsening distributary-mouth-bar and upward-fining channel-fill sequences;
5. A transgressive coastal-plain to shoreface transition comprising an erosively based sandstone with a basal pebble lag overlain by stacked, locally upward-coarsening sandstone sequences with minor mudstones and coals.

The upper-shoreface to coastal-plain sediments of broadly uniform facies are widely distributed both laterally and vertically within the Brent Group. Palynofacies evidence of varying salinities in the coastal-plain muds and the overall lithofacies distribution suggest deposition of Brent Group sandstones in a strandplain/barrier coastal setting. The northeastward progradation of a storm-wave influenced barrier and coastal plain within a deltaic setting resulted in the development of a progradational clastic wedge which is overlain by a thick unit of coastal-plain sediments.

Lithofacies

Within the Brent Group of the Northwest Hutton Field five major reservoir facies and two nonreservoir facies are identified:

1. Sublittoral sheet sandstones
2. Distributary-mouth bars and delta-front sandstones
3. Crevasse-splay sandstones
4. Distributary-channel sandstones
5. Shallow marine sandstones and claystones
6. Lagoon/bay silty claystones
7. Delta-plain overbank deposits

Sublittoral Sheet Sandstones. The sublittoral sheet sandstones are typified by the basal Broom Formation of the Brent Group. The sandstones are poorly to moderately sorted and coarse-grained, with either a sharp lower contact or a thin gradational contact with the underlying Dunlin Group. The lower section of the unit is rarely burrowed, and bioturbated sandy claystones are present. Succeeding sandstones in the sequence are parallel-laminated to low-angle cross-bedded. Deposition is inferred to have occurred by the migration of high-energy sand waves in a sublittoral setting. The depositional environment is inferred from the common burrows at the top of the sandstone and its sheet-like distribution through the Northwest Hutton Field. No marine macro- or microfossils have been found in these sandstones.

Distributary-Mouth-Bar and Delta-Front Sandstones. These sandstones are represented in the Northwest Hutton Field by the Etive and Rannoch formations, which form an overall upward-coarsening sandstone sequence (Fig. 10-16). The lower portion of the Rannoch Formation consists of locally bioturbated micaceous siltstone and claystone thinly interbedded with very fine-grained sandstone. This lower portion coarsens upward to parallel-laminated and low-angle cross-bedded, very fine-grained sandstone that locally contains micaceous laminae and rare burrows. The intermediate to upper part of the sequence has wave-induced parallel laminations, symmetrically filled scours, and local symmetrical ripples reflecting wave reworking and winnowing of the sediment. The predominance of wave-generated sedimentary structures and the high lateral continuity of the sandstone throughout the field support the interpretation of wave-dominated delta-front and coastal/barrier sedimentation.

12,711 ft

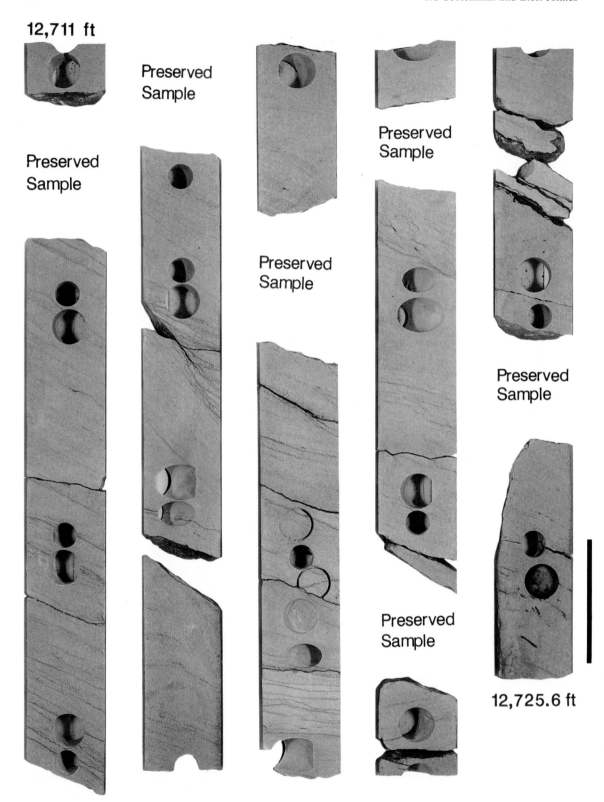

Preserved
Sample

Preserved
Sample

Preserved
Sample

Preserved
Sample

Preserved
Sample

Preserved
Sample

Preserved
Sample

Preserved
Sample

12,725.6 ft

Fig. 10-16. Core photograph showing distributary-mouth-bar sandstones from the Etive Formation in well 211/27-A22 The scale bar is 0.5 feet (15.2 cm).

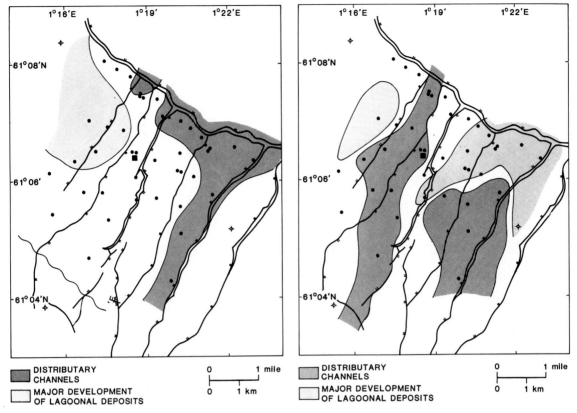

DISTRIBUTARY
CHANNELS

MAJOR DEVELOPMENT
OF LAGOONAL DEPOSITS

0   1 mile

0   1 km

DISTRIBUTARY
CHANNELS

MAJOR DEVELOPMENT
OF LAGOONAL DEPOSITS

0   1 mile

0   1 km

Fig. 10-17. Generalized distribution and orientation of distributary-channel sandstone in the Lower Ness Member.

Fig. 10-18. Distributary-channel sandstone distribution in the Upper Ness Member.

This upward-coarsening sandstone sequence is continuous into the overlying Etive Formation. The grain size typically grades from fine- to medium-grained. Within the Etive Formation burrows are rare, but local carbonaceous rip-up clasts and wood and plant fragments are common, suggesting a proximal setting at the river mouth of a delta. Whereas wave conditions were sufficient to redistribute part of the sand along strike, they did not winnow the carbonaceous mud clasts and plant debris.

Crevasse-Splay Lobe Sandstones. Crevasse-splay sandstones are found throughout the Ness Formation (Fig. 10-17). They typically coarsen upward from ripple cross-laminated very fine-grained sandstone to cross-bedded fine-grained sandstone capped by a root-mottled zone and a thick coal bed. Carbonaceous plant and woody material is common in the sandstones with burrows present locally but restricted primarily to the lower portion of the sequence. These deposits are interpreted to repre-

sent repeated lagoon/bay fill from channel breaching followed by abandonment and subsidence.

Crevasse-splay sandstones in the Lower Ness Member are individually discontinuous, but they may coalesce to form sheet sandstones which fill large areas of the lagoon/bay. In the Upper Ness Member, the crevasse-splay sandstones are thin, fine-grained and ripple cross-laminated, probably representing single-event deposition during flood stages.

Distributary-Channel Sandstones. Distributary-channel sandstones are present in the Ness Formation as the lateral equivalent of the crevasse-splay sandstones (Figs. 10-17, 10-18). Distributary-channel sandstones typically have erosional basal contacts with overlying upward-fining sequences (Fig. 10-19); they are commonly cross-bedded, and locally they contain carbonaceous mud clasts and plant material. Upper contacts may be abrupt, indicating rapid channel abandonment, or grada-

12,499 ft

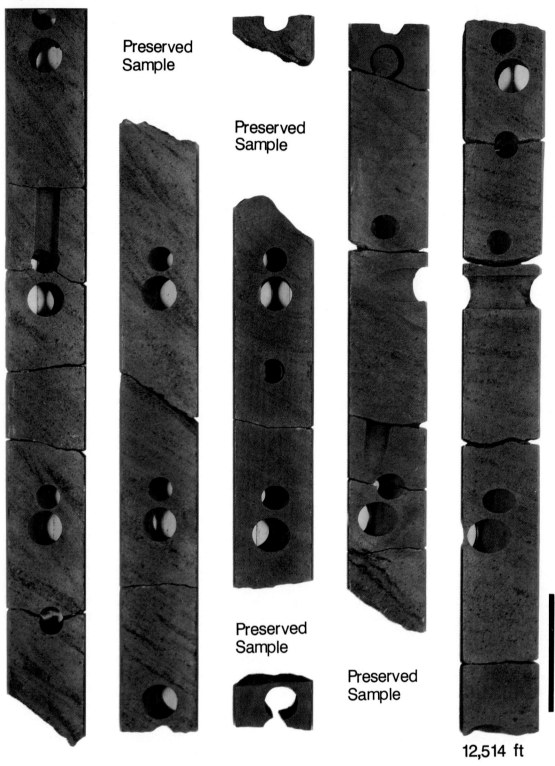

Preserved
Sample

Preserved
Sample

Preserved
Sample

Preserved
Sample

12,514 ft

Fig. 10-19. Core photograph showing distributary-channel sandstone from the Ness Formation in well 211/27-A22. The scale bar is 0.5 feet (15.2 cm).

tional and consisting of fine-grained ripple cross-laminated sandstones and siltstones, indicating gradual abandonment. Channel sandstones are often capped by a root-mottled zone.

Stacked channel-fill units consisting of amalgamated upward-fining sequences are present. The distributary-channel sandstones are laterally discontinuous along strike but may be laterally interconnected with crevasse-splay sandstones (Fig. 10-20). Distributary-channel sandstones are also present in the Etive Formation, where they locally cut into the distributary-mouth bar and give rise to an upward-fining sequence rather than the usual upward-coarsening sequence associated with the Etive Formation.

Transgressive and Regressive Marine Sandstones and Claystones. The Ness Formation is subdivided into upper and lower units by a 20-foot (6.1-m) thick shale which can be identified across the Northwest Hutton Field. The shale is wavy- to lenticular-bedded with wave-ripple cross-lamination, is locally burrowed, and contains marine palynomorphs. The presence of these structures and their widespread areal extent suggest an overall transgressive phase of deposition in a shallow marine environment. Overlying this shale is a laterally continuous sandstone which coarsens upward from the underlying silty claystone to a ripple-laminated very fine-grained sandstone in turn succeeded by horizontally stratified fine- to medium-grained sandstone. The sequence is locally interbedded with sandy and silty claystones. A rooted zone with thin coal or lagoon/bay deposits caps the sandstone sequence, which is interpreted as a regressive shelf-sand or wave-influenced distributary-mouth bar representing renewed progradation of the delta across the underlying shallow marine deposits.

Overlying the Ness Formation is a second, areally extensive shaly unit which is part of the Tarbert Formation. This shaly unit consists of horizontally laminated silty claystone interbedded with sandy claystone displaying wavy to lenticular bedding and local burrows. The burrows and local wave-induced ripple cross-lamination indicate a shallow marine depositional environment.

The Tarbert shale unit is overlain by an erosively based sandstone which coarsens and shows decreasing bioturbation upward. Grain size varies from very coarse to very fine, and both current-induced and wave-induced cross-lamination are common throughout. The Tarbert sandstone forms a continuous sheet body except in the crestal areas of the field, where several wells demonstrate that it has been removed by post-Brent erosion. This sandstone is interpreted as a regressive shelf sand.

Lagoon/Bay Silty Claystones. Interbedded with the crevasse-splay and distributary-channel sandstones of the Ness Formation are dark grey to black, laminated and burrowed silty claystones. Wavy to lenticular bedding and the presence of plant fragments on bedding planes are common, suggesting deposition as lagoon/bay sediments.

Delta-Plain Overbank Deposits. These deposits are interbedded with the distributary-channel sandstones of the Ness Formation and consist of dark grey lignitic silty claystones with common root mottling, thin coals, and occasional burrows.

Petrography

The Brent Group sandstones are predominantly quartzarenites to subfeldsarenites (classification after Folk et al., 1970) composed of framework grains of quartz, feldspar, mica, and lithic fragments. Quartz overgrowths, calcite, kaolinite, and illite are the pore-filling minerals.

Quartz is the dominant framework grain constituent, and the highest quartz content is found in the distributary-channel sandstones. Quartz grains are subangular to subround, commonly exhibiting syntaxial quartz overgrowths at the margins except where extensively cemented by carbonate. Locally, in less compacted, coarser-grained sandstones, embayed quartz margins suggest the presence of an earlier carbonate cement which has been leached.

Feldspar is the second most abundant framework grain, present in all the Brent sandstones, and is most prevalent in the sublittoral sheet sandstone of the Broom Formation. The majority of feldspars exhibits some degree of alteration ranging from clay authigenesis to partial or complete leaching.

Rock fragments are rare and occur in trace amounts in all sandstones. Mica is typically a minor constituent of most fine-grained sandstones. Higher mica contents occur in the delta-front sandstone of the Rannoch Formation. The mica is oriented parallel to bedding planes, forming laminae which severely restrict vertical permeability. Mica also commonly displays alteration to clay minerals such as kaolinite and illite. Traces of heavy minerals are present in all sandstones and where abundant are concentrated in laminae.

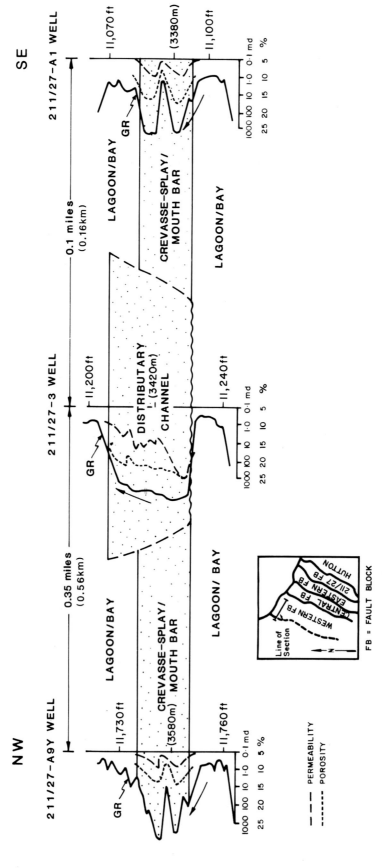

Fig. 10-20. Reservoir characteristics resulting from incision by the distributary channel into the laterally equivalent crevasse-splay or mouth-bar sandstones in the Lower Ness Member. Note the large-scale horizontal and vertical reservoir heterogeneities. Inset shows location of section.

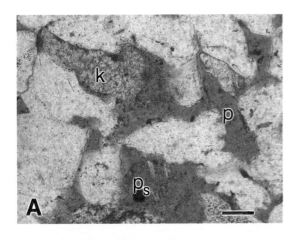

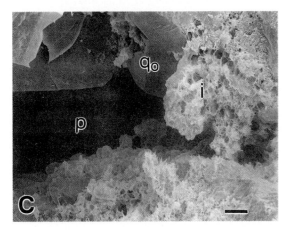
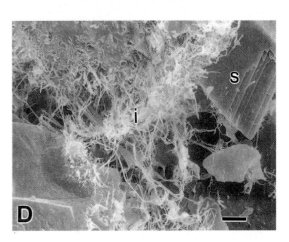

Fig. 10-21. Photomicrographs of representative samples of the Brent Group. (A) Thin section photomicrograph of a distributary-mouth-bar sandstone (Etive Formation) showing intergranular pore space (p), leached-grain secondary pore space (p_s), and a pore infilled with kaolinite (k). The surrounding grains display quartz overgrowth (k). Well 211/27-A22 at 11,937 feet (3,638 m) subsea ($\phi = 19.1\%$, k = 117 md). Scale bar is 0.1 mm. (B) SEM photomicrograph of an Upper Ness Member regressive shelf sandstone showing euhedral quartz overgrowth (q_o) and pore filling kaolinite (k). Well 211/27-A15 at 11,190 feet (3,411 m) subsea. ($\phi = 21.2\%$, k = 67.2 md). Scale bar is 0.1 mm. (C) SEM photomicrograph of pore-lining diagenetic illites (i) in coarse-grained distributary-channel sandstones. Note the quartz overgrowths (q_o) and intergranular pore space (p). Upper Ness Member well 211/27-A23 at 11,930 feet (3,636 m) subsea ($\phi = 16.5\%$, k = 853 md). Scale bar is 0.1 mm. (D) SEM photomicrograph of filamentous, pore-bridging illite (i) shows abundant microporosity resulting in good porosity (14.1%) but very low permeability (0.54 md). Siderite (s) is also present. Rannoch Formation, well 211/27-A17 at 12,188 feet (3,715 m) subsea. Scale bar is 0.005 mm.

Three types of pore space occur in the Brent Group sandstones, including intergranular pores, secondary pores, and intercrystalline clay micropores. Intergranular pores are the most significant for the storage and transmissibility of fluids (Fig. 10-21A). Distributary-channel sandstones have the highest average amount of intergranular pore space due to their coarse grain size and low clay content. This results in large interconnected, intergranular pores. These sandstones have the highest permeability and best reservoir quality.

Secondary pores result from the leaching of unstable grains in the sandstone framework and are present in all sandstones that were not cemented by early calcite. The majority of the secondary pores result from the partial to complete dissolution of feldspar grains. These pores are less important than the intergranular type because they are poorly connected. They are only effective in enhancing permeability and transmissibility when connected with intergranular pores. Only in the Broom Formation does this secondary porosity become significant in promoting reservoir quality due to the large-scale leaching of the high feldspar content of these sandstones.

The intercrystalline micropores are noneffective with respect to reservoir quality due to high tortuosity and high capillarity created by extremely small pore throats. This type of porosity is dominant in the fine-grained sandstones with high proportions of diagenetic clay, resulting in relatively high porosity but very low permeability. These micropores are associated primarily with illite and to a lesser extent kaolinite, the latter being the case with the coarse-grained Broom Formation sheet sandstones.

Diagenesis

Pore filling as a result of sandstone diagenesis is superimposed on the depositional grain-size control of reservoir quality. Diagenesis is to some extent controlled by grain size, with the finer-grained sandstones having a higher clay content and being more susceptible to diagenetic alteration. Structural depth also influences diagenetic pore filling. The relative volume of quartz and illite cements increases with depth, resulting in decreased reservoir quality on the flanks of the structure.

Pore-Filling Cements. Quartz overgrowths, calcite, siderite, illite, and kaolinite form the main pore-filling mineral phases (Scotchman et al., 1989). Volumetrically, quartz is the most important authigenic phase (Fig. 10-21B) and is ubiquitous throughout the sandstone pore system. Only in the Broom Formation sandstones that display early calcite and siderite cements is quartz cementation of lesser importance. In downdip parts of the field, quartz overgrowth cementation is very extensive and results in low porosities.

The occurrence of calcite cement (in intervals up to 6 feet (1.8 m) thick) is restricted to the marine sandstones, in particular those of the Broom and Rannoch formations. The calcite-cemented intervals are concretionary and have a limited lateral extent and erratic vertical distribution. Siderite is also restricted to marine sandstones, particularly those of the Broom Formation, where it partially or completely replaces calcite. The delta-front sandstones of the Rannoch Formation contain appreciable siderite as patchy coatings on grain margins.

Kaolinite is the second most abundant pore-filling phase and is authigenic. It is most common in the Broom Formation sandstones, where it fills the pore spaces. Its prevalence in the Broom Formation is probably due to a relatively high detrital feldspar content. In the rest of the Brent Group sandstones, kaolinite has a patchy distribution but is more prevalent in the crestal areas of the field.

Illite is a significant pore-filling phase particularly in the fine-grained sandstones and downdip in the western flank of the field, where it severely reduces reservoir quality. Illite concentrations are lowest in the distributary-channel and shelf sandstones and highest in the distributary-mouth-bar and delta-front sandstones of the Etive and Rannoch formations, respectively.

In the coarse-grained distributary-channel sandstones, illite has little effect on reservoir quality and lines pores but rarely bridges them (Fig. 10-21C). In the finer-grained sandstone facies, including the upper portions of the distributary-channel sequences and crevasse-splay sandstones, illite has a much greater abundance and is developed as a pore-bridging filamentous morphology resulting in greatly reduced reservoir quality (Fig. 10-21D). These filamentous illites contain abundant intercrystalline micropore spaces and have little effect on porosity while greatly reducing permeability. This causes the Brent Group sandstones to exhibit a widely variable porosity-permeability relationship (Fig. 10-22). Illite increases with depth, showing a marked increase below 12,000 feet (3,660 m); wells 211/27-11 and 211/27-A28 on the western flank have exceptionally poor reservoir quality.

Chlorite and dolomite occur as minor mineral phases in the Broom and Rannoch formations. Chlorite has a pore-filling or grain-coating morphology, whereas dolomite has a patchy distribution and is restricted to the flank areas of the field in the western fault blocks.

Paragenetic Sequence. The main diagenetic events which influenced the reservoir quality of the Brent Group sandstones in the Northwest Hutton Field (Scotchman et al., 1989) are listed below and illustrated in Figure 10-23.

1. Early concretionary calcite cementation of thin layers in the Broom and Rannoch formations.
2. Siderite replacement of calcite in Broom and Rannoch formations.

Fig. 10-22. Core porosity and permeability in the Brent Group sandstones. In general, permeability increases with porosity; however, there is considerable variability caused by diagenetic filamentous illite. ▷

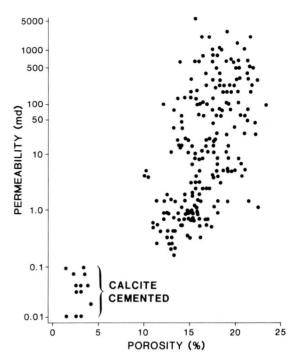

Fig. 10-23. Diagenetic sequence for the Brent Group sandstones. ▽

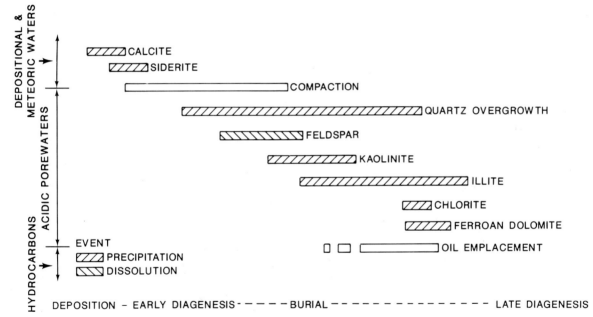

3. Development of quartz overgrowths in all sandstones not subject to calcite cementation.
4. Formation of secondary porosity by feldspar dissolution.
5. Kaolinite authigenesis.
6. Development of authigenic illite.
7. Oil emplacement.
8. Further limited illite development in the water zone in the Ness and Etive formations.

Quartz overgrowth cement forms the major porosity-occluding phase throughout the sandstones of the field. Early calcite cementation affected primarily the Broom and Rannoch formations, where it was restricted to thin intervals. This early cement inhibited the later development of the quartz overgrowths that greatly affected the other sandstones of the Brent Group. The calcite cements also protected the detrital feldspars from alteration.

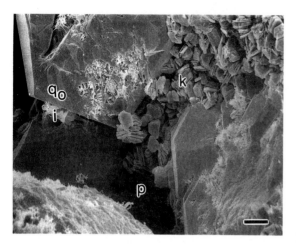

Fig. 10-24. SEM photomicrograph of a transgressive shelf sandstone, from the Lower Ness Member in well 211/27-A14 showing a pore (p) partly filled with quartz overgrowth (q_o), kaolinite (k), and illite (i) cements 11,712 feet (3,570 m) subsea. ($\phi = 19.5\%$, k = 193 md). Scale bar is 0.02 mm.

Feldspar dissolution and the development of secondary porosity occurred during the later stages of quartz cementation, with simultaneous kaolinite authigenesis occurring in crestal parts of the field. The majority of the kaolinite postdates quartz cementation (Fig. 10-21B), but a partial overlap exists in the timing of these phases. Illite cementation postdated the quartz overgrowths (Fig. 10-24), the illite having a delicate filamentous morphology emanating from the pore walls and filling the intergranular and some secondary pores. The development of illite also postdates authigenic kaolinite, which in some wells (e.g., 211/27-A28) is partially illitized.

The emplacement of oil appears to have halted diagenesis (Hancock and Taylor, 1978; Scotchman et al., 1989), causing illitization to cease in the oil-filled crestal parts of the field. Within the oil zone, the illites have a matted morphology created by the oil migration "front" passing through the reservoir (Cocker, 1986). Filamentous illites are preserved in the water-bearing zone. Illite growth appears to have continued in the water zone for a short period after oil migration into the structure (J. Cocker, pers. communication, 1988). This is particularly true in the Etive and Ness formations in wells 211/27-11 and A28, resulting in a large decrease in permeability below the oil/water contact.

Petrophysical Parameters and Reservoir Quality

The distribution of reservoir properties in the Northwest Hutton Field is strongly influenced both by depositional environment and by diagenetic factors, the latter being more important with increased depth. Depositional processes determined grain size and sorting which, in turn, dictated reservoir quality, the coarsest sands generally having the highest porosities and permeabilities. On this basis, the distributary-channel sands have the best reservoir quality, and the delta-front sandstones have the poorest quality. Grain size variations, such as the upward-fining channel sandstones or upward-coarsening deltaic sequences, control the internal reservoir quality within individual sandstone bodies. The Broom Formation sandstones are an exception because, although they are coarse-grained, they have relatively poor reservoir quality due to diagenesis (Scotchman et al., 1989).

Reservoir quality is highly variable within the sublittoral sheet sandstones of the Broom Formation. Porosity ranges from 13 to 23%, averaging 14%, with air permeabilities ranging from 10 to 100 md and averaging 17 md. In the south of the field, porosity is reduced to an average of 9% with a permeability to air of less than 1 md.

The overall upward-coarsening character of the deltaic sequences composed of delta-front sandstones overlain by distributary-mouth-bar sandstones is mirrored by an upward increase in porosity and permeability. In the Rannoch Formation, the basal siltstone and claystone form an extensive permeability barrier between the Rannoch and underlying Broom formations. Average porosity within the Rannoch Formation increases upward from 10 to 20% with a corresponding increase in air permeability from 0.1 to 100 md. A similar pattern of porosity and permeability is exhibited by the upward-coarsening deltaic sequences of the Etive Formation.

The deltaic crevasse-splay sandstones are generally poor reservoirs. Although porosity ranges from 15 to 17% and increases upward with sandstone grain size, the corresponding increase in air permeability is only from 1 to 10 md. Permeability to air decreases vertically within the distributary-channel sandstones from 100 to more than 1,000 md in the basal sandstones to between 1 and 10 md in the upper sandstones (Fig. 10-20). Porosity follows a similar trend with decreasing grain size from 20 to

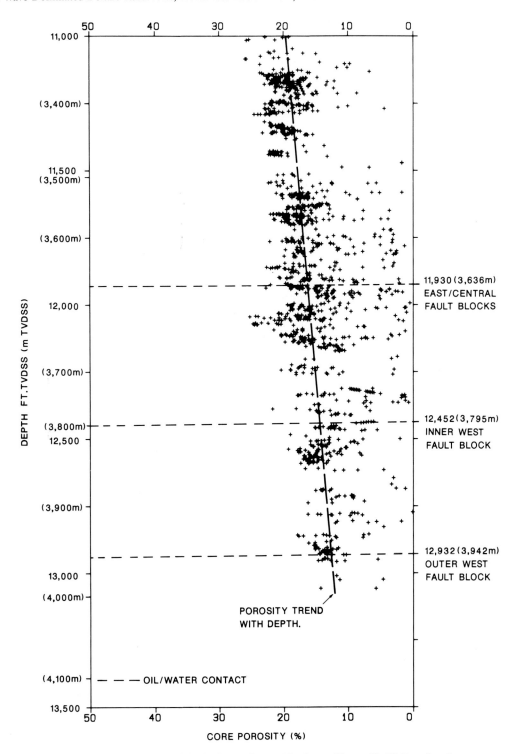

Fig. 10-25. Core porosity plotted by depth in the Brent Group, Northwest Hutton Field. Porosity shows a progressive decrease with depth in the field.

24% at the base to 15 to 20% at the top of the units. The average air permeability in the marine sandstones of the Tarbert Formation is 100 md, although values vary widely (10–300 md) across the field. Porosity is relatively uniform and averages 20%. The effect of diagenesis on reservoir quality is apparent in all reservoir facies, with a decrease in porosity from more than 20% at the crest of the structure to 12.7% on the western flank (Fig. 10-25). Permeability shows a similar trend, decreasing from 296 md in the crestal well 211/27-A1 to 1 md in the downdip 211/27-A28 well (Fig. 10-26).

Diagenetic cementation of the sandstones is an important control on reservoir quality (Scotchman et al., 1989). Most sandstones of the Etive, Ness, and Tarbert formations were not subject to early calcite cementation and were later cemented by quartz overgrowths, kaolinite, and illite. Reservoir quality in these sandstones was controlled initially by grain size but, following diagenesis, the amount of quartz overgrowth cement and illite became the controlling factor. Kaolinite is the dominant control on reservoir quality only in the Broom sandstones.

In the Broom and Rannoch formations, early calcite cementation is restricted to thin intervals where it is very detrimental to reservoir quality and results in localized vertical permeability barriers. Broom Formation reservoir quality decreases in the southwestern part of the field as the cumulative thickness of the calcite-cemented intervals increases. The least cementation occurs in the Broom Formation in the northeastern part of the field.

Reservoir quality in the field decreases down structure due to the progressive filling of the intergranular pore system by quartz overgrowths and illite cement (Fig. 10-27). With the exception of the Broom Formation, depositional texture exerts the major control on reservoir quality, the finer-grained sandstones being of poorer quality regardless of burial depth. Diagenetic cementation destroys reservoir quality at depths greater than 12,300 to 12,500 feet (3,750–3,810 m) subsea. A productive limit for the field at 12,500 feet (3,810 m) subsea, based on well 211/27-A28, can be defined for the western flank of the field.

No single oil-water contact is found in the Northwest Hutton Field. The contacts are structurally controlled with progressively higher contacts between the western and the eastern fault blocks. The various oil-water contacts may be related to the migration paths of the hydrocarbons and subsequent structural movements. The oil-water contacts within the Northwest Hutton Field are summarized in Table 10-1.

Production Characteristics

Well and Reservoir Performance

The Northwest Hutton Field came on stream in April, 1983 with well 211/27-A1 flowing at 16,000 BOPD (2.5×10^3 m³/D). Production peaked at 86,680 BOPD (1.4×10^4 m³/D) in May, 1983 when the last of the seven predrilled template wells was brought on stream. The reservoir was undersaturated and overpressured, but reservoir pressure and solution-gas flow rapidly declined with initial production (Fig. 10-28). Water injection began in February, 1984 with the conversion of well 211/27-A7, and to date eight additional wells have been converted. Artificial lift using produced gas began in October, 1984.

The field contains an estimated 670 MMBO (1.1×10^8 m³) in place with recoverable reserves of 200 to 220 MMBO (3.2–3.5×10^7 m³), a recovery factor of 30%. Additionally, it is estimated that 16 MMB (2.5×10^6 m³) of natural gas liquids and 57 BCF (1.6×10^9 m³) of gas will be recovered.

The Northwest Hutton reservoir was initially produced under primary depletion. A secondary recovery program was implemented soon after the start of production with the installation of a line-drive waterflood. The line-drive pattern was placed in the center of the field with an orientation to permit sweep in a direction parallel to the northeast-southwest trending fault system. A peripheral flood pattern was discounted because of the degradation of permeability on the flanks of the field.

The field has been difficult to produce because of its reservoir heterogeneities. During the primary depletion period, pressure declines were observed over a wide area, although these declines were areally and vertically irregular. Generally, the sheet-type shelf sandstones immediately above and below the Mid-Ness Shale showed the greatest and most consistent depletion, while the sandstones of the Broom, Rannoch, and Etive formations showed the least depletion. Pressure data collected during the secondary recovery period continue to show the large effect of the reservoir heterogeneities with zonal pressures varying from 2,000 to more than

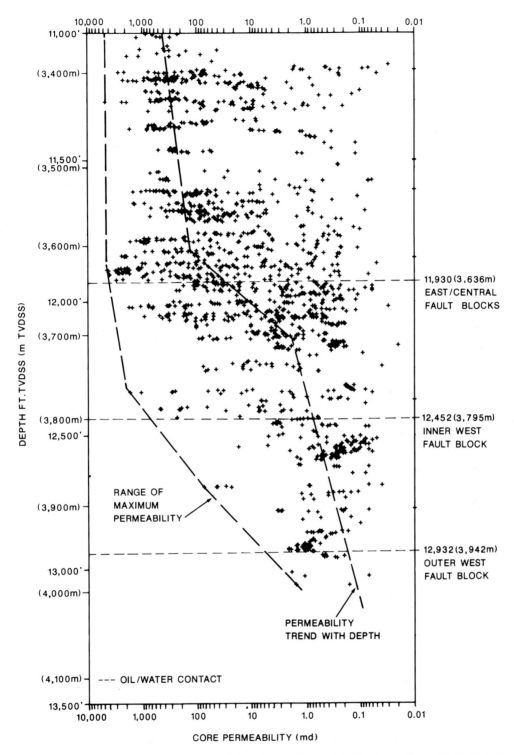

Fig. 10-26. Core air permeability variation with depth in the Brent Group of the Northwest Hutton Field. Note the rapid deterioration in permeability below about 12,000 feet (3,660 m) subsea.

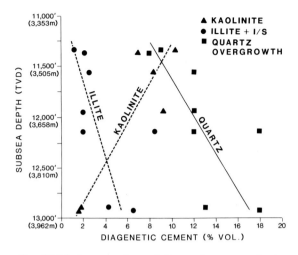

Fig. 10-27. A comparison of diagenetic cements in the Etive Formation with depth in the western fault blocks. Quartz overgrowths and illite increase with structural depth, while kaolinite decreases.

8,000 psi (1.4–5.5 × 10⁴ kPa). Attaining good reservoir sweep is difficult because small faults juxtapose some otherwise continuous sandstones against shales or other sand bodies. The severe permeability variations also give rise to thief zones which result in early water breakthrough and production of water. Consequently, well workovers are required as well as the processing of significant volumes of water.

Exploration and Production Strategy

As with any reservoir, the most important factor is the identification and accurate mapping of the structure. However, mapping trap configuration alone is insufficient. As reservoir quality is dependent on depositional characteristics, it is also important to map the major sandstones, many of which are channelized, and to understand their internal architecture and facies variability. The relation of the trap to structural depth is also critical to reservoir considerations. In this area, reservoirs below 12,000 feet (3,660 m), regardless of the depositional texture, tend to be of low quality as a product of diagenesis.

Critical factors which affect production are the faulted and compartmentalized character of the reservoir and the reservoir heterogeneity due to lithofacies and diagenetic controls on vertical and horizontal permeability variations. This results in excellent horizontal flow within the channel sandstones of a fault block but in very poor horizontal and vertical flow and drainage of the overlying tighter non-channelized sandstones. Careful siting of production and injection wells and selective completion within the reservoir are necessary to drain the lower permeability sandstones effectively. Nonselective completion allows only the higher permeability channel sandstones to be effectively drained, leaving the

Table 10-1. Fluid contacts, Northwest Hutton Field. Depths are subsea and the well where the contact was determined is noted.

Formation	Fault block		
	Western (ft)	Central (ft)	Eastern (ft)
Tarbert	12,935 ODT 211/27-8 (3,943 m)	11,635 ODT 211/27-6 (3,546 m)	11,531 ODT 211/27-A14 (3,515 m)
Ness	12,561 ODT 211/27-A2 (3,829 m)	11,849 ODT 211/27-6 (3,612 m)	11,725 ODT 211/27-A14 (3,574 m)
Etive	12,934 OWC 211/27-A28 (3,942 m)	11,813 ODT 211/27-A25 (3,601 m)	11,790 ODT 211/27-A8Z (3,594 m)
		11,897 WUT 211/27-6 (3,626 m)	
Rannoch			
Broom	12,692 ODT 211/27-A2 (3,869 m)	11,642 ODT 211/27-A19 (3,549 m)	11,450 ODT 211/27-A15 (3,490 m)

Notes:
ODT = oil-down-to
OWC = oil/water contact
WUT = water-up-to

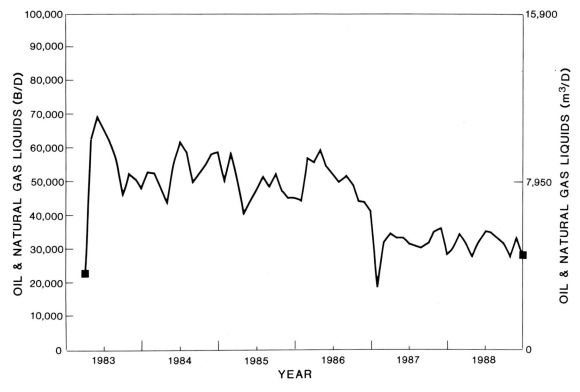

Fig. 10-28. Production decline curve for the Northwest Hutton Field from April, 1983 to December, 1988.

lower permeability sandstones relatively unswept. Production and injection well pairs must be sited within each fault block to optimize reservoir sweep with injection only into the lower permeability sandstones to prevent rapid cycling of injection water.

Conclusions

The Northwest Hutton Field Brent Group reservoirs are composed of shallow marine, wave-dominated delta-front and distributary-channel sandstones. Reservoir quality is controlled primarily by depositional facies but with a diagenetic overprint that tends not to change the relative quality of the facies. Reservoir quality is highest in the coarser-grained, clay-free, fluvial and deltaic channel sandstones which have undergone minimal diagenetic alteration. The finer-grained sandstones of the delta and shelf settings commonly contain abundant authigenic clay which has greatly reduced permeability.

Burial depth is crucial to reservoir quality, as below 12,500 feet (3,810 m) even the channel sandstones have undergone extensive quartz and clay cementation and are poor reservoirs.

Depositional and diagenetic influences have resulted in a very complex and heterogeneous reservoir. This has been complicated by two fault trends which created the structural traps but which segmented the reservoirs into four major fault blocks. Successful development and production of a field of this type are dependent on a thorough knowledge of the depositional environment, lithofacies, diagenesis, and the superimposed effects of complex faulting.

Acknowledgments. We thank Amoco and the partners in the Northwest Hutton Group (Amerada Hess, Enterprise Oil, Mobil North Sea, and Texas Eastern North Sea) for permission to publish this work. J.G. McPherson, I.D. Bryant, and an anonymous reviewer are thanked for their comments on an earlier version of the manuscript.

Reservoir Summary

Field : Northwest Hutton (UKCS Block 211/27)
Location: Northern North Sea
Operators: Amoco U.K., Enterprise Oil, Mobil North Sea, Texas Eastern, Amerada Hess
Discovery: 1975
Basin: East Shetland basin
Tectonic/Regional Paleosetting: Rift basin margin
Geologic Structure: Complex-tilted fault block
Trap Type: Fault-bounded structural closure
Reservoir Drive Mechanism: Primary depletion and water drive
  • **Original Reservoir Pressure:** 7,563 psi (5.2×10^4 kPa) Western fault block, 7,315 psi (5.0×10^4 kPa) Eastern
                                fault block at 11,500 ft (3,505 m) subsea
Reservoir Rocks
  • **Age:** Middle Jurassic, Bajocian-Bathonian
  • **Stratigraphic Unit:** Brent Group
  • **Lithology:** Fine- to medium-grained feldsarenite
  • **Depositional Environments:** Combination of shallow marine shelf and wave-dominated fluviodeltaic environ-
                                ments
  • **Productive Facies:** Distributary-channel, distributary-mouth-bar, and shelf sandstones
  • **Petrophysics**
    • **Porosity Type:** ϕ_{Total} 18% with primary 13%, secondary intergranular 5%
    • **ϕ:** Average 18%, range 8 to 24%, cutoff 12% (cores)
    • **k:** Average 99 md, range 0.1 to 2,000 md (unstressed cores)
    • **S_w:** Average 20%, range 10 to 80%, cutoff 60% (logs)
Reservoir Geometry
  • **Depth:** 11,000 feet (3,350 m). Water depth 475 feet (145 m)
  • **Areal Dimensions:** 3.7 by 5.0 miles (6.0×8.0 km)
  • **Productive Area:** 13,440 acres (5,440 ha.)
  • **Number of Reservoirs:** Multiple
  • **Hydrocarbon Column Height:** 1,400 feet (430 m)
  • **Productive Limit:** 12,400 feet (3,780 m) subsea
  • **Number of Pay Zones:** 13
  • **Gross Sandstone Thickness:** 400 feet (122 m)
  • **Net Sandstone Thickness:** 180 feet (55 m)
  • **Net/Gross:** 0.45
Hydrocarbon Source, Migration
  • **Lithology and Stratigraphic Unit:** Basinal shale, Kimmeridge Clay Formation, Upper Jurassic
  • **Time of Hydrocarbon Maturation:** Eocene
  • **Time of Trap Formation:** Late Jurassic-Early Cretaceous
  • **Time of Migration:** Early Tertiary
Hydrocarbons
  • **Type:** Oil and gas
  • **GOR:** 800:1 (Western fault block), 600:1 (Eastern fault block)
  • **API Gravity:** 35°
  • **FVF:** 1.42 (Western fault block), 1.38 (Eastern fault block)
  • **Viscosity:** NA
Volumetrics
  • **In-Place:** 670 MMBO (1.1×10^8 m³)
  • **Cumulative Production:** 83.6 MMBO (1.3×10^7 m³), 6.9 MMBNGL (1.1×10^6 m³)
  • **Ultimate Recovery:** 200 MMBO (3.2×10^7 m³)
  • **Recovery Efficiency:** 30%
Wells
  • **Spacing:** 3,900 feet (1,190 m)
  • **Total:** 48
  • **Dry Holes:** 3

Typical Well Production:
 • **Average Daily:** 1,500 BO (240 m³)
 • **Cumulative:** 4 MMBO (6.4 × 10⁵ m³)
Stimulation: Water injection

References

Badley, M.E., Price, J.D., Dahl, C.R., and Agdestein, T., 1988, The structural evolution of the northern Viking Graben and its bearing upon extensional modes of basin formation: London, Journal of the Geological Society, v. 145, p. 455–472.

Brown, S., 1984, Jurassic, *in* Glennie, K.W., ed., Introduction to the Petroleum Geology of the North Sea: Oxford, Blackwell Scientific, p. 103–131.

Brown, S., Richards, P.C., and Thompson, A.R., 1987, Patterns in the deposition of the Brent Group (Middle Jurassic) UK North Sea, *in* Brooks, J. and Glennie, K., eds., Petroleum Geology of North West Europe: London, Graham and Trotman, p. 899–914.

Budding, M.C., and Inglin, H.F., 1981, A reservoir geological model of the Brent Sands in Southern Cormorant, *in* Illing, L.V. and Hobson, G.D., eds., Petroleum Geology of the Continental Shelf of North-West Europe: London, Heyden, p. 326–334.

Challinor, A., and Outlaw, B.D., 1981, Structural evolution of the North Viking Graben, *in* Illing, L.V. and Hobson, G.D., eds., Petroleum Geology of the Continental Shelf of North-West Europe: London, Heyden, p. 104–109.

Cocker, J.D., 1986, Authigenic illite morphology: Appearances can be deceiving [abst.]: American Association of Petroleum Geologists Bulletin, v. 70, p. 575.

Death, N.G., and Schuyleman, S.F., 1981, The geology of the Magnus oil field, *in* Illing, L.V. and Hobson, G.D., eds., Petroleum Geology of the Continental Shelf of North-West Europe: London, Heyden, p. 342–351.

Deegan, C.E., and Scull, B.J., 1977, A proposed standard lithostratigraphical nomenclature for the Central and Northern North Sea: London, Report of the Institute of Geological Sciences 77/25, 35 p.

Doré, A.G., Vollset, J., and Hamar, G.P., 1985, Correlation of the offshore sequences referred to the Kimmeridge Clay Formation — Relevance to the Norwegian Sector, *in* Thomas, B.M., et al., eds., Petroleum Geochemistry in Exploration of the Norwegian Shelf: London, Graham and Trotman, p. 27–37.

Eynon, G., 1981, Basin development and sedimentation in the Middle Jurassic of the northern North Sea, *in* Illing, L.V. and Hobson, G.D., eds., Petroleum Geology of the Continental Shelf of North-West Europe: London, Heyden, p. 196–204.

Folk, R.L., Andrews, P.B., and Lewis, D.W., 1970, Detrital sedimentary rock classification and nomenclature for use in New Zealand: New Zealand Journal of Geology and Geophysics, v. 13, p. 937–968.

Goff, J.C., 1983, Hydrocarbon generation and migration from Jurassic source rocks in the E. Shetland Basin and Viking Graben of the northern North Sea: Journal of the Geological Society, London, v. 140, p. 445–474.

Hancock, N.J., and Taylor, A.M., 1978, Clay mineral diagenesis and oil migration in the middle Jurassic Brent Sand Formation: Journal of the Geological Society, London, v. 135, p. 69–72.

Morton, N., Smith, R.M., Golden, M., and James, A.V., 1987, Comparative stratigraphic study of Triassic-Jurassic sedimentation and basin evolution in the northern North Sea and north-west of the British Isles, *in* Brooks, J. and Glennie, K. W., eds., Petroleum Geology of North-West Europe: London, Graham and Trotman, p. 697–709.

Richards, P.C., and Brown, S., 1986, Shoreface storm deposits in the Rannoch Formation (Middle Jurassic), North West Hutton Oilfield: Scottish Journal of Geology, v. 22, p. 367–375.

Richards, P.C., Brown, S., Dean, J.M., and Anderson, R., 1988, Short Paper: A new palaeogeographic reconstruction for the Middle Jurassic of the northern North Sea: Journal of the Geological Society, London, v. 145, p. 883–886.

Scotchman, I.C., Johnes, L.H., and Miller, R.S., 1989, Clay diagenesis and oil migration in Brent Group Sandstones of N.W. Hutton Field: Clay Minerals, v. 24, p. 339–374.

Threlfall, W.F., 1981, Structural framework of the central and northern North Sea, *in* Illing, L.V. and Hobson, G.D., eds., Petroleum Geology of the Continental Shelf of North-West Europe: London, Heyden, p. 98–103.

Vollset, J., and Doré, A.G., 1984, A revived Triassic and Jurassic lithostatigraphic nomenclature for the Norwegian North Sea: Norwegian Petroleum Directorate Bulletin 3, p. 53.

Key Words

Northwest Hutton Field, United Kingdom, East Shetland basin, Brent Group, Middle Jurassic, Bajocian-Bathonian, wave-dominated deltas, shallow-marine shelf sandstone, shoreline sandstone, and fluviodeltaic sandstone, diagenesis, illite, petrography, North Sea, quartz, illite.

11
Deltaic Depositional Controls on Clinton Sandstone Reservoirs, Senecaville Gas Field, Guernsey County, Ohio

Brian W. Keltch, Dail A. Wilson and Paul E. Potter

BDM International Inc., Germantown, Maryland 20874; Consolidated Resources of America, Cincinnati, Ohio 45202; H.N. Fisk Laboratory University of Cincinnati, Cincinnati, Ohio 45201

Introduction

Significant reserves of oil and gas remain undiscovered in well-explored cratonic basins for several reasons. First, depositional sequences are relatively thin, often below seismic resolution, due to low subsidence rates in these basins. This can decrease the size and continuity of sandstone reservoirs, resulting in small and complex stratigraphic traps. Second, the stratigraphic detail necessary to delineate sandstone bodies requires a significant amount of well control. Finally, much of the exploration in these basins was done very early in the life of the oil industry and proceeded without the benefit of a depositional systems approach. This chapter presents methods and conclusions found useful in oil and gas development of sandstone reservoirs in a thin cratonic delta system in the Appalachian basin.

The Lower Silurian Clinton Sandstone is the most drilled and prolific hydrocarbon-bearing formation in eastern Ohio. Successful exploration for its subtle stratigraphic traps requires detailed knowledge of its depositional systems and their relationship to production. To better understand this critical relationship, sandstone thickness maps were constructed using 1,983 wells and 25 miles (40 km) of gamma-ray log cross sections, and two cores were studied in detail.

Two river-dominated, cratonic deltas, the Claysville and the Salt Fork, are present in Guernsey County, Ohio, and are typical of the small, deltaic complexes found along the eastern margin of the Clinton-Medina production trend of the Appalachian basin. From a detailed analysis of these deltaic systems, we can better interpret the economic potential of similar Clinton delta systems.

More than 1,900 wells have been drilled for Clinton development in Guernsey County since 1968. Discovery wells were eastern offsets to known production in adjacent Muskingum County. Virtually all the environments of a cratonic delta system are found in the Clinton: thin gray prodelta shale, thin distributary-mouth-bar sandstone, thick distributary-channel sandstone, interdistributary dark gray shale, and overbank shale, all of which are capped by thin marine shale and a persistent limestone. Lateral variations in deltaic subenvironments produced multistoried, coalescing, and lenticular sand bodies with abrupt lateral variations in lithology and sandstone thickness. Accurate prediction of the discontinuous sandstone bodies of the Clinton is the key to successful exploration.

First produced in 1887, the Clinton-Medina Sandstone of Silurian age is the most prolific gas and oil producer in eastern Ohio, western New York, and southern Ontario, and is one of the larger gas-bearing stratigraphic trap systems in the United States. The producing trend covers an area of approximately 27,000 square miles (6.9×10^4 km²) and lies on the northwest flank of the Appalachian basin, approximately parallel to the Cincinnati-Findlay arch and the Algonquin axis (Fig. 11-1). Guernsey County, Ohio is 531 square miles (1.3×10^3 km²) in area and lies on the eastern margin of the producing trend.

Fig. 11-1. Map showing the Clinton production trend and major structural features of the Appalachian basin and Guernsey County, Ohio.

Access to northeastern U.S. gas markets provides higher gas prices relative to more distant major gas-producing basins, and the low production rates of the Clinton are offset by a 90% well completion rate and inexpensive drilling costs.

Guernsey County was chosen for study because it has abundant well control, sits on the eastern fringe of current downdip Clinton development, and offers a variety of traps. Closely spaced well control affords an accurate definition of reservoir geometry.

Profit margins for Clinton wells preclude expensive logging techniques, petrophysical analysis, or seismic support. However, a simple gamma-ray/neutron log provides sufficient data to map sandstone bodies effectively and to determine their depositional environment to project reservoir potential. This simple, inexpensive approach has been proven an effective method for Clinton exploration and is also applied in many areas outside the Appalachian basin (i.e., the midcontinent region, the Illinois basin, and west Texas of the U.S.). These methods have application worldwide for the many small, deltaic reservoirs that occur in cratonic basins.

Regional Setting

Deposition of the Lower Silurian sequence of interbedded sandstone, siltstone, and shale occurred in response to the last stages of the Taconic Orogeny,

which formed a linear landmass southeast of the Appalachian basin. This Late Ordovician-Lower Silurian tectonism in the Appalachian basin produced a dominantly alluvial wedge of clastics which thins markedly to the west, where it intertongues with shallow-marine shales and carbonates. Lithofacies trend approximately parallel with isopach contours in an irregular arcuate pattern (Fig. 11-2).

Maximum thickness of more than 1,400 feet (430 m) occurs in northeastern Pennsylvania, northwestern New Jersey, and southeastern New York as the lower portion of the Shawangunk Conglomerate (Fisher, 1954), which is interpreted as a braided-stream deposit near the source area.

In central Pennsylvania, West Virginia, and western Virginia, the Lower Silurian Tuscarora Sandstone consists of thick-bedded, white, fine- to medium-grained quartzose sandstones cemented by silica. The Tuscarora Sandstone is generally interpreted as a distal braided-stream deposit. However, the formation has a dual origin, as many outcrops contain marine trace fossils and sedimentary structures (Cotter, 1982). Correlation of the mixed facies of the Tuscarora with the Clinton of eastern Ohio is problematic and needs study.

West of this area, in eastern Ohio and western New York, the Clinton and the Medina Sandstones form elongate sandstone bodies interbedded with shales and limestones indicating a deltaic environment (van Tyne, 1966). In central Ohio, the Brassfield Limestone is interpreted as a shallow-marine shelf carbonate (Knight, 1969). In Tennessee and northern Alabama, Lower Silurian shallow-marine deposits grade westward from the Clinch Sandstone to the shales and carbonates of the Rockwood Formation.

Thus, the lower Silurian clastic wedge is characterized by three main environmental settings (Yeakel, 1962, p. 1535): an eastern, nonmarine alluvial coastal plain; a transitional, deltaic zone; and a western, marine offshore zone. Clinton production occurs primarily in the deltaic deposits of the transitional zone.

Stratigraphy

Lower Silurian deposits in Guernsey County range from 165 to 200 feet (50.3–61.0 m) thick, rest unconformably on the Upper Ordovician Queenston Shale, and are conformably overlain by the Middle Silurian Niagaran Series of shale, limestone, and dolomite. The Brassfield Limestone, a 25-foot

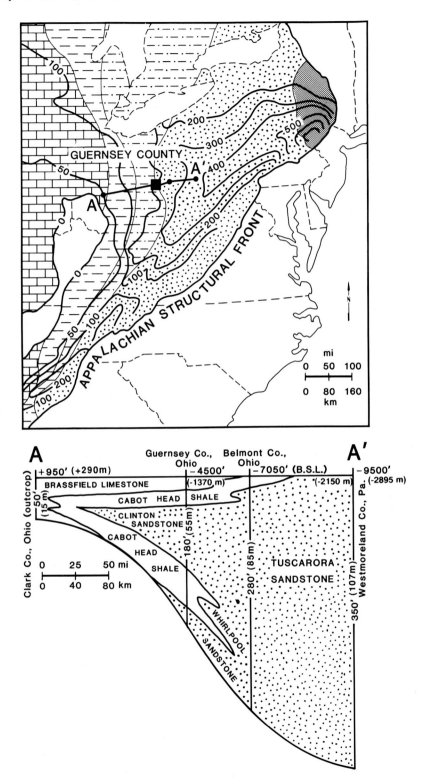

Fig. 11-2. Lithofacies and thickness (contours in feet) of Lower Silurian sediments in the Appalachian basin (after Amsden, 1955, and reprinted by permission of American Association of Petroleum Geologists; Colton, 1970, and reprinted by permission of Interscience Publishers).

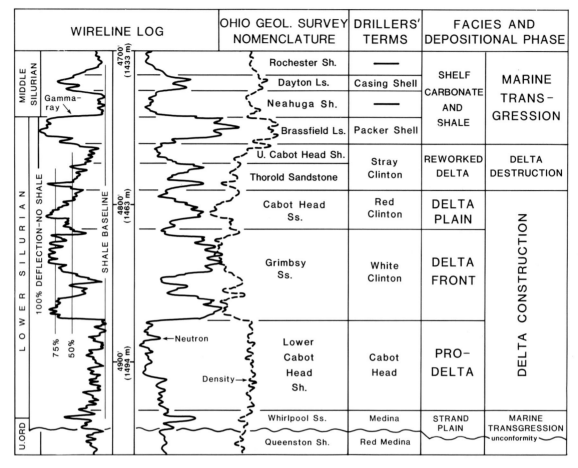

Fig. 11-3. Clinton stratigraphy and type log in Guernsey County, Ohio.

(7.6-m) thick, laterally extensive shelf carbonate, is the uppermost unit of the Lower Silurian section. The basal unit, the Whirlpool Sandstone, is a transgressive strandplain deposit 2 to 25 feet (0.6–7.6 m) thick (Knight, 1969). The reservoirs of the Clinton Sandstone lie between these two units and are separated from them by the Upper and Lower Cabot Head Shales (Fig. 11-3). Sandstones of the Clinton are divided into three units by drillers based on color variations: the upper Stray, the middle Red, and the lower White. The Ohio Geologic Survey names these units the Thorold Sandstone (Stray), the Cabot Head Sandstone (Red), and the Grimsby Sandstone (White) based on correlations with Lower Silurian outcrops in New York (Knight, 1969).

The vertical sequence of Lower Silurian lithologies in Guernsey County is an excellent example of a cratonic margin delta system. Basal sandstones of the Whirlpool were deposited as a strandplain on a Late Ordovician erosional surface. Transgression of

the Early Silurian seas ended with deposition of the prodelta mudstones of the Lower Cabot Head Shale. Deltaic progradation continued as the delta-front and delta-plain sandstones of the Grimsby and Cabot Head Sandstones were deposited. The Cabot Head Sandstone represents maximum deltaic progradation. After deposition of the Cabot Head, delta abandonment was followed by transgression and reworking of delta-plain deposits to produce the thin, littoral sandstone bodies of the Upper Cabot Head Shale and ultimately shelf limestones of the Brassfield Formation (Knight, 1969). Thus, four major depositional phases are recorded in the Clinton: an initial, marine transgression marked by a thin strandplain sandstone; a progradational sequence consisting of prodelta shale and delta-front and delta-plain reservoir sandstones; a destructional phase consisting of thin, littoral sandstones; and, finally, a marine transgression which resulted in deposition of marine shale and shelf carbonates.

A gamma-ray, compensated density-porosity, and neutron logging suite is most commonly run in Clinton wells. In areas where the Clinton produces significant water with gas, a guard induction log is also run. The gamma-ray curve is most useful for environmental interpretation, because by reacting to the clay content it most clearly differentiates sandstone from shale. Finer-grained sandstones commonly contain more matrix clay than coarser-grained sandstones; thus, the gamma- ray curve is a rough measure of grain size and helps define either an upward-coarsening or an upward-fining sequence (Hilchie, 1978).

Sandstone Trends

Isopach Maps

Isopach maps on both a regional and a local scale are essential to understand the depositional systems of the Clinton and to determine sandstone distribution and areas of major sediment input. An isopach map of the complete Lower Silurian interval (from the top of the Brassfield Limestone to the bottom of the Whirlpool Sandstone) for southeastern Ohio shows three thick lobes outlined by the 190-foot (58-m) isopach contours (Fig. 11-4). The Tuscarawas lobe is the largest and generally outlines the prolific North Canton Field described by Knight (1969). The Guernsey and Noble lobes are progressively smaller and trend to the southwest. These lobes indicate sediment dispersal patterns, define local paleodip, and outline the relative size of delta systems in the Clinton.

Clean-sandstone isopach maps show the shape and orientation of thick sandstones in Guernsey County. Thickness values for these maps are calculated from the gamma-ray curve by defining a shale baseline and a 100% shale-free line for each individual log. Prodelta shales of the Clinton sequence or Middle Silurian shales are used to define the shale baseline, and the Brassfield Limestone gamma-ray response is the 100% deflection. Thickness values are calculated by counting the cumulative thickness of sandstones with greater than 50% or 75% deflections from the shale baseline. In this way, the multiple sandstone bodies of the Clinton were reduced to one mappable thickness value. Sandstone bodies with the greatest deflections generally have relatively high porosities (10–16%) and are the best reservoirs. This simple relationship allows operators

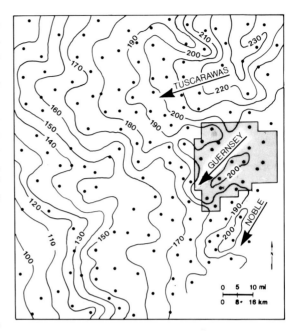

Fig. 11-4. Lower Silurian isopach map (contours in feet) in southeastern Ohio showing three major depositional lobes (arrows). Guernsey County is shaded.

in eastern Ohio to locate and project the best reservoir trends (Pees, 1987).

Care must be taken in using these maps because they combine sandstones of various depositional origin, and thus a thick sandstone trend may represent only a series of thin, poor quality reservoirs. Carefully placed log cross sections help alleviate this problem.

Cross Sections

Closely spaced cross sections were constructed to help define the depositional trends and to determine the depositional characteristics of the individual Clinton sandstone members. Fifteen cross sections using 89 well logs were constructed throughout Guernsey County. Four of these sections are shown with their environmental interpretations in Figure 11-5.

Each section is divided into five lithofacies based on the vertical sequence and gamma-ray log signatures. These lithofacies are: prodelta shale and siltstone; distributary-mouth-bar sandstone and siltstone; thin, barrier-island and sheet siltstone and shale representing reworked deltas; and marine transgressive shale and shelf limestone.

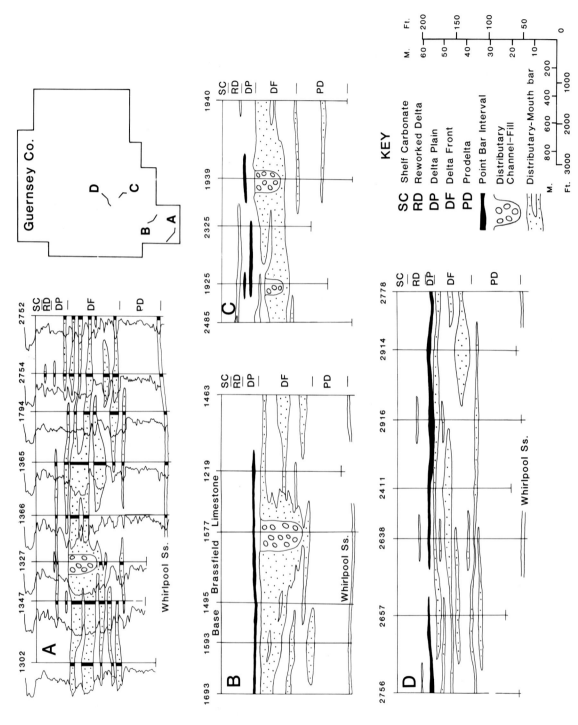

Fig. 11-5. Cross sections and environmental interpretations of the Claysville Delta. The distance between section logs was kept to one well spacing, 900 to 1,800 feet (275–550 m), where possible. Locations of cross sections are shown on the inset map.

Sections A, B, and C are strike sections through the thick sandstone of the Claysville trend in southwestern Guernsey County. Sandstone bodies are well-defined distributary-mouth bars 5,000 to 6,000 feet (1,525–1,830 m) wide, cut by distributary-channel fill sandstones less than 1,000 feet (305 m) wide. Distributary-mouth-bar sandstones are 10 to 40 feet (3.0–12.2 m) thick and fairly continuous. Distributary-channel sandstones are 50 to 60 feet (15.2–18.3 m) thick and always cut through surrounding mouth-bar deposits. Sandstone reservoirs of the delta-plain facies are 40 to 120 feet (12.2–36.6 m) below the bottom of the Brassfield Limestone with the deepest sandstones in the southwest extension of the trend. Prolific oil production in southwestern Guernsey County may be due to the low stratigraphic position (most delta-front reservoirs in the county are less than 100 feet (30.5 m) below the Brassfield Limestone) of these reservoirs. These oil-producing sandstones are thicker than the area average and are in direct contact with marine prodelta shales.

Where thick, shale-free sandstones occur, there is an increase of 6 to 10 feet (1.8–3.0 m) in the interval thickness between the Brassfield Limestone and the Whirlpool Sandstone. This increase in thickness may have the potential to delineate distributary channels on seismic data because the Brassfield and Whirlpool are good reflectors; however, current seismic technology cannot resolve such small changes at a depth of 5,000 feet (1,525 m).

Section D is a dip section located outside the thick sandstones of the Claysville Trend. Sandstone bodies in this section are interdistributary-bay and crevasse-splay deposits interbedded with siltstone and shale. Sandstones are very continuous and thicken and thin irregularly from 5 to 20 feet (1.5–6.1 m). Here the Clinton sequence produces poorly because it consist of many thin, low-permeability reservoirs.

Slice Maps

Slice maps are used to determine sandstone distribution patterns at various intervals in sandstone-shale sequences. Slice maps of 50% shale-free sandstones were constructed for four intervals in the Claysville Delta using 584 wells (Fig. 11-6). Each slice represents approximate time-equivalent intervals in the depositional history of the Claysville Delta. The choice of interval thickness is based on recurring sandstone bodies and shale breaks at about the same

stratigraphic position, in the hope of mapping the lateral extent of genetically related sandstones (Busch, 1971).

Few sandstones are found in the lowest slice, Slice 4, which is 100 to 170 feet (30.5–51.8 m) below the datum (base of the Brassfield Limestone). Gamma-ray log signatures indicate that deposits in this interval are dominantly prodelta shale and siltstone with minor, distal distributary-mouth-bar sandstone. The isopach map of this interval shows an irregular distribution of 8- to 16-foot (2.4–4.9 m) thick sandstone centered around the main distributary channel of the Claysville Delta. Distal distributary-mouth-bar sandstones in this slice mark the first progradation of delta-front sediments over prodelta shales.

A narrow, slightly meandering distributary-channel deposit, surrounded by mouth-bar deposits, is defined by Slice 3, which is 70 to 100 feet (21.3–30.5 m) below the datum. Channel-fill sandstones have blocky log signatures, are 16- to 28-feet (4.9–8.5 m) thick, and are elongate parallel to the paleodip. Thin, 8- to 16-foot (2.4–4.9 m) thick, linear sandstone bodies occur on both sides of the distributary channel and trend perpendicular to paleodip; these may be interdistributary shoal deposits.

Slice 2, 40 to 70 feet (12.2–21.3 m) below the datum, shows a distributary-channel deposit similar in orientation and thickness to the channel deposits in Slice 3, although this channel has a more fluvial character with wider and less continuous sandstone bodies. Log signatures in these channel deposits are blocky and suggest upward-fining. Sandstone bodies on both sides of the channel are interpreted as thin interdistributary shoals and crevasse subdelta deposits, some up to 20 feet (6.1 m) thick.

Reworked delta-plain sandstone and siltstone and transgressive shale are the dominant deposits of Slice 1. Sandstone bodies in this interval are rare, have thin and symmetrical log signatures, and are less than 15 feet (4.6 m) thick.

The slice isopach maps provide an accurate history of the development of the Claysville Delta. Slice 4 represents the beginning of the delta construction phase by the initial progradation of distal distributary-mouth-bar deposits over prodelta mud. Slices 2 and 3 mark the continued progradation of delta-front and delta-plain deposits, and the maximum delta progradation is represented by Slice 2. The best reservoir sandstones were deposited as the distributary channels depicted by Slices 2 and 3. Abandonment of the Claysville Delta is shown by Slice 1 and represents either decreased sediment

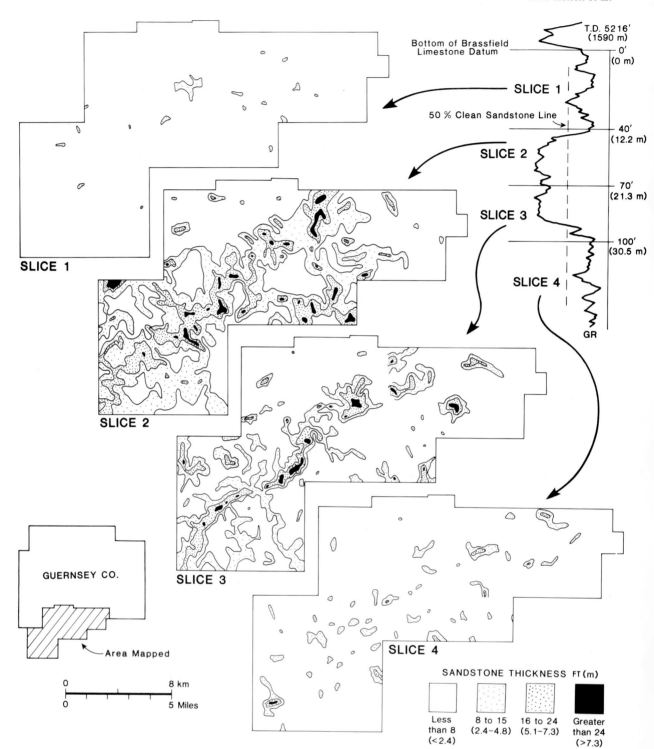

Fig. 11-6. Slice isopach maps and type log of the Claysville Delta.

supply or "delta-lobe switching" producing thin, reworked delta-front sandstone and siltstone and transgressive marine shale.

Lithofacies

Two cores demonstrate the validity of environmental interpretations of well logs by calibrating logs to lithofacies sequences. Core material in Guernsey County was not available, so the two closest cores were studied—the Chester Young No. 1/Pominex Inc., 9 miles (14.5 km) north of Guernsey County in Coshocton County, and the Ferrell No.1/National Associated Petroleum, Inc., 13 miles (21 km) to the west in Muskingum County (Fig. 11-7).

The Chester Young No. 1 well has a gamma-ray signature similar to the lower delta-front depositional environment present in northwestern Guern-

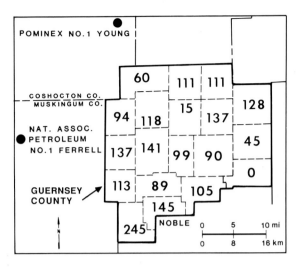

Fig. 11-7. Number of gamma-ray logs studied per township in Guernsey County and locations of cores (dark dots).

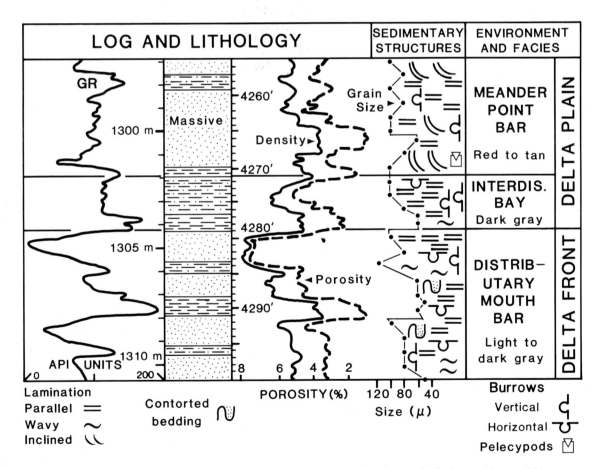

Fig. 11-8. Wireline log and core description of the Young No. 1/Pominex well, Coshocton County, Ohio.

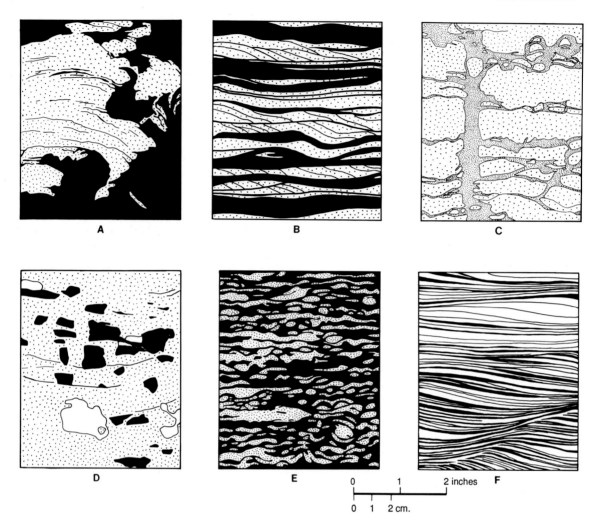

Fig. 11-9. Core drawings from the Young No. 1/Pominex: (A) 4,286 feet (1,306 m), light gray, massive sandstone slumped into underlying shale; (B) 4,288 feet (1,307 m), interbedded sandstone and shale (black) with parallel and wavy bedding; (C) 4,293 feet (1,309 m), light gray sandstone with horizontal and vertical siltstone-filled burrows; (D) 4,263 feet (1,299 m), dark red sandstone has large horizontal and vertical burrows filled with light red siltstone (black), contains light green shale clast (white); (E) 4,278 feet (1,304 m), dark gray, bioturbated shale with contorted bedding and scattered lenses of siltstone; (F) 4,255 feet (1,297 m), light red to tan cross-laminated sandstone.

sey County, which includes distal delta-front sandstones overlain by poorly developed delta-plain sandstones. Forty-five feet (13.7 m) of core were recovered from a depth of 4,255 to 4,300 feet (1,297–1,311 m) (Figs. 11-8, 11-9).

Delta-front deposits from 4,284 to 4,300 feet (1,306–1,311 m) comprise upward-coarsening sequences of light gray, very fine-grained sandstone and siltstone interbedded with light to medium gray shale. Upward-coarsening sequences are 4 to 6 feet (1.2–1.8 m) thick and consist of sandstone interbedded with lenticular shale, grading upward into massive sandstones. Bedding tends to be wavy to parallel and commonly contains soft-sediment deformation features along sandstone-shale contacts. Siltstone-filled vertical and horizontal burrows are common throughout the unit and increase in abundance with increased shale content. The upward-coarsening sequence, sedimentary structures, and bioturbation conclusively show that these sediments were deposited as distributary-mouth bars.

Distributary-mouth-bar sandstones are overlain by 12 feet (3.7 m) of dark gray shale, from 4,272 to 4,284 feet (1,302–1,306 m), interbedded with lenticular, ripple cross-laminated siltstone beds less than 0.5 inch (1.3 cm) thick. Bioturbation in these sediments was abundant and interrupts or destroys parallel laminations. Thin, repetitive interbeds of silt and shale, parallel laminations, and bioturbation suggest deposition under the low-energy conditions characteristic of interdistributary bays.

The delta-plain sequence is present in the top 17 feet (5.2 m) of the core and consists of mottled red to tan colored, very fine-grained sandstone and siltstone with common light green shale clasts. The sequence contains extensive vertical, horizontal, and U-tube burrows. Nonbioturbated sandstone zones are cross-laminated, fine upward, and contain pelecypod fragments oriented parallel to laminae. The sedimentary structures, fauna, and upward-fining grain size of these sediments indicate they were deposited on a delta plain as meander point bars.

The gamma-ray, density, and porosity curves of the Young No. 1 well accurately represent the lithologic and grain-size variations measured from the core, and sedimentary structures and faunal evidence from the core lend additional support to environmental interpretations based on log signatures. The core and the logs both show that the delta-front distributary-mouth-bar sandstones are the best reservoirs in this well, which is basinward of the thicker deltaic deposits in Guernsey County.

The Ferrell No. 1 well is located in Muskingum County approximately 14 miles (23 km) west of Guernsey County and represents a shallow-marine facies not present in the study area. The Clinton sequence in this area is composed of thin, barrier-island sandstones enveloped by interbedded marine shales and carbonates. Cores were taken from 3,900 to 3,925 feet (1,189–1,196 m), traversing a section of interbedded marine siltstone, shale, and carbonate. Unfortunately, wireline logs of this well are not available.

The upper portion of the core, from 3,900 to 3,913 feet (1,189–1,193 m), consists of light to medium gray, coarse siltstone and very fine-grained sandstone interbedded with bioturbated shale. Burrows are filled with calcite or light green siltstone. Brachiopod debris is common throughout the section. Two 6-inch (15-cm) thick, light red brachiopod and bryozoan grainstones are present in this section and indicate a high-energy carbonate shelf environment. The lower portion of the core, from 3,913 to 3,925 feet (1,193–1,196 m), consists of a single upward-coarsening sequence of dark, olive gray shale interbedded with lenticular siltstones. Bedding is parallel and uniform, except where altered by silt-filled horizontal burrows. The marine sediments and fauna found in the Ferrell No. 1 core reflect gradual basinward facies changes from deltaic sedimentation in Guernsey County to offshore shallow-marine sedimentation to the west.

Depositional Setting

Dip-oriented elongate and lobate delta-front sandstone bodies are common deposits in tectonically stable cratonic delta systems (Coleman and Wright, 1975; Brown, 1979). The gentle slopes of cratonic basin shelves result in river-dominated delta systems, because wave and tidal processes are unable to redistribute sand at the mouth of distributary channels. Distributary channels commonly change course and overlap, resulting in stacked, multiple regressive sequences.

Paleogeographic reconstruction of the prograding delta-front deposits of the Clinton in Guernsey County was accomplished by projecting facies and environmental information from cross sections and slice isopach maps along trends defined by isopach maps of 50 and 75% shale-free sandstones. Delta-plain, point-bar sandstone reservoirs were not mapped, because they are generally less than 20 feet (6.1 m) thick and are infrequent relative to thicker, delta-front sandstones. The distribution of both distributary-channel fill and mouth-bar sandstones is represented in Figure 11-10.

The Claysville Delta prograded in a southwesterly direction completely across Guernsey County as three thick distributary channels in the northeast, joining together as one channel in the southwest. Distributary channels are continuous for up to 10 miles (16 km), are cut through mouth-bar deposits, and are surrounded by thin interdistributary bay and crevasse-subdelta deposits.

In contrast to the Claysville Delta, the Salt Fork Delta prograded only to the western townships of Guernsey County as a large lobate delta composed of many smaller, discontinuous distributary channels. Continuity of distributary-channel deposits generally does not exceed 3 miles (4.8 km). Distributary-mouth-bar deposits prevail in the Salt Fork Delta. These are thinner, contain more multistory sandstone, and are more continuous than similar deposits in the Claysville Delta. The discontinuity of distributary-channel deposits and dominance of

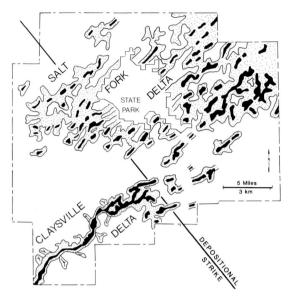

Fig. 11-10. Salt Fork and Claysville Deltas. Distributary-mouth bars are stippled, and distributary channels are shown in black.

mouth-bar deposits indicate that channel systems in this delta system were rapidly abandoned.

Production Characteristics

Clinton wells are completed in multiple, solution-gas drive reservoirs which produce oil, gas, and formation water. The relative amounts of each vary regionally. In Guernsey County, approximately 85% of Clinton production is gas (based on a BTU comparison of 6 MCFG = 1 BO), although in localized areas of the county oil accounts for the majority of production. Oil is Pennsylvania grade with an API gravity of 38 ° to 42 ° and gas is of good quality, with a BTU value exceeding 1 MBTU per cubic foot of gas (3.6×10^4 BTU/m³).

Hydrocarbons trapped in Clinton reservoirs are assumed to have been derived from the Upper and Lower Cabot Head Shales which encase the reservoir sandstones and act as both source and seal.

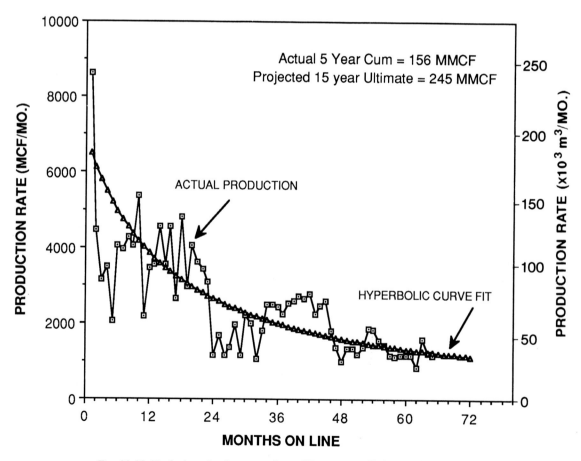

Fig. 11-11. Typical production curve for a Clinton gas well, Senecaville Gas Field.

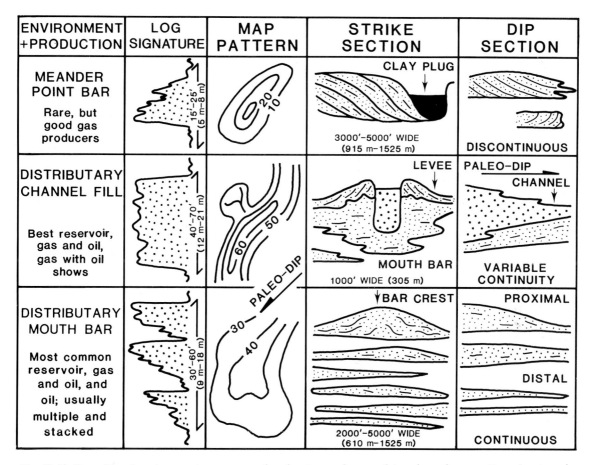

ENVIRONMENT +PRODUCTION	LOG SIGNATURE	MAP PATTERN	STRIKE SECTION	DIP SECTION
MEANDER POINT BAR Rare, but good gas producers	15'-25' (5 m-8 m)	20 / 10	CLAY PLUG ↓ 3000'-5000' WIDE (915 m-1525 m)	DISCONTINUOUS
DISTRIBUTARY CHANNEL FILL Best reservoir, gas and oil, gas with oil shows	40'-70' (12 m-21 m)	60 / 50 PALEO-DIP	LEVEE ↓ MOUTH BAR 1000' WIDE (305 m)	PALEO-DIP → CHANNEL ↓ VARIABLE CONTINUITY
DISTRIBUTARY MOUTH BAR Most common reservoir, gas and oil, and oil; usually multiple and stacked	30'-60' (9 m-18 m)	30 / 40	↓BAR CREST 2000'-5000' WIDE (610 m-1525 m)	PROXIMAL DISTAL CONTINUOUS

Fig. 11-12. Depositional environments, gamma-ray log signatures, clean-sandstone isopach map patterns (contours in feet), and strike and dip sections of Clinton reservoirs in Guernsey County.

Thus, timing and migration are assumed to be early. Unfortunately, no empirical studies have been conducted to support or disprove these assumptions.

Field size in these stratigraphic traps is determined almost wholly by the geometry and lateral extent of the reservoir sandstones, which underscores the need for a detailed understanding of Clinton depositional environments. Fields are small and range from 3 to 15 wells on 40-acre (16.2-ha.) spacing because the reservoir sandstones are thin and discontinuous. More than 150 individual fields have been identified in the two deltas in Guernsey County. The discontinuous sandstones cause highly variable production capabilities between closely spaced wells.

Formation porosity is intergranular and ranges from 2 to 16%. Permeability is low, ranging from 0.01 to 5.0 md (Knight, 1969). Thus, all wells do not produce naturally but require stimulation by hydraulic or gas fracture treatments. After stimula-

tion, formation rock pressures range from 800 to 1200 psi (5.5–8.3 × 10³ kPa).

Wells are commonly equipped with 4.5-inch (11.4-cm) casing and 1.5-inch (3.8-cm) production tubing. Fluids are cleared from the wellbore by a gas lift device called a "rabbit" or, if necessary, a pump jack is installed. Initial production rates range from 20 to 900 MCFPD (5.7–254 × 10² m³/D) including equivalent oil. Formation water production is low, ranging from 0 to 2,000 barrels (0–318 m³) per year, and generally does not interfere with gas production except in older wells with depleted reservoir pressures. Clinton production exhibits hyperbolic decline, with the majority of wells producing one-half of their ultimate production during the first three to five years. Well life in Guernsey County ranges from 5 to 25 years and averages 15 years at today's economics. A typical decline curve for a well in the Senecaville Gas Field is shown in Figure 11-11.

Facies-Related Production

Hydrocarbons are produced from multistoried and laterally discontinuous sandstone bodies deposited as distributary-mouth bars, distributary-channel fills, and delta-plain point bars. These reservoir types are interrelated in a predictable manner, each having a unique isopach pattern, log signature, relative abundance and, most importantly, production history (Fig. 11-12). Production data from 408 wells in 15 of 19 townships of Guernsey County were classified by reservoir type to provide the crucial "data link" between production and geologic factors.

Production data were obtained from well operators or the county auditor's reports. The amount of gas and oil produced during the first year and the dominant reservoir type were recorded for each well. Production during the first year was used as an equal-time comparison for each well and provides a good estimation of the lifetime production capabilities of a well.

Distributary-mouth-bar reservoirs occur in more than 74% of the wells. They are characterized by upward-coarsening log signatures and elongate isopach patterns, are 2 to 22 feet (0.6–6.7 m) thick, and exhibit poor production, with median values of 13 MMCFG (3.7×10^5 m^3) and 200 barrels (31.8 m^3) of oil in the first year of production.

Distributary-channel-fill reservoirs are superimposed on underlying mouth-bar deposits, have blocky log signatures, and form linear, narrow "shoestrings." They occur in 18% of the wells and exhibit good production with median production of 30 MMCFG (8.5×10^5 m^3) and 430 barrels (68.4 m^3) of oil in the first year. Meander point-bar reservoirs have the best production but are encountered in only 8% of the wells. These sandstones fine upward, are the most prolific gas producers with median production of 58 MMCFG (1.6×10^6 m^3) and 190 barrels (30.2 m^3) of oil, are ovoid to kidney shaped, and range up to 25 feet (7.6 m) thick (Fig. 11-13).

Senecaville Gas Field

The Senecaville Gas Field is located in south-central Guernsey County and represents the northeasternmost extension of the Claysville Delta. Development of the field began in 1971. A production map of the Senecaville Gas Field shows its elongate outline (Fig. 11-14). The best well in the field, the Dziedzic No. 1/Consolidated Resources of America, was drilled in 1980. In the first year of production, the Dziedzic No. 1 produced 153 MMCFG (4.3×10^6 m^3).

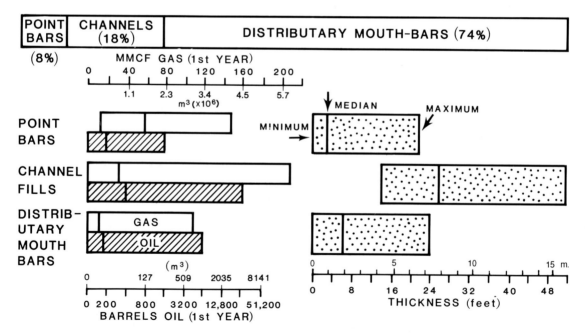

Fig. 11-13. Distribution, production, and thickness of Clinton reservoirs classified by depositional environment.

Slice maps of sandstone thickness were used to determine sandstone distribution patterns at various intervals in this sandstone-shale sequence. Slice maps of 50% shale-free sandstones (50% clean sandstones) were constructed for four intervals in the Senecaville Gas Field (Fig. 11-15). Each slice represents approximate time-equivalent intervals in the depositional history of this field. The choice of interval thickness was based on recurring sandstone bodies and shale breaks at about the same stratigraphic position, in hope of mapping the lateral extent of genetically similar sandstones (Busch, 1971). The base of the Brassfield Limestone is used as a datum because it is continuous throughout the field and Guernsey County.

Few sandstones are found in the lowest slice, Slice D, 106 to 160 feet (32.3–48.8 m) below the datum.

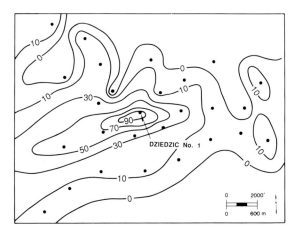

Fig. 11-14. Production map contoured on the first six months of gas production (in MMCFG) of the Senecaville Gas Field.

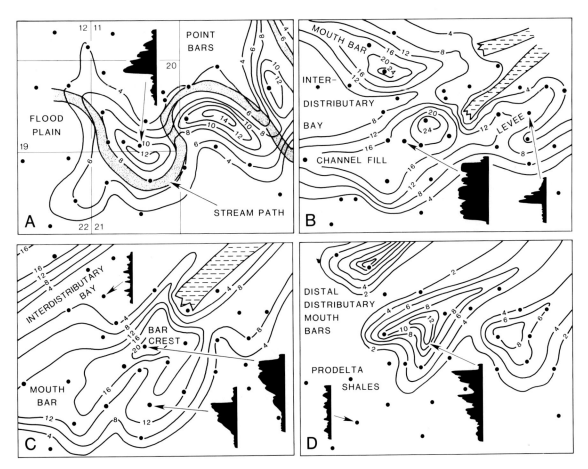

Fig. 11-15. Slice isopach maps (contours in feet), gamma-ray log signatures, and environmental interpretations of the Senecaville Gas Field. Slices A to D are described in the text.

Gamma-ray signatures indicate that deposits in this interval are dominantly prodelta shale and siltstone, with minor distal distributary-mouth-bar sandstone. The slice map of this 54-foot (16.5-m) thick interval shows three thin distributary-mouth-bar lobes less than 14 feet (4.3 m) thick.

Thicker sandstones were deposited in Slice C, 84 to 106 feet (25.6–32.3 m) below the datum, and have the elongate isopach pattern and upward-coarsening log signature of distributary-mouth bars. Maximum sandstone thickness occurs at the mouth-bar crest.

Slice B, 58 to 84 feet (17.7–25.6 m) below the datum, shows the bifurcation of the distributary channel apparent in Slice C and the development of a distributary-channel deposit of the same orientation as the original mouth-bar sandstone. The distributary-channel reservoirs have a blocky log signature and are continuous along paleodip.

Delta-plain, point-bar sandstones are the dominant reservoirs in Slice A, 40 to 58 feet (12.2–17.7 m) below the datum. These reservoirs are less than 15 feet (4.6 m) thick and have upward-fining log signatures.

Together the four slice isopach maps provide an accurate history of the development of sandstone reservoirs of the Senecaville Gas Field. Slice D represents the beginning of the delta construction phase by the initial progradation of distal distributary-mouth-bar deposits over prodelta muds. Slices C and B mark the continued progradation of delta-front distributary channels and mouth bars across the field. Slice A represents the maximum delta progradation and deposition of delta-plain deposits. The most productive reservoir sandstones were deposited as distributary channels in Slices B and C.

Exploration and Production Strategy

Our experience suggests that using the following exploration steps can establish better-than-average production in Clinton and other cratonic-margin deltaic sequences:

1. Delineate local depositional strike and dip by constructing a Lower Silurian isopach map.
2. Construct sandstone isopach maps for the entire interval using the 50% and 75% deflections of the gamma-ray log to outline gross sandstone thickness trends.
3. Construct a dense network of cross sections through the prospect area with sections along depositional strike and dip.

Fig. 11-16. Exploration and reservoir definition methodology for Clinton sandstones.

4. Divide all cross sections into four or five approximately equally thick slices based on recurring sandstone bodies and shale breaks at the same stratigraphic position. Use the log signature of each slice to identify the depositional environments (point bar, distributary-mouth bars, distributary channels, or possible beach deposits) and note the thickness of each sandstone deposit.
5. Plot the production histories on the cross sections and relate these histories to the depositional environment.
6. Carefully map the sandstone thickness and depositional environment of each slice and use all subsurface control to project thickness, environment, and production capability into the prospect area.

In summary, the above procedure effectively relates production to depositional environment based on sandstone distribution patterns and gamma-ray log signatures. As such, it effectively integrates all exploration and development components into a river-dominated delta model for the Clinton. Methods used to define these reservoir types and production characteristics are summarized in Figure 11-16.

Conclusions

Two river-dominated Clinton deltas have been studied in Guernsey County, Ohio: the Claysville and the Salt Fork. Both have elongate and lobate

sandstone bodies but differ in distribution patterns. The narrow Claysville Delta prograded completely across Guernsey County, depositing one to three distributary-mouth-bar sandstones, many of which are cut by distributary-channel-fill sandstones. The Salt Fork Delta prograded over approximately two thirds of Guernsey County and is composed of many discontinuous, distributary-mouth-bar sandstones, few of which are cut by distributary-channel sandstones.

Reservoir quality sandstones were deposited as distributary-mouth-bar and distributary-channel-fill sandstones of the Grimsby Formation (White Clinton) and as delta-plain and point-bar sandstones of the Cabot Head Formation (Red Clinton). Each reservoir type has a unique log signature, stratigraphic position, and geometry.

It is important to define and to understand the characteristics of each reservoir type because reservoir type has a dramatic influence on production. For instance, wells penetrating distributary-channel-fill and point-bar reservoirs are more than twice as productive as wells penetrating distributary-mouth-bar reservoirs.

Distributary-mouth-bar reservoirs occur in more than 70% of the wells, have upward-coarsening log signatures and elongate isopach patterns, and are 2 to 22 feet (0.6–6.7 m) thick. Distributary-channel-fill reservoirs are superimposed on underlying mouth-bar deposits, have blocky log signatures, form linear, narrow "shoestrings," and are found in 18% of the wells. Meander point-bar reservoirs have the best production but are encountered in only 8% of the wells; they fine upward, are ovoid to kidney shaped, and range up to 25 feet thick (7.6 m).

The exploration strategy used in Guernsey County should apply to many thin cratonic delta systems in other basins. Following the suggested steps should result in higher delivery rates and greater ultimate recoverable reserves.

Reservoir Summary

Field : Senecaville Gas Field
Location: Guernsey County, Ohio
Operators: Consolidated Resources of America, Enterprise Energy, and GasPro Inc.
Discovery: 1971
Basin: Appalachian basin
Tectonic/Regional Paleosetting: Cratonic basin
Geologic Structure: Homoclinal dip
Trap Type: Stratigraphic pinch-out
Reservoir Drive Mechanism: Gas solution
  • **Original Reservoir Pressure:** 1,200 psi (8.3 × 10³ kPa) at 4,500 feet (1,370 m) subsea
Reservoir Rocks
  • **Age:** Early Silurian (Llandoverian)
  • **Stratigraphic Unit:** Clinton Formation
  • **Lithology:** Well-sorted and clayey fine- to very fine-grained sublitharenites
  • **Depositional Environment:** Deltaic, river dominated
  • **Productive Facies:** Distributary-channel sandstones, point-bar sandstones, distributary-mouth bar sandstones
  • **Petrophysics**
    • **Porosity Type:** Intergranular
    • **ϕ:** Average 8%, range 2 to 16%, cutoff 4% (logs)
    • **k:** Average 0.5 md, range 0.01 to 5.0 md (from literature, unstressed air)
    • **S_w:** Average 20%, range 10 to 60%, cutoff 30% (logs)
Reservoir Geometry
  • **Depth:** 5,600 feet (1,710 m)
  • **Areal Dimensions:** 1 by 3 miles (1.6 × 4.8 km)
  • **Productive Area:** 1,900 acres (770 ha.)
  • **Number of Reservoirs:** 7
  • **Hydrocarbon Column Height:** NA
  • **Fluid Contacts:** NA
  • **Number of Pay Zones:** 3
  • **Gross Sandstone Thickness:** 185 feet (56 m)

- **Net Sandstone Thickness:** 60 feet (18 m)
- **Net/Gross:** 0.32

Hydrocarbon Source, Migration
- **Lithology and Stratigraphic Unit:** Marine shale, Lower Cabot Head Shale, Lower Silurian

Hydrocarbons
- **Type:** Gas and oil
- **GOR:** 125,000:1
- **API Gravity:** 38° to 42°, Pennsylvania grade
- **Viscosity:** NA

Volumetrics
- **In-Place:** NA
- **Cumulative Production:** 1.6 BCFG (4.5×10^7 m³) and 17.5 MBO (2.8×10^3 m³)
- **Ultimate Recovery:** 4.2 BCFG (1.2×10^8 m³) and 46 MBO (7.3×10^3 m³)
- **Recovery Efficiency:** Approximately 60%

Wells
- **Spacing:** 1,320 feet (402 m), 40 acres (16.2 ha.)
- **Total:** 28
- **Dry Holes:** 5

Typical Well Production:
- **Average Daily:** 140 MCFG (4.0×10^3 m³) and 2 BO (0.3 m³)
- **Cumulative:** 180 MMCFG (5.1×10^6 m³) and 2 MBO (318 m³)

Stimulation: Foam frac with 50,000 lb (2.3×10^4 kg) of sand is common treatment

References

Amsden, T.W., 1955, Lithofacies map of Lower Silurian deposits in central and eastern United States and Canada: American Association of Petroleum Geologists Bulletin, v. 39, p. 60–74.

Brown, L.F., Jr., 1979, Deltaic sandstone facies of the Mid-Continent, in Hyne, N.J., ed., Pennsylvanian Sandstones of the Mid-Continent: Tulsa Geological Society, Special Publication 1, 360 p.

Busch, D.A., 1971, Genetic units in delta prospecting, American Association of Petroleum Geologists Bulletin, v. 55, p. 1137–1154.

Coleman, J.M., and Wright, L.D., 1975, Modern river deltas: Variability of processes and sand bodies, in Broussard, M.L., ed., Deltas, Models for Exploration: Houston Geological Society, Texas, 555 p.

Colton, G.W., 1970, The Appalachian Basin—Its depositional sequences and their geologic relationships, in Fisher, G.W., Pettijohn, F.J., Reed, J.C., and Weaver, K.N., eds., Studies of Appalachian Geology: New York, Interscience Publishers, 460 p.

Cotter, E., 1982, Tuscarora Formation of Pennsylvania: Lewisburg, Pennsylvania, Guidebook: Society of Economic Paleontologists and Mineralogists, Eastern Section, 1982 Field Trip, 105 p.

Fisher, D.W., 1954, Stratigraphy of Medinian Group, New York and Ontario: American Association of Petroleum Geologists Bulletin, v. 38, p. 1979–1996.

Hilchie, D.W., 1978, Applied Openhole Log Interpretation for Geologists and Engineers: Department of Petroleum Engineering, Colorado School of Mines, 294 p.

Knight, W.V., 1969, Historical and economic geology of Lower Silurian Clinton Sandstone of northeastern Ohio: American Association of Petroleum Geologists Bulletin, v. 53, p. 1421–1425.

Pees, S.T., 1987, Mapping Medina (Clinton) play by lumping clean sandstones: Oil and Gas Journal, March 9, 1987, p 42–45.

Tyne, A.M. van, 1966, Progress report—subsurface stratigraphy of the pre-Rochester Silurian rock of New York: Petroleum Geologist of Appalachian Basin Symposium, Pennsylvania State University, p. 97–120.

Yeakel, L.S., 1962, Tuscarora, Juniata, and Bald Eagle paleocurrents and paleogeography in the central Appalachians: Geological Society of America Bulletin, v. 73, p. 1515–1540.

Key Words

Senecaville Gas Field, Ohio, Appalachian basin, Clinton Formation, Silurian, Clandoverian, deltaic, lenticular sandstone, slice maps, distributary-channel sandstone, distributary-mouth-bar sandstone, meander point-bar sandstone, cratonic delta.

12
Deltaic Reservoirs Of The Caño Limón Field, Colombia, South America

Michael N. Cleveland and Jorge Molina

Occidental De Colombia, Inc., Bogotá, Colombia

Introduction

The variability of sand-body architecture and reservoir quality in deltaic systems is high, and the highest variability is encountered in fluvial-dominated deltas. In general, the highest quality sands are the distributary channel sands, but these can be very problematic as reservoirs because of their highly limited lateral extent and the marked variation in the type of channel-fill possible. Some channels are filled with coarse-grained high permeability sand, whereas others are mud-plugged or filled with low reservoir quality, thin-bedded sequences of very fine-grained sand and mud. This chapter illustrates some of the variability commonly seen in fluvial-dominated deltaic reservoirs.

Discovery of the billion-barrel (1.6×10^8 m³) Caño Limón Field in 1983 established the Llanos basin of Colombia as a major oil province (Fig. 12-1). Commercial production began in December 1985, and cumulative recovery is 184 MMBO (2.9×10^7 m³). Occidental de Colombia, with a 25% working interest, operates the field. Other working interest partners are Ecopetrol (50%) and Royal Dutch Shell (25%).

Caño Limón is a structural trap with updip and lateral closure against strike-slip faults. There are 9,070 proved productive acres (3,670 ha.) within the major Caño Limón-La Yuca, Caño Yarumal, and Matanegra fault blocks (Fig. 12-2). The reservoirs are Cretaceous to Oligocene, poorly consolidated deltaic sandstones, at an average depth of 7,600 feet (2,315 m). The reservoir drive mechanism is a strong edge-water drive. River-dominated deltaic sandstones of the Lower Carbonera Formation constitute the largest reservoir. Additional reserves are present in distributary and fluvial channel sandstones of the Upper Carbonera Formation and in wave-dominated deltaic sandstones of the Lower K-1 and K-2A units. The stratigraphic seal for Caño Limón is provided by 400 to 500 feet (120–150 m) of Upper Carbonera C-4 and C-5 continental claystone and shale.

Ultimate recovery from Caño Limón Field can be maximized by controlling water influx through limited-entry perforations and balanced offtake as dictated by the mapping of reservoir heterogeneities and flow barriers. These parameters are recognized through study of the deltaic depositional environments. Although the Caño Limón deltaic reservoirs contain such heterogeneities as horizontal and vertical permeability variations and lateral fluid flow barriers, they are well organized into layers separated by impermeable shales which allow for layer-specific reservoir management (Parker, 1986).

Regional Data

The Llanos basin in eastern Colombia belongs to a series of sub-Andean sedimentary basins. This trend of basins spans the length of South America and is bounded on the east by granitic shield areas and on the west by the Andes Mountains (Fig. 12-1). Caño

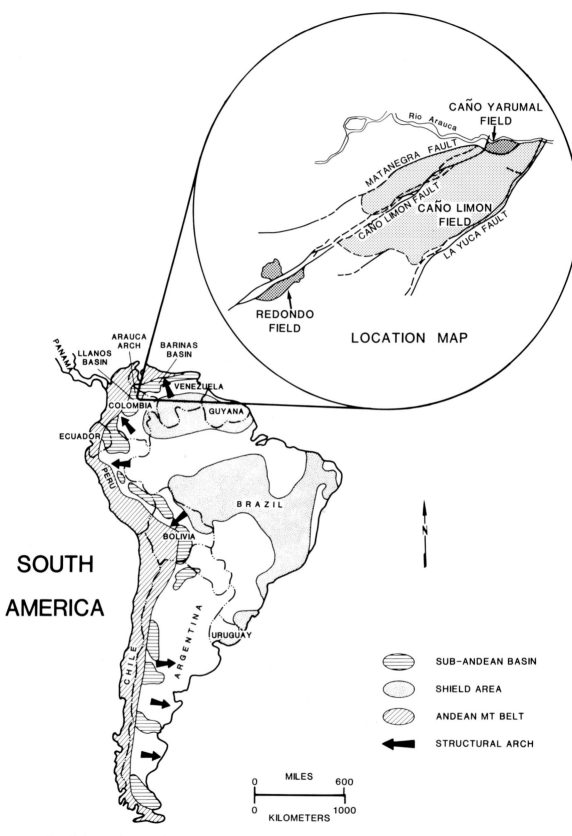

Fig. 12-1. Location map for Caño Limón Field showing the Llanos and Barinas basins and Arauca arch.

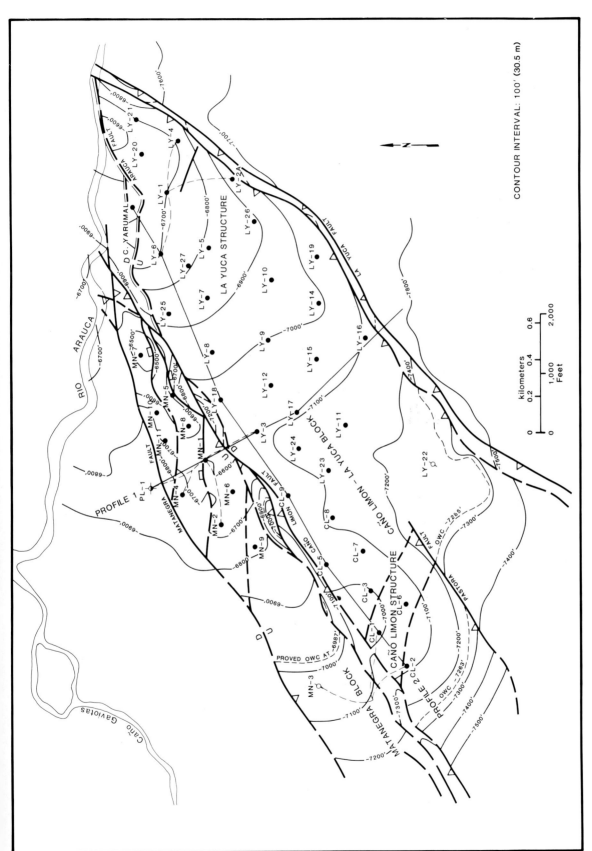

Fig. 12-2. Structural map of Caño Limón Field showing the Matanegra and Caño Limón–La Yuca fault blocks. Also shown are locations of seismic profiles Nos. 1 and 2 (Figs. 12-5, 12-6).

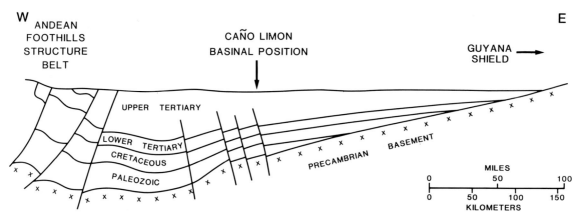

Fig. 12-3. Schematic cross section of the Llanos basin. Thickness of the Paleozoic to Tertiary section reaches 30,000 feet (9,100 m) in the western part of the basin.

Limón lies in the northern Llanos basin where the Arauca arch separates the Llanos basin in Colombia from the Barinas basin in Venezuela (Gabela, 1985).

Within the Llanos basin, 30,000 feet (9,150 m) of Paleozoic to Tertiary sediments onlap Pre-Cambrian basement (Fig. 12-3). Principal oil-productive horizons are in the Oligocene Upper Carbonera, Eocene to Oligocene Lower Carbonera, and Eocene Mirador formations and several Upper Cretaceous units (Parker, 1986). The Caño Limón accumulation occurs on the flank of the gently dipping eastern Llanos basin. Productive Upper and Lower Carbonera and Upper Cretaceous units are present at Caño Limón Field.

The structural history of the Caño Limón Field is directly related to wrench-fault movement. Major regional wrench faults are the Matanegra, Caño Limón, and La Yuca faults which transect the Caño Limón area along a northeasterly trend (Fig. 12-4).

Structure

The Caño Limón accumulation is trapped by updip and lateral closure against sealing faults. The northeast-trending Caño Limón wrench fault has probable right-lateral displacement and vertical displacement of 0 to 500 feet (0–150 m). It divides the field into two major fault blocks: the uplifted wedge-shaped Matanegra block on the northwest and the Caño Limón-La Yuca block to the southeast. The latter block contains the Caño Limón and La Yuca structural highs. Subparallel to the Caño

Limón fault are the Matanegra and La Yuca wrench faults which provide sealing lateral field boundaries and exhibit nearly vertical to reverse displacements. Updip closure is against the Arauca normal fault and other faults near the Arauca River which forms the Colombian-Venezuelan border. The Arauca fault has a vertical displacement of 10 to 350 feet (3–105 m).

The sealing nature of the wrench faults has been demonstrated by independent oil-water contacts on either side of the faults as well as interference testing between wells on either side of the Caño Limón fault. The fact that vertical displacement is zero at some points along the faults suggests that the seals are due to shale smear along the fault planes.

Syndepositional fault motion resulted in local thickening of the Lower Carbonera M1 and M3 zones. In addition, fault motion created topographically low areas such as that between the La Yuca and Caño Limón highs (Fig. 12-5) into which distributary-channel sands were deposited (Cleveland, 1988). Thickness comparisons and stratigraphic positions of fault cutouts in wells suggest that uplift of the northeastern Matanegra block occurred during deposition of the Lower Carbonera, with the Matanegra block later subjected to further faulting during deposition of the Upper Carbonera.

Downdip limits to the southwest in each block are formed by oil-water contacts (Parker, 1986). At present, the proven field area is 9,070 acres (3,670 ha.). Seismic profiles from a 3-D survey of the proven acreage illustrate strike and dip views of the field (Figs. 12-5, 12-6).

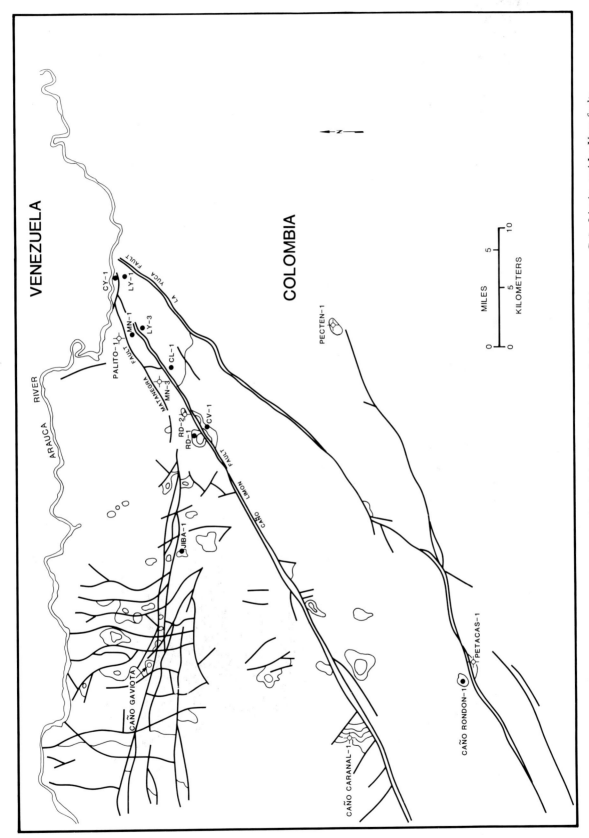

Fig. 12-4. Major northeast-trending wrench faults of the Caño Limón area include the Matanegra, Caño Limón, and La Yuca faults.

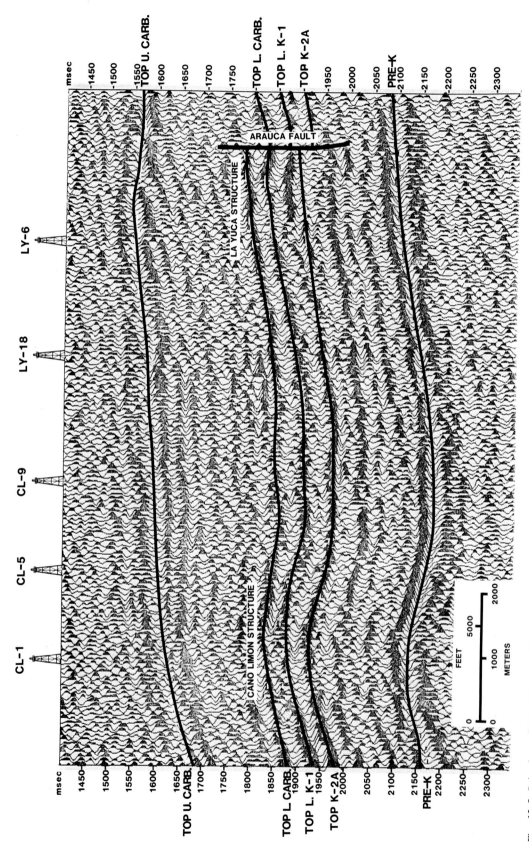

Fig. 12-5. Seismic profile 1 giving a strike view of Caño Limón. Location of the profile is noted on Figure 12-2. The Caño Limón and La Yuca structures and the Arauca normal fault were generated as the result of wrench-fault movement.

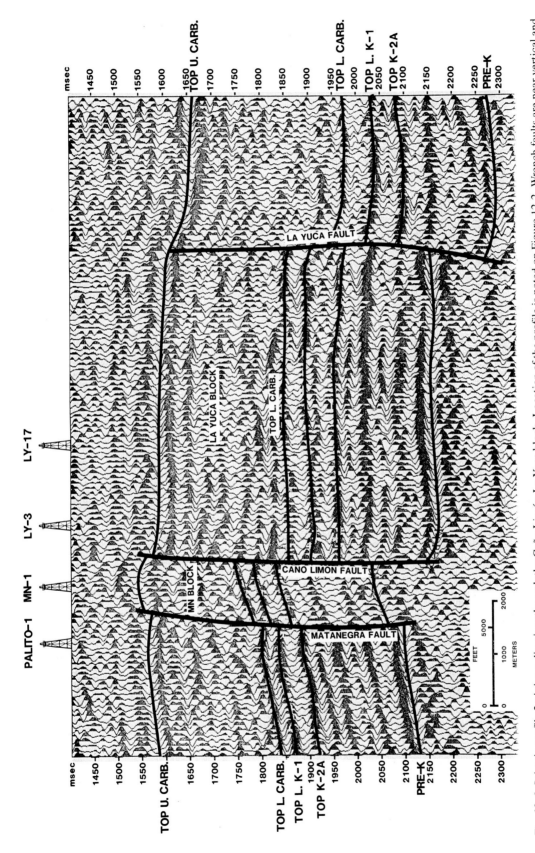

Fig. 12-6. Seismic profile 2 giving a dip view through the Caño Limón-La Yuca block. Location of the profile is noted on Figure 12-2. Wrench faults are near vertical and offset the middle Miocene Upper Carbonera through pre-Cretaceous section.

Stratigraphy

Stratigraphic nomenclature employed at Caño Limón Field is based on a framework of unconformity-bounded sequences (Fig. 12-7). The oil-bearing section comprises parts of four disconformity-separated sedimentary units with a combined thickness of 1,000 feet (305 m). Locally to regionally correlatable shales subdivide the units into productive zones. Overall, the stratigraphic sequence records a major regressive episode. The sequence progresses upward from Cretaceous shallow marine wave-dominated deltaic deposits through the Lower Carbonera river-dominated deltaic complex and ends with Upper Carbonera channel and floodplain deposits.

Sediments encountered at Caño Limón were derived from the Guyana shield to the east, while sediments found in the western Llanos basin were derived from sources to the west. Thus, some time-equivalent units on opposite sides of the basin have different names. Stratigraphic nomenclature for the four productive units is described below from oldest to youngest.

The Cretaceous (Senonian) K-2A unit, which is subdivided into the K-2A1, K-2A2, K-2A3, and K-2A4 zones at Caño Limón, has an average thickness of 245 feet (75 m). The Palito Shale separates K-2A from underlying K-2B sandstone and is a regional barrier to vertical fluid flow. The K-2A unit is age-equivalent to the Chipaque Formation which crops out in the Eastern Cordillera of the Andes to the west. Figure 12-8 is a net oil-sandstone isochore map for the K-2A unit.

The Cretaceous (Senonian) Lower K-1 unit is subdivided into the K-1A, K-1B, K-1C, K-1D, K-1E, and K-1F zones. It is an average of 313 feet (95 m) thick in the Caño Limón-La Yuca block and 247 feet (75 m) thick in the Matanegra block. The La Yuca Shale, which forms the lower part of the Lower K-1, forms a regional barrier to vertical flow. The Lower K-1 is correlatable with the Guadelupe Group present in the western Llanos basin. The Upper K-1 is not found at Caño Limón Field but is present to the west. Figure 12-9 is a net oil-sandstone isochore map for the Lower K-1.

The late Eocene to Oligocene Lower Carbonera Formation is subdivided into the M-1, M-2, M-3, and M-4 zones. It has an average thickness of 280 feet (85 m) in the Caño Limón-La Yuca block and 215 feet (66 m) in the Matanegra block. The Guafita Shale forms a regional barrier to vertical flow within the Lower Carbonera, separating the M-1 zone from the M-2 zone. The Lower Carbonera was previously referred to as the "Mirador Formation" at Caño Limón. However, the Mirador Formation, for which the type locality is found in the Eastern Cordillera of the Andes, has recently been shown to be slightly older than Lower Carbonera and does not extend into the Caño Limón area (Bogotá-Ruiz, 1988). Figure 12-10 is a net oil-sandstone isochore map for the Lower Carbonera Formation.

The early Oligocene to middle Miocene Upper Carbonera Formation is divided into the C-1 through C-5 members. Although these members are continuous across the Llanos basin, individual channel sandstones within the formation are discontinuous. The Upper Carbonera has an average thickness of 1,500 feet (460 m) in the Caño Limón area. The Oligocene C-5 member and the lower part of the C-4 member, with a combined thickness of 150 to 200 feet (46–61 m), are productive in the field and are subdivided into the C-4A, C-4B, C-5A, and C-5B zones. Figure 12-11 is an Upper Carbonera net oil-sandstone isochore map.

Lower Carbonera reservoir continuity between the Caño Limón and La Yuca areas is illustrated in cross section A-A' (Fig. 12-12). Excellent reservoir continuity in the Lower Carbonera has been demonstrated by detection of pressure interference in updip La Yuca wells from producing Caño Limón wells. A common oil-water contact is shared by Lower Carbonera and Lower K-1 reservoirs, whereas the thinner and more discontinuous Upper Carbonera sandstones have locally independent, multiple oil-water contacts.

Cross section B-B' (Fig. 12-12) shows that the Matanegra block has been uplifted 300 feet (91 m) vertically across the Caño Limón fault. Independent oil-water contacts on either side of the fault are evidence of the fault's sealing nature.

Hydrocarbon Characteristics

Crude Oil Typing

Caño Limón crude oils are aromatic intermediate to paraffinic naphthenic oils (Fig. 12-13) according to the Tissot and Welte classification (1978). They are of moderate density and viscosity, with a sulfur content of 0.41%. Additional properties are listed in the Reservoir Summary.

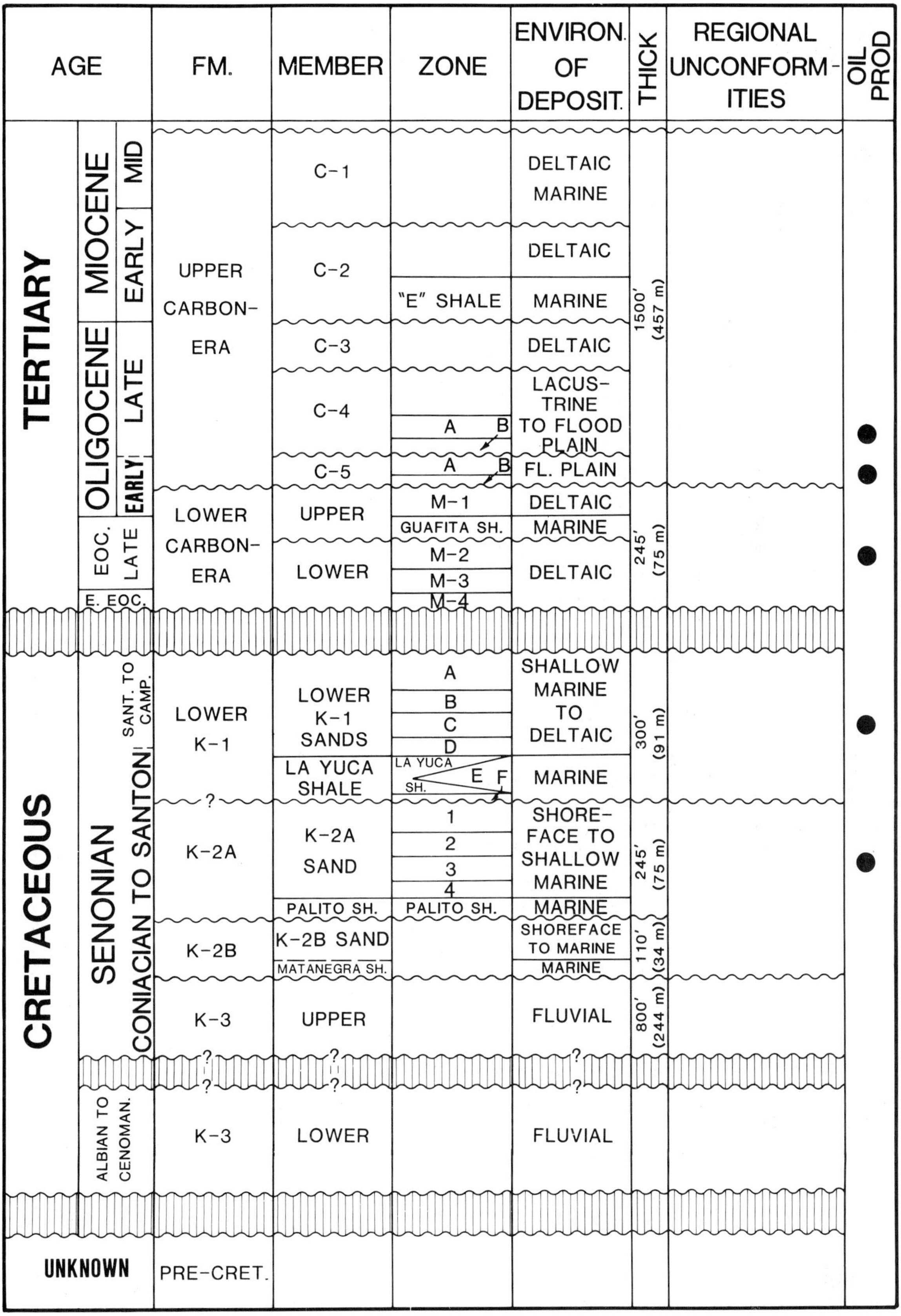

AGE			FM.	MEMBER	ZONE	ENVIRON. OF DEPOSIT.	THICK	REGIONAL UNCONFORMITIES	OIL PROD.
TERTIARY	MIOCENE	MID / EARLY	UPPER CARBONERA	C-1		DELTAIC MARINE	1500' (457 m)		
				C-2		DELTAIC			
					"E" SHALE	MARINE			
	OLIGOCENE	LATE		C-3		DELTAIC			
				C-4		LACUSTRINE TO FLOOD PLAIN			
					A — B				●
		EARLY		C-5	A — B	FL. PLAIN			●
	EOC.	LATE	LOWER CARBONERA	UPPER	M-1	DELTAIC	245' (75 m)		●
					GUAFITA SH.	MARINE			
				LOWER	M-2	DELTAIC			
					M-3				
	E. EOC.				M-4				
CRETACEOUS	SENONIAN	CONIACIAN TO SANTON. / SANT. TO CAMP.	LOWER K-1	LOWER K-1 SANDS	A	SHALLOW MARINE TO DELTAIC	300' (91 m)		●
					B				
					C				
					D				
			?	LA YUCA SHALE	LA YUCA SH. — E F	MARINE			
			K-2A	K-2A SAND	1	SHOREFACE TO SHALLOW MARINE	245' (75 m)		●
					2				
					3				
					4				
				PALITO SH.	PALITO SH.	MARINE			
			K-2B	K-2B SAND		SHOREFACE TO MARINE	110' (34 m)		
				MATANEGRA SH.		MARINE			
			K-3	UPPER		FLUVIAL	800' (244 m)		
			?	?	?	?			
		ALBIAN TO CENOMAN.	K-3	LOWER		FLUVIAL			
	UNKNOWN		PRE-CRET.						

Fig. 12-7. Caño Limón stratigraphic column based on unconformity-bounded sequences. The Upper and Lower Carbonera, Lower K-1 and K-2A are oil-bearing at Caño Limón. The Late Miocene to Recent section is not shown.

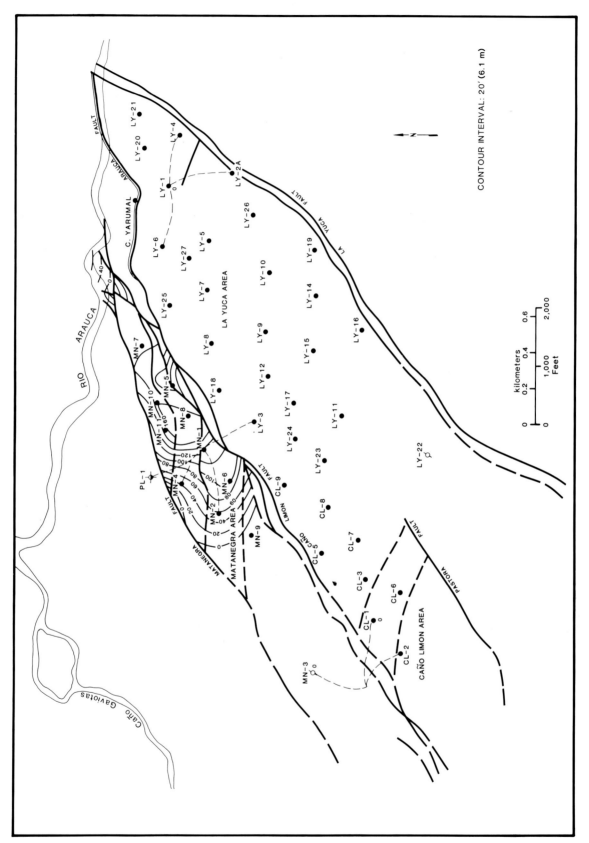

Fig. 12-8. K-2A net oil-sandstone isochore map showing that K-2A oil is confined to the Matanegra area.

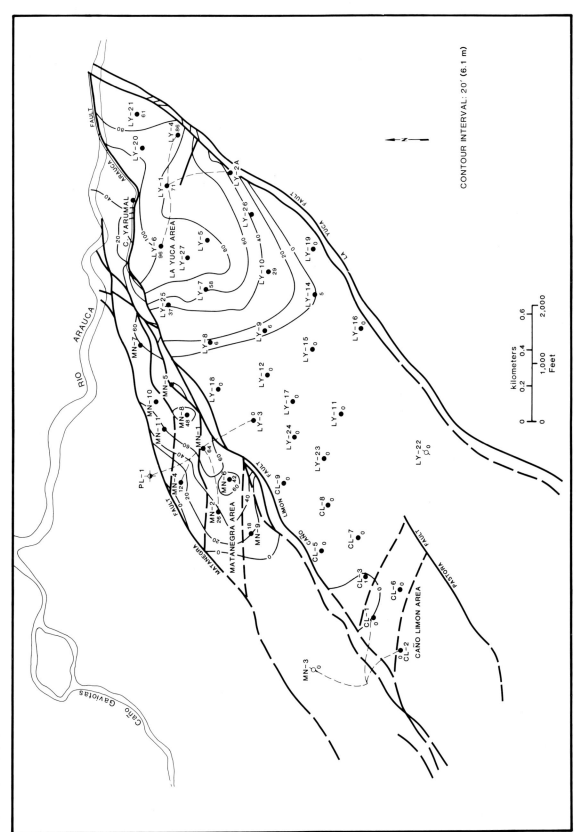

Fig. 12-9. Lower K-1 net oil-sandstone isochore map showing that Lower K-1 oil is confined primarily to the Matanegra and updip La Yuca areas.

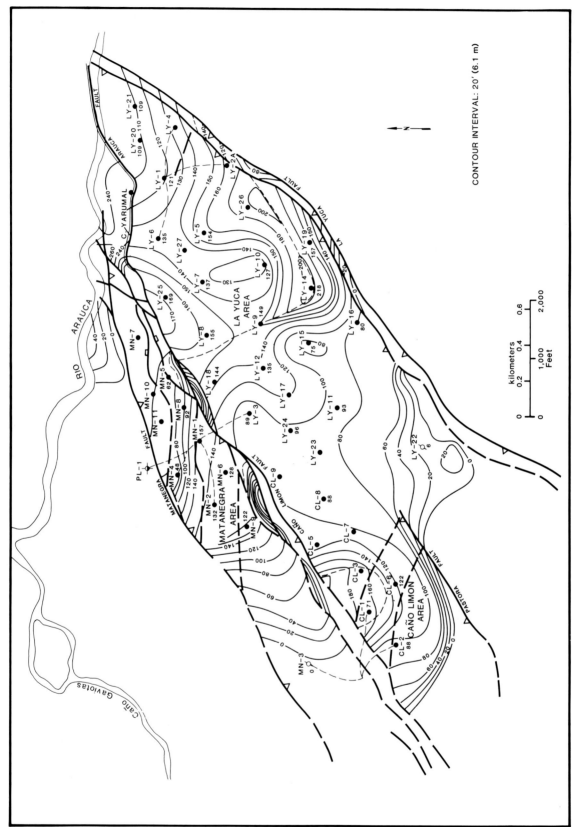

Fig. 12-10. Lower Carbonera net oil-sandstone isochore map showing a broad distribution of Lower Carbonera oil. Thicknesses locally exceed 200 feet (61 m).

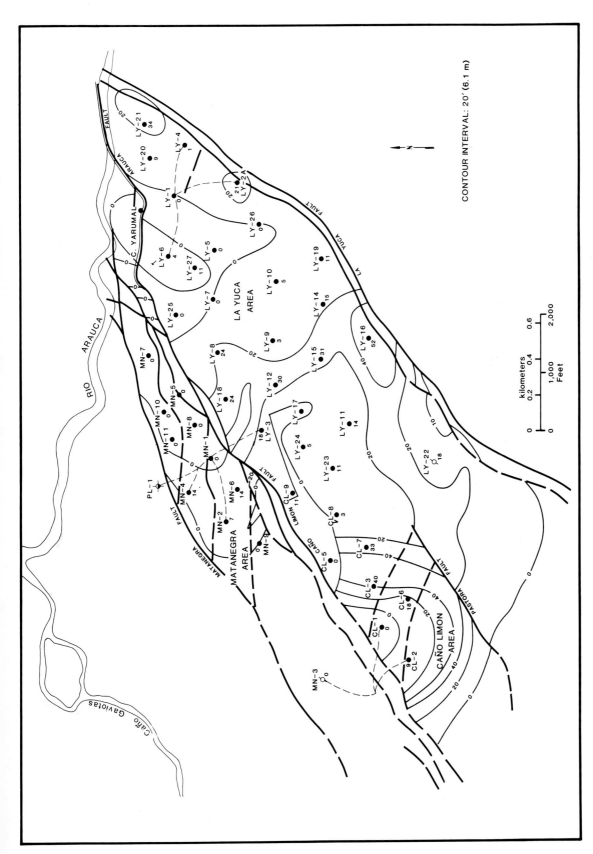

Fig. 12-11. Upper Carbonera net oil-sandstone isochore map showing greatest thickness occurs in downdip Caño Limón and La Yuca areas.

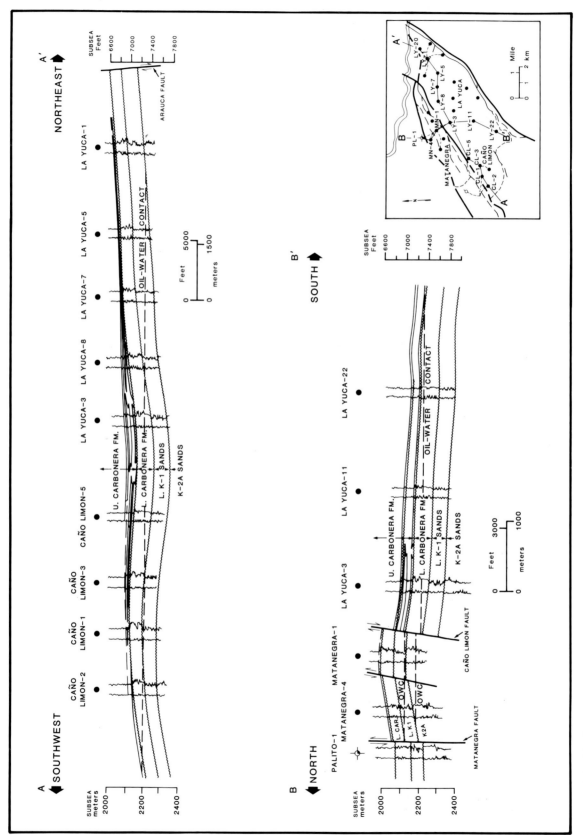

Fig. 12-12. Structural cross sections A-A' and B-B' showing reservoir continuity, fault offsets, and multiple oil-water contacts in Caño Limón Field.

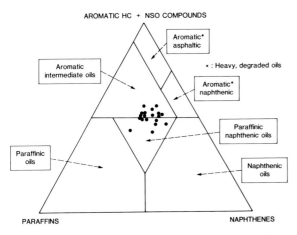

Fig. 12-13. Caño Limón crude oils are classified as aromatic intermediate to paraffinic naphthenic oils.

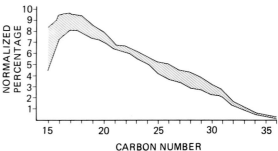

Fig. 12-14. C15+ normal paraffins versus normalized percentage plot for 13 Caño Limón oil samples confirms that all oils were derived from the same, mostly marine, source facies.

Isotopic and isoprenic compositions and biomarker patterns, which are matched by oils at similar depths, suggest that all crude oils were probably generated from the same, mostly marine-derived kerogen. A plot of C15+ normal paraffins versus normalized percentage for thirteen Caño Limón area oils clearly confirms that the oils belong to the same family and were generated from the same, mostly marine-derived, source facies (Fig. 12-14).

Source Rocks, Migration Model, and Pathways

Marine-derived organic source matter from which Caño Limón crude oils could have been generated is present only in Late Cretaceous and Paleozoic outcrops to the west. Although abundant organic-rich Tertiary zones are present within the Llanos basin, these rocks contain primarily immature terrestrially derived kerogen. Subsurface Cretaceous sections in the Llanos basin contain fair to good source potential, but Caño Limón core and cuttings analyses indicate these sediments have immature (Ro = 0.48), terrestrially derived kerogen.

The generally accepted thoughts on generation and migration of Llanos oils hold that these oils were probably expelled by Cretaceous source beds. The generative basinal area was located along the present-day Eastern Cordillera, with subsequent east-southeastward oil migration through carrier beds and fault avenues (Fig. 12-15). Long distance oil migration of approximately 100 miles (161 km) is

implied if Caño Limón oils were derived from the Late Cretaceous source rocks which are presently exposed.

Lopatin analyses indicate that oil generation and migration probably occurred during late Oligocene to early Miocene deposition of the Upper Carbonera Formation. Thus, trapping structures were present and filled prior to regional tilting which began during deposition of the middle Miocene Leon Formation.

The structural development and oil migration history of the Caño Limón Field is illustrated in two schematic cross sections (Fig. 12-16). The trends of gravity and other geochemical properties versus depth vary by both zone and area. For example, original pretilt (pre-middle Miocene) accumulations at the Caño Limón and La Yuca structural highs are interpreted to have each had distinct depth versus gravity trends. Subsequent structural tilting caused remigration of lower gravity oil from the Caño Limón structure into the La Yuca structure. At comparable depths, oil gravities are lower at La Yuca than at Caño Limón.

Oil-Water Contacts

Original free-water contacts were determined using repeat formation tester (RFT) pressure-gradient plots derived from wells drilled before the commencement of production. For example, a plot of Matanegra-6 RFT data illustrates how the intersection of the oil and water gradients demonstrates independent oil-water contacts in the Lower K-1 and K-2A units (Fig. 12-17). Within each fault block, separate oil-water contacts can be demonstrated for the Upper Carbonera (3 or more contacts), the

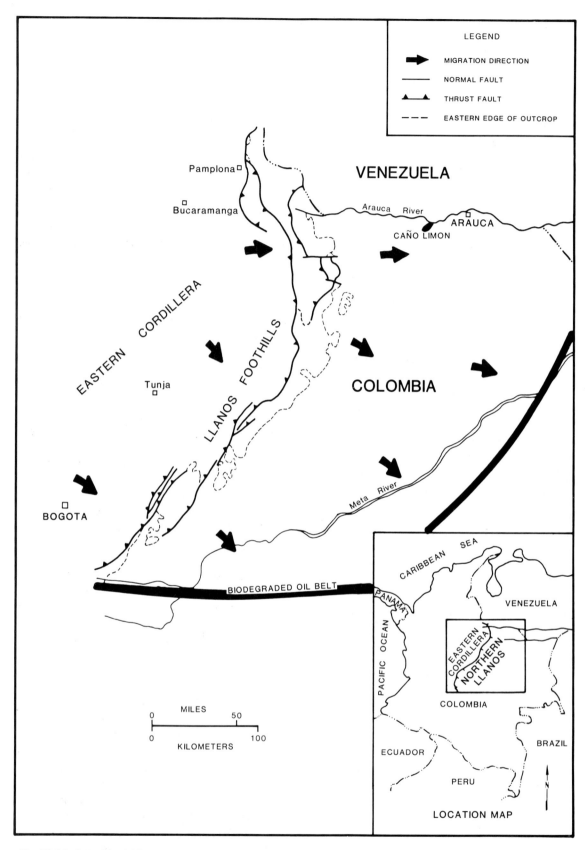

Fig. 12-15. Caño Limón Cretaceous migration pathways. Source rocks were located in the present-day Llanos foothills and oil migrated to the east.

296

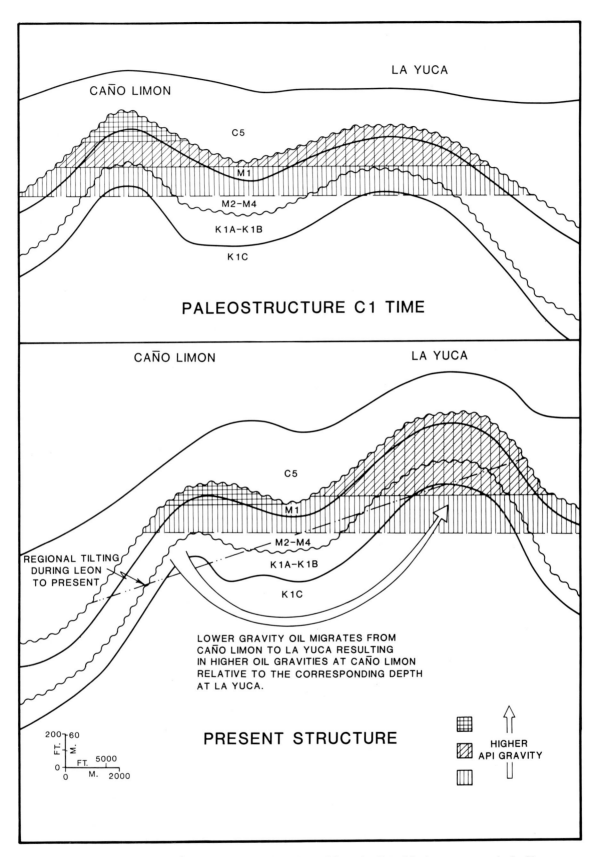

Fig. 12-16. Schematic cross sections showing how oil remigrated from the Caño Limón structure to the La Yuca structure in response to regional tilting.

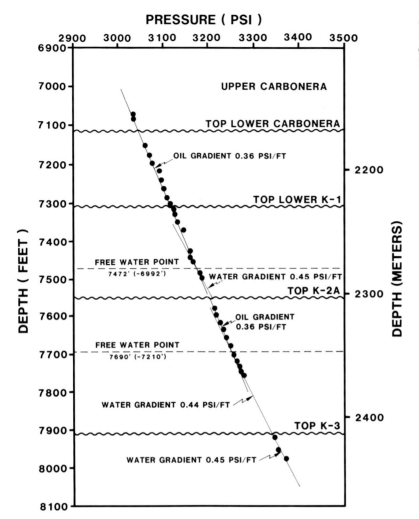

Lower Carbonera/Lower K-1 (1 contact), and the K-2A (1 contact). The Cretaceous K-2A has the deepest oil-water contact and is productive only in the Matanegra block.

Mapping of fluid distribution required interpretation of oil-water contacts from logs and core shows in addition to RFT data. The Lower Carbonera/Lower K-1 oil-water contact map reveals a tilted surface subparallel to the geologic structure and dipping an average 0.38° southeast (Fig. 12-18). This tilt suggests that a dynamic regional aquifer system, with southeasterly flow, is in contact with the oil accumulation. Active hydrodynamic aquifers are now recognized throughout the Llanos basin (Molina, 1988). Potentiometric analyses in conjunc-

tion with regional water resistivity, geochemical data, and regional stratigraphic studies sustain this interpretation (Molina, 1988). Field performance also indicates a very dynamic aquifer within the Caño Limón block. After 148 MMBO (2.4 × 10[7] m³) of production, pressure was quickly restored at Caño Limón during field shut-in indicating strong aquifer support.

Field Development

Up to 1988, 46 wells had been drilled on approximately 247-acre (100-ha.) spacing, of which 43 are producers, two wells are used solely for pressure

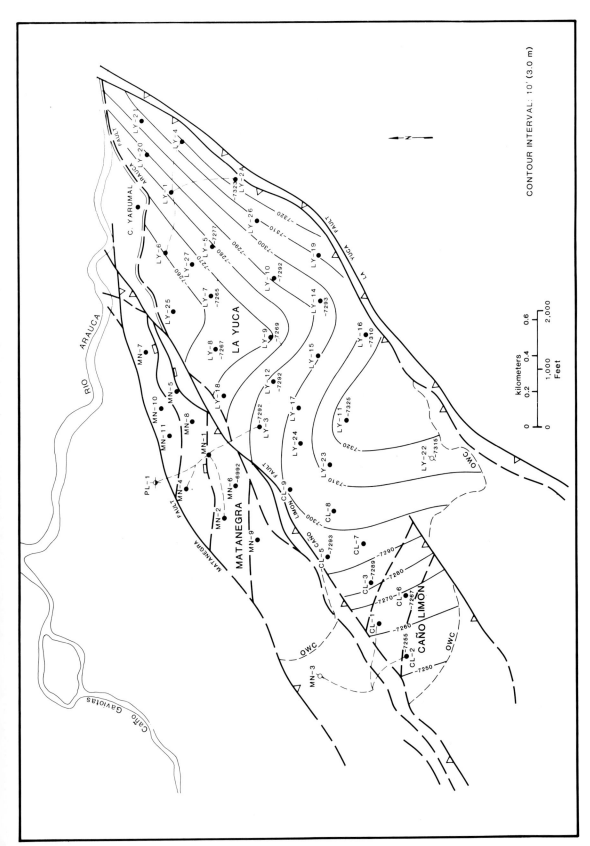

Fig. 12-18. Free-water level map showing tilted oil-water contact.

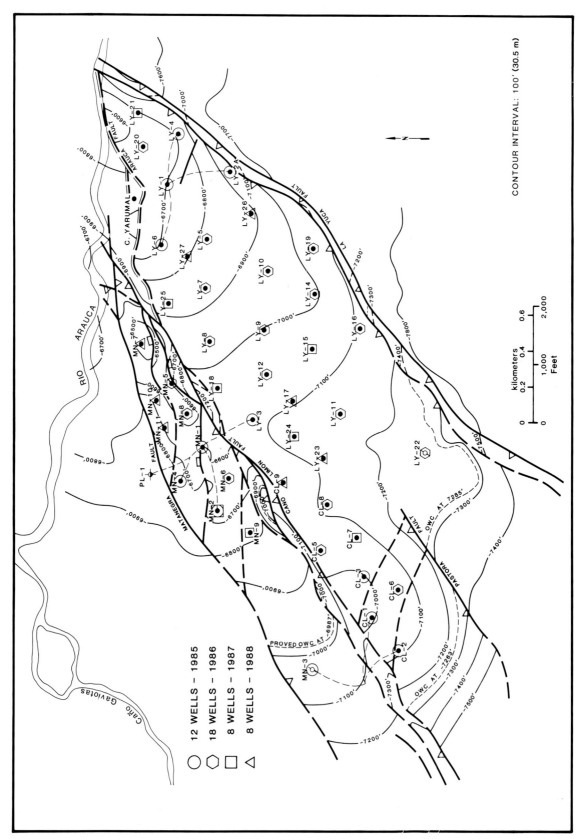

Fig. 12-19. Map of the well drilling sequence by year. A total of 46 wells have been drilled.

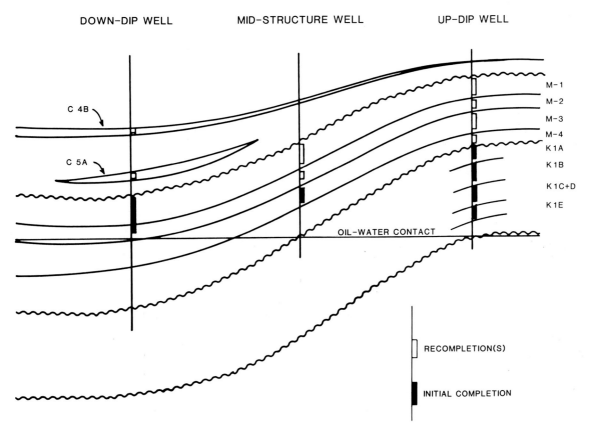

DOWN-DIP WELL   MID-STRUCTURE WELL   UP-DIP WELL

C 4B

C 5A

OIL-WATER CONTACT

M-1
M-2
M-3
M-4
K 1A
K 1B
K 1C+D
K 1E

☐ RECOMPLETION(S)

■ INITIAL COMPLETION

Fig. 12-20. Caño Limón completion schematic showing that zones close to the oil-water contact are completed first and overlying zones are completed as the contact rises.

monitoring, and one was drilled to replace a previous oil producer with mechanical problems. Twelve wells were drilled during 1985, 18 in 1986, 8 in 1987, and 8 in 1988 (Fig. 12-19). Most wells were initially completed in the Lower Carbonera Formation. Eight completions include the K-1A or K-1B zone, three updip wells in the La Yuca area are completed in just the Lower K-1 zone, and four wells in the Matanegra block are completed in the K-2A zone.

A schematic shows how zones closer to the oil-water contact are completed first and overlying zones are completed as the water front advances (Fig. 12-20). This selective completion and recompletion program assists reservoir management by balancing fluid offtake from the various reservoir layers and is a dominant element in achieving uniform water influx and a high sweep efficiency.

Well test rates range from 3,000 to 10,000 BOPD (477–1,590 m³/D) depending on the completed interval. Individual wells have produced at rates of more than 20,000 BOPD (3,180 m³/D).

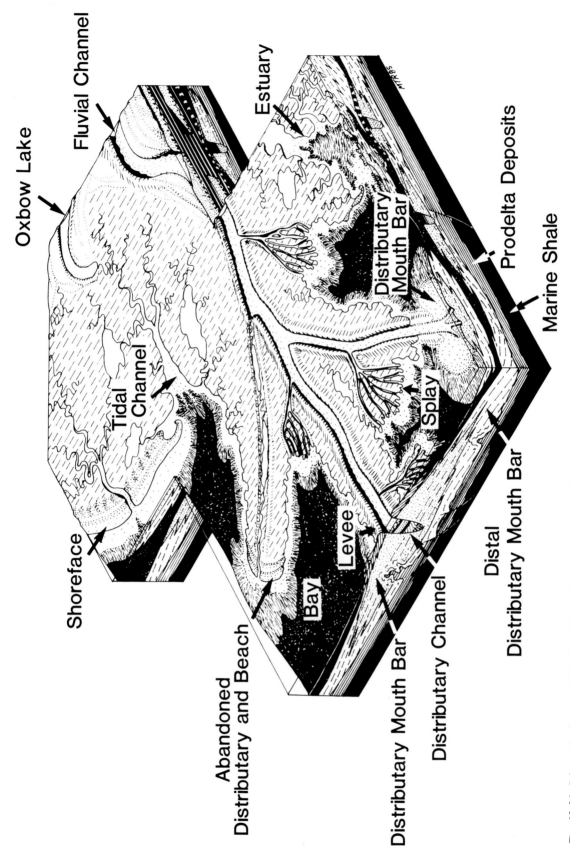

Fig. 12-21. Schematic diagram of the Caño Limón depositional system. K-2A reservoirs are shoreface sandstones. Lower K-1 reservoirs are shoreface and tidal-channel sandstones. Lower Carbonera reservoirs are mainly distributary-channel sandstones. Upper Carbonera reservoirs are fluvial-channel sandstones.

Reservoir Characterization

Depositional Settings

Sediments of the productive section at Caño Limón Field were derived from the granitic Guyana shield to the southeast. The reservoir sandstones are quartzarenites comprising predominantly subangular to angular quartz grains.

The Caño Limón Field depositional model is illustrated in Figure 12-21. The interpretations are based on 2,226 feet (679 m) of cores from 11 wells in the field. Each producing formation represents a different depositional environment. The Cretaceous K-2A reservoir consists of wave-dominated deltaic sandstone having high continuity parallel to the paleocoastline. The Cretaceous K-1 sediments include marine shale with interbedded wave-dominated deltaic sandstone, tidal-channel sandstone, and inner-shelf sandstone. The Lower Carbonera is interpreted as a sequence of river-dominated deltaic sandstone, and the Upper Carbonera as continental, lower to upper delta-plain claystone. Each of these environments is discussed in detail in the following sections.

Cretaceous K-2A. Lower-shoreface sandstone comprises 56% of Cretaceous K-2A reservoirs, and upper-shoreface sandstone comprises the remainder. A shallow marine origin for both facies is indicated by glauconite concentrations that can exceed 10%. Lower-shoreface sandstone is very fine- to fine-grained and is of poorer reservoir quality due largely to bioturbation (Fig. 12-22A). Burrows are frequently clogged with kaolinite. Upper-shoreface sandstone is fine-grained but not bioturbated, and uniform cross-bedding and thin clay laminae have been preserved (Fig. 12-22B). Shoreface sandstone reservoirs are noteworthy for extensive coastline-parallel continuity and the absence of significant barriers to lateral fluid movement.

Two kinds of local barriers to vertical fluid movement in the K-2A reservoir are present: (1) marine and interdistributary-bay shale, and (2) sandstone cemented with calcite derived from the dissolution of adjacent bivalve shell beds.

Cretaceous Lower K-1. The marine Lower K-1 unit contains a higher percentage of nonreservoir rock (66%) than the K-2A unit (24%). Nonreservoir Lower K-1 deposits are primarily marine and interdistributary-bay shale, with secondary shell beds and calcite-cemented sandstone as in the K-2A. Reservoir deposits consist of marine inner-shelf (Fig. 12-22C), shoreface, and tidal-channel sandstones. The sandstones are all very fine- to fine-grained with minor glauconite. Kaolinite occurs in varying amounts and reduces reservoir permeability. It occurs as burrow fillings where it can form lateral barriers in otherwise fairly continuous sandstone. The only high quality K-1 reservoirs are those with few or no burrows.

Interbedded marine and interdistributary-bay shales separate the Lower K-1 vertically into six sandstone members. The lowermost 60 to 100 feet (18–30 m) of Lower K-1 is the continuous marine La Yuca Shale, which separates the Lower K-1 and K-2A into different hydrodynamic regimes with separate oil-water contacts.

Lower Carbonera. The Eocene Lower Carbonera Formation is the main Caño Limón reservoir, having the greatest number of permeable sandstones and 74% of total oil in place. Deltaic distributary-channel sandstone comprises 68% of the Lower Carbonera reservoirs. Channel-fill deposits exhibit sharp, erosional basal contacts with basal coarse sandstone and conglomerate and moderate- to high-angle cross-bedding (Fig. 12-22D). Moderate- to poorly sorted, upward-fining sequences are present but are often obscured or incomplete due to channel stacking or amalgamation.

An additional 24% of Lower Carbonera reservoir rocks are low-energy channel-fill sandstone, which typically overlie or grade laterally into distributary-channel sandstone. The low-energy channel-fill sandstone is fine-grained, well sorted, and ripple cross-laminated. Occasional bioturbation indicates shallow marine or brackish water conditions (Fig. 12-22E).

An abrupt termination of channel flow is represented by channel abandonment sediments. Interlaminated claystone with minor very fine- to fine-grained sandstone filled the stagnant channel depressions. Sandier portions are often burrowed. Channel abandonment fills are mostly nonreservoir units, forming localized barriers to vertical flow which extend only as far as the filled channel depression.

Distributary-mouth bars constitute only 6% of the Lower Carbonera reservoirs. This low proportion is due to the bars having been partially to completely eroded by the distributary channels. Gradational contacts with underlying prodelta shale and abrupt erosional contacts with overlying distributary channel-fill sandstone are characteristic (Fig. 12-22F). The bar sandstone is fine-grained and

moderately well sorted, with upward-decreasing clay content and increasing grain size. Low-to high-angle cross-bedding and wavy parallel laminations, locally burrowed, are present.

Of minor significance (2% of the the Lower Carbonera reservoirs) are estuary-fill, crevasse-splay, and overbank deposits. These are typically interbedded sandstone and shale sequences with fair to poor reservoir quality.

Nonreservoir rocks include flood-plain claystone and marine, interdistributary-bay, and prodeltaic shales. Marine shale is dark gray, parallel laminated, fissile, and regionally continuous. The best example is the approximately 18-foot (5.4-m) thick Guafita Shale which separates the M-1 and the M-2 Lower Carbonera members. This shale acts as an effective regional seal to fluid movement. Interdistributary-bay shale is dark gray, subparallel to wavy laminated, and often highly burrowed with interlaminated siltstone and sandstone. The silts and sands were deposited in the bays when nearby channels flooded. Interdistributary-bay shale occasionally forms field-wide barriers to fluid flow. Prodelta shale is dark gray and interlaminated with fine-grained sandstone. These thicken and coarsen upward, grading into distributary-mouth bars. Floodplain claystone is light gray with occasional red and ochre mottling. It may contain thin coal interbeds and root tubes lined with carbonaceous material. These continental deposits represent the maximum progradational phase of delta building. They can serve as effective barriers to vertical fluid movement but are locally cut by distributary or fluvial channels.

Upper Carbonera. The Upper Carbonera is composed primarily of floodplain claystone which serves as the seal for Lower Carbonera reservoirs. This formation also contains some relatively discontinuous fluvial or distributary-channel sandstones that have good reservoir properties. These channel sandstones

have thicknesses and reservoir properties comparable to those of the Lower Carbonera (Cleveland, 1988).

The depositional settings were also interpreted using wireline logs after correlation with cores. Since the differences between many depositional settings are gradational, fixed log cutoffs for particular environments were impractical. Correlations show, however, that: (1) gamma-ray values increase with higher clay content and (2) deep resistivity values increase with grain size in oil-bearing rocks. For example, distributary-channel sandstone exhibits very low gamma-ray readings and, when oil-bearing, exhibits very high resistivities in addition to typical abrupt basal and gradational upper contacts. Relationships of depositional environments to petrophysical parameters are shown for La Yuca-6 in Figure 12-23.

Core-to-log correlations have been used to determine depositional environments throughout the entire field in a series of stratigraphic sections (Fig. 12-24). These were used to construct structural sections and fence diagrams (Fig. 12-25) which have been invaluable in understanding uneven water advance. For example, continuous and highly permeable channel sandstones shown in the sections are swept at a faster rate than are other sandstone types. Detailed correlations of higher and lower quality reservoirs throughout the field provide a basis to balance offtake by locating more drainage points in the lower quality sandstones.

Thicknesses of individual reservoir units such as channel sandstones have been mapped based on the cross sections and paleocurrent directions derived from dipmeters (Fig. 12-26). The mapped channel sandstones often trend southwest, parallel to major faults, but trend more northerly to the east of the Caño Limón high. Movement along the faults is believed to have created topographically low areas which localized channel deposition (Cleveland, 1988).

Fig. 12-22. Core photographs representative of the Caño Limón sandstones. The scale bar is 2 inches (5 cm). (A) Lower-shoreface sandstones are very fine- to fine-grained and glauconitic. Burrows filled with clay result in mottled oil stains. (B) Upper-shoreface sandstones are fine-grained and glauconitic. Because they are unburrowed, uniform cross-bedding and thin laminae are preserved. (C) Burrowed inner-shelf/lower-shoreface sandstone. (D) Distributary-channel sandstones contain basal coarse-grained sand and have moderate- to high-angle cross-bedding. (E) Low-energy channel-fill sandstones are fine-grained, well sorted, and argillaceous. (F) Distributary-mouth-bar sandstones are fine-grained, well sorted, and laminated.

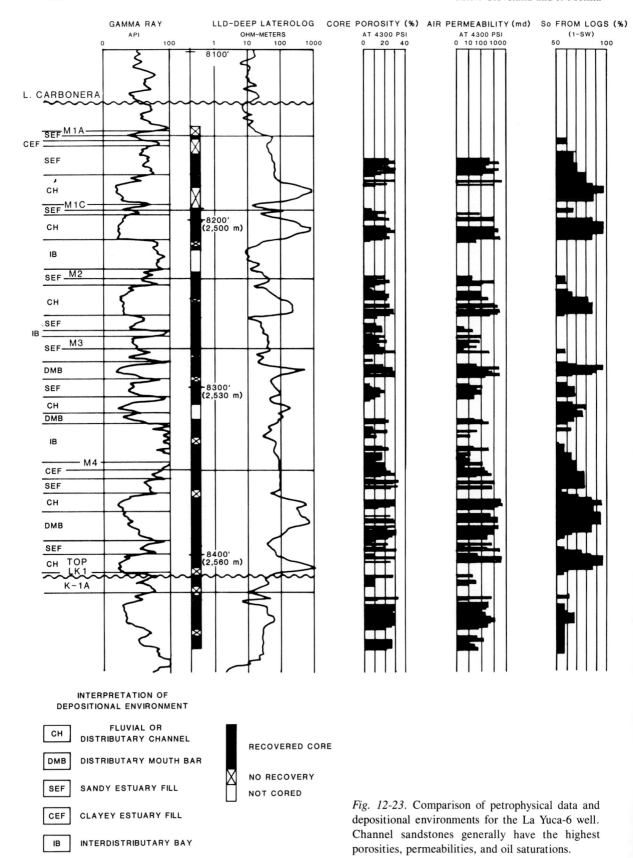

Fig. 12-23. Comparison of petrophysical data and depositional environments for the La Yuca-6 well. Channel sandstones generally have the highest porosities, permeabilities, and oil saturations.

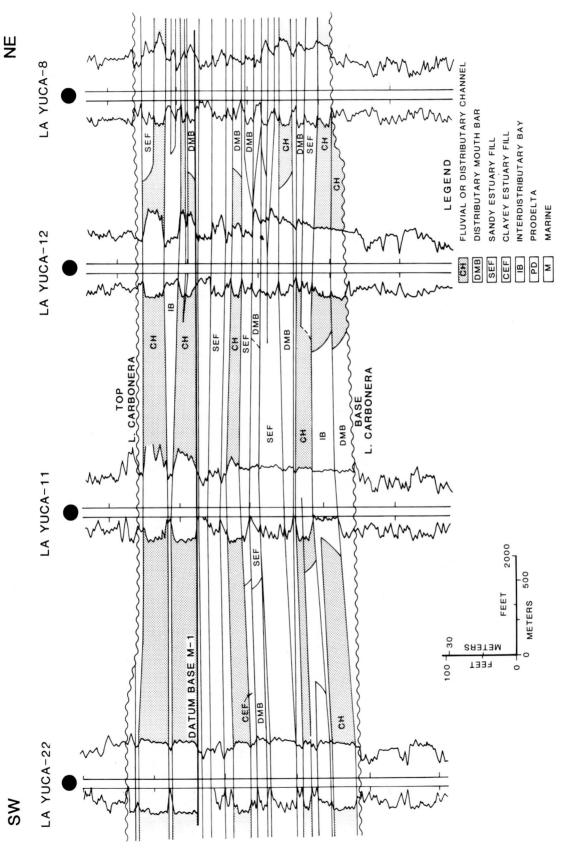

Fig. 12-24. Stratigraphic correlations showing environments of deposition for the Lower Carbonera Formation. Channel sandstones are the highest quality reservoirs.

M.N. Cleveland and J. Molina

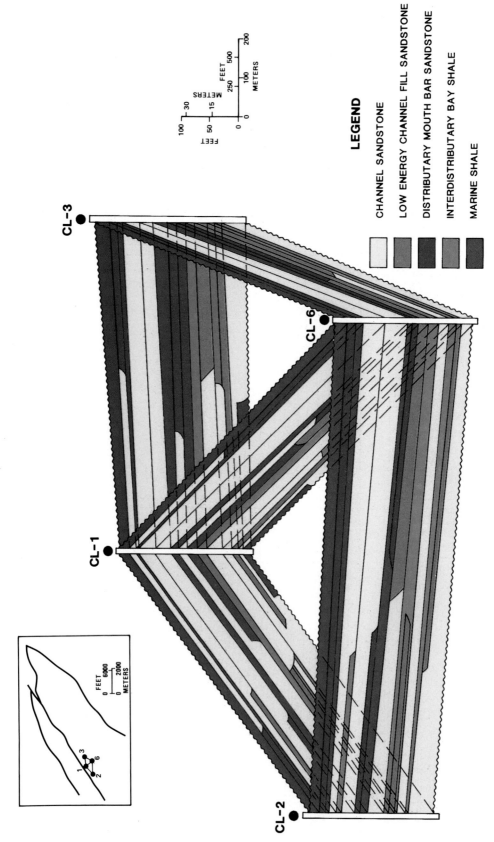

Fig. 12-25. A fence diagram shows the importance of depositional facies to water advance in the reservoir. Channel sandstones often have appreciably higher permeabilities than do the other reservoir facies and are therefore swept much faster.

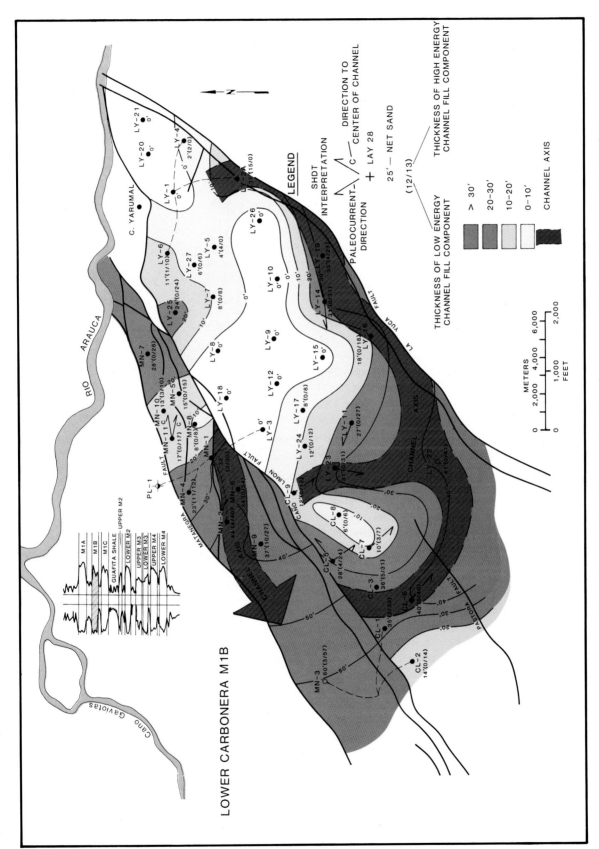

Fig. 12-26. Channel-sand isopach map developed from stratigraphic cross sections and stratigraphic dipmeter interpretations.

Table 12-1. Reservoir parameters by environment.

Environment	Weighted average porosity[1]	Weighted average clay volume[1]	Weighted average water saturation[2]
Distributary Channel	25.7%	6.2%	16.4%
Low Energy Channel Fill	22.5%	19.6%	30.8%
Distributary-Mouth Bar	23.8%	13.1%	27.0%
Upper Shoreface	23.7%	12.3%	22.9%
Lower Shoreface	22.3%	21.5%	34.3%

[1] Total net sand.
[2] Net pay only.

Petrophysical Properties

The porosity of the productive facies averages 24% and ranges from 12 to 32%. The economic cutoff is set at 14%. This porosity, determined from logs and stressed conventional cores, is all primary intergranular porosity. Permeability for the productive facies averages 1,450 md and ranges from 10 to 8,000 md. The economic limit is approximately 10 md.

Permeability measurements have been routinely made on conventional core plugs, but obtaining accurate measurements is problematic. The measured values are low when compared to permeabilities derived from well-test buildup data. The more permeable sandstones show the greatest discrepancy in values. Study has demonstrated that, during the measurement process, the fine particles migrate to the screens at the ends of core plugs, resulting in reduced permeability values. It is possible that the movable fine particles have been introduced by the invasion of drilling solids into the highly permeable sandstones. In addition, the permeability of some sandstones from well-test buildup may exceed 10,000 md, which suggests that nonelastic deformation may occur in cores when the most permeable rocks are brought to the surface.

Average reservoir parameters have been determined for the principal depositional environments (Table 12-1). There is no significant variation between environmental facies. However, clay volume and water saturation are significantly lower in the

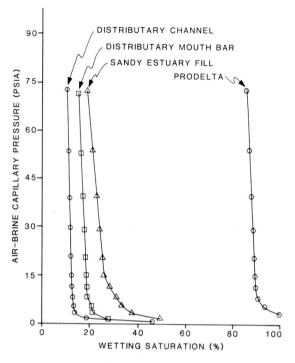

	DEPTH (FT)	STRESSED PERMEABILITY (md)	STRESSED POROSITY (%)
⊖—⊖—⊖	7732.6 (2356.9 m)	3104	28.6
⊟—⊟—⊟	7739.7 (2359.1 m)	3563	25.2
△—△—△	7743.5 (2360.2 m)	342	22.5
⊘—⊘—⊘	7540.1 (2298.2 m)	186	21.7

Fig. 12-27. Air-brine capillary pressure curves for four samples from the Caño Limón-3 well representing the range of depositional environments. The measurements suggest that oil-water transition zones will be thinner in high-quality distributary-channel sandstones than in low-quality prodelta sandstones.

distributary channel facies. This lowered water saturation value reflects a coarser grain size. Coarser grains have a lower percentage of surface area per unit volume than do finer grains and so exhibit a proportional decrease in bound or connate water.

Water saturations for the productive facies average 30% but range from 5 to 70%. The cutoff is 50 to 70%, depending on the zone. Oil saturations average 23% but vary according to the zone. Capillary-pressure curves have been derived from core-plug measurements and plotted by environment (Fig. 12-27). Four samples from Caño Limón-3

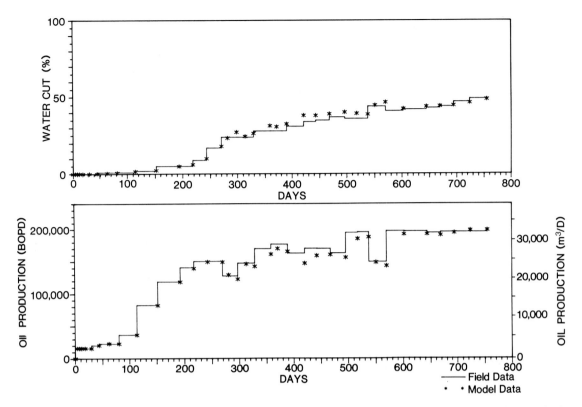

Fig. 12-28. Plots of water cut and oil rate versus reservoir simulator results for the first 750 days of Caño Limón production.

well provide examples representative of the different depositional environments. These case studies suggest that the oil-water transition zones will be thinner in high-quality distributary-channel sandstone than in low-quality prodelta sandstone.

Reservoir Simulation and Management

Future field performance and ultimate oil recovery were simulated using structural, isopach, and petrophysical property maps of reservoir layers, based on the mapping of individual reservoir environments. Channel maps were especially useful in constructing permeability maps, since channel sandstones are the most permeable (Cleveland, 1988).

A history match of actual and simulated field water cuts and oil production rates for the first 750 days is shown in Figure 12-28. Computer simulation results indicate that oil recovery efficiency for Caño Limón will be 54% with ultimate recovery of more than one billion barrels (1.6×10^8 m³).

Caño Limón well performance indicates that geological heterogeneity has a significant effect on the pattern of water invasion in the reservoir (Fig. 12-29). Geological factors affecting sweep efficiency include the following:

1. Areal permeability variation in reservoir zones is related to areal variations in depositional environments.
2. Vertical permeability and layering distributions within each zone are related to depositional environments.
3. Vertical permeability within individual sandstones is affected by discontinuous shales or other discontinuous permeability features.
4. Very high-capacity flow paths of probable limited extent and volume occur within portions of the reservoir. Examples are thin, high permeability layers such as narrow channel-fill sandstones, faults, or combinations of these.
5. The magnitude of vertical permeability varies across the apparently continuous mappable shales separating major reservoir zones.

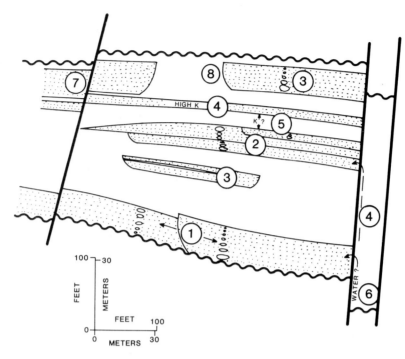

Fig. 12-29. Schematic depiction of Caño Limón geological heterogeneities which will affect water invasion. The keys are: (1) areal permeability variation; (2) vertical permeability distribution within a zone; (3) vertical permeability distribution within a sandstone; (4) high permeability flow paths; (5) vertical permeability in shales; (6) water influx sources; (7) communication across internal faults; and (8) lateral barriers.

6. Facies variations and fault-plane channeling of water control water influx location.
7. Facies distribution controls communication across faults and between different sand units.
8. Lateral barriers within sandstone layers include channel edges, abandoned-channel claystone plugs, and kaolinite-filled burrows.

Detailed cross sections and maps, based in part on interpretations of depositional environment, are used in balancing oil offtake from the various reservoir layers and in maintaining an even water-front advance. Balancing decisions include planning well locations and designing selective completion programs. Well recompletions to perforate new intervals or to exclude high water-cut intervals in previously completed wells may be recommended. Decision making is especially influenced by recognition of high permeability zones and lateral and horizontal barriers which will affect fluid flow within the reservoir.

Exploration and Production Strategy

Caño Limón Field was drilled and discovered on the basis of seismically interpreted structural closure at Lower Carbonera, Lower K-1, and K-2A levels. The bounding faults are easily recognized seismically. Eocene and Cretaceous reservoir sandstones are ubiquitous in the Llanos basin. Exploration strategy in the Llanos, thus far, involves careful seismic definition of structural traps. Factors which may influence whether traps are oil-bearing include the following:

1. Is the trap faulted? The fact that Caño Limón Field contains several productive levels separated by sealing shales suggests that the field may have been charged via faults. Faults which are sealing today may have been conduits when active.
2. Could the trap have been flushed by the dynamic aquifer?
3. Did the trap form prior to oil migration?

Caño Limón development has included the interpretation of a 3-D seismic survey acquired on a closely spaced grid, 82 by 164 feet (25 × 50 m). The seismic data have been used to pinpoint fault locations, to construct maps, and to select well locations. In addition, seismic data have been used to identify stratigraphic layers within the Upper Carbonera C-5 member. This horizon readily lends itself to seismic stratigraphic interpretation because it contains discontinuous, relatively high-quality sandstones encased in claystones. Seismic stratigraphic interpretation of other horizons has not been practical or necessary due to high sandstone/shale ratios and good sandstone continuity.

Optimum development of deltaic reservoirs like those of Caño Limón depends on the recognition and mapping of lateral and vertical fluid barriers and reservoir heterogeneities. Maximum sweep efficiency cannot be realized by draining the sandstones from only crestal producers, in spite of the strong edge-water drive. Stratigraphic complexities result in flow-unit complications, and drainage points are located with respect to the mapped reservoir heterogeneities. For example, infill wells drilled later can be located on each critical side of easily mapped local lateral barriers. The realization of maximum sweep efficiency requires a balanced (perforation-deferred) fluid offtake from low-structural, mid-structural, and crestal wells in a manner that allows uniform water influx.

Conclusions

Caño Limón Field is a faulted structural trap with at least four separate oil accumulations. The Cretaceous K-2A unit is productive only in the Matanegra block where it has the deepest oil-water contact. The Eocene to Oligocene Lower Carbonera and Cretaceous Lower K-1 units share a common oil-water contact. The Oligocene Upper Carbonera sandstones are discontinuous and exhibit multiple oil-water contacts.

Each of the four units belongs to a well-characterized deltaic facies association. Proper development of Caño Limón depends on coordination of well location and completion programs with recognition of the deltaic reservoir heterogeneities and flow barriers. The fundamental goal of this development strategy is to maximize ultimate recovery through maintenance of a balanced offtake and uniform water advance.

Acknowledgments. We thank Occidental for supporting this work and also Ecopetrol, Royal Dutch Shell, and Repsol (part owner of Occidental de Colombia) for approving its publication. The original core descriptions and definitions of depositional environments were done by R.W. Tillman of the Cities Service Technology Center in Tulsa, Oklahoma. All subsequent work was carried out by the Production Geology staffs of Occidental de Colombia and Occidental Bakersfield. The comments of J.H. Barwis, J.G. Bryant, B.J. Casey, J. Karlo, J.G. McPherson, C.A. Parker, and J.R.J. Studlick improved this work and are greatly appreciated.

Reservoir Summary

Field : Caño Limón
Operator: Occidental de Colombia (Ecopetrol, Royal Dutch Shell — partners)
Discovery: 1983
Location: Northeastern Colombia, South America
Basin: Llanos basin
Tectonic/Regional Paleosetting: Foreland basin
Geologic Structure: Faulted anticline
Trap Type: Fault-bounded structural closure
Reservoir Drive Mechanism: Edge-water drive
 • **Original Reservoir Pressure:** 3,213 psi (2.2 × 10⁴ kPa) at 7,100 feet (2,165 m) subsea
Reservoir Rocks
 • **Age:** Late Cretaceous, Eocene, Oligocene

- **Stratigraphic Unit:** K-2A and Lower K-1 units, Lower Carbonera and Upper Carbonera formations
- **Lithology:** Quartzarenite, very fine- to coarse-grained sandstones
- **Depositional Environments:** Deltaic and shallow marine
- **Productive Facies:** Distributary-channel sandstones and shoreface sandstones
- **Petrophysics**
  - **Porosity Type:** ϕ total $= 23$ to 25%, all primary intergranular
  - **ϕ:** Average 23 to 25%, range 12 to 32%, economic cutoff 14% (logs and stressed cores)
  - **k:** Average 1,450 md, range 10 to 8,000 md, economic limit is approximately 10 md (stressed cores)
  - **S_w:** Average 21 to 42%, range 5 to 70%, cutoff is 50 to 70% depending on zone (logs)
  - **S_{or}:** 17 to 29%, depending on zone (cores)

Reservoir Geometry
- **Depth:** 7,600 feet (2,315 m)
- **Areal Dimensions:** 3 by 7.5 miles (5 × 12 km)
- **Productive Area:** 9,070 acres (3,675 ha.)
- **Number of Reservoirs:** 4+
- **Hydrocarbon Column Height:** 700 feet (215 m) (combined reservoirs)
- **Fluid Contacts:** Oil/water contacts at 7,300 feet (2,225 m) subsea, 7,200 feet (2,195 m) subsea, 6,990 feet (2,130 m) subsea, and 6,980 feet (2,128 m) subsea
- **Number of Pay Zones:** 4
- **Gross Sandstone Thickness:** Average-Upper Carbonera 75 feet (23 m), Lower Carbonera 235 feet (72 m), Lower K-1 252 feet (77 m), K-2A 221 feet (67 m)
- **Net Sandstone Thickness:** Average-Upper Carbonera 7 feet (2.1 m), Lower Carbonera 144 feet (44 m), Lower K-1 86 feet (26 m), K-2A 168 feet (51 m)
- **Net/Gross:** Average-Upper Carbonera 0.23, Lower Carbonera 0.61, Lower K-1 0.34, K-2A 0.76

Hydrocarbon Source, Migration:
- **Lithology and Stratigraphic Unit:** Late Cretaceous shales
- **Time of Hydrocarbon Maturation:** Late Oligocene to Early Miocene
- **Time of Trap Formation:** Early Oligocene
- **Time of Migration:** Late Oligocene to Early Miocene

Hydrocarbons
- **Type:** Oil
- **GOR:** 8 SCF/STB (1.4 m³/m³)
- **API Gravity:** 29°
- **FVF:** 1.05
- **Viscosity:** 4 cP (4.0 × 10³ Pa·s) at 207°F (97°C) and 3,200 psi (2.2 × 10⁴ kPa)

Volumetrics
- **In-Place:** 1.94 billion STB (3.1 × 10⁸ m³)
- **Cumulative Production:** 184 MMBO (2.9 × 10⁷ m³)
- **Ultimate Recovery:** 1.05 billion MMBO (1.7 × 10⁸ m³)
  Recovery Efficiency: 54%

Wells
- **Spacing:** 3000 feet (915 m), 247 acres (100 ha.)
- **Total:** 43 producers
- **Dry Holes:** 1

Typical Well Production:
- **Average Daily:** 4,800 BO (760 m³)
- **Cumulative:** No wells have been abandoned yet. The largest production to date comes from the La Yuca-7 well, which has produced 10 MMBO (1.6 × 10⁶ m³).

Other: Caño Limón is the largest oil field in Colombia.

References

Bogotá-Ruiz, J., 1988, Contribución al conocimiento estratigráfico de la cuenca de Los Llanos (Colombia): III Simposio Bolivariano, Memorias, Tomo I, p 308– 346.

Cleveland, M.N., 1988, The role of geological core interpretation in reservoir management of Caño Limón field, Colombia: Tercer Congreso Colombiano del Petróleo, Memorias, Tomo I, p. 165–182.

Gabela, V.H., 1985, Campo Caño Limón, Llanos Orientales de Colombia: II Simposio Bolivariano, Publicaciones, Tomo I, p. 1–29.

Molina, J., 1988, Condiciones hidrodinámicas de Los Llanos septentrionales: Tercer Congreso Colombiano del Petróleo, Memorias, Tomo I, p. 183–198.

Parker, C.A., 1986, Caño Limón reservoir properties suggest high recovery factor: Oil and Gas Journal, May 12, p. 55–58.

Tissot, B.P., and Welte, D.H., 1978, Petroleum Formation and Occurrence: Berlin, Springer-Verlag, 538 p.

Key Words

Caño Limón Field, Columbia, Llanos basin, Upper Carbonera Formation, Lower Carbonera Formation, Late Cretaceous, Eocene, Oligocene, deltaic, shallow marine, wave-dominated delta, river-dominated delta, repeat formation testers, tiled oil-water contact, dipmeters, wrench faults.

Estuarine/Barrier Environments

13

Fluvial-Estuarine Valley Fill at the Mississippian-Pennsylvanian Unconformity, Main Consolidated Field, Illinois

Richard H. Howard and Stephen T. Whitaker

Illinois State Geological Survey, Champaign, Illinois 61820

Introduction

Approximately 4 billion barrels of oil (6.4×10^8 m³) have been produced in the Illinois basin since oil was discovered there more than a century ago (Fig. 13-1). Production has come largely from structural traps at depths of less than 5,500 feet (1,675 m) that collectively contain more than sixty different pay zones ranging in age from Ordovician to Pennsylvanian (Fig. 13-2). Despite the maturity of the basin, significant discoveries could come from numerous plays that remain to be tested.

One such play involves the Mississippian-Pennsylvanian unconformity that separates the Kaskaskia and Absaroka sequences (Sloss, 1963) in the Illinois basin (Fig. 13-2). The unconformity in the subcrop is characterized by an anastomosing pattern of paleovalleys ranging up to 300 feet (91 m) deep and to more than 20 miles (32 km) wide (Fig. 13-3). Although the nature and evolution of this paleovalley network have been studied (Siever, 1951; Bristol and Howard, 1971; Howard, 1979a, 1979b), the character, distribution, and depositional environments of the sediments that buried this surface have been documented only locally (Davis et al., 1974; Pryor and Potter, 1979). Because these paleovalleys contain Pennsylvanian coarse-grained, discontinuous sandstone and dark gray shale, they provide the potential for hydrocarbon traps throughout the paleovalley network. Published discussions of hydrocarbon reservoir potential associated with the paleovalley network, however, have been limited to only two reports (Shiarella, 1933; Howard and Whitaker, 1988).

Of the nearly 500 million barrels of oil (8.0×10^7 m³) produced from Pennsylvanian reservoirs in the Illinois basin, more than 75% has come from fields along the La Salle anticlinal belt (Fig. 13-1). Although approximately 40% of this production has come from sandstones lying at or near the base of the Pennsylvanian System (Swann and Bell, 1958), it is not clear how much production can be attributed to sandstones deposited within the unconformity paleovalleys. This uncertainty is due to a lack of understanding regarding detailed sub-Pennsylvanian paleotopography.

The vast majority of Pennsylvanian tests have been based on structural prospects with no regard given to the sub-Pennsylvanian paleovalley network. Consequently, reservoirs found in these paleovalleys were discovered more by chance than by intent. This chapter examines the results of such a chance encounter with a basal Pennsylvanian (Caseyville) reservoir within the giant Main Consolidated Field near Hardinville, Crawford County, Illinois. The geologic insight is presented here as a guide for reservoir exploration and development within the sub-Pennsylvanian paleovalley network in the Illinois basin.

Discovery and Production History

Despite the fact that the study area is located in the oldest and one of the most densely drilled oil-producing areas in Illinois, the existence there of a

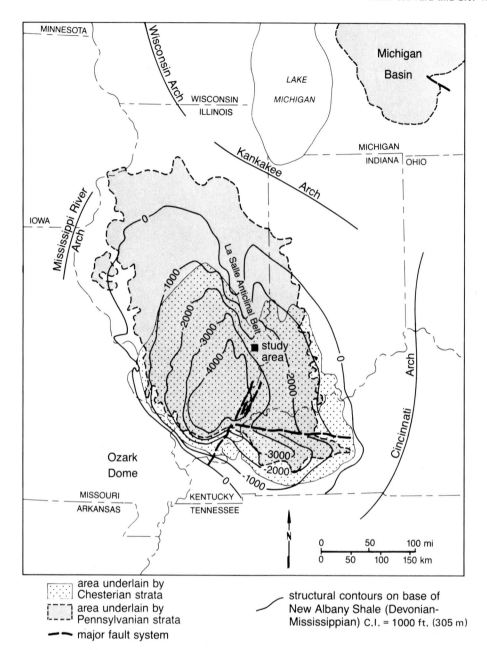

Fig. 13-1. Regional and tectonic setting of the Illinois basin showing the regional structure on the base of the Devonian-Mississippian New Albany Shale, the areas underlain by Pennsylvanian and Mississippian Chesterian strata, and the study area.

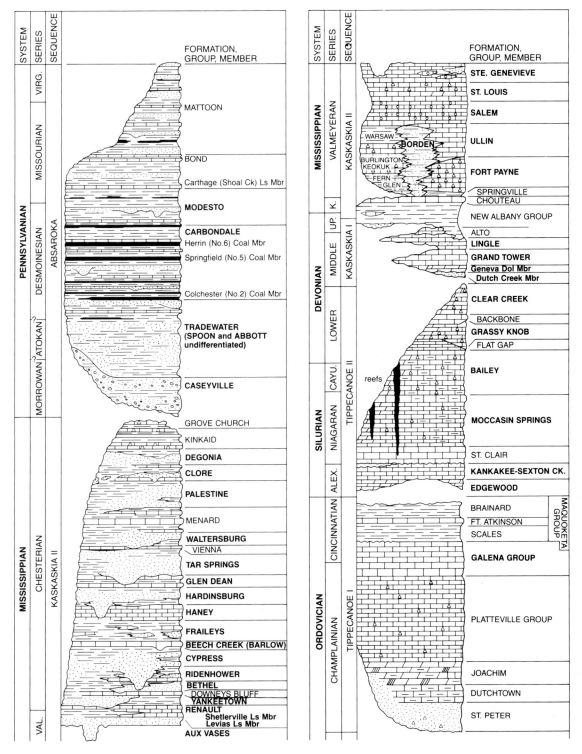

Fig. 13-2. Generalized geologic column of southern Illinois. Rocks that underlie the St. Peter Sandstone are not shown, and formations or members that contain pay zones are shown in bold type. The names Alexandrian, Cayugan, Upper Devonian, Kinderhookian, Valmeyeran, and Virgilian are abbreviated as Alex., Cayu., Up., K., Val., and Virg., respectively. Variable vertical scale.

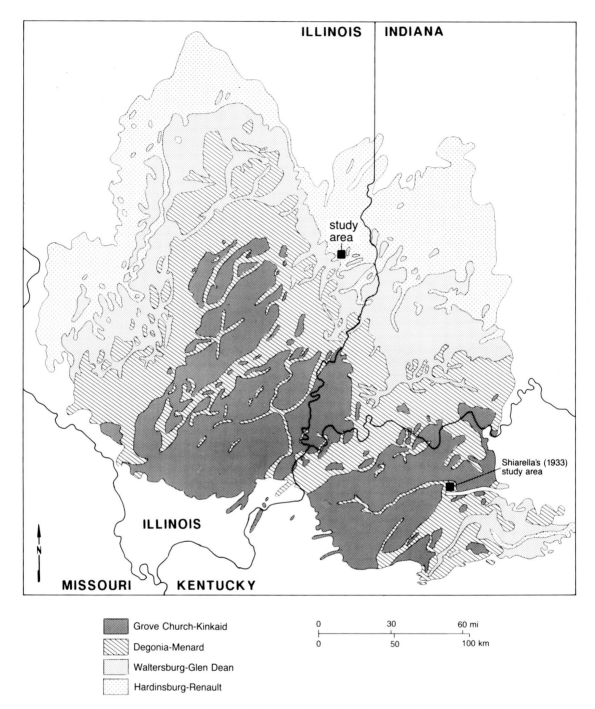

■ (dark)	Grove Church-Kinkaid
▨ (hatched)	Degonia-Menard
□ (light stipple)	Waltersburg-Glen Dean
▫ (stipple)	Hardinsburg-Renault

```
0              30            60 mi
|----|----|----|----|----|----|
0         50          100 km
```

Fig. 13-3. Paleogeologic map of the sub-Pennsylvanian Chesterian surface in the Illinois basin showing an anastomosing erosional pattern at the Mississippian-Pennsylvanian unconformity (modified from Howard and Whitaker, 1988).

hydrocarbon-bearing conglomeratic sandstone lens within a buried paleovalley at the Mississippian-Pennsylvanian unconformity went unrecognized for more than 70 years. This lack of recognition was due principally to the exploration focus of structural traps in shallower Pennsylvanian and deeper Mississippian (Chesterian and Valmeyeran) reservoirs along the La Salle anticlinal belt (Fig. 13-1). Recognition of the paleovalley was also delayed by miscorrelation of a basal Pennsylvanian sandstone, which comprises part of the valley fill, with Mississippian sandstone that occurred at the same depth adjacent to the paleovalley. The potential for miscorrelation and the topographic relief of the unconformity surface at Hardinville are shown in Figure 13-4.

Oil production from a basal Pennsylvanian reservoir within the Caseyville Formation was established two miles (3.2 km) east of Hardinville in 1955, when a Mississippian McClosky (Ste. Genevieve) test on the northeast flank of the Hardinville Anticline, the Miracle and Wooster No. 1 Richart (SW NW NE Sec. 36, T6N-R13W), was plugged back and completed at a rate of 120 BOPD (19.1 m³/D) (Fig. 13-5). Subsequent exploitation resulted in three north and south, ten-acre (4.1-ha.) offsets which were all productive. The exploration significance of the oil-bearing basal Pennsylvanian sandstone reservoir situated along the floor of a deeply incised paleovalley was not recognized.

In 1974 the basal Pennsylvanian reservoir was again fortuitously encountered, this time on the anticline's southwest flank, one and one-half miles (2.4 km) southwest of the 1955 discovery well. Energy Resources of Indiana (E.R.I.), however, mistook the reservoir in their No. 1 Richart Heirs well (NE NE SE Sec. 2, T5N-R13W) for Chesterian sandstones that were expected at about the same depth. Consequently, the paleovalley was not recognized in this well. Finally, in 1976, the sequence was correctly identified in the E.R.I. No. 1 Due Heirs II (NE SW SW Sec. 1, T5N-R13W), which resulted in the drilling and completion of 16 oil wells in the southern portion of the accumulation (Fig. 13-5).

Recoverable oil from the 16 producing wells in the southern portion of the pool is estimated to be 1.4 MMBO (2.2×10^5 m³). An estimated 100 MBO (1.6×10^4 m³) should be recovered from the four wells in the northern portion of the pool. Two additional wells, the No. 1-A Richart Heirs (NE NE SE Sec. 2, T5N-R13W) and the No. 1-A Coulter Heirs (SE SE NE Sec. 2, T5N-R13W), are shut-in gas wells. Although Energy Resources of Indiana estimated that a gas cap in the southern portion of the reservoir originally contained 277 MMCFG (7.8×10^6 m³), produced gas has been flared due to the lack of a sales market.

Reservoir and Trap Characteristics

Reservoir Geometry

The configuration and apparent topographic relief of the sub-Pennsylvanian drainage pattern across the study area are interpreted from an isopach map of Pennsylvanian strata beneath the Colchester (No. 2) Coal (Fig. 13-6). This map shows a south-southwest-trending, one-mile (1.6 km) wide, 200-foot (61-m) deep, bifurcating valley crossing the study area. The paleotopographic relief shown in Figure 13-6 is somewhat muted due to compaction of Pennsylvanian shale that underlies the coal. Additionally, divergence between the unconformity surface and the Colchester Coal along the steeply dipping southwest flank of the anticline creates the illusion of deeper valley incision there than along the anticlinal crest.

A three-mile (5-km) long, slightly arcuate, 0.25- to 0.5-mile (0.4–0.8-km) wide conglomeratic sandstone body, ranging up to 45 feet (13.7 m) thick, occurs along the western side of the valley near its junction with a somewhat smaller valley that contains a similar sand body (Figs. 13-5, 13-6). The basal Pennsylvanian sandstone in the main valley is overlain by an extensive, impermeable shale which forms an effective top seal (Fig. 13-7). The basal sandstone in the smaller paleovalley 0.75 mile (1.2 km) to the west is overlain by sandstones which preclude trap integrity.

The apparent continuity of the basal sandstone in the main valley is illustrated in Figure 13-8. This longitudinal transect reveals that (1) the sandstone is overlain by a thick shale which forms the top seal, (2) the sandstone intertongues with shale at its southern margin and pinches out to the north, (3) a gas cap existed in the southern portion of the reservoir at about 800 feet (244 m) subsea and at about 790 feet (241 m) subsea in the northern portion, and (4) an oil-water contact existed at 875 feet (267 m) subsea in the southern part of the sandstone and at about 815 feet (248 m) subsea in the northern part.

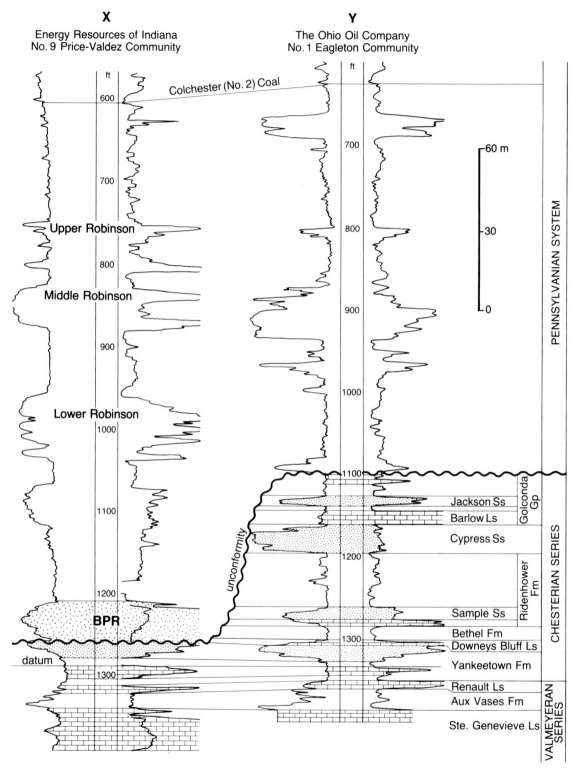

Fig. 13-4. Cross section X-Y (location shown in Fig. 13-5) shows paleotopographic relief at the sub-Pennsylvanian unconformity. Note the potential for miscorrelation due to structural relation of the basal Pennsylvanian reservoir (BPR) sandstone on the valley floor to adjacent Mississippian sandstones (modified from Howard and Whitaker, 1988).

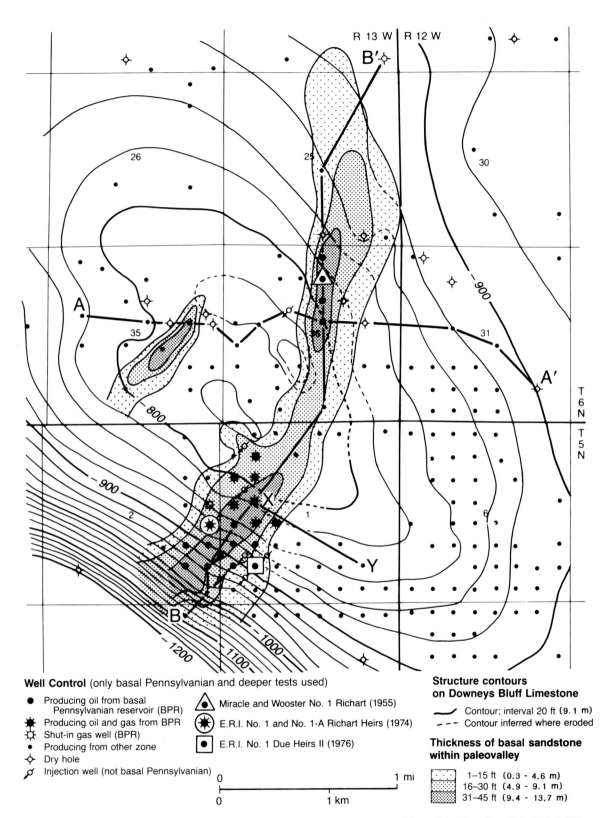

Well Control (only basal Pennsylvanian and deeper tests used)

- ● Producing oil from basal Pennsylvanian reservoir (BPR)
- ✸ Producing oil and gas from BPR
- ☼ Shut-in gas well (BPR)
- • Producing from other zone
- ✧ Dry hole
- ⌀ Injection well (not basal Pennsylvanian)

- ▲ Miracle and Wooster No. 1 Richart (1955)
- ✴ E.R.I. No. 1 and No. 1-A Richart Heirs (1974)
- ▣ E.R.I. No. 1 Due Heirs II (1976)

Structure contours on Downeys Bluff Limestone

╱ Contour; interval 20 ft (9.1 m)
--- Contour inferred where eroded

Thickness of basal sandstone within paleovalley

- 1–15 ft (0.3 - 4.6 m)
- 16–30 ft (4.9 - 9.1 m)
- 31–45 ft (9.4 - 13.7 m)

0   1 mi
0   1 km

Fig. 13-5. Isopach thickness map of the basal sandstone superimposed on the Downeys Bluff Limestone structure map. The basal Pennsylvanian reservoir (BPR) was discovered by Miracle and Wooster's No. 1 Richart on the northeast flank of the Hardinville Anticline in 1955. The southern portion of the BPR was first encountered in 1974 by the E.R.I. No. 1 Richart Heirs but was not recognized until the drilling of the E.R.I. No. 1 Due Heirs II in 1976. Locations of cross sections X-Y (Fig. 13-4), A-A′ (Fig. 13-7), and B-B′ (Fig. 13-8) are shown (modified from Howard and Whitaker, 1988). Note: Well symbols reflect well status reported by the operator and may not indicate the precise nature of produced hydrocarbons.

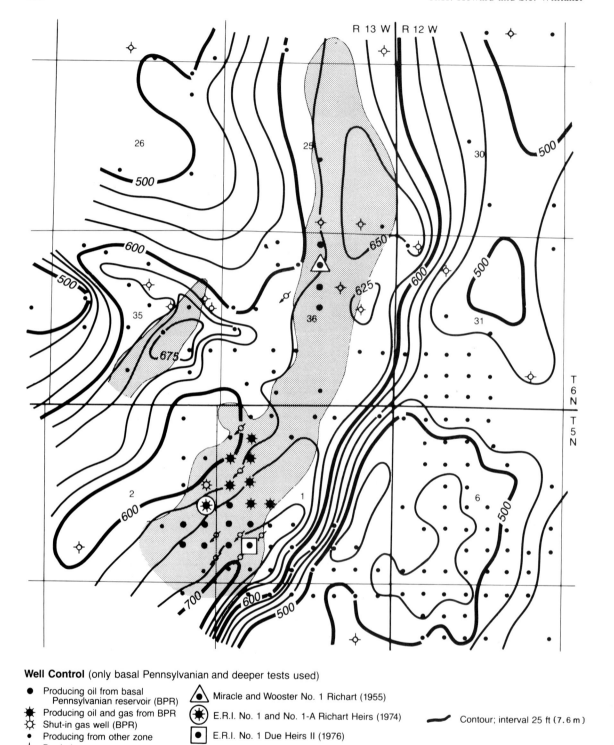

Well Control (only basal Pennsylvanian and deeper tests used)

- ● Producing oil from basal
  Pennsylvanian reservoir (BPR)
- ✳ Producing oil and gas from BPR
- ☆ Shut-in gas well (BPR)
- · Producing from other zone
- ◇ Dry hole
- ⌀ Injection well (not basal Pennsylvanian)

- △ Miracle and Wooster No. 1 Richart (1955)
- ✳ E.R.I. No. 1 and No. 1-A Richart Heirs (1974)
- ▣ E.R.I. No. 1 Due Heirs II (1976)

Contour; interval 25 ft (7.6 m)

0 ——————————— 1 mi
0 ——————————— 1 km

Fig. 13-6. Thickness map of Pennsylvanian strata below the Colchester (No. 2) Coal showing the configuration and apparent paleotopographic relief of the sub-Pennsylvanian surface. A south-southwest-trending bifurcating valley that crosses the study area contains basal sandstones (shaded areas). Note that the basal Pennsylvanian reservoir lies along the western side of the main valley floor (modified from Howard and Whitaker, 1988).

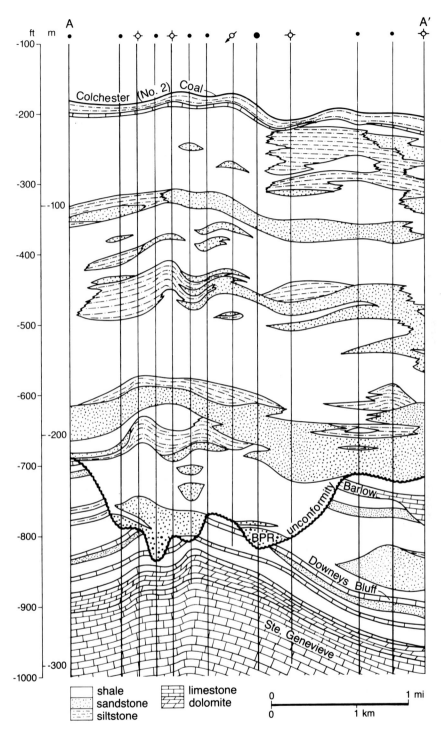

Fig. 13-7. Cross section A-A' (location shown in Fig. 13-5) shows structural and stratigraphic relationships of Chesterian and Pennsylvanian strata below the Colchester Coal along the Hardinville anticline. Note that Pennsylvanian shale covers the basal sandstone in the main valley and forms the top seal (modified from Howard and Whitaker, 1988).

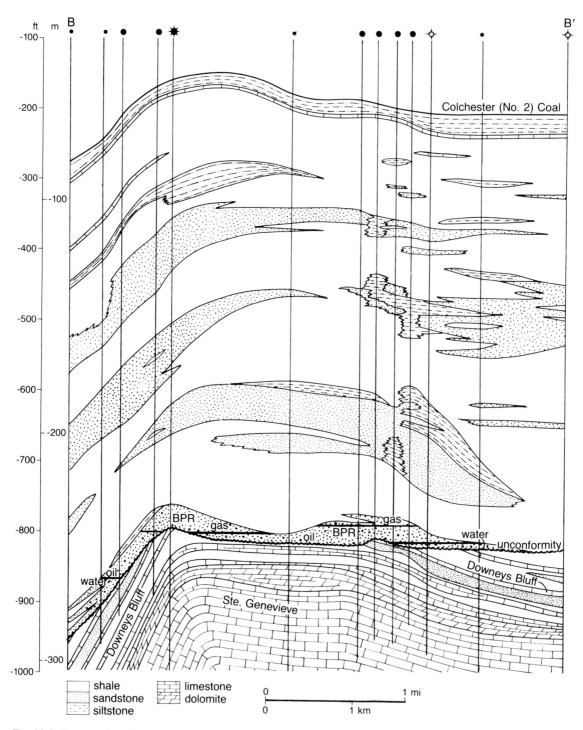

Fig. 13-8. Cross section B-B′ (location shown in Fig. 13-5) oriented along the main paleovalley and showing the variable thickness of the basal Pennsylvanian reservoir across the Hardinville anticline. The reservoir intertongues with shale and siltstone at its southern end and pinches out to the north. Note the lower gas-oil and oil-water contacts in the southern portion (modified from Howard and Whitaker, 1988).

These differences in the gas-oil and oil-water contacts within the reservoir will be discussed in a later section on oil migration.

Trapping Mechanism

Although the present structural configuration at Hardinville provided the inducement for exploration in the area, structure is not the primary factor in the development of the trap of the basal Pennsylvanian reservoir. Indeed the position of this reservoir on the Hardinville anticline masks the paleogeomorphic (stratigraphic) nature of the trap. The pinchout of porous sandstone that is overlain by impermeable shale within the paleovalley (Fig. 13-8) provided the necessary components for stratigraphic entrapment within the sandstone regardless of its fortuitous orientation draped over an anticline. The field is thus a structurally focused stratigraphic trap.

Reservoir Lithofacies

In 1977 E.R.I. cored the No. 1-A Richart Heirs, located 50 feet (15.2 m) south of the No. 1 Richart Heirs (NE NE SE Sec. 2, T5N-R13W), recovering 20 feet (6.1 m) from the basal sandstone and 40 feet (12.2 m) from subjacent Chesterian strata (Figs. 13-6, 13-9).

The basal foot (0.3 m) of the sandstone, from 1,248 to 1,247 feet (380.4–380.1 m), is separated from the rest of the reservoir by a sharply defined contact highlighted by a thin, undulating clay parting. This lowest part consists of tan-colored, well-sorted, medium-grained quartz sandstone with some thin layers of dark gray, medium-grained, angular to subrounded, well-cemented quartz sandstone (Fig. 13-10). Planar crossbeds have an apparent dip of 25° and are highlighted by dark gray, silicified bedding surfaces and at least one planar clay parting. The contact with underlying siltstone of the Chesterian Ridenhower Formation is sharp.

From 1,247 to 1,238 feet (380.1–377.3 m), the core consists of generally light gray, poorly sorted quartz sandstone (with minor feldspar, chert, and rock fragments) and ranges in grain size from medium sand to granules, with a modal size of coarse to very coarse sand. Indistinct, low-angle planar bedding lacks vertical grain-size segregation. Numerous shale chips up to 2 inches (5 cm) in diameter are scattered throughout this portion of the basal sandstone (Fig. 13-11) but are most abundant

in the lower 3 feet (0.9 m), from 1,247 to 1,244 feet (380.1–379.2 m). Limonite staining is pervasive.

From 1,238 to 1,235.4 feet (377.3–376.5 m), the reservoir exhibits poorly defined upward-fining sequences that are mineralogically similar to the lower portion of the core except that there are no large shale chips. The upward-fining sequences become slightly better defined in the uppermost foot (0.3 m) of this interval (Fig. 13-12) and are terminated at a sharp contact with an overlying layer of shale.

At 1,235.2 feet (376.5 m) is a 2.5-inch (6.4-cm) thick layer of dark gray shale that contains some very fine-grained sand laminae (Fig. 13-13). This shale bed can be traced by its wireline-log response to several nearby wells. The presence of a few acritarchs (rare in the Pennsylvanian), which are considered to have affinity with planktonic algae, indicates that this shale was probably deposited under marine conditions (R. Peppers, personal communication, 1988). The upper contact of this shale with overlying sandstone is slightly undulatory and sharply defined.

The upper third of the sandstone in this core, from 1,235.2 to 1,229.2 feet (376.5–374.7 m), comprises several upward-fining sequences which display more pronounced vertical grain-size segregation than those observed below the shale unit. These sequences range from 0.2- to 1.5-feet (0.06–0.46 m) thick (Fig. 13-14) and consist of dark gray to light gray quartz sand with minor feldspar, chert, and rock fragments. The sandstone is angular to subrounded, mostly poorly sorted, and cemented with silica and calcite. Grain size varies from coarse pebbles, sand, or granules at the base of each sequence to coarse to fine sand at the top. The basal contact of each sequence is well defined, may be pyritic, and may be slightly inclined. Several shale chips up to 2 inches (5 cm) in diameter occur at 1,235 feet (376.4 m).

Depositional Setting

The close of Chesterian (latest Mississippian) time was marked by a global drop in sea level (Vail et al., 1977) of more than 300 feet (91 m). The ancient Michigan River system (Swann, 1963), which during Chesterian time had repeatedly supplied the near-equatorial Illinois basin area (Rowley et al., 1985) with siliciclastics, became the agent of removal of much of the Chesterian sedimentary

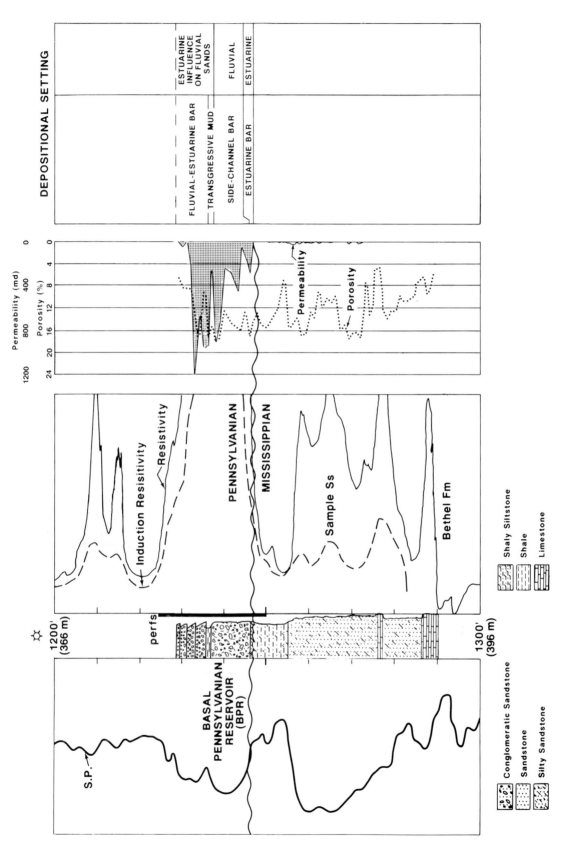

Fig. 13-9. Induction log of bottom portion of E.R.I. No. 1-A Richart Heirs (see Fig. 13-5). Lithology, porosity, and permeability of a 60-ft (18.3-m) core, including 20 ft (6.1 m) of basal Pennsylvanian reservoir and 40 ft (12.2 m) of subjacent Chesterian strata, are depicted; the interpreted depositional setting of the Pennsylvanian portion of the core is shown (modified from Howard and Whitaker, 1988). Fm = formation, Ss = sandstone.

Fig. 13-11. Coarse-grained, poorly sorted, subangular to subrounded quartz sandstone containing a high percentage of clay clasts and lacking size grading. The abundance of clay material and limonite cement in the lower basal Pennsylvanian reservoir has adversely affected permeability. Core from E.R.I. No. 1-A Richart Heirs at a depth of 1,246 feet (379.8 m). Scale bar is 1 inch (2.5 cm).

Fig. 13-10. The basal foot (0.3 m) of the basal Pennsylvanian reservoir contains a medium-grained, well-sorted, angular to subrounded, well-cemented quartz sandstone with planar beds at an apparent dip of 25°. Dark, silicified streaks highlight the bedding surfaces. This sandstone is evidence of an earlier (estuarine?) deposit along the valley floor that was partially scoured, then buried by a subsequent stage of fluvial sedimentation. An undulating shale parting marks the abrupt contact with overlying coarse-grained sandstone of the basal Pennsylvanian reservoir (Fig. 13-11). Core from E.R.I. No. 1-A Richart Heirs at a depth of 1,247 feet (380.1 m). Scale bar is 1 inch (2.5 cm).

Fig. 13-12. Sandstone similar to that in Figure 13-11 except that it is virtually free of clay clasts. The permeability in this portion of the basal Pennsylvanian reservoir is high (about 900 md). Core from E.R.I. No. 1-A Richart Heirs at a depth of 1,236 feet (376.7 m). Scale bar is 1 inch (2.5 cm).

Fig. 13-13. Part of a 2.5-inch (6.4-cm) thick shale bed that is correlatable to a number of adjacent wells by wireline logs. The presence of acritarchs in this shale indicates that it is of marine origin. Core from the E.R.I. No. 1-A Richart Heirs at a depth of 1,235.2 feet (376.5 m). Scale bar is 1 inch (2.5 cm).

Fig. 13-14. Two upward-fining sequences typical of the upper 8 feet (2.4 m) of cored basal Pennsylvanian reservoir. Very coarse-grained, subangular to subrounded quartz sandstone at the base of each sequence grades upward to medium-grained, subangular to subrounded quartz sandstone. Occasional quartz pebbles, feldspar grains, and clay clasts are scattered throughout. The lower part of the graded-bedded sandstone has very good permeability due to lack of cements and clays; however, the uppermost portion (shown in this figure) is tightly cemented and impermeable (see Fig. 13-9). Core from the E.R.I. No. 1-A Richart Heirs at a depth of 1,229.2 feet (374.7 m). Scale bar is 1 inch (2.5 cm).

record (Fig. 13-15A). Notwithstanding evidence of a semiarid climate (Howard, 1979a), the area's erosional history was apparently punctuated by periodic flooding (Howard and Whitaker, 1988). The southwest-flowing rivers (present coordinates) that carried coarse sand and gravel from eastern Canada to Arkansas during earliest Pennsylvanian time (Garner, 1974) incised the emergent portions of the Mississippian surface as they followed the southwestwardly retreating sea. Prograding deltas that accompanied this marine regression were themselves eventually dissected much as those along the southwest Louisiana continental shelf during late Quaternary time (Suter et al., 1987). The resulting topography comprised an anastomosing pattern of valleys cut into Mississippian strata, with interfluves capped by remnants of deltaic deposits or by Mississippian bedrock.

Uplift of the La Salle anticlinal belt (Fig. 13-1) during Late Mississippian and Early Pennsylvanian time accelerated the rate of erosion across the associated Hardinville anticline. The river that flowed southward through the Hardinville area consequently created a relatively steep-walled valley (Fig. 13-15B). Fluvial sands deposited within the valley represent amalgamated sequences deposited by separate flood cycles.

As the northward-advancing Pennsylvanian sea entered the paleovalley at Hardinville, the depositional environment changed from fluvial to estuarine. It is difficult to determine the extent of estuarine influence on deposition in a setting such as this (Smith, 1987, p. 87), since fluvial events continued to affect the environment. For instance, the basal, well-sorted, medium-grained sandstone observed in the No. 1-A Richart Heirs core suggests deposition in an estuarine environment. Apparently, however, a subsequent influx of fluvial sediment, caused either by a modest regression or by a period of high river flow, established another sandbar over the eroded remnants of the estuarine deposit. The latter bar, which stretches for about 3 miles (5 km) along the western side of the main valley (Fig.

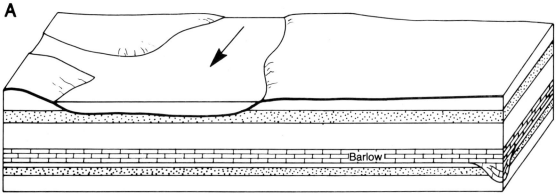

LATE MISSISSIPPIAN
Rivers flow across low-relief plains.

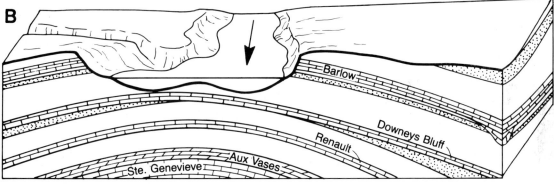

LATEST MISSISSIPPIAN
River incision caused by drop in base level and gradual uplift of anticline.

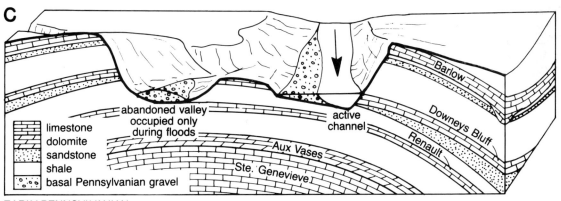

EARLY PENNSYLVANIAN
Gravelly sandbars were deposited along the paleovalleys at the end of each period of high flow. The fluvial conditions were ultimately superceded by an estuarine environment resulting from the transgressing Pennsylvanian sea.

Fig. 13-15. Evolution of paleovalleys at the Mississippian-Pennsylvanian unconformity and subsequent sandbar deposition across the Hardinville anticline during Pennsylvanian time (modified from Howard and Whitaker, 1988).

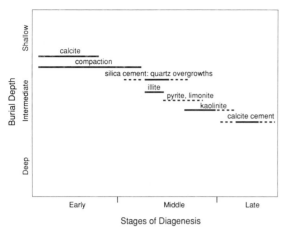

Fig. 13-16. Order of diagenetic events in the basal Pennsylvanian reservoir based on petrographic analyses of core samples from the E.R.I. No. 1-A Richart Heirs.

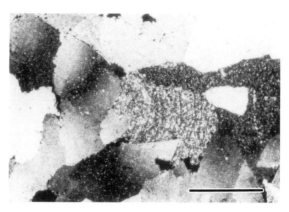

Fig. 13-17. Thin-section photograph showing large crystals of calcite cement displaying undulose extinction that are evidence of an early stage of cementation which partially filled intergranular void space. Subsequent compaction of the sandstone stressed this calcite cement and resulted in its characteristic undulose-extinction pattern. This early cement was observed only in the top 3 feet (0.9 m) and the bottom 1 foot (0.3 m) of the cored sandstone. From the E.R.I. No. 1-A Richart Heirs at a depth of 1,229.2 feet (374.7 m). Scale bar is 1 mm.

13-15C), appears to be a side-channel bar that typifies deposition in straight, narrow valleys on relatively gentle slopes (Miall, 1977, p. 3–5). It seems likely that the lower 9 feet (2.7 m) of this bar, consisting of conglomeratic sandstone lacking graded bedding, was deposited during a period of high fluvial flow.

When marine conditions again dominated the valley, a thin bed of estuarine/marine mud was deposited over the bar as indicated by the shale at 1,235.2 feet (376.5 m) in the core. This shale is more continuous over a larger portion of the bar than are any of the other shales noted from wireline logs and appears to have recorded the ultimate transformation of the valley into an estuary.

A final influx of fluvial sediment into the estuary is indicated by distinct upward-fining sequences deposited above the estuarine/marine shale (Fig. 13-14). The rising sea then completely dominated the environment in the area and deposited 100 feet (30.5 m) of marine shale over the sandbar (Figs. 13-7, 13-8). This last episode provided the requisite seal to trap the hydrocarbons that later migrated into the reservoir. The breaching of this shale by younger porous strata eliminated the potential for entrapment in the basal sandstone in section 35, T6N-R13W (Fig. 13-7).

Petrography and Diagenesis

Petrographic analyses of the E.R.I. No. 1-A Richart Heirs core reveal a prevalence of calcite, silica,

pyrite, and limonite cements within the basal Pennsylvanian reservoir. In addition, varying amounts of kaolinite and illite/smectite are present. The interpreted order of diagenetic events is illustrated in Figure 13-16.

The first period of cementation is thought to be represented by a calcite cement that exhibits an undulose extinction pattern (Fig. 13-17). This cement is found only in the basal foot (0.3 m) and top 3 feet (0.9 m) of the cored sandstone, where it partially filled void space before continued compaction stressed the cement and resulted in its characteristic extinction pattern.

The early calcite cement in the bottom foot (0.3 m) of the sandstone could have been introduced during the sandbar's early exposure to calcium-carbonate-rich waters emanating from adjacent Mississippian strata or from an early infusion of Pennsylvanian marine waters. Early calcite cement in the top 3 feet (0.9 m) of the core may represent the influence of the marine/estuarine waters that flooded the paleovalley soon after deposition of the sand body and penetrated only the upper portion of the bar.

Shortly after initial calcite cementation began, compaction of the sand caused some pressure dissolution of the metaquartzite grains that compose most of the reservoir rock. A period of silica cemen-

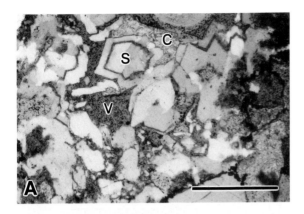

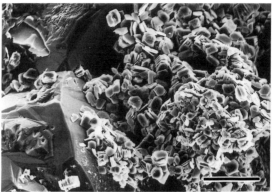

Fig. 13-19. SEM photomicrograph showing books of authigenic kaolinite coating grains and reducing permeability. From the E.R.I. No. 1-A Richart Heirs at a depth of 1,244 feet (379.2 m). Scale bar is 0.025 mm.

Fig. 13-18. (A) Thin section photograph showing euhedral crystals of silica cement and quartz overgrowths (s) partially filling an intergranular void (v). A thin layer of clay within silica cement outlines an earlier stage of crystal growth and suggests an interruption in silicification. Note the calcite cement (c). From the E.R.I. No. 1-A Richart Heirs at a depth of 1,229.2 feet (374.7 m). Scale bar is 0.5 mm. (B) SEM photomicrograph showing euhedral crystals of silica cement and quartz overgrowths partially filling an intergranular void. Note sparse kaolinite books on crystal faces. From the E.R.I. No. 1-A Richart Heirs at a depth of 1,247.5 feet (380.2 m). Scale bar is 0.05 mm.

of minor amounts of illite or illite/smectite, the breakdown of fine-grained detrital feldspar produced authigenic kaolinite, which is particularly abundant in the finer-grained portions of the reservoir (Fig. 13-19). A second period of calcite cementation continued to fill pore space between quartz overgrowths and any previous cements (Figs. 13-18A, 13-20). This calcite cement exhibits a normal, flat extinction pattern and thus can be differentiated from the earlier calcite cement.

Reservoir Quality

Petrographic analyses indicate that cementation is a major agent influencing the reservoir's heterogeneity.

tation and quartz overgrowth formation began, which partially filled the remaining void space throughout the reservoir (Figs. 13-18A, 13-18B). Relationships with neighboring cements suggest that precipitation of pyrite and limonite, particularly abundant near partially decomposed shale clasts, occurred after silicification. Although present throughout the core, the concentration of pyrite and limonite is greatest in the lower 9 feet (2.7 m) of the unit.

Clays have also formed in varying amounts throughout the sandstone. Following the formation

Fig. 13-20. SEM photomicrograph showing that calcite cement has occluded much of the intergranular porosity. From the E.R.I. No. 1-A Richart Heirs at a depth of 1,229.2 feet (374.7 m). Scale bar is 0.5 mm.

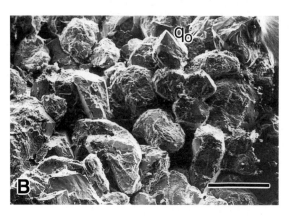

Fig. 13-21. (A) Thin section showing excellent porosity and permeability in the basal Pennsylvanian reservoir. The general lack of shale clasts and relative distance below the tightly cemented upper portion of the sandstone have resulted in a lack of clays and in sparse cementation. From E.R.I. No. 1-A Richart Heirs at a depth of 1,236.3 feet (376.8 m). Scale bar is 0.5 mm. (B) SEM photomicrograph showing relatively clean, open pore throats typical of the coarser-grained portions of the sandstone. Note the quartz overgrowth (q_o). From E.R.I. No. 1-A Richart Heirs at a depth of 1,236.3 feet (376.8 m). Scale bar is 0.5 mm.

Finer-grained layers within the reservoir, notably the upper portion of each upward-fining sequence, have been more thoroughly cemented due to the smaller pore spaces originally present. Indeed, the upper 4 to 7 feet (1.2–2.1 m) of the reservoir are impermeable due to cementation. In contrast, the coarser-grained portions of the upward-fining sequences in the upper part of the reservoir were insufficiently cemented to severely hinder permeability and thus are the best reservoirs (Figs. 13-9, 13-21A, 13-21B). In the lower portion of the sandstone, poor sorting, abundant shale clasts,

and the formation of cements and clays have decreased permeability but have not totally occluded porosity.

In summary, because the textural fabric of the basal Pennsylvanian reservoir that was inherited from the depositional and early postdepositional conditions controlled the early porosity and permeability of the sandstones, the present-day reservoir quality is largely a function of subsequent diagenetic processes, but is facies related.

Hydrocarbon Migration into the BPR

Hydrocarbons in Pennsylvanian reservoirs within the Illinois basin were apparently generated from the Devonian-Mississippian New Albany Shale (Barrows and Cluff, 1984). Maturation studies indicate that the onset of oil generation in the New Albany began during Permo-Triassic time and was limited to the central, deeper portion of the basin (Oltz and Crockett, 1989). Hydrocarbon migration through Mississippian strata into Pennsylvanian rocks probably began during the Mesozoic (Bethke et al., in press) and would have been facilitated by paleovalleys that transected hydrocarbon-bearing Mississippian units.

As hydrocarbons migrated into the reservoir at Hardinville, a gas cap began to form in the highest structural position along the southern part of the reservoir. The gas cap continued to expand until it encompassed enough of the structurally highest portions of the reservoir to act as a barrier to further oil migration between the southern and northern portions (Fig. 13-8). Oil continued to fill the southern part of the BPR but was prevented by the gas cap from spilling over into the northern part to completely fill the trap.

Production Characteristics

Although recoverable reserves from the BPR were estimated at 1.5 MMBO (2.4×10^5 m³), certain recovery practices have adversely affected production and resulted in a recovery of only about 500 MBO (8.0×10^4 m³). Flaring of produced gas has reduced reservoir pressure and allowed expansion of the oil column into the pore spaces formerly occupied by the gas cap. In addition, commingling of production from the reservoir with that from younger Robinson sandstones drastically reduced recovery from the basal sandstone and precluded any ability to obtain accurate production values for the individual reservoirs (Larry Whitmer, personal communication, 1988).

Relatively high concentrations of iron in basal sandstone brines reacted chemically with Robinson brines to form precipitates of iron oxide and iron sulfide that rapidly reduced permeability of the reservoir near the wellbore. This problem is illustrated by a production curve (Fig. 13-22A), which shows a dramatic drop in production upon the initiation of commingling in 1977. In 1980, some of these wells were recompleted to only the deeper interval using a diesel oil frac with iron sequestering agents and clay stabilizers. This practice improved subsequent production. The contrast between Figure 13-22B, which shows production from a lease where commingling did not occur, and Figure 13-22A dramatically manifests the deleterious effect of commingled production.

An attempt to waterflood the field began in 1978 using water largely composed of brines produced from Robinson reservoirs. This effort was not effective due to the permeability degradation and, in part, due to probable inherent heterogeneities within the reservoir caused by cementation.

Optimum recovery of hydrocarbons from basal Pennsylvanian sandstone reservoirs similar to the one discussed here requires drilling and completion practices that prevent formation damage like that experienced at Hardinville. Additionally, the likelihood of compartmentalization due to both cementation and probable clay drapes requires that careful consideration be paid to well spacing patterns and to subsequent reservoir management.

Exploration and Production Strategy

The discovery and ultimate exploitation of the basal Pennsylvanian reservoir at Hardinville, in the Main Consolidated Field, were due to an accidental encounter with a previously unrecognized paleovalley. The position of this reservoir on the Hardinville anticline was indeed fortunate but did not readily enable geologists to note the paleogeomorphic (stratigraphic) component of the trap.

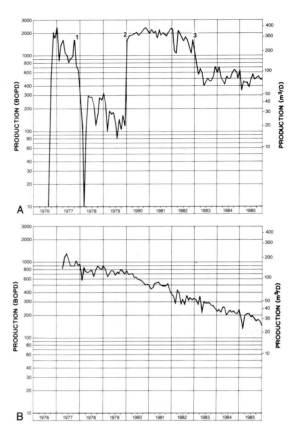

Fig. 13-22. (A) Production curve from the 3-well Valdez lease (Sec. 1, T5N-R13W) shows the effects of commingling of Robinson sandstones with the basal Pennsylvanian reservoir. (1) Precipitation of iron oxides and iron sulfides reduced permeability upon initiation of commingling in 1977. (2) In 1980, wells in the basal Pennsylvanian reservoir were recompleted without commingling. (3) Waterflooding of the basal Pennsylvanian reservoir using treated Robinson brines began in 1982. (B) Production curve from the Price-Valdez No. 10, which has not been commingled, reveals a more gradual and stable decline.

The distribution of discontinuous reservoirs along the floors of paleovalleys at the Mississippian-Pennsylvanian unconformity is not dependent on present structural position. Consequently, a prime concern should be the delineation of the paleovalley network and its relationship to surrounding lithologic units.

The locations of medium to large paleovalleys are revealed by Bristol and Howard's (1971) paleogeologic map of the sub-Pennsylvanian Chesterian surface in the Illinois basin, a generalized version of which is shown in Figure 13-3. From this regional

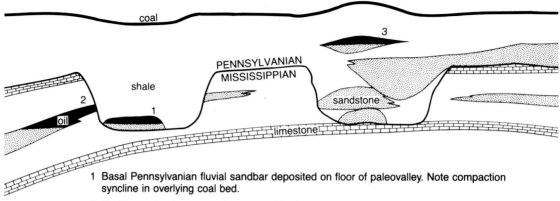

1 Basal Pennsylvanian fluvial sandbar deposited on floor of paleovalley. Note compaction syncline in overlying coal bed.

2 Pre-Pennsylvanian reservoir truncated by impermeable shale valley fill.

3 Anticline resulting from compactiơn of shales around stacked sandstones deposited within and above the paleovalley.

Fig. 13-23. Diagram of types of stratigraphic traps associated with paleovalleys and their sedimentary fill (after Howard and Whitaker, 1988).

map, areas can be identified for further detailed mapping. Structure maps of the closest overlying coal have proven to be effective in locating buried paleovalleys (Howard and Whitaker, 1988). Buried paleovalleys are indicated on such maps by narrow, sublinear depressions of the coal bed caused by compaction of shale within the buried paleovalley (Fig. 13-23). The numerous coal tests and mine maps available in the basin should prove to be especially useful even in areas where oil drilling and logged holes are sparce.

Another indicator of the buried paleovalley configuration is the thickness of Pennsylvanian strata between the closest mappable overlying coal bed and the Mississippian-Pennsylvanian unconformity (Fig. 13-6). However, as the unconformity itself is not always easily picked from log data, sample examinations are commonly needed to differentiate finer-grained Mississippian sediments from coarser-grained Pennsylvanian sediments.

Seismic data may be useful in locally locating buried paleovalleys, but the accuracy of these data depends on their quality and the reflection coefficient between sedimentary rocks within the paleovalley and the surrounding strata. A shale-filled valley surrounded by dense, relatively tight, high-velocity limestone or sandstone may show a characteristic seismic anomaly. Seismic studies could probably not delineate the entire course of a paleovalley, however, because some lithologies occurring in and along a paleovalley might not pro-

duce adequate acoustic impedance contrast to produce good seismic reflection.

Other kinds of hydrocarbon traps have been created by the paleovalley network in addition to the one described in this chapter (Fig. 13-23). A truncation trap (type 2 in Fig. 13-23) was recently discovered a few miles north of the study area and demonstrates that the relationship of a paleovalley to subcropping lithologies should also be examined for potential trap geometries.

Conclusions

Most hydrocarbon exploration and development in the Illinois basin has been based on the search for structural traps, mainly anticlines. Consequently, stratigraphic traps have received far less attention than they warrant.

This chapter suggests that stratigraphic traps associated with the sub-Pennsylvanian paleovalley network can be delineated through conventional mapping methods and possibly locally by high-resolution seismic acquisition. Primary targets of exploration within these paleovalleys should be discontinuous, coarse-grained, Pennsylvanian sandstone reservoirs. Additional traps associated with the paleovalley network include truncated reservoirs sealed by shale-filled valleys and reservoirs in compaction anticlines above stacked sandstones in these valleys.

Acknowledgments. We thank Zakaria Lasemi for preparing samples for SEM analysis and for his advice during the interpretation of the resulting photomicrographs. We also thank Bryan Huff for preparing the thin sections used in this study. James H. Clement and Joseph R.J. Studlick provided constructive reviews of the manuscript, which are gratefully acknowledged.

Reservoir Summary

Field: Main Consolidated
Location: Southeastern Illinois
Operator: Energy Resources of Indiana, Shakespeare Oil Company
Discovery: 1955, 1976
Basin: Illinois basin
Tectonic/Regional Paleosetting: Intracratonic basin
Geologic Structure: Domal anticline
Trap Type: Stratigraphic pinch-out across anticlinal axis
Reservoir Drive Mechanism: Depletion
  • **Original Reservoir Pressure:** 380 psi (2.6×10^3 kPa) at 820 ft (250 m) subsea
Reservoir Rocks
  • **Age:** Early Pennsylvanian (Bashkirian)
  • **Stratigraphic Unit:** Caseyville Formation
  • **Lithology:** Medium-grained to pebbly sandstone, quartzarenite
  • **Depositional Environment:** Fluvial-estuarine
  • **Productive Facies:** Elongate channel-bar sandstone
  • **Petrophysics**
    • **Porosity Type:** Intergranular (total = 15%; 15% primary, 0% secondary)
    • **ϕ:** Average 15%, range 6 to 25%, cutoff NA (unstressed cores)
    • **k:** Average 670 md, range 0 to 3,570 md, cutoff NA (unstressed cores)
    • **S_w:** Average 30%, range 10 to 55%, cutoff 45% (logs)
    • **S_{or}:** NA
Reservoir Geometry
  • **Depth:** 1,250 feet (380 m)
  • **Areal Dimensions:** 2.0 by 0.5 miles (3.2×0.8 km)
  • **Productive Area:** 300 acres (120 ha.) oil; 120 acres (50 ha.) gas
  • **Number of Reservoirs:** 1
  • **Hydrocarbon Column Height:** 105 feet (32 m)
  • **Fluid Contacts:** Oil-water contacts at 875 feet (266.7 m) subsea in the southern portion of the reservoir and 815 feet (248.4 m) subsea in the northern portion of reservoir; gas-oil contacts at 800 feet (243.8 m) subsea in southern portion and 790 feet (240.8 m) subsea in the northern portion
  • **Number of Pay Zones:** 1
  • **Gross Sandstone Thickness:** Up to 45 feet (13.7 m)
  • **Net Sandstone Thickness:** 19 feet (5.8 m)
  • **Net/Gross:** 0.40
Hydrocarbon Source, Migration
  • **Lithology & Stratigraphic Unit:** Basinal shale, New Albany Group, Late Devonian
  • **Time of Hydrocarbon Maturation:** Permo-Triassic
  • **Time of Trap Formation:** Pennsylvanian
  • **Time of Migration:** Triassic to Recent
Hydrocarbons
  • **Type:** Oil, gas
  • **GOR:** Up to 25,000:1
  • **API Gravity:** 36°
  • **FVF:** 1.15
  • **Viscosity:** NA

Volumetrics
  • **In-Place:** 4.6 MMBO (7.3×10^5 m³)
  • **Cumulative Production:** 500 MBO (8.0×10^4 m³), estimated due to commingling with other zones
  • **Ultimate Recovery:** 1.5 MMBO (2.4×10^5 m³); 277 MMCFG (7.8×10^6 m³)
  • **Recovery Efficiency:** 33%
Wells
  • **Spacing:** 660 feet (201 m), 10 acres (4.1 ha.)
  • **Total:** 24
  • **Dry Holes:** 2
Typical Well Production
  • **Average Daily:** 70 BO (11 m³)
  • **Cumulative:** 70 MBO (1.1×10^4 m³)

References

Barrows, M.H., and Cluff, R.M., 1984, New Albany Shale Group (Devonian-Mississippian) source rocks and hydrocarbon generation in the Illinois Basin, *in* Demaison, G.R. and Murris, R.J., eds., Petroleum Geochemistry and Basin Evaluation: American Association of Petroleum Geologists Memoir 35, p. 111–138.

Bethke, C.M., Reed, J.D., Barrows, M.H., and Oltz, D.F., in press, Long range petroleum migration in the Illinois Basin, *in* Leighton, M. W., ed., Cratonic Basins: American Association of Petroleum Geologists Petroleum Basin Series, v. 5.

Bristol, H.M., and Howard, R.H., 1971, Paleogeologic map of the sub-Pennsylvanian Chesterian (Upper Mississippian) surface in the Illinois Basin: Illinois State Geological Survey, Circular 458, 14 p.

Davis, R.W., Plebuch, R.O., and Whitman, H.M., 1974, Hydrology and geology of deep sandstone aquifers of Pennsylvanian age in part of the western coal field region, Kentucky: Kentucky Geological Survey Series X, Report of Investigations 15, 26 p.

Garner, H.F., 1974, The Origin of Landscapes: London, Oxford University Press, p. 651–659.

Howard, R.H., 1979a, The Mississippian-Pennsylvanian unconformity in the Illinois Basin–Old and new thinking, *in* Palmer, J.E. and Dutcher, R.R., eds., Depositional and structural history of the Pennsylvanian System of the Illinois Basin: Guidebook for Field Trip 9, Part 2, Ninth International Congress on Carboniferous Stratigraphy and Geology, Illinois State Geological Survey, p. 34–43.

Howard, R.H., 1979b, Carboniferous cyclicity related to the development of the Mississippian-Pennsylvanian unconformity in the Illinois Basin, *in* Proceedings, Eighth International Congress on Carboniferous Stratigraphy and Geology, v. 5, p. 32–41.

Howard, R.H., and Whitaker, S.T., 1988, Hydrocarbon accumulation in a paleovalley at Mississippian-Pennsylvanian unconformity near Hardinville, Craw-

ford County, Illinois: A model paleogeomorphic trap: Illinois State Geological Survey, Illinois Petroleum 129, 26 p.

Miall, A.D., 1977, A review of the braided river depositional environment: Earth-Science Reviews, v. 13, p. 1–62.

Oltz, D.F., and Crockett, J.E., 1989, Focusing future exploration in a mature basin: Maturation and migration models integrated with timing of major structural events in Illinois [abst.]: American Association of Petroleum Geologists Bulletin, v. 73, p. 1037.

Pryor, W.A., and Potter, P.E., 1979, Sedimentology of a paleovalley fill: Pennsylvanian Kyrock Sandstone in Edmonson and Hart Counties, Kentucky, *in* Palmer, J.E. and Dutcher, R.R., eds., Depositional and structural history of the Pennsylvanian System of the Illinois Basin: Guidebook for Field Trip 9, Part 2, Ninth International Congress on Carboniferous Stratigraphy and Geology, Illinois State Geological Survey, p. 49–62.

Rowley, D.B., Raymond, A., Parish, J.T., Lottes, A.L., Scotese, C.R., and Ziegler, A.M., 1985, Carboniferous paleogeomorphic, phytogeographic, and paleoclimatic reconstructions, *in* Phillips, T.L. and Cecil, C.B., eds., Paleoclimatic Controls on Coal Resources of the Pennsylvanian System of North America: International Journal of Coal Geology, v. 5, p. 7–42.

Shiarella, N.W., 1933, Typical oil producing structures in the Owensboro Field of western Kentucky: Kentucky Bureau of Mineral and Topographic Survey, Series VII, Bulletin 3, 14 p.

Siever, R., 1951, The Mississippian-Pennsylvanian unconformity in southern Illinois: American Association of Petroleum Geologists Bulletin, v. 35, p. 542–581.

Sloss, L.L., 1963, Sequences in the cratonic interior of North America: Geological Society of America Bulletin, v. 74, p. 93–114.

Smith, D.G., 1987, Meandering river point bar lithofacies models: Modern and ancient examples compared, *in* Ethridge, F.G. and Flores, R.M., eds., Recent Developments in Fluvial Sedimentology: Society of Economic Paleontologists and Mineralogists Special Publication 39, p. 83–91.

Suter, J.R., Berryhill, H.L., Jr., and Penland, S., 1987, Late Quaternary sea-level fluctuations and depositional sequences, southwest Louisiana continental shelf, *in* Nummedal, D., Pilkey, O.H., and Howard, J.D., eds., Sea Level Fluctuation and Coastal Evolution: Society of Economic Paleontologists and Mineralogists Special Publication 41, p. 199–219.

Swann, D.H., 1963, Classification of Genevievian and Chesterian (Late Mississippian) rocks of Illinois:

Illinois State Geological Survey, Report of Investigations 216, 91 p.

Swann, D.H., and Bell, A.H., 1958, Habitat of oil in the Illinois Basin, *in* Weeks, L.G., ed., Habitat of Oil: American Association of Petroleum Geologists, p. 447–472.

Vail, P.R., Mitchum, R.M., and Thompson, S., III, 1977, Seismic stratigraphy and global changes of sea level, part 4: Global cycles of relative changes of sea level, *in* Payton, C.E., ed., Seismic Stratigraphy—Application to Hydrocarbon Exploration: American Association of Petroleum Geologists Memoir 26, p. 83–97.

Key Words

Main Consolidated Field, Illinois, Illinois basin, Caseyville Formation, Early Pennsylvanian, Bashkirian, fluvial-estuarine, Absaroka Sequence, anastomosing paleovalley network, Kaskaskia Sequence, Mississippian-Pennsylvanian unconformity, paleogeomorphic (stratigraphic) trap, paleovalley fill, channel-bar sandstone.

14

Continuity and Performance of an Estuarine Reservoir, Crystal Field, Alberta, Canada

John E. Clark and Gerald E. Reinson

Westcoast Petroleum Ltd., Calgary, T2P 0T8 Alberta; Geological Survey of Canada, Calgary, T2P 0T8 Alberta

Introduction

Performance of the reservoir in Crystal Field is directly related to the distribution of depositional lithofacies. Rock types consistent with an estuarine-fill episode have been described and delineated, then used to explain anomalous production characteristics. The result of such an exercise is a greater understanding of the dependence of flow-unit geometry on lithofacies distribution within the reservoir.

The Viking Formation in Crystal Field, Alberta, Canada, contains a linear sandstone-conglomerate deposit which attains a thickness of 100 feet (30.5 m) and is aligned in a north-south direction. The thickness and alignment trends of this reservoir sand body contrast sharply with other established Viking sandstone reservoirs, which are generally much thinner and oriented more in a northwest-southeast direction (Fig. 14-1). Crystal Field is unique in comparison to other Viking oil fields in that the producing reservoir consists of two distinct, hydrodynamically separated oil pools (Fig. 14-2). The stratigraphically higher Crystal Viking "H" pool partially overlaps the larger Viking "A" pool, which contains 95% of the 36.5 MMBO (5.8 × 10⁶ m³) field reserves.

The occurrence of two separate pools and the variability in reservoir capacity, continuity, and performance trends of the "A" pool are controlled directly by depositional factors (Reinson et al., 1988). The depositional model of prograding estua-

rine valley fill under transgressive conditions readily explains the presence of two separate oil pools and also accounts for some of the reservoir behavior, especially with respect to spatial continuity and comparative well performance within the "A" pool. The purpose of this chapter is to document reservoir production behavior and relationships of specific reservoir lithofacies, especially in response to enhanced recovery operations.

Crystal Field, the most recent significant Viking oil find in Alberta, was discovered in 1978. The discovery well (6-7-46-3 W5M) encountered a thick but apparently low-porosity sandstone interval which was not tested or cored. It was evaluated as gas-bearing and was cased for later completion. A year later, another well (3-8-46-3 W5M) was drilled east of the discovery well (Fig. 14-2). It yielded 1,320 feet (402 m) of oil on a drill-stem test and was completed as an oil well. This prompted reexamination of the 6-7 well, and it was subsequently completed for oil. The next well was drilled in 1981 on the eastern limit of the present field at location 13-5-46-3 W5M. This also produced oil at moderate rates, and development drilling began in earnest.

One hundred and twenty-five wells have been drilled in the Crystal Field: 84 are producing oil wells, 12 are dry holes, and 29 have been converted to water injection wells in the "A" pool waterflood project. It is estimated that the ultimate recovery will be enhanced to 34.6 MMBO (5.5 × 10⁶ m³), representing 34% of original oil in place of the "A" pool.

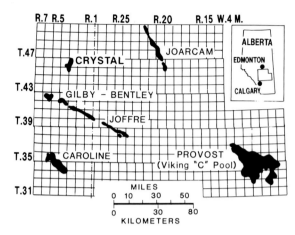

Fig. 14-1. Location of Crystal Field relative to other Viking fields in south-central Alberta. From Reinson, Clark, and Foscolos (1988) and reprinted by permission of American Association of Petroleum Geologists.

Regional Framework

Setting

The Lower Cretaceous Crystal Viking Field of south-central Alberta is in the Cretaceous Western Interior basin of North America (Fig. 14-1). This large foreland basin stretched from the Arctic Ocean to the Gulf of Mexico and, by late Albian time, was totally flooded by the shallow epicontinental Western Interior seaway (Williams and Stelck, 1974; Caldwell, 1984). The basin formed by downward flexure of the ancient continental lithospheric margin in response to passive thrust-sheet loading of supracrustal rocks during evolution of the fold-thrust belt of the eastern Cordillera (Price, 1981; Caldwell, 1984).

Tectonic activity along the marginal orogenic belt, although intermittent and of varying intensity, persisted throughout the Cretaceous. This activity provided both the source and the depositional gradient required to transport these clastics to the seaway. The size and shape of the foreland basin changed through time due mainly to the varying degree of crustal shortening within the marginal orogen and the distribution and volume of sedimentary fill (Caldwell, 1984). The fill consists primarily of thick wedges of westerly derived sandstone which alternate with sequences of marine shale (Stott, 1982). Individual clastic wedges are thought to record cycles of major tectonic activity to the west, resulting in high rates of sediment availability and

rapid progradational input to the adjacent basin. By mid-late Albian time, this previously dominant continental sedimentation pattern was interrupted by a major global rise of sea level which inundated the entire basin.

It is widely recognized that several major sea-level fluctuations affected the Western Interior seaway during the Early Cretaceous (Hancock, 1975; Kauffman, 1977; Vail, et al., 1977; Weimer, 1984). Caldwell (1984) suggests that two major transgressive-regressive cycles in the interior plains reflect global eustatic sea-level fluctuations. Further, the two major cycles contain several transgressive-regressive "subcycles" within their transgressive phases, which Caldwell attributes to local and/or regional tectonic controls. Recently, several workers have documented the presence of these higher frequency sea-level movements in the form of stacked progradational sequences, unconformities, and incised-valley and fill sequences within Viking equivalent strata in western Canada (Reinson et al., 1988; Leckie and Reinson, 1989) and in Montana and Wyoming (Weimer, 1983). Reinson, Clark, and Foscolos (1988) have demonstrated the occurrence of sequence and parasequence boundaries within the Viking Formation in Crystal Field. Such depositional breaks have led to very diversified stratigraphic sequences in near juxtaposition.

Stratigraphy

The nomenclature and stratigraphic position of the Viking Formation have been summarized by Beaumont (1984), Reinson, Clark, and Foscolos (1988), and Leckie and Reinson (1989). The Viking Formation of south and central Alberta is overlain by the Lower Colorado shale and underlain by the Joli Fou Formation (also a shale). All three units constitute the lower part of the Colorado Group of late Albian age. The Viking ranges from 45 to 100 feet (13.7–30.5 m) in thickness over most of south-central Alberta, thickening southwestward and southward to more than 190 feet (58 m). To the northeast, the formation becomes very thin, and the sandstones tend to become finer-grained and very silty, until they finally "shale-out."

In the Crystal Field area, the Viking Formation varies from 80 to 125 feet (24.4–38.1 m) thick and is unconformable with the overlying Colorado shale. The basal contact can be either unconformable or conformable with the underlying Joli Fou Formation, depending on the depth of channel

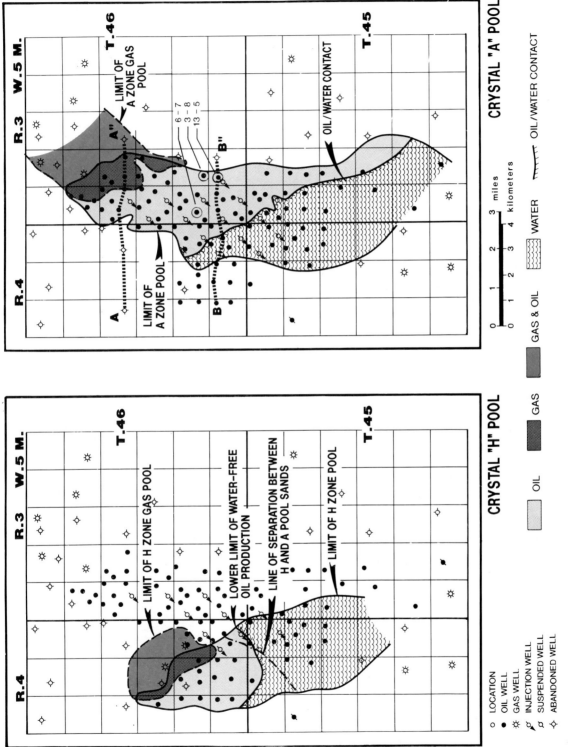

Fig. 14-2. Crystal Field, "A" and "H" pool limits and well control. From Reinson, Clark, and Foscolos (1988) and reprinted by permission of American Association of Petroleum Geologists.

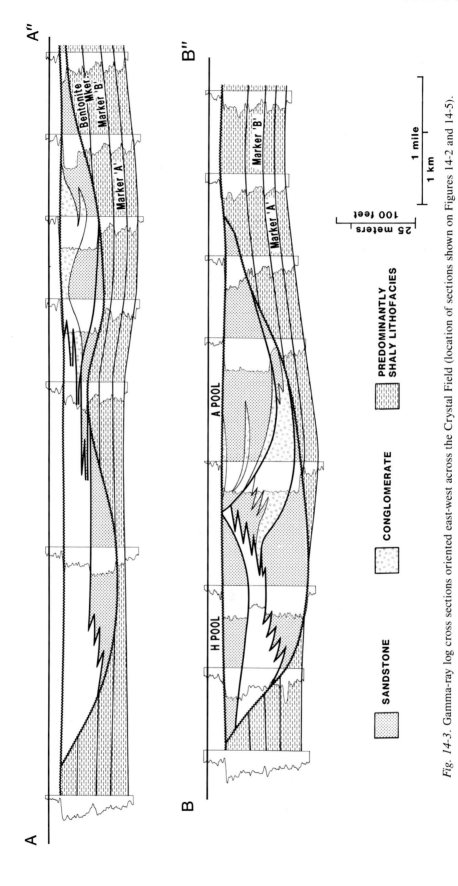

Fig. 14-3. Gamma-ray log cross sections oriented east-west across the Crystal Field (location of sections shown on Figures 14-2 and 14-5).

incisement (Fig. 14-3 and Reinson et al., 1988). The stratigraphic section illustrated in Figure 14-4 is a typical Viking log section from Crystal Field showing an unconformable upper contact, a conformable lower contact, and an unconformable surface at the base of the reservoir body within the formation.

Reservoir Characterization

Depositional Setting and Facies Geometry

The Crystal Field reservoir is a south-north trending sandstone and conglomerate deposit that is thickest in the central area and bifurcates northward (Fig. 14-5). Reinson, Clark, and Foscolos (1988) documented the depositional setting and geometry of the reservoir, and it is summarized briefly here.

The Crystal reservoir sand body is interpreted to be a multistage estuarine tidal channel-bar complex which is flanked by laterally equivalent bay-fill "muddy" sediments (Figs. 14-3, 14-6). Mapping indicates that the reservoir sand body comprises three successive but partially superimposed channel depositional cycles ("A" pool) and an upper less channelized and lower-energy depositional event which forms a shallow channel-bar deposit ("H" pool).

Four depositional settings have been recognized: (1) cyclical shelf-to-shoreface, (2) tidal channel, (3) estuary bay-fill, and (4) upper transgressive shelf. The sandstone body sits in a valley which was incised into the older cyclical shelf-to-shoreface deposits during a major basin-wide drop in sea level. Coarse-grained fluvial sand and gravel were supplied to the system during valley downcutting and then reworked to varying degrees during successive channel depositional stages. The channel depositional cycles (stages) record progressive estuarine valley fill under conditions of rising sea level with each stage representing a stillstand during an overall transgressive event. Corresponding subtidal shaly facies were deposited in the adjacent estuarine bay, with each channel stage modifying the estuary margin on the eastern side while intertonguing with equivalent bay-fill sediments to the west and northwest (Fig. 14-3). The final episode of estuary fill is represented by a shallow channel-bar complex of the "H" pool. After the estuary filled, sand deposition probably continued across the entire area. The uppermost deposits, however, were removed by the rapid transgression of the Lower Colorado sea,

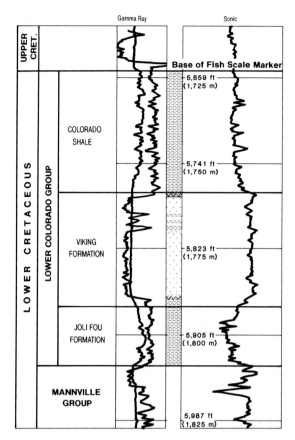

Fig. 14-4. Typical well-log response of the Viking Formation and enclosing stratigraphic units in the Crystal Field area. The well location is 6-1-46-4 W5M.

which left a thick succession of shale across the entire basin. Thus, the Crystal sandbody is bounded by unconformities and encased in predominantly muddy facies of the lower Viking Formation and overlying Colorado shale.

Lithofacies

Five lithofacies, exhibiting a wide range of reservoir parameters, have been recognized within the Crystal reservoir sandbody (Figs. 14-7, 14-8; Table 14-1). These are conglomerate, medium-grained sandstone, conglomeratic (granular) sandstone, fine-grained laminated sandstone, and fine-grained shaly sandstone. Each of these lithofacies corresponds to a specific subenvironment within the overall tidal-channel complex of each successive depositional stage. The geometric relationships of these lithofacies are extremely complex; this is reflected in the high degree of heterogeneity within the reservoir

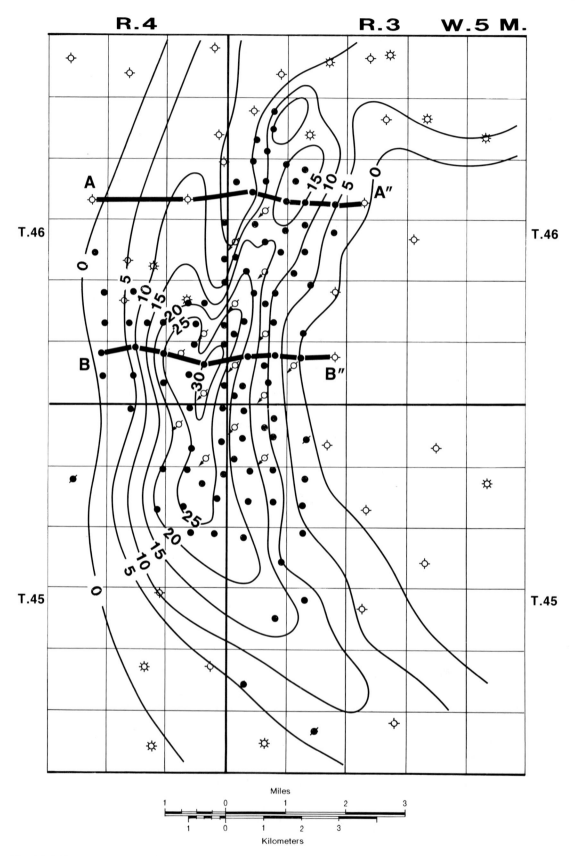

Fig. 14-5. Net porous sandstone isopach map of the reservoir body (sandstone and conglomerate) in Crystal Field. Contour interval is 16.4 feet (5 m). From Reinson, Clark, and Foscolos (1988) and reprinted by permission of American Association of Petroleum Geologists.

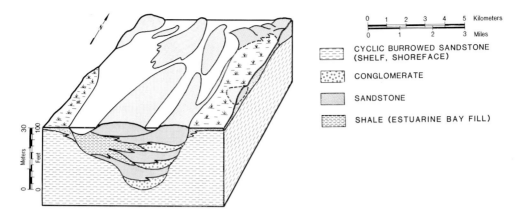

Fig. 14-6. Conceptual block diagram of the depositional setting of the channelized estuarine valley at Crystal Field.

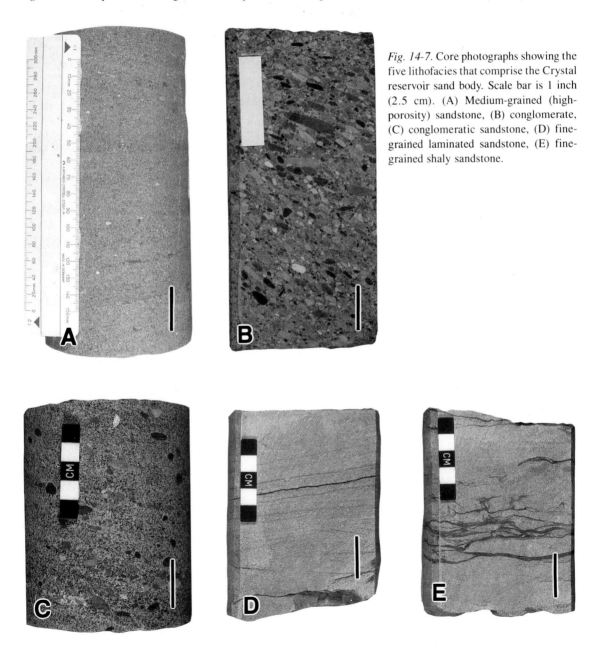

Fig. 14-7. Core photographs showing the five lithofacies that comprise the Crystal reservoir sand body. Scale bar is 1 inch (2.5 cm). (A) Medium-grained (high-porosity) sandstone, (B) conglomerate, (C) conglomeratic sandstone, (D) fine-grained laminated sandstone, (E) fine-grained shaly sandstone.

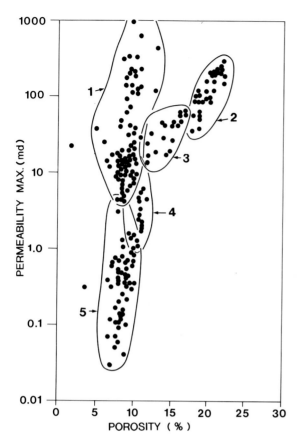

PERMEABILITY MAX. (md)

POROSITY (%)

(Fig. 14-9), which bears directly upon reservoir continuity and performance (discussed later in this chapter).

Conglomerate. This lithofacies consists of true conglomerates—they are clast-supported with variable amounts of sand size matrix. Clast size ranges from granule to medium pebble size. Some beds are moderately to well-sorted granular conglomerates, whereas others are dominated by moderately to poorly sorted layers with granules constituting from 30 to 50% of the rock. Clasts are generally discoidal in shape, subround to round, and consist primarily of black and buff chert or dark gray siliceous siltstone. Graded and imbricate bedding are also common. Contacts with underlying strata are always sharp or erosive, the latter type being most pronounced at the base of the sand body.

Fig. 14-8. Porosity-permeability (maximum) cross-plot (from conventional core analysis data) illustrating the five lithofacies that comprise the reservoir. (1) Conglomerate, (2) medium-grained sandstone, (3) conglomeratic sandstone (4) fine-grained laminated sandstone, (5) fine-grained "shaly" sandstone. Data are from the 2-1-46-4 and 4-1-46-4 W5M wells.

Table 14-1. Principal characteristics of Crystal Field reservoir lithofacies.

Lithofacies	Texture	Detrital mineralogy	Principal diagenetic features	Porosity (%) [permeability (md)]	Reservoir quality
Conglomerate	Granular to pebbly; poorly to well-sorted	Chert arenite	Moderate compaction; minor silica cement	5–15 [10 to > 1,000]	Excellent
Conglomeratic sandstone	Bimodal; poorly sorted	Chert arenite	Moderate compaction; minor silica cement	10–20 [10–100]	Excellent
Medium-grained sandstone	Predominantly medium-grained; well-sorted; no detrital matrix	Quartz sublitharenite	Moderate compaction; moderate silica cement	<17 [20–400]	Excellent
Fine-grained laminated sandstone	Moderately to well-sorted; some detrital matrix	Quartzose litharenite	Abundant silica cementation; moderate compaction	7–12 [1.0–10]	Marginal to good
Fine-grained shaly sandstone	Moderately to well-sorted; abundant detrital matrix as shaly partings	Quartzose litharenite	Abundant silica cementation; moderate compaction	5–10 [<0.1–1.0]	Poor

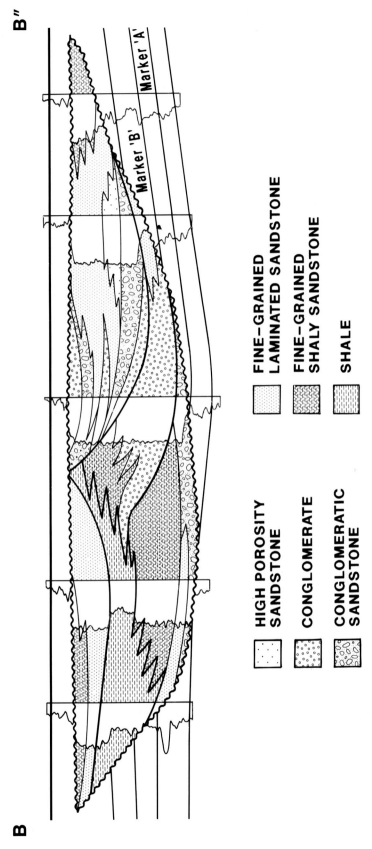

Fig. 14-9. Portion of cross section B-B″ of Figure 14-3, showing the intricate relationships of the five lithofacies across the reservoir body.

The conglomerate lithofacies forms the most productive part of the reservoir because of consistently high permeabilities (200–2,000 md) despite porosities which are generally less than 10% (Fig. 14-8). No evidence in any of the cores suggests the existence of lithologically controlled permeability barriers between the conglomerate and any of the abutting sandstone lithofacies. Thus, the conglomerate acts as a natural conduit through which hydrocarbons can migrate from juxtaposed highly porous, but less permeable sandstones.

Conglomeratic (Granular) Sandstone. Intermediate to the medium-grained sandstone and conglomerate lithofacies is a suite of rock types characterized by bimodal grain size distributions. This lithofacies includes fine- to medium-grained sandstone with pebbles and granules. The conglomeratic sandstone constituents vary from 10% to as much as 50% in some beds. Clasts may occur either as isolated floating grains or as alternating interbedded sandstone and conglomerate layers.

Medium-Grained Sandstone. This lithofacies is distinguished by porosity which is commonly greater than 18% and permeability which ranges from 50 to 500 md (Fig. 14-8). These buff- to light-brown sandstones range from fine- to coarse-grained but are usually fine- to medium-grained and well sorted. They are characterized by weathered "chalky" buff-colored chert grains that occur either as floating granules or as very coarse sand grains in medium-grained sandstone. This lithofacies displays faint to distinct cross-bedding but most commonly occurs in massive beds. It is apparent from the porosity and permeability values that this sandstone lithofacies constitutes the best reservoir rock type within the entire "A" pool.

Fine-Grained Laminated Sandstone. These sandstones generally are light olive-gray to light brown-gray and are moderately to well sorted within specific fine- to medium-grained intervals. Fine-grained sandstones are most common, but medium-grained layers are also present. Sedimentary structures include planar laminae, cross-bedding and mud-layer couplets as well as indistinctly bedded and massive intervals. Iron mudstone pebble clasts commonly occur at the base of cross-bed sets, and biogenic structures are present in the form of isolated *Skolithos* and *Ophiomorpha* burrows.

Fine-Grained "Shaly" Sandstone. This facies is characterized by a light buff-gray to light gray color, abundant shale partings, and biogenic structures (*Planolites, Paleophycus, Skolithos, Teichichnus* (?)) associated with the shaly zones. Sandstone beds display wispy laminae as well as discontinuous and disrupted laminated structures, and mud-draped ripple cross-laminae. Stylolites are also common. Rocks of this category grade into the low-porosity intervals of the fine-grained to laminated sandstone (Fig. 14-8) and generally exhibit low porosity and permeability values. This lithofacies is the poorest reservoir rock of the entire sand body.

The fine-grained laminated and fine-grained shaly lithofacies together form the bulk of the Crystal Field reservoir (Fig. 14-10). These fine-grained rocks contribute very little to the productive capacity of the reservoir relative to the other rock types.

Petrography and Diagenesis

The five reservoir facies are very similar to other Viking sandstones and conglomerates described by Reinson and Foscolos (1986). Characteristic mineralogical and diagenetic features are listed in Table 14-1.

Using the sandstone classification of Folk and others (1970), the conglomerate and conglomeratic sandstone lithofacies are classified as chert arenites, with chert grains being the dominant framework component of the detrital fraction (Fig. 14-11). The principal diagenetic features of the conglomeratic lithofacies are compaction and silica cementation as syntaxial quartz overgrowths. Primary intergranular porosity was not greatly reduced by these processes. In fact, the chert-rich intervals display more mechanical and chemical compactional effects than do the quartz-rich intervals because, although silica is present, fewer nucleation sites (i.e., monocrystalline quartz grains) are accessible for syntaxial overgrowth cementation (Reinson and Foscolos, 1986).

The fine-grained shaly and fine-grained laminated rock types are classified as quartzose litharenites, whereas the medium-grained sandstone facies is a quartz sublitharenite. All three display varying degrees of chemical compaction and syntaxial quartz overgrowths. Silica cementation is quite significant in the two finer-grained units as it occludes much of the primary intergranular pore space (Fig. 14-11).

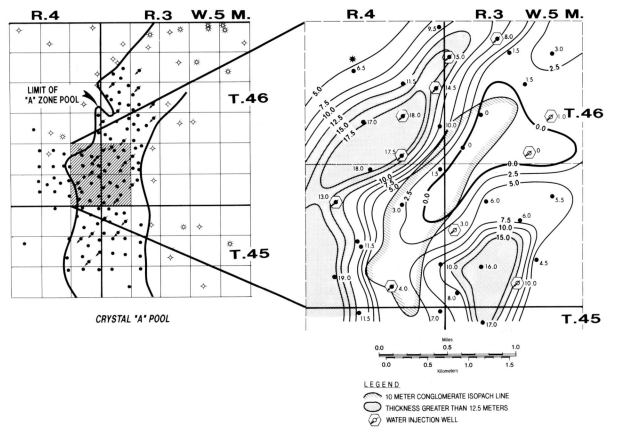

Fig. 14-10. Enlarged map of a four-section area in the center of Crystal Field showing isopachs of the combined thickness of fine-grained laminated and fine-grained shaly lithofacies (contour interval = 2.5 meters (8.2 ft)). The outline of the most productive area of the reservoir is represented by the 10-meter (32.8-ft) conglomerate isopach line. Note the arrangement of water injection wells along the boundaries of the conglomerate pod.

As illustrated in Figure 14-8, porosity, and particularly permeability values, are governed primarily by depositional texture. Conglomeratic lithofacies display the highest permeabilities and the fine-grained lithofacies the lowest. The porosity-permeability distribution and petrographic analysis strongly suggest that, relative to depositional texture, diagenetic factors have been minimal in governing reservoir quality and performance.

Production Characteristics

Oil-Water Relationships

The Crystal Viking reservoir lies in an oil/water transition zone as complex as the facies relation-ships identified in the reservoir (Figs. 14-9, 14-12). Since the central portion of the reservoir (Fig. 14-10) is occupied predominantly by the two most productive lithofacies (medium-grained sandstone and conglomerate), discussion of the fluid distribution in this area is necessary.

Valley-fill sedimentation, where channels have cut into preexisting porous sediment, may result in the reservoir having several oil-water contacts in an apparently continuous reservoir (Harms, 1966). This situation exists in the Crystal "A" pool, where water-free oil production is obtained from highly porous, medium-grained sandstones which are substantially downdip from wells which produce only water from finer-grained sandstones. Such a distribution of fluid productivity is attributable (Berg, 1975) to pore-throat sizes being generally larger in

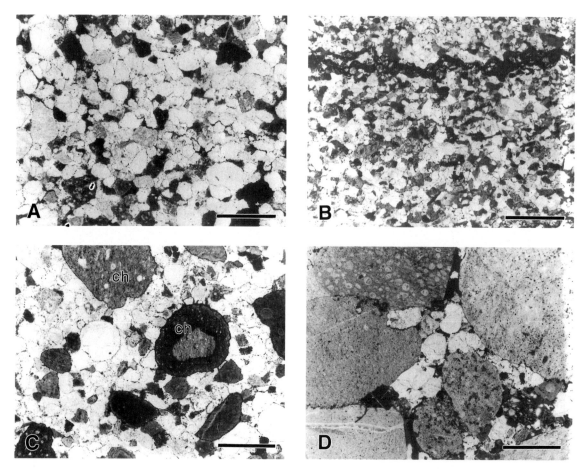

Fig. 14-11. Photomicrographs of reservoir lithofacies. Scale bar is 1.00 mm. (A) Medium-grained sandstone. Note the quartz-rich content of this facies and the relatively high porosity (black areas). (B) Fine-grained "shaly" sandstone. Note the fine-grained nature and relatively tight grain packing as well as the shaly parting. (C) Conglomeratic sandstone. The bimodality of this lithofacies is evident with large chert grains (ch) "floating" in a mosaic of fine-grained quartz. (D) Conglomerate. Large chert grains with sutured boundaries form intergranular pore space. Pore spaces are partially filled by fine- to medium-grained quartz and chert grains which form the matrix.

the medium-grained facies than in the finer-grained rock, resulting in lower capillary pressures and lower oil-water contacts in that facies type. Since the Crystal "A" pool is in a transition zone, oil-water contacts may be defined only for an individual lithofacies and are difficult to establish for the whole reservoir.

Water saturations calculated from well logs and confirmed by special core analysis range from a low of 20% in conglomerates near the gas cap to 100% in fine-grained laminated sandstones in the southern end of the reservoir. Where coarse-grained sandstones and conglomerates occupy the channel, they may still produce water-free oil although offset by water-saturated fine-grained sandstones.

For wells above the oil-water contact in similar reservoirs, residual oil saturation and grain size are inversely proportional. That is, the finer-grained, less permeable rock types tend to have higher residual oil saturations. Such a relationship exists at the Crystal "A" pool (Fig. 14-13) but with exceptions. Reservoir units located on the western margin of the pool which have relatively constant permeabilities are observed to have oil-saturated and water-saturated zones without obvious barriers between them. In some examples, thin clay drapes between

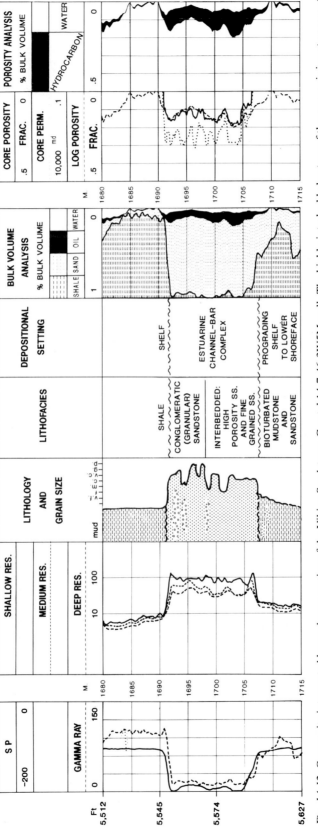

Fig. 14-12. Composite interpreted log and core section of the Viking Sandstone, Crystal 14-7-46-3W5M well. The highly interbedded nature of the reservoir is not apparent from the gamma-ray or induction logs.

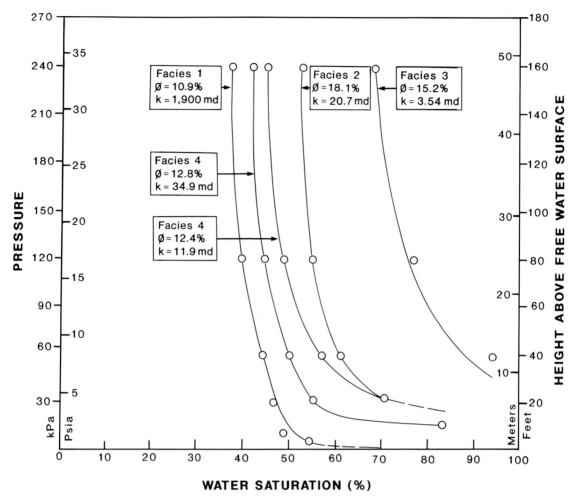

Fig. 14-13. Plot of air-brine capillary pressure versus water saturation for different facies. Facies are identified in Figure 14-8.

these zones may have restricted oil migration. In others, the capillary pressures, estimated to be only 3 to 4 psi (21–28 kPa) for fine-grained sandstones (Harms, 1966), may have been insufficient in the smaller pore throats to permit oil migration in this strongly water-wet rock (Morgan and Gordon, 1970). Zones of lower residual oil saturation are common in the fine-grained laminated and "shaly" sandstone lithofacies, where planar cross-bedding and bedding-plane discontinuities are more abundant and pore-throat size distributions more complex.

Facies-Reservoir Performance Relationships

The stacking of succeeding estuarine tidal-channel cycles has led to a complex facies architecture

which directly governs permeability heterogeneities and fluid behavior within the reservoir. Each depositional cycle compounds the internal complexity to the extent that adjacent wells within the channel boundaries may contain almost entirely different reservoir facies sequences (Fig. 14-9). This leads to highly variable production even for closely spaced wells. Productive wells and efficient injectors correspond to lithofacies characteristic of tidal-channel deposits (conglomerate, conglomeratic sandstone, and medium-grained sandstone) particularly in areas where these channels are superimposed. In contrast, marginally productive wells (and poor injectors) have encountered fine-grained laminated and "shaly" sandstones representative of the estuarine subtidal-bar complex, deposited marginal to the channel axis (Fig. 14-10). Deterioration in reservoir continuity is also evident in areas where successive channel stage events diverge (Reinson et al., 1988).

Most (85%) of the wells in the "A" pool have been fully cored in the Viking. These cores were carefully examined to detect barriers to fluid flow at channel-scour surfaces or lithofacies boundaries. Despite the apparent absence of such barriers, the pool behaves as a parallel series of adjacent reservoirs. Permeable zones exhibit relatively poor flow laterally across lithofacies. Fluids preferentially follow the south-to-north orientation of depositional strike.

Production from the highly permeable conglomerates in the central portion of the "A" pool exceeds that from any of the other lithofacies. These conglomerates are somewhat discontinuous and frequently abut granular and high-porosity sandstones. The conglomerates have permeabilities of up to 2,500 md but have relatively low porosities and hydrocarbon pore volumes. Reserves in the conglomerates are quickly depleted if not recharged from adjacent oil-bearing zones. Lower flow capacities and recharge capabilities in the sandstones on the eastern and northern sides of the main conglomerate deposit (Fig. 14-10) have resulted in pressure drawdown and the formation of a secondary gas cap at the northern end of the conglomerate trend. This cap is separate and unrelated to the pool gas cap.

Since the lateral and vertical distribution of lithofacies is very complex, selection of completion intervals should proceed with caution. Perforation of the entire porous interval should be avoided, particularly in water injection wells. It is most important that pressure communication be established between the highest quality lithofacies. Temperature log surveys run on fully perforated injection zones in some wells have shown water to be moving into a lithofacies different from that of the offsetting producing wells. This results in a compromise in pressure maintenance in some areas.

Waterflood Response

Water injection in the "A" pool was initiated in 1984 on a modified 9-spot pattern on 80-acre (32.4-ha.) spacing. Pattern distribution among the 62 producers and 29 injectors was arranged to minimize premature water breakthrough in the permeable conglomerate facies (Fig. 14-10). Wells encountering thick fine-grained sandstone lithofacies (facies 4 and 5, Fig. 14-8) were converted to injection wells to form a "line-drive" pressure fence along the boundaries of conglomerate and high-porosity sandstone lithofacies.

Although production response to injection was almost immediate and sustained (Fig. 14-14), it appears that the performance of the wells producing from the conglomerate was not influenced as greatly by the pressure maintenance as were those producing from the sandstone facies. This is suggested by the formation of a minor secondary gas cap in the northern portion of the conglomerate deposit, indicating relatively poor pressure maintenance in this area. This was further indicated by declining water cuts and productivity (Fig. 14-15). In contrast, production from the sandstone lithofacies was consistent with predicted behavior, displaying a gradual decline in oil production as a consequence of gradually increasing water cuts (Fig. 14-15). Recently, production between the producing facies has been balanced by injecting water directly into the conglomerate facies to avoid the deleterious effects of interfacies boundaries.

Exploration and Production Strategy

Although Crystal Field lies in one of the most intensively drilled regions of the province, most production is derived from younger strata, and relatively few wells have penetrated the Viking Formation. Thus, Viking paleogeography in this region is poorly defined and little understood, especially at the level required to delineate more Crystal-type reservoirs. Exploration is further hampered by the low seismic reflection coefficient between the sandstone and the encasing marine shale. Mapping of the valley-fill sediments that indicate proximity to estuaries and/or tidal channels appears to be the most viable technique for finding reservoirs analogous to that of Crystal Field.

Reservoir performance of Crystal Field has been shown to be governed primarily by external geometry and internal facies distribution. The external architecture of each succeeding channel environment is overprinted on the lithofacies distribution generated at the subenvironment level in the preceding channel stage. Resultant variations in depositional texture, manifested in the form of lithofacies variability, ultimately control lateral and vertical permeability heterogeneities within the reservoir. Diagenetic factors are minimal in governing reservoir quality and performance.

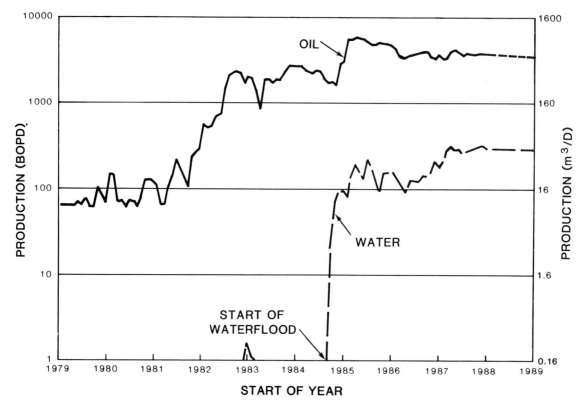

Fig. 14-14. Production plot for the main Crystal "A" pool waterflood project. Water breakthrough was almost instantaneous along the pool margins following the start of injection in November, 1984.

Care should be exercised in planning enhanced recovery schemes in similar reservoirs to ensure that the facies distribution is identified and mapped. Barriers to fluid flow at lithofacies boundaries are likely to be abundant and subtle although not observed in available cores.

Conclusions

Crystal Field represents one of the most significant reservoirs discovered in the Lower Cretaceous Viking Formation. Most Viking reservoirs are stratigraphically controlled shoreface or shelf bars.

In contrast, Crystal Field has resulted from valley incisement and subsequent estuarine-fill events. Consequently, the field comprises a complex layering of flow units and lithofacies which must be carefully identified and mapped to maximize reservoir performance.

Acknowledgments. We thank Westcoast Petroleum Ltd. for the permission to publish this work and Millie L'Amarca, Jim Kotyk and Pat Goeres for their help in assembling it. Ted Beaumont and Ray Rahmani provided valuable comments in their reviews of the manuscript.

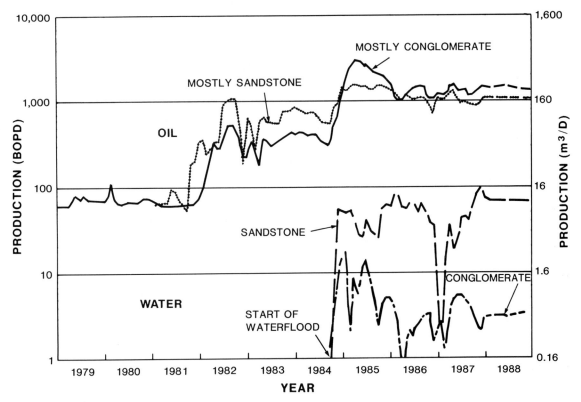

Fig. 14-15. Production plots for wells producing from conglomerate and sandstone lithofacies. For the conglomerate reservoir, water breakthrough occurred early in the waterflood program, with watercuts declining coincidentally with productivity. For sandstone reservoirs, the response to water injection was immediate and sustained.

Reservoir Summary

Field : Crystal
Location: South-central Alberta, Canada
Operators: Westcoast Petroleum, Bumper Oil & Gas
Discovery: 1978
Basin: Western Interior (Alberta) basin
Tectonic/Regional Paleosetting: Marginal to a shallow epicontinental seaway (foreland basin)
Geological Structure: Homoclinal dip
Trap Type: Stratigraphic pinch-out
Reservoir Drive Mechanism: Partial water drive (now under waterflood)
 • **Original Reservoir Pressure:** 1,490 psi (1.0×10^4 kPa) at 2,500 feet (760 m) subsea
Reservoir Rocks
 • **Age:** Early Cretaceous, late Albian
 • **Stratigraphic Unit:** Viking Formation
 • **Lithology:** Fine- and medium-grained sandstone and conglomerate; quartzose sublitharenite to chert arenite
 • **Depositional Environment:** Estuary
 • **Productive Facies:** Estuarine channel-fill complex
 • **Petrophysics:**
  • **Porosity Types:** ϕ total = 10.5%; 10.0% primary intergranular, 0.5% secondary microporosity
  • **ϕ:** Average 10.5%, range 6.0 to 25%, cutoff 6.0% (cores/logs)
  • **k:** Average 200 md, range 0.01 to 2,500 md, cutoff 0.5 md (cores)

- S_w: Average 45%, range 20 to 100%, cutoff 65% (logs)
- S_{or}: Average 10%, range 0 to 55%, cutoff > 0% (cores)

Reservoir Geometry
- **Depth:** 5,750 feet (1,750 m)
- **Areal Dimensions:** 9 by 4 miles (14.5 × 6.4 km)
- **Productive Area:** 14,500 acres (5,875 ha.)
- **Number of Reservoirs:** 2
- **Hydrocarbon Column Height:** 236 feet (71.9 m)
- **Fluid Contacts:** OWC at 2,700 feet (820 m) subsea; GOCs at 2,464 feet (751 m) and 2,588 feet (789 m) subsea
- **Number of Pay Zones:** 1
- **Gross Sandstone Thickness:** 10 to 105 feet (3.0–32.0 m)
- **Net Sandstone Thickness:** 10 to 100 feet (3.0–30.5 m)
- **Net/Gross:** Approximately 0.95

Hydrocarbon Source, Migration
- **Lithology & Stratigraphic Unit:** Basinal shale, Joli Fou Formation (?)
- **Time of Hydrocarbon Maturation:** Unknown
- **Time of Trap Formation:** Cretaceous
- **Time of Migration:** Cretaceous (?)

Hydrocarbons
- **Type:** Oil
- **GOR:** 450 SCF/STB (79.2 m³/m³)
- **API Gravity:** 39°
- **FVF:** 1.23
- **Viscosity:** 2.345 cP (2.3 × 10³ Pa·s) at 77°F (25°C)

Volumetrics
- **In Place:** 102 MMBO (1.6 × 10⁷ m³)
- **Cumulative Recovery:** 12.0 MMBO (1.9 × 10⁶ m³)
- **Ultimate Recovery:** 34.6 MMBO (5.5 × 10⁶ m³)
- **Recovery Efficiency:** 34%

Wells
- **Spacing:** Nominal interwell distance: 660 feet (200 m); 80-acre (32.4-ha.)
- **Total:** 125
- **Dry Holes:** 12

Typical Well Production
- **Average daily:** 75 BO (11.9 m³)
- **Cumulative:** 200 MBO (3.2 × 10⁴ m³)

References

Beaumont, E.A., 1984, Retrogradational shelf sedimentation; Lower Cretaceous Viking Formation, central Alberta, in Tillman, R.W. and Seimers, C.T., eds., Siliciclastic Shelf Sediments: Society of Economic Paleontologists and Mineralogists Special Publication 34, p. 163–177.

Berg, R.R., 1975, Capillary pressure in stratigraphic traps: American Association of Petroleum Geologists Bulletin, v. 59, p. 932–956.

Caldwell, W.G.E., 1984, Early Cretaceous transgressions and regressions in the southern interior plains, in Stott, D.F. and Glass, D.J., eds., The Mesozoic of Middle North America: Canadian Society of Petroleum Geologists Memoir 9, p. 173–204.

Folk, R.L., Andrews, P.B., and Lewis, D.W., 1970, Detrital sedimentary rock classification and nomenclature

for use in New Zealand: New Zealand Journal of Geology and Geophysics, v. 13, p. 937–968.

Hancock, J.M., 1975, The sequence of facies in the Upper Cretaceous of northern Europe compared with that in the Western Interior, in Caldwell, W.G.E., ed., The Cretaceous System in the Western Interior of North America: Geological Association of Canada Special Paper 13, p. 83–118.

Harms, J.C., 1966, Stratigraphic traps in a valley fill, western Nebraska: American Association of Petroleum Geologists Bulletin, v. 50, p. 2119–2149.

Kauffman, E.G. 1977, Geological and biological overview: Western Interior Cretaceous Basin, in Kauffman, E.G., ed., Cretaceous Facies, Faunas, and Paleoenvironments across the Western Interior Basin: Rocky Mountain Association of Geologists, v. 14, p. 76–99.

Leckie, D.A., and Reinson, G.E., 1989, Effects of middle to late Albian sea level fluctuations in the Cretaceous

Interior Seaway, Western Canada, *in* Caldwell, W.G.E., ed., Evolution of the Western Interior Basin: Geological Association of Canada Special Paper (in press).

Morgan, J.T., and Gordon, D.T., 1970, Influence of pore geometry on water-oil relative permeability: Journal of Petroleum Technology, v. 22, p. 1199–1209.

Price, R.A., 1981, The Cordilleran foreland thrust and fold belt in the southern Canadian Rocky Mountains, *in* McClay, K. and Price, N.J., eds., Thrust and Nappe Tectonics: Geological Society, London Special Paper 9, p. 427–448.

Reinson, G.E., and Foscolos, A.E., 1986, Trends in sandstone diagenesis with depth of burial, Viking Formation, southern Alberta: Bulletin of Canadian Petroleum Geology: v. 34, p. 126–151.

Reinson, G.E., Clark, J.E., and Foscolos, A.E., 1988, Reservoir geology of Crystal Viking Field, Lower Cretaceous estuarine tidal-channel-bay complex, south-central Alberta: American Association of Petroleum Geologists Bulletin, v. 72, p. 1270–1294.

Stott, D.F., 1982, Lower Cretaceous Fort St. John Group and Upper Cretaceous Dunvegan Formation of the foothills and plains of Alberta, British Columbia, District of Mackenzie and Yukon Territory: Geological Survey of Canada Bulletin 328, p. 124.

Vail, P.R., Mitchum, R.M., and Thompson, S., 1977, Seismic stratigraphy and global changes of sea level, *in* C. E. Payton, ed., Seismic Stratigraphy—Applications of Hydrocarbon Exploration: American Association of Petroleum Geologists Memoir 26, p. 83–97.

Weimer, R.J., 1983, Relation of unconformities, tectonics and sea level changes, Cretaceous of the Denver Basin and adjacent areas, *in* Reynolds, M.W. and Dolly, E.D., eds., Mesozoic Paleogeography of West-Central United States: Society of Economic Paleontologists and Mineralogists Rocky Mountain Section, Symposium 2, p. 359–376.

Weimer, R.J., 1984, Relation of unconformities, tectonics, and sea-level changes, Cretaceous of Western Interior, U.S.A., *in* Schlee, J.S., ed., Interregional Unconformities and Hydrocarbon Accumulation: American Association of Petroleum Geologists Memoir 36, p. 7–35.

Williams, G.D., and Stelck, C.R., 1974, Speculations on the Cretaceous paleogeography of North America, *in* Caldwell, W.G.E., ed., The Cretaceous System in the Western Interior of North America: Geological Association of Canada Special Paper 13, p. 1–20.

Key Words

Crystal Field, Alberta, Canada, Western Interior (Alberta) basin, Viking Formation, Early Cretaceous; late Albian, estuarine, channel sandstone, waterflood, diagenesis, flow-unit geometry, conglomerate, oil-water transition zone.

15

Wave-Influenced Estuarine Sand Body, Senlac Heavy Oil Pool, Saskatchewan, Canada

Brian A. Zaitlin and Bruce C. Shultz

Esso Resources Canada, Ltd., Calgary, Alberta Canada, T2P 0H6; PanCanadian Petroleum Ltd., Calgary, Alberta Canada, T2P 2S5

Introduction

A review of several petroleum-producing basins has shown that significant volumes of hydrocarbons are contained in fluvial and estuarine facies deposited within incised paleovalley systems (Table 15-1). Few of these studies investigated the relationships between estuarine facies, their reservoir characteristics, and their production behavior. The absence of such information is at least in part attributable to the incomplete understanding of estuarine stratigraphy within paleovalley systems. The aims of this chapter, therefore, are (1) to present an overview of the facies distribution and diagnostic criteria of estuarine systems, with particular emphasis upon the wave-dominated estuarine complex, and (2) to relate the spatial distribution and vertical stacking of estuarine facies to trapping mechanism, reservoir characteristics, and production behavior.

The reservoir that will serve as the case study is the Senlac Heavy Oil Pool (hereafter termed the Senlac Pool), located in west-central Saskatchewan, Canada (Fig. 15-1). The Senlac Pool was previously interpreted as an estuarine sand plug located at the mouth of an incised paleovalley system (Zaitlin and Shultz, 1984). Additional study has resulted in a more detailed understanding of the relationship between facies architecture, reservoir characteristics, and reservoir behavior. However, since no comprehensive suite of stratigraphically applicable estuarine facies models now exists (Frey and Howard, 1986; Zaitlin, 1987), we will briefly summarize the estuarine depositional model as determined from the study of modern estuarine systems.

The Estuarine Model

An estuary was defined by Pritchard (1967) as "a semi-enclosed coastal body of water having a free connection with the open sea and within which seawater is measurably diluted by fresh water of river origin." The majority of classifications of modern estuarine systems are based on salinity variations (Rochford, 1951; Pritchard, 1967; Biggs, 1978). Salinity can remain depressed for considerable distances beyond the limit of active deposition; in addition, tidal flux can exert significant influence on depositional patterns headward of the limit of measurable salinity (Rochford, 1951; Journeau and Latouche, 1981; Zaitlin, 1987). Because it is difficult to resolve the effects of salinity variations on ancient sequences, geologists tend to define estuarine sequences as "the deposits found within drowned paleovalleys" (Curray, 1969). To resolve this ambiguity, Zaitlin (1987) proposed that an **estuarine complex** would include "all tidally influenced deposits, seaward from the limit of tidal marine influence to the major facies change with normal marine deposits, occurring within a constrained (paleo-) valley setting, affected by the mixing of marine and fresh waters."

An important characteristic of estuaries is that a significant percentage of their sedimentary fill con-

Casebooks in Earth Science
Sandstone Petroleum Reservoirs
Eds.: Barwis/McPherson/Studlick
© 1990 by Springer-Verlag New York, Inc.

Table 15-1. Examples of selected hydrocarbon-bearing formations/groups with interpreted estuarine facies as a major reservoir type.

Location	Formation/group	Age	References
Western Canada	Paddy Fm.	Upper Cretaceous	Leckie, 1988
			Leckie and Reinson, in press
	Viking Fm.	Middle Cretaceous	Reinson and others, 1988
	Lloydminster Fm.	Lower Cretaceous	Zaitlin and Shultz, 1984
	McMurray Fm.	Lower Cretaceous	Flach and Mossop, 1985
	General Petroleum and Rex Fms.	Lower Cretaceous	O'Connel and Benns, 1988
	Halfway Fm.	Triassic	Unpublished Data
	Kiskatinaw Fm.	Permo-Pennsylvanian	Unpublished Data
	Granite Wash	Devonian (?)	Unpublished Data
USA	Mesaverde Gp.	Upper Cretaceous	Lorenz and Rutledge, 1985
	Muddy Fm	Lower Cretaceous	Stone, 1972
			Mitchell, 1978
	Morrowan Fm.	Lower Pennsylvanian	Emery and Sutterlin, 1986
	Tyler Fm.	Lower Pennsylvanian	Maughan, 1984
			Barwis, this volume
	Simpson Gp.	Middle Ordovician	Esslinger, 1983
India	Gujarat Fm.	Eocene	Raju and Rao, 1975
Australia	Latrobe Gp.	Tertiary	Sloan, 1987
Libya	Marada Fm.	Miocene	El-Hawat, 1980

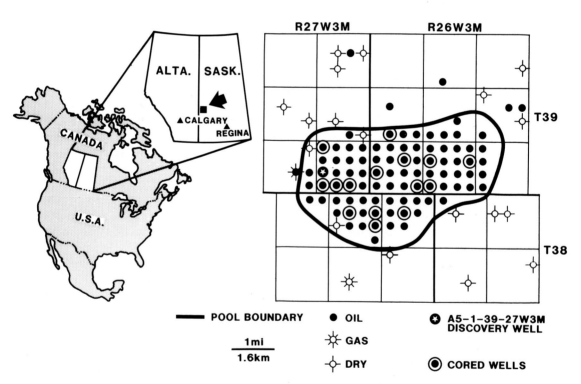

Fig. 15-1. Distribution of cored wells and location of the discovery well in the Saskoil-Gulf Senlac Pool. Arrow in inset map shows the location of the Senlac area (solid square) in west-central Saskatchewan. ALTA. = Alberta; SASK. = Saskatchewan.

RIVER

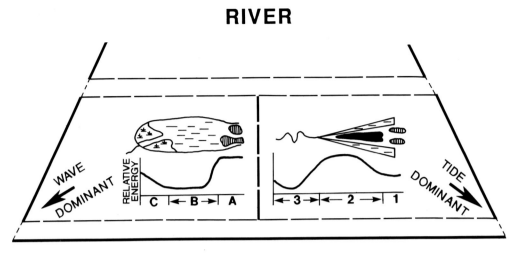

WAVE   TIDE

Fig. 15-2. Conceptual ternary diagram exhibiting the spatial distribution in plan view of depositional zones for wave- and tide-dominated estuarine complexes. Wave-dominated estuarine complexes are divisible (after Roy, 1984) into three zones: Zone A—seaward sand plug (vertical lines); Zone B— central basin (horizontal dashes); and Zone C (marsh symbols). The relative energy dissipated across each zone is shown below the inset diagrams. Tidally dominated estuarine complexes (after Dalrymple and Zaitlin, 1989) are fringed by intertidal to supratidal fine-grained deposits (horizontal dashes), and the axial facies is divisible into Zone 1—elongate tidal sandbars (horizontal lines); Zone 2—high-energy sand flats (solid); and Zone 3—tidal-fluvial deposits.

sists of sediment supplied from seaward of the incised valley mouth. This material is transported both along shore and onshore by nearshore coastal marine processes, until those processes (i.e., waves or tides), operating at the estuary's mouth, transport the material headwards into the lower reaches of the drowned valley. Sediment may also be supplied from the hinterland to the upper reaches of the estuary by rivers. Estuarine deposition is thus characterized by sediment filling the valley from two opposing directions: headward of the estuarine zone, valley fill is composed of a variety of fluvial deposits; at the seaward end of the estuarine zone, the incised valley fill consists of more open-marine deposits. The estuarine portion of the incised valley is unique, however, as it is the only area in the valley characterized by sedimentary fill sourced from opposing directions (land and sea). The estuarine portion of the incised valley is influenced by the interaction between marine and fluvial depositional processes redistributing this sediment. The resulting depositional response to this interaction is a complex longitudinal facies distribution along the axis of the valley, with possible compositional changes due to mixed

provenance. These variations may have a major control on reservoir quality.

Investigations of modern estuarine facies (Clifton, 1982; Roy, 1984; Chappell and Woodroffe, 1985; Frey and Howard, 1986; Boyd et al., 1987; Zaitlin, 1987; Dalrymple and Zaitlin, 1989; Dalrymple et al., in press) have documented the nature of incised-valley fill. They have also considered a variety of sediment types (sand- vs. mud-dominated), sediment sources (headward vs. seaward), depositional process (wave-, tide-, or fluvial-dominated) and sea-level conditions (relative sea-level rise, fall, or stillstand; single or multiple cycles) in relation to the final incised-valley fill.

Estuarine complexes are characterized by a predictable facies architecture (Roy, 1984; Zaitlin, 1987; Dalrymple et al., in press). Variations in facies architecture, in the preserved vertical sequence, and in diagnostic sedimentological features result from the relative dominance of marine processes (waves versus tides) responsible for redistribution of sediment at the estuary mouth. A wave-dominated estuarine model (Fig. 15-2) developed from estuaries along the New South Wales coast of

Australia (Roy, 1984) and the eastern shore of Nova Scotia (Boyd et al., 1984) exhibits the following threefold zonation: (a) an outer estuarine sand plug, (b) a central basin, and (c) an inner tidal-fluvial zone. In contrast, tidally dominated estuaries similar to those studied in the Bay of Fundy (Dalrymple, 1977; Knight, 1977, 1980; Zaitlin, 1987; Dalrymple et al., in press), Western Australia (Wright et al., 1973, 1975), Queensland (Cook and Mayo, 1977), and the Northern Territory of Australia (Chappell and Woodroffe, 1985) are divisible into (a) a fringing intertidal to supratidal flat, and (b) an axial zone divisible longitudinally landwards into elongate bars, high-energy sand flats, and inner tidal-fluvial facies.

Variations in facies characteristics between wave- and tide-dominated estuarine complexes, particularly those near the estuary mouth (i.e., wave-dominated estuarine sand plug versus tidally dominated elongate bar-sand flats), can therefore be used to determine the relative position of any point along the estuarine portion of an incised valley system. These variations can also be used to infer regional paleogeographic changes along a shoreline or to predict the reservoir characteristics and continuity within an individual estuarine sand body. A brief summary of the spatial distribution, reservoir potential, and sealing characteristics of sandy, wave-dominated estuarine deposits is presented below. Detailed descriptions of tide-dominated estuarine complexes can be found in Cook and Mayo (1977), Chappell and Woodroffe (1985), Zaitlin (1987), Dalrymple and Zaitlin, (1989); and Dalrymple and others (in press).

Sandy Wave-Dominated Estuarine Systems

The distribution of facies within sandy, wave-dominated estuarine systems is controlled both by the effective dissipation of wave energy at the mouth of the complex and by the shape of the incised valley (Roy, 1984). Dissipation of wave energy and sand-plug development at the mouth of the system have been shown to differ slightly between shallow and deeply incised valley systems (Roy, 1984). The distinction between deep and shallow wave-dominated estuaries most probably depends on the actual depth of the valley relative to the amount of the sediment available to fill the system. Any estuary mouth would be "deep" at the time of initial flooding but will become "shallow" as additional sand is transported into the mouth of the system. As a result, subaqueous tidal-delta deposits initially develop at

the mouth of deeply incised valley systems as the amount of available sediment is insufficient to fill up the valley mouth during the initial flooding. In such a system, the tidal delta is effective in only the partial dissipation of the marine processes (i.e., wave energy and tidal circulation), thus allowing for the free interchange of marine and fresh waters between the fluvial and the open-marine zones. In contrast, as the incised valley becomes shallow due to increased sediment input relative to its effective depth, a subaerial barrier or spit complex may develop across the mouth of the embayment. This restricts the free interchange of marine and fluvial waters, except through tidal inlets, and effectively dampens the wave energy between the inner estuary and the outer, open-marine area.

Wave-dominated estuarine complexes developed within shallow valley systems are divisible, after Roy (1984), into three major facies belts (Zones A, B, and C – Fig. 15-2), as follows:

Zone A consists of a wave-dominated composite sand plug (Fig. 15-3) across which wave energy is dissipated (Fig. 15-2). The facies geometry within a composite sand plug is dependent on the interplay of sediment supply and sea-level history. The sand plug is divisible into an inner transgressive core and an outer progradational sand body. The transgressive core consists of a barrier shoreface and/or spit complex cut by multiple, laterally migrating tidal inlets; the resulting vertical sequence is dominated by upward-fining tidal inlet deposits capped by upward-coarsening shoreface sequences. The transgressive core is developed during the initial stages of flooding as the result of transport of sediment headwards into the mouth of the valley. As the rate of sea-level rise decreases or as sediment supply is increased to the mouth of the system, sediment cannot be driven headward through the tidal inlet at sufficiently fast rates, thus resulting in the progradation of a sand body accreted onto the seaward side of the transgressive core. This progradational shoreline is dominated by an upward-coarsening shoreface sequence which may be dissected by a few relatively stationary tidal inlets with tidal deltas prograding into the central basin.

Zone B is a central basin zone containing the lowest-energy intertidal to shallow subtidal deposits in the complex (Figs. 15-2, 15-4). Zone B comprises fringing intertidal to supratidal flats and an axial subtidal area containing "lagoonal," interdistributary-bay and/or fluvial deposits. These Zone B facies are deposited behind the barrier con-

Fig. 15-3. An example of a modern Zone A sand plug displaying barrier, inlet, and tidal-delta deposits from the New South Wales coast of Australia. (Photo courtesy of A. Short, University of Sydney)

Fig. 15-4. An example of modern central basin and tidal-fluvial delta deposits from the New South Wales coast of Australia. (Photo courtesy of A. Short, University of Sydney)

structed in Zone A but are sourced predominantly from updip fluviodeltaic environments.

Zone C is the innermost zone consisting of tidal-fluvial channels and interdistributary-bay and overbank (delta-plain) deposits (Figs. 15-2, 15-4). Zone C sediments are transported from above the limit of tidal influence and are reworked to varying degrees by the tidal processes operating in the headward reaches of the estuarine system. Reworking by waves is minimal.

Reservoir Potential of Wave-Influenced Estuarine Complexes

The composite sand plug (Zone A) commonly has the best reservoir potential of the wave-dominated estuarine complex. The sand plug is usually oriented parallel to the coastline, its lateral extent being controlled by the geometry of the valley. The sand plug is restricted to the valley mouth along deeply embayed coastlines but may be more laterally continuous in low-relief coastal-plain settings where the sand bodies form more typical barrier-island/lagoonal systems (Rehkemper, 1969; Hayes, 1979; Davis and Hayes, 1984).

Facies of the tidal-fluvial zone (Zone C) may have moderate reservoir potential associated with tidally modified distributary channel-fill deposits oriented parallel to depositional dip. However, these "bay-head deltas" are typically immature and texturally and mineralogically very muddy because of their protected location and lack of reworking, so even the

channel-fill and distributary mouth-bar sequences can be poor reservoirs (McEwen, 1969).

Finer-grained deposits are common within Zones B and C and are represented by interdistributary-bay, lagoonal, and coastal-plain facies. These facies have a variable spatial distribution and display poor reservoir quality because of their generally finer grain size. Fine-grained deposits within Zones B and C form effective lateral seals for reservoirs in Zones A and C. In addition, the quiet water, anoxic "lagoonal" facies of Zone B may, under certain conditions, be an excellent source of hydrocarbons (e.g., the Cretaceous Paddy Formation of western Canada; personal communication, J. Allan and S. Creaney, 1988). Top and bottom seals for wave-dominated estuarine reservoirs are controlled by the vertical stacking of backstepping parasequences due to either the stepped rise in relative sea level or fluctuations in sediment input. In a transgressive or backstepping sequence, the superposition of open-marine mudstones over the sand plug forms the seal. With progradation, the development of coastal-plain deposits over the reservoir sands forms the topseal.

Senlac Heavy Oil Pool

Reservoir Characteristics and Available Data

The Senlac Pool was discovered in 1980 with the drilling of Gulf-Saskoil Senlac A5-1-39-27W3M (Fig. 15-5). Production is from the Lloydminster

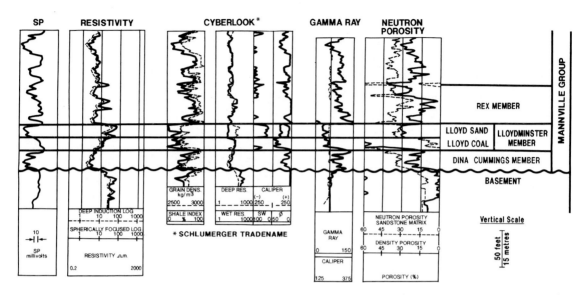

Fig. 15-5. Characteristic log responses from Gulf et al. Senlac A5-1-39-27W3M discovery well. LLOYD = Lloyd-minster.

Member of the Lower Cretaceous (Aptian to lower Albian) Mannville Group. Operatorship of the pool was transferred from Gulf Canada Resources to Saskatchewan Oil and Gas Corporation (Saskoil) in 1983. The Senlac Pool is estimated to contain 84.3 million barrels (1.3×10^7 m³) of 13° to 15° API gravity oil in place (OIP). A 7.5% recovery factor is predicted to yield an ultimate 6.4 million barrels

(1.0×10^6 m³) of oil on primary recovery. The pool is presently on primary production but is being evaluated as a possible candidate for an in situ combustion (fire flood) enhanced oil recovery (EOR) project.

The data base from the Senlac Pool includes 98 wells (84 producing) on 40-acre (16.2-ha.) spacing distributed over an area of 4,818 acres (1,951 ha.)

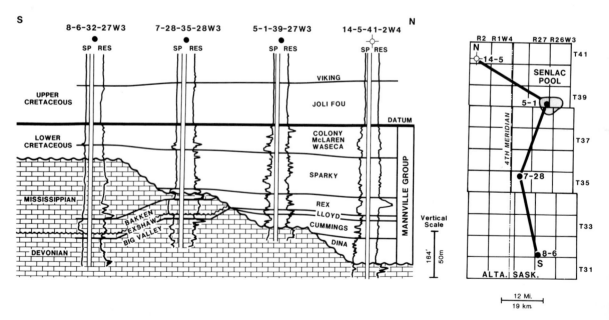

Fig. 15-6. Stratigraphic nomenclature and north-south cross section through the Senlac Pool in west-central Saskatchewan. LLOYD = Lloydminster Formation; ALTA. = Alberta; SASK. = Saskatchewan.

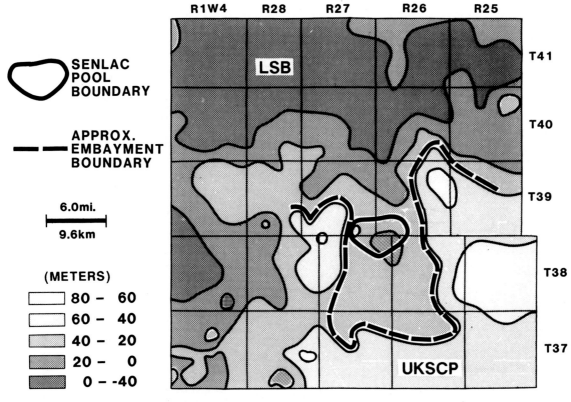

Fig. 15-7. Third-order structural residual map of the sub-Cretaceous erosional surface for the Senlac Pool and surrounding area contoured in meters. The embayment measures approximately 15 miles (24 km) N-S by 4 miles (6.4 km) E-W. LSB = Lloydminster sub-basin; UKSCP = Unity-Kindersley-Swift Current platform.

(Fig. 15-1). Fourteen of the wells that were cored penetrated the Lloydminster Member. All wells from the pool were examined and all cores were logged in detail. Surrounding wells in a three-township area (108 mi²; 275 km²) were also incorporated into the study in order to establish the Senlac Pool's general paleogeographic context.

Regional Stratigraphic and Structural Framework

The regional stratigraphic framework for west-central Saskatchewan is presented in Figure 15-6. Devonian and Mississippian carbonates and clastics are truncated by a regionally extensive sub-Cretaceous erosional surface. Overlying this unconformity is a complex northward-thickening wedge of flat-lying (Lower Cretaceous) Mannville Group terrestrial to shallow-marine clastic sediments. The Mannville Group is overlain by regionally extensive marine shale of the Upper Cretaceous Joli Fou Formation.

The Western Canada Sedimentary basin has undergone repetitive eustatic fluctuations in sea level. Tectonic activity to the west resulted in varying sediment supply to the basin; together these factors result in a composite stacking of transgressive-regressive cycles throughout the basin. Lower Cretaceous strata were deposited in a number of discrete mappable sub-basins controlled by the irregular sub-Cretaceous paleotopography (Ranger, 1983). The Senlac area is situated at the southern limit of one such area (Fig. 15-7), termed the Lloydminster sub-basin (Zaitlin and Shultz, 1984). The Senlac area straddles the boundary between this sub-basin and the northern edge of the Unity-Kindersley-Swift Current platform (Christopher, 1980; hereafter termed the Swift Current platform).

The Mannville Group, divided into nine members by Nauss (1945) and Vigrass (1977), is 590 to 755 feet (180–230 m) thick within the Lloydminster sub-basin and abruptly thins to 230 to 328 feet (70–100 m) over the Swift Current platform (Christopher, 1980). The Mannville Group isopach map exhibits thick-

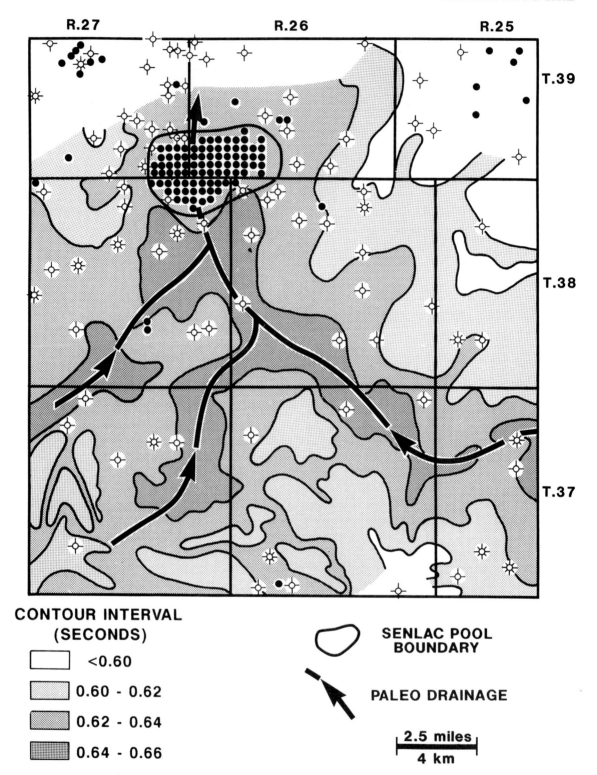

**CONTOUR INTERVAL
(SECONDS)**

<table>
<tr><td>□</td><td><0.60</td></tr>
<tr><td>▦</td><td>0.60 - 0.62</td></tr>
<tr><td>▦</td><td>0.62 - 0.64</td></tr>
<tr><td>▦</td><td>0.64 - 0.66</td></tr>
</table>

**SENLAC POOL
BOUNDARY**

PALEO DRAINAGE

2.5 miles
4 km

Fig. 15-8. Structural seismic interpretation on the sub-Cretaceous unconformity. Contour interval is 0.02 seconds using two-way travel time. Arrows indicate interpreted paleovalley trends and inferred paleodrainage directions.

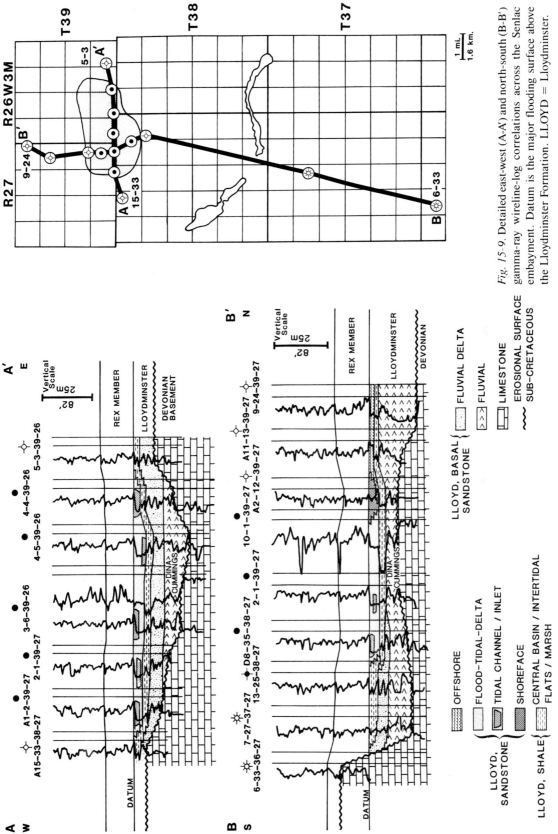

Fig. 15-9. Detailed east-west (A–A') and north-south (B–B') gamma-ray wireline-log correlations across the Senlac embayment. Datum is the major flooding surface above the Lloydminster Formation. LLOYD = Lloydminster.

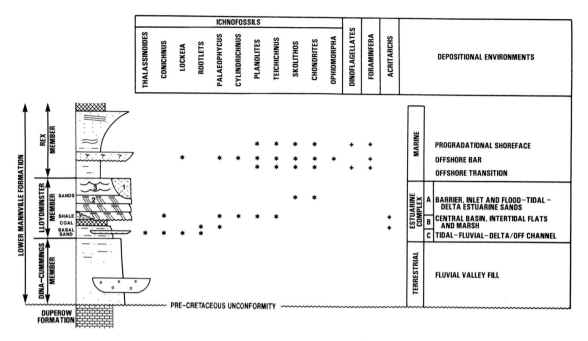

Fig. 15-10. Idealized vertical sequence of the Lower Mannville Group in the Senlac area. Ichnofossils (*) identified by G. Pemberton and micropaleontology (+) identified by Robertson Research and C. Vervoloet. A, B, and C refer to zones defined in Figure 15-2 for a wave-dominated estuarine complex. Numbers 1, 2, and 3 refer to sand types described in the chapter.

ening of more than 650 feet (200 m) in areas interpreted as incised paleovalleys associated with the sub-Cretaceous erosional surface. The third-order residual and structure maps (Figs. 15-7, 15-8) are observed to mimic closely the inferred paleotopography at the time of deposition. The residual map clearly shows the occurrence of a paleotopographic embayment in the edge of the Swift Current platform in the Senlac area. This incised paleovalley system is 15 miles (24 km) north-south by 4 miles (6.4 km) east-west and opens northwards into the Lloydminster sub-basin. This feature has been termed the Senlac embayment (Zaitlin and Shultz, 1984). Detailed correlations (Fig. 15-9) and third-order residual and structural mapping on the sub-Cretaceous unconformity (Figs. 15-7, 15-8) exhibit a localized depression with 65 to 130 feet (20–40 m) of topographic relief in the Senlac area. The depression is interpreted to represent a northward-draining paleovalley system on the sub-Cretaceous erosional surface at the edge of the Swift Current platform (Fig. 15-8). From detailed correlations, this depression can be shown to contain the localized development of Dina-Cummings and Lloydminster Members

(Fig. 15-9). These units are absent from the surrounding paleotopographic highs.

Reservoir Characterization

Stratigraphy and Facies Description

Core-controlled correlations permit the development of a composite stratigraphic section of the Senlac embayment fill (Fig. 15-10). The documentation and interpretation of lithofacies in the Senlac area have been presented in Zaitlin and Shultz (1984) and are summarized briefly below. The composite vertical section constructed suggests a slow continuous sea-level rise with a resultant limited headward translation of facies belts. The composite section does not take into consideration local erosional events within the system.

Dina-Cummings Member. The undifferentiated Dina-Cummings Member is in erosional contact with underlying carbonates of the Devonian Duperow Formation (Figs. 15-6, 15-9, 15-10). The Dina-Cummings Member, where present, forms an

Fig. 15-11. Core photographs of Saskoil-Gulf Senlac B4-5-39-26W3M. Top of core is to upper right, base of core is to bottom left. Lloydminster basal sandstone (bs) is overlain by the carbonaceous shales of the Lloydminster coal (coal), which is overlain by Lloydminster shale (S) and the Lloydminster sand (sand). Rex Member siltstone and shale (rex) overlie Lloydminster Member deposits. Scale bar is 6 inches (15 cm).

Fig. 15-12. Lloydminster basal sandstone (Gulf-Saskoil Senlac 1-6-39-26W3M) displaying carbonaceous root casts and a possible *Thalassinoides* burrow. Scale bar is 1 inch (2.5 cm).

overall upward-fining sequence, up to 100 feet (30 m) thick, composed of interbedded calcareous-cemented, cross-bedded sandstone and siltstone. The finer-grained facies contain pyritized rootlets but have yielded no recognizable micropaleontologic or palynologic material. This unit is interpreted to have been deposited as part of a fluvial-dominated valley-fill sequence prior to the flooding of the valley during Lloydminster time.

Lloydminster Member. The Lloydminster Member is divisible into four informal units (Figs. 15-10, 15-11): the basal sandstone, the Lloydminster coal, the Lloydminster shale, and the Lloydminster sands (Figs. 15-9, 15-10).

Basal sandstone. The basal sandstone ranges in thickness up to 16 feet (4.9 m) and appears to be in gradational contact with the underlying Dina-Cummings Member. This unit is characterized by calcite-cemented sandstone organized into repetitive upward-fining cycles ranging from 0.3 to 3.3 feet (0.1–1 m) thick. Each upward-fining cycle is composed of unstratified to cross-bedded sandstone, with ripple cross-laminae and flaser beds which grade into carbonaceous, rooted siltstone (Fig. 15-12). The siltstone contains an ichnofossil assemblage composed of *Palaeophycus herberti*, *Conichnus*, *Lockeia*, and *Thalassinoides* and has yielded a few undiagnostic acritarchs. Thin-section

and core analyses reveal low porosity ($< 5\%$) and no effective permeability (< 0.01 md). The basal sandstone is situated toward the south or headward reaches of the estuarine system (Fig. 15-9) and is interpreted to have been deposited as part of the sandy, tidal-fluvial and off-channel environments in the inner reaches of the estuarine complex (Fig. 15-13). The unit has little reservoir potential and, where higher permeability or porosity is present, is water-bearing due to the gentle southwest structural dip in the Senlac area.

Lloydminster Coal. The Lloydminster coal ranges in thickness up to 13 feet (4.0 m) and overlies the basal sandstone (Figs. 15-9, 15-10). The coal is interbedded with dark, carbonaceous shale and siltstone. The siltstone contains pinstripe bedding, lenticular and flaser bedding, and isolated rootlets. A sparse ichnofossil assemblage dominated by *P. herberti* is present. The coal and interbedded siltstone and shale are interpreted to represent the fringing upper intertidal to supratidal facies. The coal forms both within Zone C of the inner tidal-fluvial environments and within the Zone B central basin.

Lloydminster Shale. The Lloydminster shale occurs locally throughout the Senlac embayment and, where present, directly overlies the Lloydminster coal (Figs. 15-9, 15-10). The shale is less than 2.5 feet (0.8 m) thick and contains well-developed pinstripe, lenticular, and tidal bedding features (Fig. 15-14 A–D). The shale has yielded a few acritarchs and contains a suite of recognizable restricted shallow subtidal to intertidal ichnofossils (Fig. 15-10). The Lloydminster shale is interpreted to represent a variety of central basin-fringing tidal-flat, and shallow subtidal subenvironments (Zone B, e.g., lagoon, interdistributary bay, and prodelta).

Lloydminster Sands. The Lloydminster sands are developed near the mouth of the Senlac embayment. Based upon the examination of primary sedimentary structures, spatial distribution, production data, and reservoir characteristics, the hydrocarbon-saturated Lloydminster sands are divisible into three distinct sand types (Fig. 15-14 E–G). Subtle differences in the gamma-ray log signature, directly correlatable to the cored intervals, are discernible for each type. Porosity, permeability, and water/oil saturations also differ among the three types (Fig. 15-15).

Type 1 sands range in thickness up to 13 feet (4.0 m) and average 5 feet (1.5 m) thick. They are organized into 2-inch to 5-foot (5–150 cm) thick, repetitive upward-fining cycles. Each upward-fining cycle begins with a sharp, erosional basal contact (with or

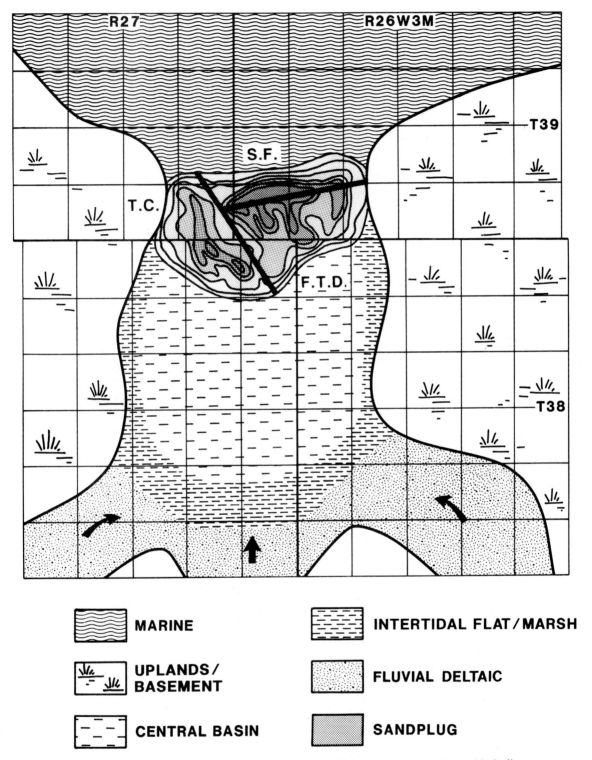

MARINE

UPLANDS / BASEMENT

CENTRAL BASIN

INTERTIDAL FLAT / MARSH

FLUVIAL DELTAIC

SANDPLUG

Fig. 15-13. Distribution of depositional environments during Lloydminster time in the Senlac embayment. S.F. = shoreface deposits; T.C. = tidal-channel deposits; F.T.D. = flood-tidal-delta deposits. Heavy black lines separate depositional environments. Arrows indicate paleodrainage directions in the fluviodeltaic environments.

B.A. Zaitlin and B.C. Shultz

Fig. 15-14. Examples of the intertidal to shallow subtidal Lloydminster shale (scale bar is 1 inch (2.5 cm)): (A) Pinstripe-laminated to lenticular-bedded shale and very fine-grained sandstone exhibiting *Teichichnus* burrows; (B) bioturbated shale exhibiting well-developed *Conichnus* and *Planolites* burrows and some intact zones of pinstripe lamination; (C) *Teichichnus* and *Planolites* burrows within massive shale; (D) well-developed fine pinstripe lamination within carbonaceous shale with *Teichichnus* and *Palaeophycus* burrows. Examples of Lloydminster sands: (E) upward-fining, oil-saturated, cross-bedded to ripple cross-laminated fine-grained sands to rippled siltstone (Type 1 sands); (F) upward-coarsening bioturbated silty shale to cross-bedded oil-stained sands (Type 2 sands); (G) Wavy-bedded fine-grained sands (Type 3 sands).

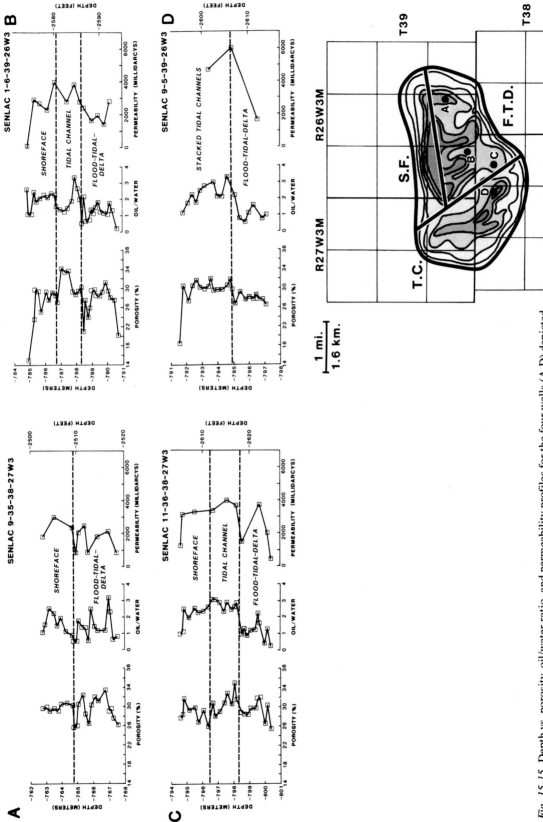

Fig. 15-15. Depth *vs.* porosity, oil/water ratio, and permeability profiles for the four wells (A-D) depicted in the inset isopach map of the Lloydminster sands (see Fig. 15-17). T.C. = Tidal-channel deposits; S.F. = Shoreface deposits; F.T.D. = Flood-tidal-delta deposits.

 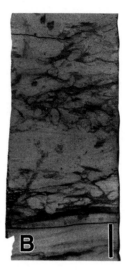 

Fig. 15-16. (A) Bioturbated, starved ripple to lenticular bedded shale of the Rex Member, indicative of offshore transition to shelf deposits overlying Lloydminster sands. (B) Bioturbated shale and silty sandstone of the Rex Member, indicative of lower shoreface to offshore transition deposits. (C) Wave-ripple bedding within shale and sandstone of the Rex Member, indicative of lower shoreface deposits. (Scale bars are 1 inch (2.5 cm)).

without a gravel or shell lag) and overlain by cross-bedded, fine- to very fine-grained sandstone and ripple cross-laminated siltstone. Porosity values range from 25 to 30%, and permeability values are on the order of 3,000 to 4,000 md.

Type 2 sands range in thickness up to 21 feet (6.4 m) and are organized into upward-coarsening cycles that are 0.8 to 3.3 feet (0.25–1.0 m) thick. The lower part of each cycle generally consists of a bioturbated shale overlying a sharp contact and then coarsens upward into current-ripple cross-laminated siltstone and cross-bedded sandstone. *Chondrites* and *Skolithos lineraris* are present in the lower portions of each cycle. Porosity and permeability are variable, with maximum values on the order of 27 to 31% and 1,000 to 4,000 md developed near the top of each cycle. The basal bioturbated shale forms laterally restricted permeability barriers within the sand plug.

Type 3 sands range in thickness from 2 to 25 feet (0.6–7.6 m). They are fine- to medium-grained, undulatory, wavy-bedded, and cross-bedded and are organized into well-developed upward-coarsening cycles. Porosities range from 29 to 31%, and permeabilities are 2,000 to 3,500 md.

Rex Member. The Rex Member ranges in thickness from 16 to 33 feet (4.9–10.0 m) and directly overlies the Lloydminster Member (Figs. 15-9, 15-10). The Rex Member forms a well-developed upward-

coarsening sequence, characterized by a basal carbonaceous pyritic shale (Fig. 15-16A). It is overlain by bioturbated shale, siltstone, and sandstone (Fig. 15-16B), with interbedded, 3 to 10 feet (0.9–3.0 m) thick, low-angle, wavy to hummocky cross-bedded sandstone (Fig. 15-16C). This upward-coarsening sequence is capped by a coal or carbonaceous shale. The lower shale contains a diverse open-marine ichnofossil assemblage and yields dinoflagellate cysts (Fig. 15-10).

Lloydminster Sands Distribution and Facies Interpretation. The Lloydminster gross-sandstone isopach map is presented in Figure 15-17 and describes a sand plug consisting of a 23-foot (7.0-m) thick lenticular sand body near the mouth of the embayment (Fig. 15-13). The sand body displays a north-south asymmetric profile, with the northern margin sloping steeply northward and the southern face tapering gently toward the south (Fig. 15-17). The Lloydminster sand undergoes a southward facies change, interfingering with the subtidal facies of the Lloydminster shale (Figs. 15-9, 15-10). Northward from the mouth of the paleovalley, the sands are not present and are replaced by open-marine shale and siltstone. To the east and west, the sand plug grades into fringing intertidal deposits of the Lloydminster shale, which in turn onlap the edge of the paleovalley.

Three distinct facies identified as the deposits of tidal channel (facies T.C.), flood-tidal delta (facies F.T.D), and shoreface (facies S.F.) are recognized within the sand plug (Figs. 15-13, 15-17).

The tidal-channel sequence is developed within a northwest-southeast trend along the western edge of the pool. Interpretation of core data and the gamma-log signature indicates that the sequence is dominated by Type 1 sands but may be capped to the north by a thin veneer of Type 2 and/or Type 3. Based on the dominant upward-fining profile, lack of bioturbation, and spatial distribution, this facies is interpreted to comprise composite, stacked tidal-channel deposits that cut across the sand plug.

The flood-tidal delta sequence is dominated by Type 2 sands and exhibits a number of north-south Type 1 sand trends. This facies ranges in thickness from 0 to 21 feet (0–6.4 m) and thins towards the south. The environmental interpretation is based upon its variable upward-coarsening profile, the thickness of the Type 2 sands, and the spatial distribution with respect to other facies.

Shoreface deposits are dominated by Type 3 sands, occur in an east-west trend, and display thicknesses extending to more than 23 feet (7.0 m). Their environmental interpretation is based upon their upward-coarsening profile, absence of crosscutting tidal-inlet fills, and sedimentary structures. The sequence formed on the seaward side of the Zone A sand plug.

Composition of the Lloydminster sands. The Lloydminster sands are dominated by quartz and chert (~85%), with accessory feldspar, carbonaceous material, rock fragments, and glauconite. No major compositional differences occur between the Type 1, 2, and 3 sands. A clay matrix (< 8% by weight) composed of kaolinite, minor illite, smectite, and traces of smectite-illite mixed-layer clays is present preferentially within Type 2 sands associated with the flood-tidal delta. The higher water saturations associated with the flood-tidal-delta facies are thought to be due to this higher interstitial clay content. This S_w criterion can be employed in uncored wells as an indicator of Type 2 sands.

Depositional Model And Trapping Mechanism

The Lower Mannville Group in the Senlac area is interpreted to represent the complex fill of an incised paleovalley system which evolved strati-

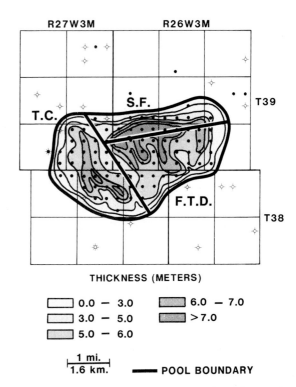

Fig. 15-17. Lloydminster gross-sandstone isopach map of the Senlac Pool. S.F. = shoreface; T.C. = tidal-channel; and F.T.D. = flood-tidal-delta.

graphically from fluvial through estuarine to open-marine deposits (Zaitlin and Shultz, 1984). Deposition is believed to have occurred during slow sea-level rise to stillstand conditions (Lloydminster Member), followed by a rapid sea-level rise and landward facies translation (Rex Member). The superposition of marine mudstone (Rex Member) over the Lloydminster Member sandstone forms the top seal of the Senlac reservoir (Figs. 15-9, 15-10).

The estuarine fill of the Lloydminster Formation in the Senlac area is characterized by facies deposited in three major environments (Fig. 15-13): (1) an outer composite sand plug composed of three distinct sand facies forming the reservoir; (2) a fine-grained central basin; and (3) an inner fluvially dominated, tidally modified delta, which grades headward into fluvial deposits. Figure 15-13 summarizes the interpreted paleogeography of the Senlac embayment at the time of Lloydminster deposition. Figure 15-3 is a modern example of a wave-dominated barrier sand plug from the east coast of Australia that is considered to be an analog to the Senlac area. Figure 15-10 summarizes the

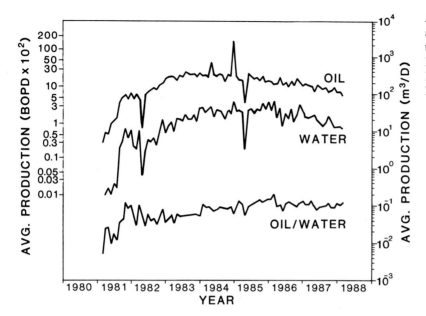

Fig. 15-18. Group summary of averaged fluid production from the Lloydminster sands, Senlac Pool. Data from the Alberta Energy Resources Conservation Board.

the idealized vertical sequence of the fill of the Senlac incised paleovalley.

The Lloydminster sand is a 3.5 by 2 mile (5.6 × 3.2 km) lenticular sandstone body situated near the mouth of the paleovalley (Fig. 15-13). The pool is a pure stratigraphic trap and is divisible into three major components represented by tidal-channel, flood-tidal-delta, and barrier/shoreface environments. The reservoir is encased within a variety of impermeable shales: (1) an updip seal formed by contemporaneous, open-marine shaly facies to the north (i.e., seaward) of the embayment mouth; (2) a top seal provided by the offshore deposits of the overlying Rex parasequence; (3) lateral seals to the east and west consisting of fringing intertidal-flat deposits; and (4) to the south by central estuary fine-grained deposits. Geochemical analysis of the biodegraded oil indicates that it was sourced from the Mississippian Exshaw Formation located to the southwest of the Senlac area (personal communication, S. Creaney and J. Allan, 1988) and subsequently migrated updip to the Senlac stratigraphic trap.

Production Characteristics

The Senlac Pool was discovered in 1980 with development drilling continuing until the end of 1986. Daily production peaked during late 1983 (Fig. 15-18), remained steady at a rate greater than 1,250 BOPD (200 m³/D) between 1983 and 1985, and has

slowly declined to 950 BOPD (150 m³/D). Cumulative production exceeds 3.7 MMBO (5.9 × 10⁵ m³) (Fig. 15-19, inset).

Figure 15-20 presents the daily oil production averaged over the second 3,000 hours of production for each well in the pool. The wells were averaged by this method to eliminate erratic production during the initial period of well completion and production and to present a common datum from which to compare production across the reservoir. Daily oil production per well varies across the pool from less than 25 barrels (4.0 m³) to more than 50 barrels (8.0 m³).

Production rates are controlled by lithofacies. A strong correlation exists between the northwest-southeast-trending zone producing more than 50 BOPD (8.0 m³/D) along the western edge of the sand plug, and the area interpreted as having been cut by tidal channels. In addition, the east-west-trending shoreface (Fig. 15-13) is characterized by production on the order of 37 BOPD (5.9 m³/D) (Fig. 15-19). Production is erratic in the southeast portion of the pool in the area dominated by flood-tidal-delta deposits. The majority of the wells in this area are producing less than 25 BOPD (4.0 m³/D) except in areas that are interpreted to be tidal channels, where rates are 50 BOPD (8.0 m³/D). In areas where the relationship between depositional facies and production appears inconsistent, variations can be attributed to the differing percentages of each sand type in a particular well.

Fig. 15-19. Plot of cumulative oil production vs. time for three typical examples of wells completed in tidal-channel, shoreface, and flood-tidal-delta lithofacies. Inset—total cumulative oil production vs. time for the Senlac Pool. For well locations, see Figure 15-21. Data from the Alberta Energy Resources Conservation Board.

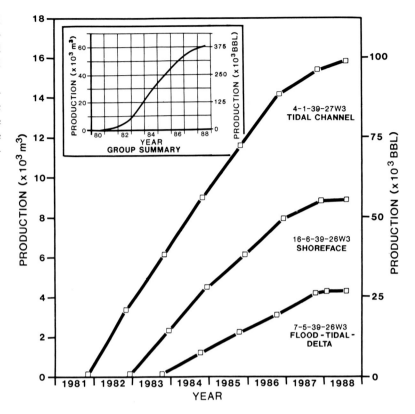

Figures 15-19 and 15-21 show the cumulative daily oil production for representative wells completed in sandstones of tidal channel, shoreface, and flood-tidal delta environments. Line breaks in Figure 15-21 signify temporary production shut-ins, most commonly due to well maintenance. The wells were chosen to exhibit the typical porosity, permeability, and water/oil ratios associated with each of the major depositional facies (Fig. 15-15). The tidal-channel sandstones average 50 to 63 BOPD (8.0–10.0 m³/D); shoreface sandstones average between 37 and 50 BOPD (5.9–8.0 m³/D); and the tidal-delta sandstones average between 12 and 25 BOPD (1.9–4.0 m³/D). Cumulative production data for each depositional facies are shown in Figure 15-19 and vary significantly between facies. Thus, it is clear that the subfacies of the estuarine depositional environment

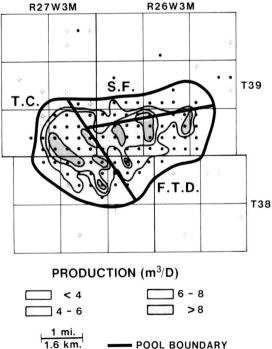

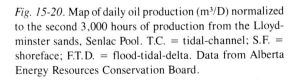

Fig. 15-20. Map of daily oil production (m³/D) normalized to the second 3,000 hours of production from the Lloydminster sands, Senlac Pool. T.C. = tidal-channel; S.F. = shoreface; F.T.D. = flood-tidal-delta. Data from Alberta Energy Resources Conservation Board.

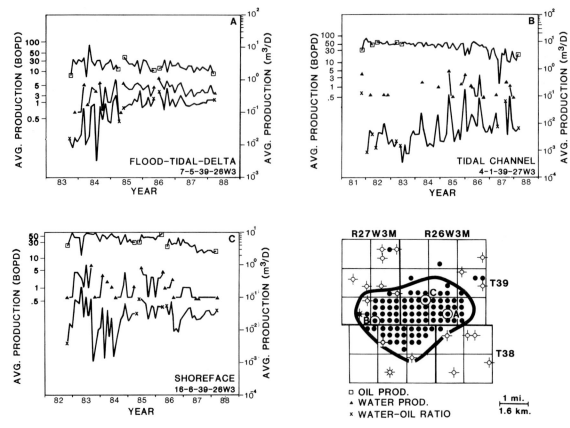

Fig. 15-21. Daily fluid production vs. time for typical examples of wells completed in flood-tidal-delta (A), tidal-channel (B), and shoreface (C) facies. Map shows well locations. Data from the Alberta Energy Resources Conservation Board.

provides the primary control to production characteristics of the Senlac Pool.

In addition, channel and shoreface sandstones produce increased volumes of water with time, whereas the oil/water ratio appears constant in wells which produce from flood-tidal-delta deposits (Fig. 15-21). This is confirmed by the variation between initial versus current water cuts (Figs. 15-22, 15-23). The initial water cut was based on the first three months average production for each well (Fig. 15-23). Water saturation is generally homogeneous across the pool; however, slightly higher water cuts exist in the southeastern portion of the pool. The current water cut is based on the last three months (January- March, 1988) average water production (Fig. 15-23). Higher volumes of water are being produced in the pool area interpreted to be dominated by the tidal-channel facies. A less obvious east-west trend associated with the shoreface deposits can also be detected by comparing Figures 15-22 and 15-23. Areas dominated by flood-tidal-delta deposits have

not yielded large volumes of water. The tidal-channel and shoreface water cuts indicate a depleting reservoir in the vicinity of the wellbores. In these wells, water has begun to flow preferentially with respect to the oil. Although pressure data from these wells are not available to the authors, we believe that water is not being coned from the possible waterleg in the southeastern portion of the pool; rather it is the residual water in the vicinity of the wellbore which is being produced as the reservoir begins to deplete.

Exploration And Production Strategy

Exploration Strategy

Successful exploration for estuarine reservoirs within paleovalley systems depends on the ability to identify the distribution of incised paleovalleys in an area and determine the spatial distribution of reservoir facies within these systems.

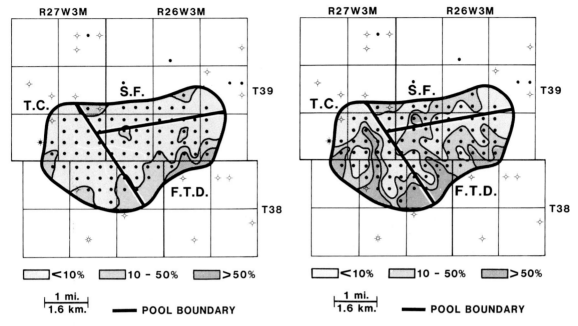

Fig. 15-22. Map of average initial water cut based on the first three months of production for each well in the Senlac Pool. T.C. = tidal-channel; S.F. = shoreface; F.T.D. = flood-tidal-delta. Data from the Alberta Energy Resources Conservation Board.

Fig. 15-23. Map of average current water cut based on the final three months of production, Senlac Pool. T.C. = tidal-channel; S.F. = shoreface; F.T.D. = flood-tidal-delta. Data from the Alberta Energy Resources Conservation Board.

To locate incised valley systems, it is critical to document the nature of the irregular paleotopographic surface and the direction of paleodrainage. A variety of techniques can be employed to map such a paleotopographic surface: seismic structural mapping; simple structural mapping of an erosional surface from well logs; and residual mapping of structural surfaces in areas that have undergone postdepositional structural tilting. Other methods that can be utilized to identify incised paleovalley systems include detailed isopach mapping to locate anomalously thick stratigraphic sections confined to paleotopographic lows; downward shift in facies; and detailed correlations to identify truncation of regionally extensive markers, thus defining the incision (Weimer, 1984).

Reservoir facies can be predicted with detailed core analysis calibrated to wireline logs. Because comprehensive estuarine facies models are not available to evaluate these cores (Frey and Howard, 1986; Zaitlin, 1987), the opportunity exists for effective depositional models to be constructed from subsurface data using log correlations and high-resolution seismic stratigraphy.

A regional understanding of parasequence stacking patterns can be applied within the paleovalley to predict the spatial distribution of reservoir facies throughout the system. Once the transgressive and regressive limits of the facies within a well-documented paleovalley have been determined, these can be extrapolated as fairways parallel to depositional strike; the intersection of these fairways with other paleovalley systems are areas of possible exploration potential.

Production Strategy

The ultimate aim of all development strategies is to maximize hydrocarbon recovery at the lowest unit cost in the shortest possible time. This can be achieved by reducing the number of wells required for primary production and by more efficient use of injectors during enhanced recovery operations. Both require a complete understanding of the reservoir facies architecture and its effect on fluid movement.

As demonstrated in this chapter, estuarine deposits contain a variety of possible reservoir facies. Wave-dominated estuarine systems can contain significant reservoir potential in the Zone C tidal-fluvial channels located at the head of the estuarine system (e.g., Crystal Viking Pool, Reinson et al., 1988; Reinson and Clark, this volume) or in the composite estuarine sand plug of Zone A. In each case, a different production strategy must be developed to account for the differences in reservoir geometry.

Zone C reservoirs are long, sinuous stacked channel-fill sandstones which may or may not be in reservoir communication. Zone A reservoirs, although in communication, are formed by a complex association of facies, each with different permeability and porosity characteristics. Evaluation of the production characteristics across the Senlac Pool indicated that lithofacies distribution is the primary control on reservoir behavior. Better production characteristics associated with the tidal-channel facies is a function of internal homogeneity, better lateral continuity, higher effective permeabilities, and coarser grain size. The upward-coarsening profiles of the shoreface deposits are characterized by textural inhomogeneities and variations in vertical porosity and permeability which result in decreased reservoir continuity. The flood-tidal-delta facies are also composed of repetitive coarsening-upward sequences and have the poorest production behavior. This poor production is interpreted to be a function of more internal permeability barriers, greater internal inhomogeneity, a higher proportion of fines, and lower reservoir continuity.

This complex arrangement of reservoir subfacies has an important control on fluid production response. As an example, Senlac Pool was characterized by an initial water cut that was uniform across the pool; however, after seven years of primary production, water cuts in the reservoir have become significantly more variable and will impact the combustion attributes for a potential fire flood of the pool.

Conclusions

The Saskoil Gulf-Senlac Heavy Oil Pool produces from the Lloydminster Member of the Lower Cretaceous Mannville Group. The distribution of facies within the Lloydminster Member in the Senlac area is interpreted to represent a wave-dominated sandy estuarine complex comprising a tidal-fluvial delta at the headward end, a middle central basin, and an estuarine sand plug at the seaward end. The Senlac reservoir is developed within the estuarine sand plug and is a composite feature dominated by tidal-channel, flood-tidal-delta, and shoreface sandstones. The reservoir body is a stratigraphic trap encased in fine-grained open-marine, central basin, and intertidal-flat deposits. The best production in the reservoir is associated with tidal-channel deposits; poorer production performance is associated with the shoreface deposits, and flood-tidal-delta deposits display the poorest production capabilities. Knowledge of the reservoir complexities of estuarine deposits and an understanding of the spectrum of estuarine depositional systems can significantly improve both exploration and production results.

Acknowledgments. We thank Esso Resources Canada, Ltd., PanCanadian Petroleum Ltd., Gulf Canada Resources, and Saskatchewan Oil and Gas Corporation for permission to publish this study. We also thank R.W. Dalrymple, D.P. James, and D.A. Leckie for their helpful comments after critically reading the manuscript. Thanks goes to C. Heath of PanCanadian Petroleum Ltd. and Esso Resources Canada, Ltd. for drafting and photographic assistance. Special thanks to M. Ranger, S.G. Pemberton, and F. Haidl for their earlier contributions to the Senlac study and to S.G. Pemberton for identification of trace fossils. We appreciate the ideas of P. Roy, B.G. Thom, and A. Short (University of Sydney, Australia) who provided insights into wave-dominated estuarine systems.

Reservoir Summary

Field: Senlac Heavy Oil Pool
Operators: Gulf Canada, Saskoil (Saskatchewan Oil & Gas)
Discovery: 1980
Location: West-central Saskatchewan, Canada
Basin: Western Canada sedimentary basin
Tectonic/Regional Paleosetting: Marginal to shallow epicontinental seaway (foreland basin)
Geologic Structure: Homoclinal dip

Trap Type: Stratigraphic pinch-out
Reservoir Drive Mechanism: Solution-gas drive
  • **Original Reservoir Pressure:** NA
Reservoir Rocks
  • **Age:** Early Cretaceous (Aptian to lower Albian)
  • **Stratigraphic Unit:** Lloydminster Member, Mannville Group
  • **Lithology:** Fine- to medium-grained sublitharenites
  • **Depositional Environment:** Wave-dominated estuarine complex
  • **Productive Facies:** Tidal-channel, shoreface, and tidal-delta of estuarine sand body
  • **Petrophysics**
    • **Porosity Type:** Intergranular
    • **ϕ:** Tidal Channel—Average 27.5%, range 25 to 30% (cores/logs); Shoreface—Average 29.0%, range 27 to 31% (cores/logs); Flood-Tidal Delta—Average 30.0%, range 29 to 31% (cores/logs)
    • **k:** Tidal Channel—Average 3,000 md, range 3,000 to 4,000 md (cores); Shoreface—Average 2,500 md, range 1,000 to 4,000 md (cores); Flood-Tidal Delta—Average 2,700 md, range 2,000 to 4,000 md (cores)
    • **S_w:** 16 to 37% (logs)
Reservoir Dimensions
  • **Depth:** 2,540 to 2,640 feet (775–805 m)
  • **Areal Dimensions:** 3.5 by 2.0 miles (5.6 × 3.2 km)
  • **Productive Area:** 4,818 acres (1,951 ha.)
  • **Number of Reservoirs:** 1
  • **Hydrocarbon Column Height:** NA
  • **Fluid Contacts:** NA
  • **Number of Pay Zones:** 1
  • **Gross Sandstone Thickness:** 10 to 23 feet (3.0–7.0 m)
  • **Net Sandstone Thickness:** 10 to 23 feet (3.0–7.0 m)
  • **Net/Gross:** Approximately 1.0
Source Rocks
  • **Stratigraphic Unit:** Basinal shales of the Exshaw Formation (Mississippian)
  • **Time of Maturation:** Late Cretaceous (?)
  • **Time of Trap Formation:** Early Cretaceous
  • **Time of Migration:** Late Cretaceous (?)
Hydrocarbons
  • **Type:** Oil
  • **API Gravity:** 13 to 15°
  • **FVF:** 1.11
  • **Viscosity:** 1,245 to 3,959 cP (1.2–4.0 × 10^6 Pa·s) at 77°F (25°C)
Volumetrics
  • **In-Place:** 84.3 MMBO (1.3 × 10^7 m^6)
  • **Cumulative Recovery:** 3.7 MMBO (5.9 × 10^5 m^3)
  • **Ultimate Primary Recovery:** 6.4 MMBO (1.0 × 10^6 m^3)
  • **Primary Recovery Efficiency:** 7.5%
Wells
  • **Spacing:** 1,320 feet (400 m), 40 acres (16.2 ha.)
  • **Total:** 94 (84 presently producing)
  • **Dry Holes:** 7
Typical Well Production:
  • **Cumulative Production:** NA
  • **Average Daily:** Tidal Channel—50 to 63 BOPD (8.0–10.0 m^3/D); Shoreface—37 to 50 BOPD (5.9–8.0 m^3/D); Flood-Tidal Delta—12 to 25 BOPD (1.9–4.0 m^3/D)

References

Biggs, R.B., 1978, Coastal bays, in Davis, R.A., Jr., ed., Coastal Sedimentary Environments: New York, Springer-Verlag, p. 69–100.

Boyd, R., Bowen, A.J., and Hall, R.K., 1987, An evolutionary model for transgressive sedimentation on the eastern shore of Nova Scotia, in FitzGerald, D.M. and Rosen, P.S., eds., Glaciated Coasts: New York, Academic Press, p. 87–114.

Chappell, J., and Woodroffe, C.D., 1985, Morphodynamics of northern rivers and floodplains, in Bardsley, K.N., Davie, J.D.S., and Woodroffe, C.D., eds., Coasts and Tidal Wetlands of the Australian Monsoon Region: NARV Monograph, A.N.U. Press, Canberra, p. 85–96.

Christopher, J.E., 1980, The Lower Cretaceous Mannville Group of Saskatchewan, a tectonic overview, in Lloydminster and Beyond—Geology of Mannville Hydrocarbon Reservoirs: Saskatchewan Geological Society Special Paper 5, p. 4–32.

Clifton, H.E., 1982, Estuarine deposits, in Scholle, P.A. and Spearing, D., eds., Sandstone Depositional Environments: American Association of Petroleum Geologists Memoir 31, p. 121–141.

Cook, P.J., and Mayo, W., 1977, Sedimentology and Holocene History of a Tropical Estuary (Broad Sound, Queensland): Australian Bureau of Mineral Resources Bulletin 170, 206 p.

Curray, J.R., 1969, History of continental shelves, in Stanley, D.J., ed., The New Concepts of Continental Sedimentation: Application to the Geologic Record: American Geological Institute, p. JC-6-1-JC-6-7.

Dalrymple, R.W., 1977, Sediment dynamics of macrotidal sandbars, Bay of Fundy [Ph.D. thesis]: Hamilton, Ontario, McMaster University, 635 p.

Dalrymple, R.W., and Zaitlin, B.A., 1989, Tidal sedimentation in the macrotidal Cobequid Bay-Salmon River Estuary, Bay of Fundy: Canadian Society of Petroleum Geologists' Field Guide to the Second International Research Symposium on Clastic Tidal Deposits, August 22–25, Calgary, Alberta, 84 p.

Dalrymple, R.W., Knight, R.J., Zaitlin, B.A., and Middleton, G.V. (in press), Dynamics and facies model of a macrotidal sand bar complex, Cobequid Bay-Salmon River Estuary (Bay of Fundy): Sedimentology.

Davis, R.A., Jr., and Hayes, M.O., 1984, What is a wave-dominated coast?: Marine Geology, v. 60, p. 313–329.

El-Hawat, A.S., 1978, Carbonate-terrigenous cyclic sedimentation and paleogeography of the Mirada Formation (middle Miocene), Sirte Basin: Symposium on the Geology of Libya, p. 427–448.

Emery, M., and Sutterlin, P.G., 1986, Characterization of a Morrowan sandstone reservoir, Lexington Field, Clark County, Kansas: Shale Shaker, v. 37, p. 18–33.

Esslinger, B.A., 1983, Facies and depositional environment of the Simpson Group (Middle Ordovician) of central Oklahoma [M.S. thesis]: Columbia, Missouri, University of Missouri, 187 p.

Flach, P.D., and Mossop, G., 1985, Depositional environments of the Lower Cretaceous McMurray Formation, Athabasca Oil Sands, Alberta: American Association of Petroleum Geologists Bulletin, v. 69, p. 1195–1207.

Frey, R.W., and Howard, J.D., 1986, Mesotidal estuarine sequences: A perspective from the Georgia Bight: Journal of Geology, v. 70, p. 737–753.

Hayes, M.O., 1979, Barrier island morphology as a function of tidal and wave regime, in Leatherman, S.P., ed., Barrier Islands—From the Gulf of St. Lawrence to the Gulf of Mexico: New York, Academic Press, p. 1–27.

Journeau, J.M., and Latouche, C., 1981, The Gironde Estuary: Stuttgart, E. Schweizerhart'sche Verlogsbuchhandlung, (Nagele U. Obermiller), 115 p.

Knight, R.J., 1977, Macrotidal Sediments, Bedforms and Hydraulics, Cobequid Bay (Bay of Fundy, Nova Scotia) [Ph.D. thesis]: Hamilton, Ontario, McMaster University, 693 p.

Knight, R.J., 1980, Linear sand bar development and tidal current flow in Cobequid Bay, Bay of Fundy, Nova Scotia: Geological Survey of Canada Paper 80-10, p. 123–152.

Leckie, D.A., 1988, Sedimentology and sequences of the Paddy and Cadotte Members along the Peace River: Canadian Society of Petroleum Geologists Field Guide to Sequences, Stratigraphy and Sedimentology: Surface and Subsurface Technical Meeting, Calgary, Alberta, 78 p.

Leckie, D.A., and Reinson, G.E., in press, Effects of Middle to Late Albian sea level fluctuations in the Cretaceous Interior Seaway, Western Canada, in Caldwell, W.G.E. and Kauffman, E., eds., Evolution of the Western Canada Interior Basin: Geological Association of Canada Special Paper.

Lorenz, J.C., and Rutledge, A.K., 1985, Facies relationships, reservoir potential of Ohio Creek interval across the Piceance Creek Basin: Oil and Gas Journal, v. 83, p. 91–96.

Maughan, E.K., 1984, Paleogeographic setting of Pennsylvanian Tyler Formation and relation to underlying Mississippian rocks in Montana and North Dakota: American Association of Petroleum Geologists Bulletin, v. 68, p. 178–195.

McEwen, M.C., 1969, Sedimentary facies of the modern Trinity Delta, in Cankford, R.R. and Rogers, J.J.W., eds., Holocene Geology of the Galveston Bay Area: Houston Geological Society, p. 53–77.

Mitchell, G., 1978, Greive Oil Field; a Lower Cretaceous estuarine deposit, in Boyd, R.G., Boberg, W.W., and Olsen, G.W., eds., Resources of the Wind River Basin: Wyoming Geological Association Guidebook 30, p. 147–165.

Nauss, A.W., 1945, Cretaceous Stratigraphy of the Vermillion Area, Alberta, Canada: American Association of Petroleum Geologists Bulletin, v. 29, p. 1605–1629.

O'Connell, S.C., and Benns, G.W., 1988, The geology of the Mannville Subgroup in the west-central Lloydminster Heavy Oil Trend, Alberta: Alberta Research Council Open File Report E.O. 1988-1, 34 p.

Pritchard, D.W., 1967, What is an estuary? Physical viewpoint, in Lauff, G.D., ed., Estuaries: Washington, D.C., American Association for the Advancement of Science, p. 3–5.

Raju, A.T.R., and Rao, M.R., 1975, Depositional environment of oil bearing Eocene sands in Anklesvar Field, Cambay Basin, India: Geological Society of India Journal, v. 16, p. 165–175.

Ranger, M.J., 1983, The paleotopography of the sub-Cretaceous erosional surface in the western Canada basin [abst.], in The Mesozoic of Middle North America Conference, Program and Abstracts: Canadian Society of Petroleum Geologists, p. 118.

Rehkemper, L.J., 1969, Sedimentology of Holocene estuarine deposits, Galveston Bay, in Lankford, R.R., and Rogers, J.J.W., eds., Holocene Geology of the Galveston Bay Area: Houston Geological Society, p. 12–52.

Reinson, G.E., Clark, J.E., and Foscolos, A.E., 1988, Reservoir geology of Crystal Viking field, Lower Cretaceous estuarine tidal channel-bay complex, south-central Alberta: American Association of Petroleum Geologists Bulletin, v. 72, p. 1270–1294.

Rochford, D.J., 1951, Studies in Australian estuarine hydrology, in Introductory and Comparative Features: Australian Journal of Marine and Freshwater Research, v. 2, p. 1–116.

Roy, P.S., 1984, New South Wales estuaries: Their origin and evolution, in Thom, B.G., ed., Coastal Geomorphology in Australia: Sydney, Academic Press, p. 99–121.

Sloan, M.W., 1987, Flounder—A complex intra-Latrobe oil and gas field: Australian Petroleum Exploration Association Journal, v. 27, p. 308–317.

Stone, W.D., 1972, Stratigraphy and exploration of the Lower Cretaceous Muddy Formation, North Powder River Basin, Wyoming and Montana: The Mountain Geologist, v. 9, p. 355–378.

Vigrass, L.W., 1977, Trapping of oil at the intra-Mannville (Lower Cretaceous) disconformity in the Lloydminster area, Alberta and Saskatchewan: American Association of Petroleum Geologists Bulletin, v. 61, p. 1010–1028.

Weimer, R.J., 1984, Relation of unconformities, tectonics and sea level changes, Cretaceous of Western Interior, U.S.A., in Sclee, J.S., ed., Interregional Unconformities and Hydrocarbon Accumulation: American Association of Petroleum Geologists Memoir 36, p. 7–35.

Wright, L.D., and Coleman, J.M. 1973, Variations in morphology of major river deltas as functions of ocean wave and river discharge regime: American Association of Petroleum Geologists Bulletin, v. 57, p. 370–398.

Wright, L.D., Coleman, J.M., and Thom, B.G., 1973, Processes of channel development in a high-tide range environment: Cambridge Gulf-Ord River Delta, Western Australia: Journal of Geology, v. 81, p. 15–41.

Wright, L.D., Coleman, J.M., and Thom, B.G., 1975, Sediment transport and deposition in a macrotidal river channel: Ord River, Western Australia, in Cronin, R.E., ed., Estuarine Research, Vol. II: New York, Academic Press, p. 309–322.

Zaitlin, B.A., 1987, Sedimentology of the Cobequid Bay—Salmon River Estuary, Bay of Fundy, Canada [Ph.D. thesis]: Kingston, Ontario, Queen's University, 393 p.

Zaitlin, B.A., and Shultz, B.C., 1984, An estuarine-embayment fill model from the Lower Cretaceous Mannville Group, west-central Saskatchewan, in Stott, D.F. and Glass, D.J., eds., Mesozoic of Middle North America: Canadian Society of Petroleum Geologists Memoir 9, p. 455–469.

Key Words

Senlac Heavy Oil Pool, Saskatchewan, Canada, western Canada Sedimentary basin, Lloydminster Member, Mannville Group, Early Cretaceous; Aptian-lower Albian, estuarine facies, tidal-channel deposits, flood-tidal-delta deposits, incised paleovalley systems, in situ combustion, fire flood, EOR, heavy oil, progradational-shoreface deposits.

16
Flood-Tidal Delta Reservoirs, Medora-Dickinson Trend, North Dakota

John H. Barwis

Shell Oil Company, Houston, Texas 77210

Introduction

Barrier island depositional systems are particularly challenging as exploration and development objectives because they span a very wide range of sizes and shapes depending on their tectonic setting and temporal position within a sea-level cycle. Sands in a barriered shoreline commonly grade along depositional strike into wave-dominated, sandy deltas and strand plains. Where rates of subsidence and sediment supply are high, such as on passive margins or on the thrustbelt sides of foreland basins for example, relative sea-level rise is accompanied by rapid shoreline aggradation, which results in thick, multistoried sand bodies with uniform strike orientations. Regional sandstone/shale ratios are often so high in these stacked intervals that trap integrity requires structural closure. Where subsidence rates are relatively low and the result is a single-story regressive sequence, preservation of even one entire barrier island can provide enough reservoir for a giant hydrocarbon accumulation.

Conversely, because transgressive barriers on relatively stable margins produce thin, discontinuous sand bodies, the stratigraphic components of traps in this setting are far more important. Because reservoirs may be relatively small and may display any number of orientations, exploration and development in these situations carry correspondingly higher reservoir risks. The keys to recognizing and developing reservoirs in stable-margin transgressive barriers are to combine detailed facies analyses of individual reservoirs and seals with broad-scale sequence analysis of the entire basin, and to recognize the role played by sea-level changes in determining facies relationships and sand-body architecture at both scales. This chapter describes reservoirs in these types of less predictable barrier subenvironments, using stratigraphic traps within the Tyler Formation of North America's Williston basin as examples.

Regional Setting: Medora-Dickinson Trend

The Medora-Dickinson trend is a string of relatively small fields (Table 16-1) which produce oil from thin, lenticular sandstones that were deposited along a microtidal coastline. Reservoirs in these fields originated as tidal-channel point bars and flood-tidal deltas, and are preserved as thin, linear to podlike sandstone bodies configured in a wide variety of structurally influenced stratigraphic traps. These reservoirs represent the remnants of a transgressively modified, sand-poor barrier island system deposited on a relatively stable cratonic margin.

Location and Reserves

The Medora-Dickinson area is east of the Cedar Creek anticline in the southern Williston basin, and consists of eight fields which define a 40-mile

Table 16-1. Summary of oil fields producing from the Tyler Formation in the Medora-Dickinson trend, Billings and Stark Counties, North Dakota. All production is from barrier-related lithofacies; Madison carbonate production is excluded.

Field	Current operator	Year	Cum (MBO)	EUR (MBO)	Area (ac)	Area (ha.)		Richness (BO/ac)	Success ratio
Dickinson	Conoco	1958	23,141	24,000	10,700	4,330		2,240	0.65
Fryburg	Amerada-Hess	1954	15,241	16,000	9,380	3,800		1,710	0.66
Medora	Amerada-Hess	1964	6,724	6,900	3,240	1,310		2,130	0.44
Zenith	Columbus et al.	1968	4,996	5,300	2,320	940		2,280	0.35
Bell	Samedan et al.	1982	1,727	3,000	1,900	770		1,580	0.50
South Heart	S. Heart Co.	1973	1,191	1,300	930	380		1,400	0.78
Green River	Cenex	1974	1,003	1,020	620	250		1,760	0.25
Heart River	Samson	1978	66	70	580	230		120	0.38
Total			54,089	57,590	29,670	12,010	Avg.	1,940	0.54

Year: Year of discovery.
Cum: Cumulative production through September, 1988 (North Dakota Division of Oil and Gas).
EUR: Estimated ultimate recovery based on decline curves.
Area: Productive acreage, not unitized area, based on net-pay maps.
Richness: Estimated ultimate recovery per acre of productive area.
Success ratio: Number of productive wells divided by the sum of all wells drilled, excluding injection wells and wells drilled within the field for deeper objectives.

(64-km) long oil-producing trend that spans Billings and Stark Counties in southwestern North Dakota (Fig. 16-1). Oil production is from two horizons: Tyler Formation sandstones (Pennsylvanian) and Madison Group carbonates (Mississippian). Cumulative Tyler production (March, 1989) from this trend since the first field's discovery in 1954 is about 54 million barrels (8.6×10^6 m³) of low-sulfur, high-gravity oil. Estimated ultimate recovery of Tyler oil from the Medora-Dickinson trend, one of several producing areas in the region, is about 58 million barrels (9.2×10^6 m³). Total proven Tyler reserves recoverable in all of North Dakota and Montana are estimated at 150 million barrels (2.4×10^7 m³), with many of the more than 40 fields now undergoing secondary recovery operations. Depositional origin determined reservoir continuity and trap style in virtually all of these fields. Associated exploration and production problems are exemplified by Medora Field, at the western end of the Medora-Dickinson trend.

Medora Field History

Amerada Petroleum Company discovered Medora Field in 1964 when they drilled a simple dome in an attempt to extend known production at Fryburg and Dickinson fields. The discovery well, the deeper primary objective of which had been Madison Group carbonates, was completed just below 7,800 ft (2,380 m) in the Tyler Formation, where oil shows had been encountered during drilling. The first test was drilled on the domal crest and flowed at an initial potential of 499 BOPD (79.3 m³/D) with no water from a 16-foot (4.9-m) thick sandstone. An adjacent crestal confirmation well penetrated the objective section less than 4,000 feet (1,220 m) away and only 8 feet (2.4 m) deeper but encountered only 3 feet (0.9 m) of nonporous sandstone. In 12 of 23 additional wells drilled within the structural closure at Medora Field, this sandstone was absent, tight, or tested water with minor amounts of oil.

Similar development problems were experienced by other operators along the trend, where both stepout and infill wells often penetrated either poor quality reservoir sandstones or no sandstone at all. Just over half of all attempted field development wells in the trend were completed as producers. Cumulative oil production in these fields, which is an indication of reservoir size, spans a wide range (Table 16-1). As shown by Figure 16-1, stratigraphic components of these traps are evidenced by fields which produce from structural positions ranging from simple closure to homoclinal dip, including the axes of gentle synclines.

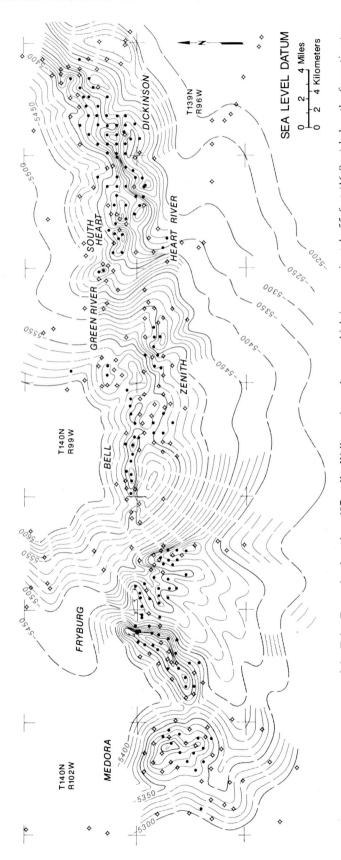

Fig. 16-1. Structure contour map of the Tyler Formation based on 407 wells. Well symbols apply to this formation only; many of the dry holes produce oil from deeper intervals. The contoured horizon is the base of the top seal for the uppermost reser-voir sandstone, which is approximately 55 feet (16.8 m) below the formation top. Contour interval is 10 feet (3.0 m).

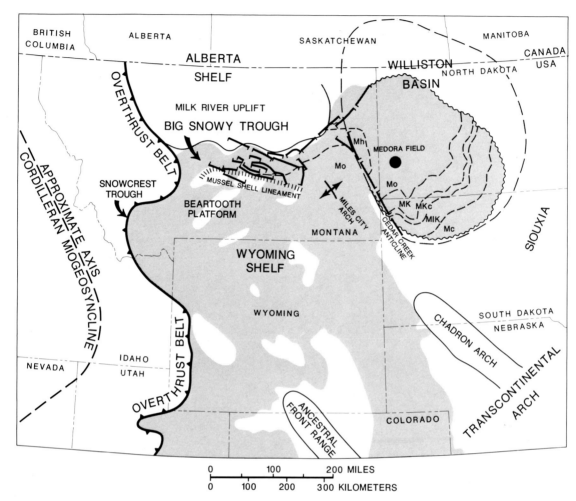

Fig. 16-2. Principal paleogeographic and tectonic elements affecting Tyler deposition. Present-day limits of the Tyler Formation are shown in gray. Note the Tyler subcrop in Williston basin. Mc: Charles Fm.; Mlk: Lower Kibbey Sandstone; Mkc: Kibbey "lime"; Mk: Upper Kibbey Sandstone; Mo: Otter Fm.; Mh: Heath Fm. Modified after Maughan, 1984 and reprinted by permission of American Association of Petroleum Geologists.

Regional Geology

Tectonic Setting

The Williston basin, now a roughly elliptical, gentle structural depression, was a very shallow intracratonic basin during the Carboniferous, part of a region of broad basins and low uplifts bordering the Canadian Shield that was dominated until the Early Meramecian (Visean) by carbonate shelf sediments (Rose, 1976; Roberts, 1979; Smith and Gilmour, 1979). Subsequent tectonism is recorded by several Mississippian structural elements. Most important of these was a shallow seaway (Willis, 1959), a graben complex of possible aulacogenic origin

(Maughan, 1984) which connected the Williston basin to the Cordilleran miogeosyncline (Fig. 16-2). This feature, the Big Snowy trough (Peterson, 1981), provided the trunk-drainage pathway for the entire Williston basin catchment area when it was subaerially exposed during a mid-Carboniferous sea-level lowstand.

The Cedar Creek structure (Fig. 16-2), a northwest-oriented faulted anticline, played a major role in Carboniferous deposition. Folding and faulting influenced facies patterns along the anticline throughout the Mississippian (Gerhard et al., 1982; Clement, 1986), eventually culminating in erosion of uppermost Mississippian sediments from the paleostructural crest (Maughan, 1975), west of the bounding fault. Subaerial exposure continued into the Morrowan, as evidenced by Tyler rocks being thin to absent in the same area, but thickening abruptly eastward across the fault. Subsequent movement is indicated by similar abrupt eastward thickening across the Cedar Creek structure of Permian (Opeche) and Triassic (Spearfish) redbeds and evaporites. During Cretaceous-Paleocene reactivation, movement on this steeply dipping basement fault was reversed and the eastern limb of the fold was upthrown and remains the high block today.

Tyler Formation thickness patterns reveal pre-Tyler erosional furrows that are roughly parallel to the Cedar Creek anticline (Fig. 16-3), and that coincide with the thickest Tyler fluvial sequences. These furrows suggest structural control of river systems that eroded Mississippian clastic and carbonate sediments and probably preferentially localized drainage along basement faults (Thomas, 1974; Brown, 1978; Sturm, 1983).

Orientations of structural elements suggest a regional wrench tectonic regime (Brown, 1978), as does the configuration and deformational history of the Cedar Creek anticline (Clement, 1986). The Cedar Creek anticline and its contemporary folds and faults in the Williston basin are roughly parallel to the Frontrange and Central Kansas uplifts, which also exhibit components of mid-Carboniferous transcurrent movement. These similarities may reflect a northern expression of the large-scale distributive shear which Kluth and Coney (1981) proposed as the cause of Pennsylvanian foreland block uplifts throughout the mid-continent region and attributed to Mississippian collision of the North and South American plates.

Stratigraphy

Carboniferous stratigraphic relationships in the Williston basin and Big Snowy trough were determined by correlations of wireline logs from more than 2,400 wells and by interpretation of about 4,000 line-miles (6,400 km) of reflection seismic data. Lithologic calibration of logs and interpretation of Tyler depositional environments in the Medora-Dickinson area were guided by examination of slabbed cores from 35 wells. These observations were supplemented by published descriptions from an additional 151 cored wells and by 14 outcrops of the Tyler in the Big Snowy trough in central Montana.

The Tyler Formation overlies Mississippian carbonate and clastic rocks with angular unconformity throughout most of North Dakota (Fig. 16-4). Uppermost Mississippian rocks comprise the Big Snowy Group, a single third-order depositional sequence whose base was defined by Sloss (1950) as the onset of the final Paleozoic second-order eustatic cycle, the Absaroka sequence. Because the Madison-Kibbey boundary does not represent a major unconformity, the base of the Absaroka sequence more logically belongs at the major hiatus represented by the Tyler-Big Snowy contact (Edwin Maughan, 1989, personal communication). The Big Snowy Group (Maughan, 1984) records a general upward deepening from sabkha redbeds, carbonates and gypsum (Kibbey) to shallow-marine shales and limestones (Otter) to organic-rich, open-marine shales and limestones (Heath) (Williams, 1983), the uppermost section of which is upward shoaling. Shallow-water marine carbonates of the Alaska Bench Limestone overlie the fluvial lacustrine and shallow-marine sediments of the Tyler throughout the region (Fig. 16-4).

Tyler strata are readily discernible by wireline-log correlation in the Williston basin because structural deformation has been minor, and because Tyler lithologies contrast markedly with underlying and overlying units. Contacts with units in the Big Snowy Group and the Alaska Bench Limestone are characterized by abrupt changes in gamma radiation (Fig. 16-5) and resistivity, and differences in thicknesses of the strata provide evidence of the erosional unconformity at the Big Snowy-Tyler contact.

Tyler rocks record the depositional response to a mid-Carboniferous sea-level fall and subsequent rise, a single chronostratigraphic sequence on which higher frequency cyclicity was superimposed. In addition to the transgression represented by the overlying Alaska Bench Limestone, at least ten regionally correlative base-level rises are evident within the Tyler. Eight of these surfaces are shown as dashed lines in Figure 16-5. In the lower part of the Tyler, they usually correspond to the contact between paleosol horizons and overlying shales. In the upper Tyler, they are represented by marine limestones. This higher-order cyclicity has been attributed to periodic tectonism (Willis, 1959) and to glacial eustasy (Meissner et al., 1984). Only three cycles have been recognized previously, however (Foster, 1956; Willis, 1959; Maughan, 1984; Sturm, 1987), probably due to generally poor outcrop qual-

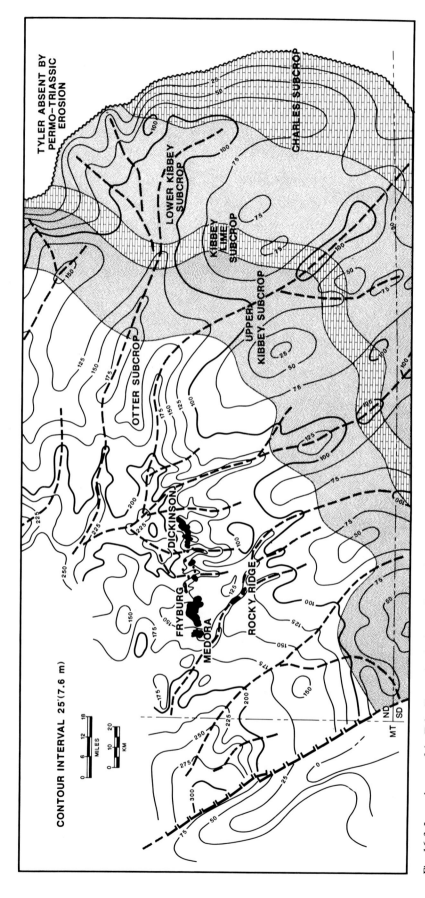

Fig. 16-3. Isopach map of the Tyler Formation in the southern Williston basin, based on approximately 1,200 wells. Subcropping Mississippian units are shaded. Dashed lines represent probable Morrowan drainage pattern, inferred from the isopach pattern, subcrop pattern, and locations of Tyler fluvial sequences.

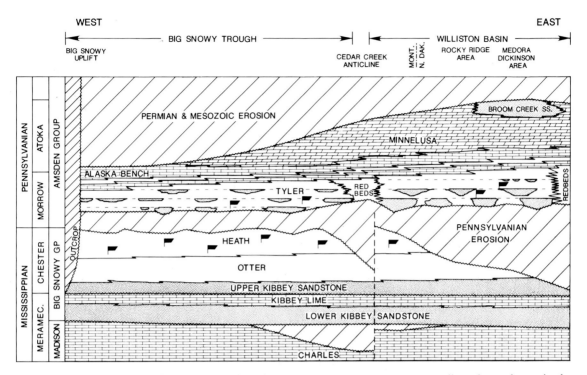

Fig. 16-4. Time-stratigraphic chart of mid-Carboniferous rocks in the Williston basin. Subcrop of formations underlying the Tyler are shown on Figure 16-3. Flags represent hydrocarbon source rocks. The Medora-Dickinson reser- voirs are the uppermost, small sandstone lenses in the Tyler at the right side of the chart. All Tyler sandstone locations are diagrammatic only.

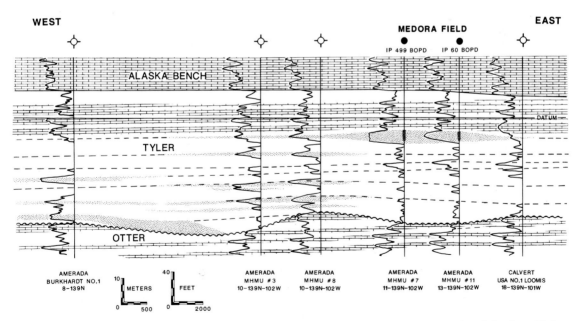

Fig. 16-5. Stratigraphic cross section through Medora Field. Note the upper reservoir sandstone and its pay zone. Dashed lines are basin-wide rises in relative sea level. Datum is a thin, transgressive shale about 20 feet (6.1 m) above the pay section. See Figure 16-7 for location of section.

ity and the paucity of cores sufficiently long to have penetrated all ten cycles.

The duration of these cycles is unknown. Big Snowy and Amsden Group faunas (Willis, 1959; Maughan, 1984) suggest that the Tyler Formation may represent the entire Morrowan (Bashkirian) record in the Williston basin. Using a Morrowan duration of 8 m.y. (Ross and Ross, 1987), a cycle in the Tyler would represent about 0.8 m.y., which is of the order of the Early Carboniferous eustatic periodicity documented by Ross and Ross (1988).

Hydrocarbon Source Rocks

Isotopic analyses of Williston Basin oils and source rocks indicate that Medora-Dickinson reservoirs are charged locally from Tyler shales and not from the Heath Formation, which is the source rock for Tyler oils in Montana (Williams, 1974). Figure 16-4 depicts these relationships between source rock and reservoir. Maughan (1984) suggested that Tyler oils in North Dakota may have originated from an intraformational source or from the Heath Formation in eastern Montana. However, burial history and heat-flow studies done by the author indicate that Tyler oils were not generated in North Dakota until the Oligocene. Structural reconstructions suggest that Paleocene and later hydrocarbon migration to the area around Medora-Dickinson was only from the center of the basin and not from the Big Snowy trough. Tyler oils were thus generated from intraformational source rocks in North Dakota, as originally suggested by Willis (1959).

Tyler source rocks consist of thinly bedded to laminated, brown to black calcareous shales deposited in lagoonal and shallow shelf environments. Because some of the richly organic, dark shale beds coarsen upward within only a few feet (tens of centimeters) to paleosol horizons (mottled red and green mudstone containing root casts), deposition of these source rocks must have occurred in extremely shallow water. Total organic carbon content in these rocks, which are largely limited to the center of the basin, ranges as high as 9%. Basin-wide patterns of Tyler oil shows match the limits of dark, radioactive Tyler shales which grade southward and eastward into redbeds (as mapped by Ziebarth, 1972). The distribution of these dark shales also corresponds roughly to the only area which reached the thermal maturity necessary for oil generation (Dow, 1974). Kerogen types reflect overall marine transgression

by a change from humic macerals (vitrinite) near the fluviodeltaic base of the sequence to relatively more lipid material (alginite) in the upper, more lagoonal and marine sections.

Reservoir Characterization

Seismic Expression

Seismic modeling shows that sandstone bodies in the Tyler that are 15 to 20 feet (4.6–6.1 m) thick at depths of about 7,000 to 8,000 feet (2,130–2,440 m) are acoustically detectable at a bandwidth of 10 to 50 Hz. At these frequencies the lenticular sand bodies appear as small positive amplitude anomalies in an otherwise low-velocity shale section (Fig. 16-6). However, seismic exploration for Tyler shoreline sandstones is hampered by two problems. First, the thickness of the Tyler Formation brackets the tuning thickness for most existing conventional reflection seismic data in the Williston basin. Second, in many places where barrier-island sandstones are absent, the interval is occupied by thin lagoonal and shelf limestones. Both of these factors introduce reflections that can spuriously suggest the presence of reservoir rock.

Geometry

An isopach map of the uppermost sandstone body at Medora Field shows that the reservoir is a sigmoidal to linear pod oriented north-south (Fig. 16-7). Elsewhere along the trend, lenticular sandstones are lobate to linear in plan view, consisting of coalesced lenses that are discontinuous at scales ranging from 2,000 feet to more than 4 miles (610 m-6 km).

Long axes of these sandstone lenses occur at all angles to depositional strike. Their ultimate updip lateral seals are at the edges where they feather into surrounding shale beds. Very few productive wells have been drilled within 2,000 feet (610 m) of this pinch-out. In Medora and Fryburg fields, for example, no wells with less than 7 feet (2.1 m) of sandstone were productive. In Dickinson Field, of the 36 updip wells in which sandstones are 4 feet (1.2 m) or less in thickness, 10 wells produced oil whereas, in the other 26 wells, the sandstones had insufficient porosity and permeability to produce. Detailed correlations suggest that the Tyler sandstone sequences penetrated in these dry holes are connected with the main reservoir.

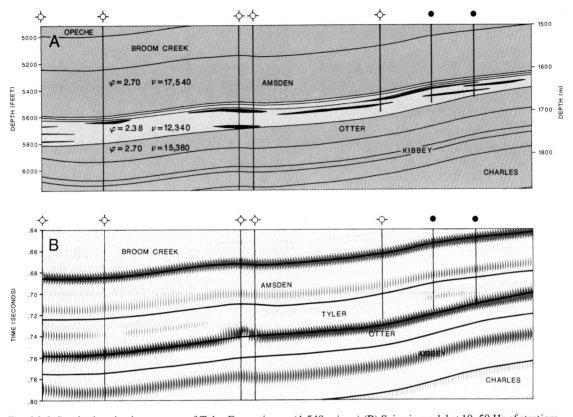

Fig. 16-6. Synthetic seismic response of Tyler Formation lenticular sandstones. (A) Dip-oriented geological cross section through Dickinson Field, showing velocities and densities used for the seismic model. Tyler sandstone properties used were ρ = 2.60 g/cm³, υ = 14,920 ft/sec.

(4,548 m/sec.) (B) Seismic model at 10–50 Hz of stratigraphy shown in (A). Note presence of a small, right-kicking event at location of field, and "pull-up" of basal reflection. Also note presence of a spurious right-kicking event between dry holes where no sandstone is present.

Depositional Setting

Most earlier lithofacies work in the region recognized general upward-deepening in the Tyler, but described Medora-Dickinson reservoirs as representing environments such as stable shelf or beach (Willis, 1959), fluvial (Ziebarth, 1962, 1972), or deltaic distributary channel (Grenda, 1977). Land (1976, 1979), in a significant reinterpretation of depositional strike, proposed that the Medora-Dickinson trend represents not a river system but a chain of barrier islands. This interpretation was supported by isopach and facies interpretations of Sturm (1983, 1987), who incorporated evidence for high-frequency sea-level fluctuations proposed by Roux and Schindler (1973) by suggesting that these barriers formed by longshore transport of deltaic sand during the youngest of three Tyler eustatic

cycles. Maughan (1984, his Fig. 7) mapped the Medora-Dickinson trend as a westerly oriented, dominantly fluvial meander belt, but suggested that other uppermost Tyler sandstones may have originated as "barrier island sands."

Tyler isopach and subcrop patterns indicate that the Medora-Dickinson trend defines depositional strike. The dominant trend of Tyler thickness contours is roughly concentric to the paleostructural center of the basin. Fingerlike reentrants in these contours are subperpendicular to this concentric trend and also to the subcrop belts of underlying Mississippian rocks (Fig. 16-3). These reentrants define the locations of alluvial valleys and contain the major fluvial sequences throughout the basin. The direction of flow in these alluvial valleys probably represents the Pennsylvanian depositional dip. The Medora-Dickinson trend, however, is approxi-

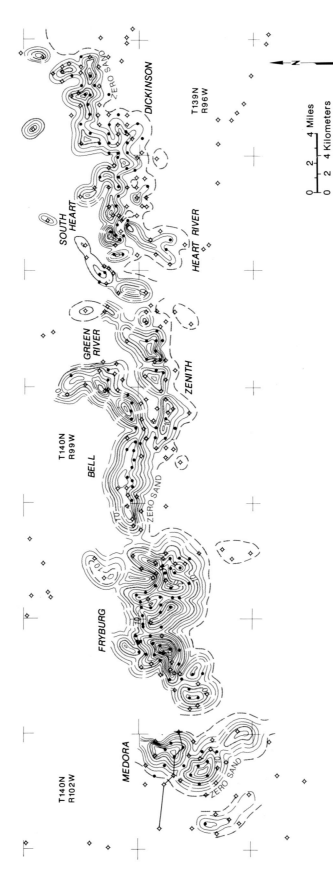

Fig. 16-7. Isopach map of the uppermost Tyler sandstone in the Medora-Dickinson trend. Contour interval is 2 feet (0.6 m). Tics mark township corners. See Figure 16-5 for section across Medora Field.

mately parallel with the margin of the basin and intersects the fluvial trends at acute angles. Sandstone lenses in the upper part of the Tyler along this trend cannot be deltaic because they are stratigraphically above any fluvial channel-fill sequence in any updip well. Furthermore, they could not have been strike-fed by longshore transport from adjacent deltas (Sturm, 1983), because nowhere along strike do these reservoirs correlate to sandstones with characteristics of deltaic facies.

Medora-Dickinson productive sandstones represent tidal-channel and flood-tidal delta deposits that overlie and intertongue with back-barrier lagoonal mudstone (Fig. 16-8). The overlying, diversely fossiliferous dark shale and limestone are evidence for shallow shelf deposition following continued sea-level rise. The barrier islands themselves were removed by shoreface erosion as the shoreline migrated updip, leaving only beheaded back-barrier deposits. Two separate lines of stratigraphic evidence support this transgressive origin, rather than genesis by northward progradation and longshore transport as proposed by Sturm (1983). First, Medora-Dickinson sandstone sequences are abruptly based and do not display the upward coarsening exhibited by the shoreface portions of all modern regressive barriers. Second, detailed correlations show that the northward offlapping stratal patterns demonstrated by Sturm (1983) are partially the result of a correlation option, enhanced by a possibly inconsistent datum. Facies characteristics provide additional evidence for a transgressive origin, as discussed below.

The coalescing, lobate forms of Medora-Dickinson reservoirs (Fig. 16-7) derive from the discontinuous nature of back-barrier sand bodies. Modern flood-tidal deltas produce thin sheet sands which may interfinger along strike (Fig. 16-9) as a result of inlet formation and migration (Barwis and Hayes, 1978). Where shapes of flood-tidal deltas are constrained by tidal-creek bathymetry, they become dip-oriented and grade geomorphically into tidal-channel point bars. Because channel migration rates are much lower in tidal than in fluvial systems, tidal point bars generally do not evolve into dip-oriented, fully developed sandy-meander belts, but often remain as single bars surrounded by mud (Barwis, 1978). Clay drapes and lags of bioclastic material form potential cross-channel impediments to fluid migration and reservoir drainage because they are concordant with the inclined accretion surfaces of the migrating point bars.

Many modern microtidal shorelines display pronounced variations in back-barrier sand-body geometry that range from small, isolated lenses to sheetlike forms. Pronounced changes in size and shape commonly occur within less than 10 miles (16 km), as is the case with the Medora-Dickinson trend. Regardless of shape, the crests of most modern flood-tidal deltas and tidal point bars are commonly vegetated (Fig. 16-9).

The sandstone beds in the Tyler flood-tidal deltas are intensely bioturbated, whereas the enclosing pyritic shales are well laminated. This suggests that climate and nutrient supply supported a community of infaunal deposit feeders which inhabited only the sand bodies and not the surrounding anoxic lagoonal muds. Primary physical sedimentary structures in the thicker sandstones have been largely destroyed, resulting in an homogenized fabric similar to that described in detailed studies of a microtidal flood-tidal delta along the subtropical Texas coast by Israel and others (1987).

The linear geometry of Bell Field (Fig. 16-7), together with its slightly downdip position, suggests that it may represent an inlet-fill sequence. Wireline-log curves from wells in Bell Field demonstrate the abrupt (i.e., erosional) bases common to all inlet-fill deposits.

Lithofacies

The Tyler Formation of the Williston basin and adjacent Big Snowy trough comprises varicolored shales and siltstones, thin quartzose sandstones, and minor coals, lime mudstones and evaporites deposited in fluviodeltaic to very shallow marine environments. Lowermost Tyler lithofacies represent fluvial-channel and overbank deposits (Sturm, 1983; Maughan, 1984), which drape a regional angular unconformity (Figs. 16-3, 16-4) often marked by a paleosol horizon (caliches, nodular ironstones) developed on underlying Mississippian units. Basal sandstones are gray, and medium to fine grained with conglomeratic, disconformable bases. Cross stratification is ubiquitous, with set thicknesses decreasing upward from medium to small scale. Shale interbeds often display mudcracks and root casts. Sandstones are organized as dip-oriented, upward-fining sequences from 10 to 50 feet (3–15 m) thick, the lateral equivalent of gray to red mudstones containing abundant plant fossils and paleosols. In many places, the lower Tyler comprises two stacked channel-fill sequences.

A

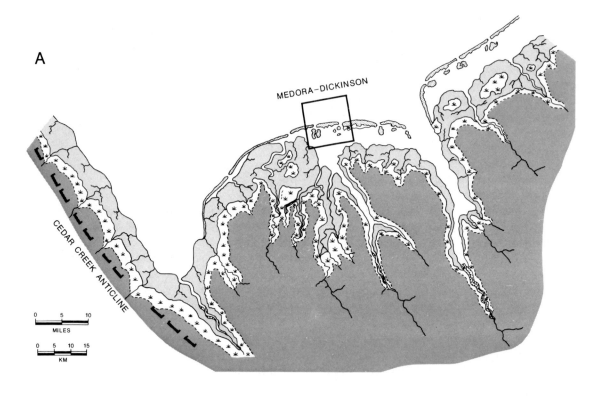

B

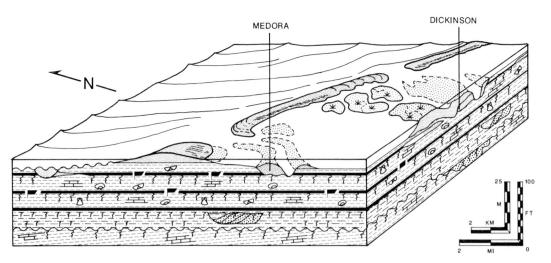

Fig. 16-8. Tyler transgressive-shoreline depositional model. (A) Regional paleogeography during deposition of Medora-Dickinson reservoirs. Rising relative sea level causes stream alluviation and transgressive reworking of downdip deltas. (B) Local block diagram for the Medora-Dickinson area (shown as box in A). Heavy lines are hydrocarbon source rocks. The unit below the unconformity is the Otter Formation.

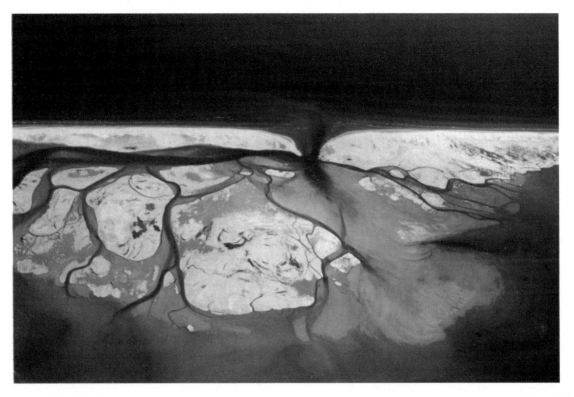

Fig. 16-9. Aerial photograph looking seaward at microtidal coastline near Shishmaref, Alaska. Coalesced, flood-tidal-delta sand bodies result from the migration of the inlet from left to right. Note vegetation on older tidal deltas. Barrier at left is about 1000 feet (305 m) wide. Photo by C. H. Ruby.

The fluvial sequences are overlain by and correlate laterally to dark gray shale (Fig. 16-10). This shale coarsens and becomes redder upward, passing into rooted and mudcracked siltstone. In places the siltstone is a paleosol or contains thin, nodular anhydrites or coals. Paleosols are abruptly overlain by ostracod-rich dark shales which define transgressive surfaces. Three such upward-coarsening sequences and their associated transgressive shales can be correlated from wireline logs over much of southern North Dakota. The lowermost of these surfaces caps the youngest of the two basal fluvial sequences. The upper two are separated by a 30-foot (9.1-m) thick zone of thin, dark gray to black, micritic limestone interbedded with dark gray to black shale containing an abundant, low-diversity faunal assemblage. This carbonate interval correlates westward into the Big Snowy trough (Fig. 16-4) to the Bear Gulch Limestone, a marine lime mudstone which contains the world's most abundant and diverse assemblage of Carboniferous fossil vertebrates (Williams, 1983).

Facies identified by the study of cores suggest that the sandstone bodies shown in Figures 16-4 and 16-5 represent the bases of back-barrier tidal channels, probably the feeder systems for flood-tidal deltas. They lie disconformably on dark shales and mudstones containing an abundant, low-diversity ostracod-brachiopod faunal assemblage. These shales are characterized by thin, upward-coarsening and upward-reddening sequences capped by paleosol horizons containing root casts and nodular, iron-rich mudstones (Figs. 16-10, 16-11A) and are interpreted as the record of salt marsh encroachment over a subtidal lagoon.

The dark, organic-rich fossiliferous shale that overlies the uppermost rooted zone (Figs. 16-10, 16-11B) is disconformably overlain by the lenticular sandstones which produce along the Medora-Dickinson trend. These sandstone lenses comprise white, very fine-grained, very well-sorted quartz-arenites. Bedding is thick at their disconformable base and thins upward to less than an inch (2.5 cm) at the top. Flaser and lenticular beds in the upper few feet (tens of centimeters) are the only primary physical sedimentary structures visible in cores (Fig. 16-11C). Large-scale cross-stratification is

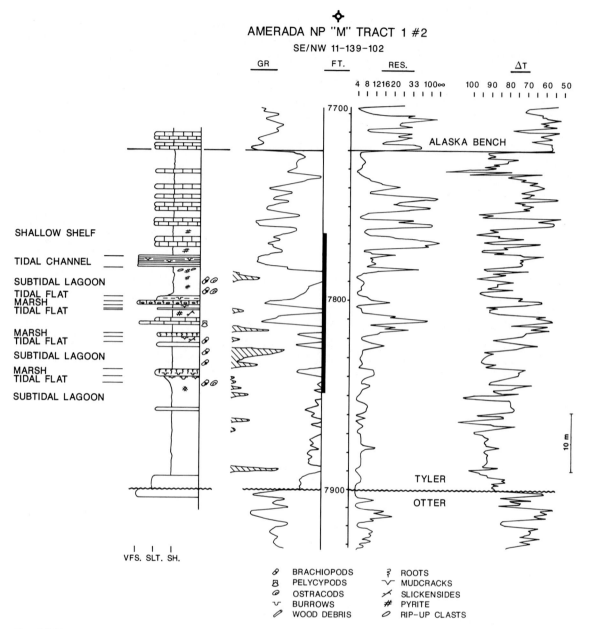

Fig. 16-10. Core-calibrated type log of the Tyler Formation in Medora Field. See Figures 16-7 and 16-16 for the field and well locations, respectively. The cored interval is shown by the vertical bar in the depth track. Other lithologies were identified in cores from adjacent wells.

absent but may have been erased by bioturbation. Most of the reservoir is so biogenically homogenized that individual trace fossils are difficult to discern (Fig. 16-11D). Thin, fossiliferous claystone partings may represent neap-tide deposits. Body fossils are present only as finely comminuted shell debris.

The sandstone lenses are overlain by laminated dark shale which is thinly interbedded with argillaceous lime mudstone, ostracod-phylloid algal boundstone, and occasional mollusk grainstones (Sturm, 1987). Biogenic sedimentary structures are rare. This shale grades upward into the thick carbonate section of the Alaska Bench Limestone (Figs. 16-4, 16-5).

Fig. 16-11. Core photographs from wells in the Medora-Dickinson trend. (A) Rooted nodular ironstone (paleosol horizon); (B) bioturbated argillaceous lime mudstone from backbarrier lagoon; (C) argillaceous ostracod-brachiopod packstone from shallow shelf; (D) laminated lagoonal shale (hydrocarbon source rock) immediately under the reservoir; white laminae are ostracod wacke-stones; (E) ripple cross-laminated sandstone near the edge of reservoir (porosity is 13%, permeability is 5 md); (F) bioturbated sandstone near the top of reservoir. See Figure 16-10 for relative stratigraphic position. Scale bar is 1 inch (2.5 cm).

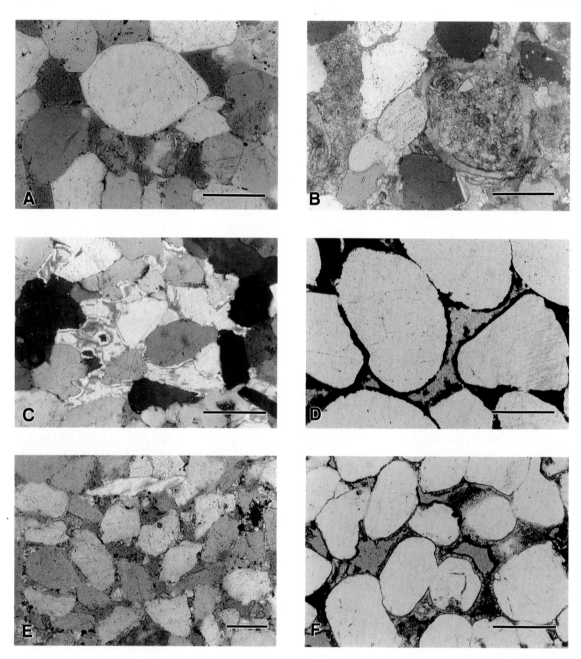

Fig. 16-12. Thin-section photomicrographs showing porosity types in Tyler barrier-related sandstones. (A) Clean, porous quartzarenite with minor quartz overgrowths; (B) ostracodal quartz wacke near feather edge of reservoir; (C) anhydrite-cemented quartzarenite; (D) hematite-cemented quartzarenite; (E) leached calcite-cemented quartzarenite; (F) clay-coated quartz grains in reservoir. Black is residual oil. Scale bar is 0.25 mm.

Petrography and Diagenesis

Composition varies and is relative to position within the lenticular sandstone reservoir rocks. Where the sandstone is thicker than 6 to 8 feet (1.8–2.4 m), well-rounded, fine- to very fine-grained quartz composes 90 to 98% of the framework grains (Fig. 16-12A).

The remainder consists largely of broken and abraded brachiopod, pelycypod, and ostracod fragments. The bioclastic fraction is up to 50% of some isolated lentils within the sandstone. Chert grains, rock fragments, and heavy minerals are present only in trace amounts. Detrital clays compose from 1 to 10% of the primary minerals. Near the thinner edges

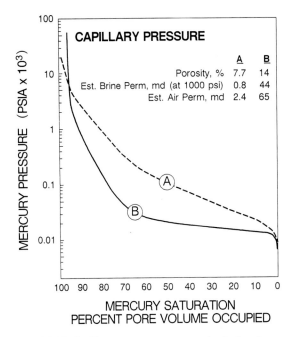

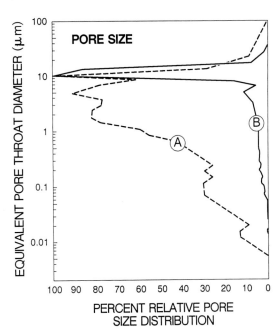

Fig. 16-13. Capillary pressure curves and calculated pore-size distribution in the Medora Field reservoir. A is a tight sandstone just below the top seal; B is from the oil pay zone 6 feet (1.8 m) lower (7,780 feet (2,371.3 m) and 7,786 feet (2,373.2 m), respectively).

of reservoirs, the proportional volume of clay matrix increases to 25% (Fig. 16-12B). Throughout the reservoir, the detrital clay content is locally high in swirls or blotches associated with burrow traces.

Several types of cement are distributed preferentially within the reservoir. Near the topseals, porosity is often completely plugged by calcite or anhydrite (Fig. 16-12C). Cement-filled intergranular porosity is as high as 40% in these zones because cementation occurred before compaction, during pedogenesis (Land, 1979). Early, pore-lining hematite cement (Fig. 16-12D) is also related to the thin soils which capped these tidal deltas. These zones of early cementation are seldom more than 3 feet (0.9 m) thick. Pyrite replacement of anhydrite or calcite has contributed to minor porosity reduction in patchy zones associated with ostracod shell hash in shale laminae (Sturm, 1983).

Quartz overgrowths are common but destroy porosity only in small, irregular patches where the sandstone was originally very clean (Fig. 16-12E). Some reservoir-quality sandstones (Fig. 16-12F) exhibit porosity preserved from overgrowth cementation by detrital clay coatings (Sturm, 1983); others exhibit secondary porosity as etched remnants of precompaction calcite cement. Both clay-protected and calcite-etched porosities are concentrated in oil-saturated zones, but in no readily discernible pattern. Below oil-water contacts, pore-lining authigenic kaolinite and illite commonly reduce porosity by 5 to 10% absolute. Sturm (1983) suggested that these clays were concentrated in water-wet zones where they nucleated on crystal faces of quartz overgrowths, indicating late-stage growth (i.e., after oil migration).

Petrophysics

Porosity ranges from 1 to 22% (avg. 12%) and, as shown by the capillary pressure curves (Fig. 16-13) and histogram (Fig. 16-14), is bimodally distributed. Poor quality reservoir rocks (avg. 5.5% porosity in Fig. 16-14) comprise two types of sandstone. One type consists of clean quartzarenite (less than 5% matrix) with porosity occluded by quartz, calcite, or anhydrite (Figs. 16-12C, 16-12E). This is the tightest rock in the Medora-Dickinson trend, with permeability between 0.01 and 0.3 md (Fig. 16-15). These sandstones are the permeability equivalents of the sandstones with up to 25% detrital clay which characterize the updip sandstone pinchouts. The other type also consists of "clean" quartzarenite, but with a component of porosity occlusion by authigenic clay. This type of sandstone, shown by triangles in Figure 16-15, plots in the high end of the

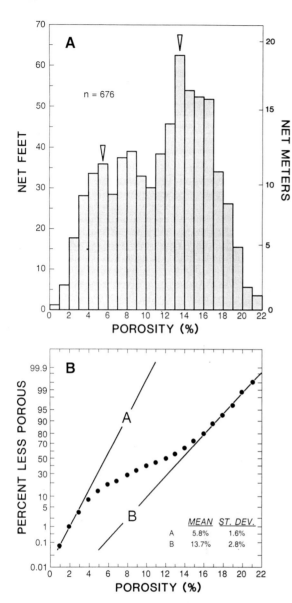

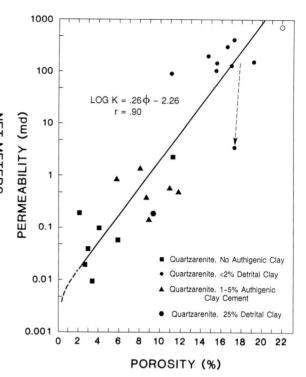

Fig. 16-15. Stressed air porosity (% bulk volume) versus permeability measurements in Dickinson sandstone cores. The arrow points to quartzarenite with meniscus hematite cement that has reduced pore connectivity but not significantly reduced porosity. The dashed line is the suggested porosity-occlusion path resulting from soil formation. The open circle (uppermost right) is the average value reported by Amerada Hess for friable sandstone in the central channel at Medora Field (not included in regression).

Fig. 16-14. (A) Histogram of core-measured porosity (% bulk volume) in the Medora-Dickinson area. Each porosity value is assumed to represent one foot of reservoir. Triangles indicate modes of two porosity populations. (B) Porosity (% bulk volume) data from (A) expressed as cumulative frequency. Porosity distribution can be described as the sum of two normally distributed populations (straight lines). These populations correspond to the two data clusters in Figure 16-15.

low-porosity cluster and has permeability between 0.2 and 0.3 md.

The highest quality reservoir rocks average about 14% porosity (Fig. 16-14) and are also quartzarenites. They display the same range of cements observed in the low-porosity zones. The best have an average porosity of about 16% and permeability between approximately 100 and 400 md (Fig. 16-15). Some values of up to 23% and 750 md occur in the more friable sandstones (Alex Chaky, 1989, personal communication). The detrital clay matrix content in all of the permeable sandstones is less than 2%. Early oil migration, rather than clay coating of grains as suggested by Sturm (1983), appears to have been the primary mechanism for porosity preservation in these highest reservoir-quality sandstones, because many oil-saturated samples with porosity

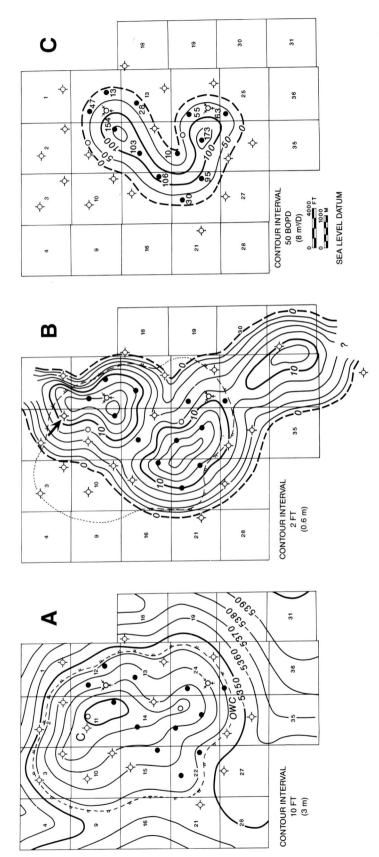

Fig. 16-16. Structural versus facies control of production rates in Medora Field. Sections are 640 acres (259.2 ha.). (A) Structure contour map of the top of the Tyler producing sandstone. Note dry holes within the structural closure and above the inferred oil-water contact. "C" in sec. 11 is the cored well shown in Figure 16-10. Open circles are dry-hole development wells completed after the waterflood was initiated. (B) Thickness map of Tyler reservoir sandstone. Note that all wells with more than 8 feet (2.4 m) of sandstone above the oil-water contact are producers with one exception (arrow, discussed in text). The dotted line is the oil-water contact inferred from the structural spill point in A. (C) Map of preflood production rates. Note that the highest rate, 173 BOPD (27.5 m³/D), was from a well more than 100 feet (30 m) below the structural crest, and was from a slightly thinner pay section than the crestal well. Note the general pronounced mimicking of total sandstone thickness shown in B.

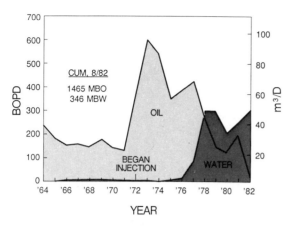

Fig. 16-17. Decline curve for the discovery well in Medora Field, with daily rates averaged by year.

greater than 20% are completely clay-free. Porosity preservation by clay coatings was noted in quartz-arenites with 12 to 15% porosity. However, none of the sandstones in this porosity range were sampled for permeability.

Production Characteristics of Medora Field

Preflood production rates are proportional to reservoir thickness (Fig 16-16). Only wells penetrating more than 8-foot (2.4-m) thick sandstone beds have produced oil, despite the fact that the entire sandstone body was filled to the structural spill, a 54-foot (16.5-m) oil column. The dry hole in Section 2 (arrow in Fig. 16-16B) penetrated thick sandstone above the oil-water contact but is nonproductive, and it is unknown whether this part of the reservoir is isolated by a local permeability barrier (e.g., a point-bar clay drape) or whether the reservoir was damaged during drilling. The productive area represents only 52% of the 6,150 acres (2,490 ha.) within the structural closure. The two best wells had initial potentials of 400 to 500 BOPD (64–80 m³/D) as shown in Figure 16-16. Neither well was acidized or fracture-treated. These wells are good performers compared to other Medora-Dickinson wells; their initial production rates have been exceeded by only 6% of all producing wells in the trend. Their preflood average production rates (154 and 173 BOPD (24.5 and 27.5 m³/D) nearly equal the average initial potential production of 190 BOPD (30.2 m³/D) for all completions in similar facies along the trend.

No initial rates for barrier-related sandstones in the trend exceeded 800 BOPD (127 m³/D).

More than half of the wells in Medora Field were acid stimulated, which caused later waterflood problems by creating differential wettabilities to oil. The poorest wells, those with initial potentials of less than 100 BOPD (15.9 m³/D), were improved by 20 to 30 BOPD (3.2–4.8 m³/D) with a saltwater-fracture treatment using sand and glass proppant. Wells in the thickest portion of the reservoir required no stimulation.

The field was developed on 320-acre (130-ha.) spacing and produced by depletion drive from 12 wells whose average performance during eight years ranged from 10 to 173 BOPD (1.6–27.5 m³/D). Primary recovery is estimated to have been about 13% of the 24.8 MMB (4.0 × 10⁶ m³) OIP, based on calculations employing material balance, decline curves, and an initial water saturation of 20%. After water injection was initiated in 1970, average per-well production rates increased to as much as 526 BOPD (83.6 m³/D). Total production for the reservoir increased from 580 BOPD (92.2 m³/D) to 2,200 BOPD (350 m³/D). Water cuts exceeded 50% in 1977 and the field has since been in steep production decline (Fig. 16-17). Transient pressure analyses indicate less efficient drainage in the east-west direction (Alex Chaky, personal communication, 1989), which probably is the result of retardation of fluid flow across the tidal-channel sand body due to accretion bed drapes of claystone or cemented shell hash.

Ultimate waterflood oil is estimated at 120% of the primary recovery, with total estimated ultimate recovery at 6.9 MMBO (1.1 × 10⁶ m³). The three best wells in the field each produce from 14 net feet (4.3 m) of pay, and each will yield almost 1.5 MMBO (2.4 × 10⁵ m³). Ultimate recoveries for each of the other nine wells are as low as 11 MBO (1,750 m³) and average 547 MBO (8.7 × 10⁴ m³). A planned CO_2 flood could recover an additional 8 to 11% of the OOIP, a maximum of about 2.7 MMBO (4.3 × 10⁵ m³) (Alex Chaky, 1989, personal communication).

Exploration and Production Strategy

Tyler sandstones are most prospective where they parallel the youngest Pennsylvanian southern shore-line of the paleo Williston basin. Barrier remnants may also occur downdip along older paleoshorelines that were formed during earlier phases of the regional transgression. Barrier-remnant reservoirs will be hydrocarbon charged only where they are intercalated within dark, organic-rich shales or lie along updip migration paths from those shales. Isolated, small zones of recoverable oil are likely to be undrained due to permeability barriers resulting from claystone drapes.

Without the ability to detect sandstone less than 15 feet (4.6 m) thick on a seismic grid with line spacings less than 1 mile (0.6 km), evaluation of exploration and development economics for these types of reservoirs must use probabilities of success which are consistent with the average sizes of Tyler sandstone bodies. Based on the ratio of the productive area to trend area along the Medora-Dickinson shoreline, a randomly drilled well in a similar setting would have a 20% chance of discovery, assuming that the shoreline trend could be located in the first place. The probability of drilling successful development wells ranges from about 25 to 75%, only slightly higher in some instances than expected for an exploration well.

Medora-Dickinson sandstones, like many cratonic-margin shoreline sands, are mineralogically mature, so no special drilling or completion techniques need be considered to avoid damage to reservoir framework grains. However, acid treatment may decrease waterflood efficiencies by altering matrix mineralogy and creating differential permeability to water. Hydraulic fracturing has been demonstrated to improve Tyler well performance in the thin, diagenetically altered "rinds" near the edges of reservoirs. Because natural water drives are very unlikely in these small, isolated sandstone lenses, pressure maintenance must be implemented early in the life of a field and should be designed for long-term waterflooding.

Conclusions

1. The potential reservoirs generated by transgressive barrier systems on stable margins are thin and discontinuous. These reservoirs are distributed unevenly along depositional strike and may be individually oriented in any direction.
2. Traps are stratigraphic, so simple seismic structural mapping alone is inadequate for exploration. Individual sandstone bodies must be seismically imaged or predicted from detailed facies analysis using well control.
3. The main risks to field development are sandstone body discontinuity, poor reservoir quality in thinner zones, and porosity occlusion by paleosols.

Acknowledgments. I thank Jim Clement for introducing me to Williston basin stratigraphy and for sharing his knowledge of Montana and North Dakota geology. Helpful ideas were gained from Don Beard, David Childers, and John Hastings of Shell Western E&P, who performed companion studies and provided insights to petrography, outcrops, and cores, respectively. Jennifer Thompson provided kerogen analyses and valuable interpretations of their link to depositional environment. I also benefited from discussions with Steve Sturm on Tyler paleogeography and with Alex Chaky of Amerada Hess on the development of Medora Field. I greatly appreciate the incisive review of Ed Maughan of the U.S. Geological Survey. This work was done at Shell Western E&P Inc., to whom I am grateful for permission to publish.

Reservoir Summary

Field : Medora
Location: Billings County, North Dakota
Operator: Amerada Hess Petroleum
Discovery: 1964
Basin: Williston basin
Tectonic/Regional Paleosetting: Intracratonic basin
Geologic Structure: Simple dome
Trap Type: Stratigraphic pinch-out on simple dome
Reservoir Drive Mechanism: Depletion
  • **Original Reservoir Pressure:** 3,511 psi (2.4 × 10⁴ kPa) at 5,350 feet (1,631 m) subsea

Reservoir Rocks
- **Age:** Pennsylvanian, Bashkirian
- **Stratigraphic Unit:** Tyler Formation
- **Lithology:** Quartzarenite
- **Depositional Environments:** Barrier island - lagoon system
- **Productive Facies:** Back-barrier tidal channel, flood-tidal delta
- **Petrophysics**
  - **Porosity Type:** Intergranular
  - **ϕ:** Average 12%, range 2 to 22%, cutoff 8% (cores)
  - **k:** Average 90 md, range 0.1 to 750 md (brine, stressed cores)
  - **S_w:** Average 18%, range 17 to 22% (cores)

Reservoir Dimensions
- **Depth:** 7,766 feet (2,367 m)
- **Areal Dimensions:** 2.3 by 3.0 miles (3.7 × 4.8 km)
- **Productive Area:** 3,240 acres (1,310 ha.)
- **Number of Reservoirs:** 1 (excluding Mississippian carbonates)
- **Hydrocarbon Column Height:** 52 feet (15.8 m)
- **Fluid Contacts:** Oil-water contact at 5,320 feet (1,622 m) subsea
- **Number of Pay Zones:** 1
- **Gross Sandstone Thickness:** 14 feet (4.3 m) maximum
- **Net Sandstone Thickness:** 14 feet (4.3 m) maximum
- **Net/Gross:** 1.0

Source Rocks
- **Lithology:** Lagoonal shale (intraformational), Tyler Formation
- **Time of Hydrocarbon Maturation:** Oligocene-Miocene
- **Time of Trap Formation:** Paleocene
- **Time of Hydrocarbon Migration:** Oligocene-Miocene

Hydrocarbons
- **Type:** Oil (low sulfur, parafinic with high pour point)
- **GOR:** 230:1
- **API Gravity:** 34°
- **FVF:** 1.125
- **Viscosity:** NA

Volumetrics
- **In-Place:** 24.8 MMBO (3.9×10^6 m³)
- **Cumulative Production:** 6.8 MMBO (1.08×10^6 m³)
- **Ultimate Recovery:** 7.1 MMBO (1.13×10^6 m³) after waterflood
- **Recovery Efficiency:** 29% after waterflood

Wells
- **Spacing:** 3,750 feet (1,145 m); 320 acres (129.6 ha.)
- **Total:** 24
- **Dry Holes:** 12

Typical Well Production:
- **Average Daily:** 73 BO (11.6 m³), range 10 to 154 BO (1.6–24.5 m³)
- **Average Cumulative:** 550 MBO (8.8×10^4 m³), range 15 to 1,500 MBO (2.4–240×10^3 m³)

Stimulation: Saltwater frac with 2,000 to 10,000 lb (0.9–4.5×10^3 kg) of sand and glass proppant improves flow rate by 20 to 30%. Frac pressure gradients are 0.58 to 0.85 psi/ft (13.1–19.2 kPa/m)

References

Barwis, J.H., 1978, Sedimentology of some South Carolina tidal creek point bars, and a comparison with their fluvial counterparts, *in* Miall, A.D., ed., Fluvial Sedimentology: Canadian Society of Petroleum Geologists Memoir 5, p. 129–160.

Barwis, J.H., and Hayes, M.O., 1978, Regional patterns of modern barrier island and tidal inlet deposits as applied to paleoenvironmental studies, *in* Ferm, J.C. and Horne, J.C., eds., Carboniferous Depositional Environments in the Appalachian Region: Dept. Geology, University of South Carolina, American Association of Petroleum Geologists Field Course Guidebook, p. 472–498.

Brown, D.L., 1978, Wrench-style deformation patterns associated with a meridional stress axis recognized in Paleozoic rocks in parts of Montana, South Dakota and Wyoming, *in* Williston Basin Symposium: Montana Geological Society, 24th Annual Conference, p. 17–34.

Clement, J.H., 1986, Cedar Creek: A significant paleotectonic feature of the Williston Basin, *in* J.A. Peterson, ed., Paleotectonics and Sedimentation in the Rocky Mountain Region, United States: American Association of Petroleum Geologists Memoir 41, p. 213–240.

Dow, W.G., 1974, Application of oil-correlation and source rock data to exploration in Williston Basin: American Association of Petroleum Geologists Bulletin, v. 58, p. 1253–1262.

Foster, D.I., 1956, N.W. Sumatra Field (Montana), *in* Judith Mountains-Central Montana: Billings Geological Society, 7th Annual Field Conference Guidebook, p. 116–123.

Gerhard, L.C., Anderson, S.B., LeFever, J.A., and Carlson, C.S., 1982, Geological development, origin, and energy mineral resources of Williston Basin, North Dakota: American Association of Petroleum Geologists Bulletin, v. 66, p. 989–1020.

Grenda, J.C., 1977, Paleozoology of cores from the Tyler Formation (Pennsylvanian) in North Dakota, U.S.A. [Ph.D. thesis]: Grand Forks, North Dakota, University of North Dakota, 338 p.

Israel, A.M., Ethridge, F.G., and Estes, E.L., 1987, A sedimentologic description of a microtidal, flood-tidal delta, San Luis Pass, Texas: Journal of Sedimentary Petrology, v. 57, p. 288–300.

Kluth, C.F., and Coney, P.J., 1981, Plate tectonics of the ancestral Rocky Mountains: Geology, v. 9, p. 10–15.

Land, C.B., 1976, Stratigraphy and petroleum accumulation, Tyler sandstones (Pennsylvanian), Dickinson area, North Dakota (abst.): American Association of Petroleum Geologists Bulletin, v. 60, p. 1401–1402.

Land, C.B., 1979, Tyler sandstones (Pennsylvanian), Dickinson area, North Dakota—a 24 million barrel soil-zone stratigraphic trap (abs): American Association of Petroleum Geologists Bulletin, v. 63, p. 485.

Maughan, E.K., 1975, Montana, North Dakota, northeastern Wyoming, and northern South Dakota, *in* McKee, E.D. and Crosby, E.J., eds., Paleotectonic Investigations of the Pennsylvanian System in the United States: U.S. Geological Survey Professional Paper 853, Part I, p. 279–293.

Maughan, E.K., 1984, Paleogeographic setting of Pennsylvanian Tyler Formation and relation to underlying Mississippian rocks in Montana and North Dakota: American Association of Petroleum Geologists Bulletin, v. 68, p. 178–195.

Meissner, F.J., Woodward, J., and Clayton, J.L., 1984, Stratigraphic relationships and distribution of source rocks in the greater Rocky Mountain region, *in* Woodward, J., Meissner, F.J., and Clayton, J.L., eds., Hydrocarbon Source Rocks of the Greater Rocky Mountain Region: Denver, Rocky Mountain Association of Geologists, p. 1–30.

Peterson, J.A., 1981, General stratigraphy and regional paleostructure of the western Montana overthrust belt, *in* Tucker, T.E., ed., Field Conference and Symposium Guidebook, Southwest Montana: Montana Geological Society, p. 5–35.

Roberts, A.E., 1979, Northern Rocky Mountains and adjacent plains regions, *in* McKee, E.D. and Crosby, E.J., eds., Paleotectonic Investigations of the Mississippian System in the United States: U.S. Geological Survey Professional Paper 1010, p. I-221–I-247.

Rose, P.R., 1976, Mississippian carbonate shelf margins, western United States: U.S. Geological Survey Journal of Research, v. 4, p. 449–466.

Ross, C.A. and Ross, J.R.P., 1987, Late Paleozoic sea levels and depositional sequences, *in* Ross, C.A. and Haman, D., eds., Timing and Depositional History of Eustatic Sequences: Constraints on Seismic Stratigraphy: Cushman Foundation for Foraminiferal Research, Special Publication 24, p. 137–149.

Ross, C.A. and Ross, J.R.P., 1988, Late Paleozoic transgressive-regressive deposition, *in* Wilgus, C.K., Hastings, B.S., Ross, C.A., Posamentier, H., Van Wagoner, J., and Kendall, C.G.St.C., eds., Sea-Level Changes: An Integrated Approach: Society of Economic Paleontologists Mineralogists Special Publication 42, p. 227–248.

Roux, W.F., Jr., and Schindler, S.F., 1973, Late Mississippian cyclothems of the Heath Formation, western North Dakota (abst.): American Association of Petroleum Geologists Bulletin, v. 57, p. 961.

Sloss, L.L., 1950, Paleozoic sedimentation in Montana area: American Association of Petroleum Geologists Bulletin, v. 34, p.423–451.

Smith, D.L., and Gilmour, E.H., 1979, The Mississippian and Pennsylvanian (Carboniferous) Systems in the United States—Montana, *in* McKee, E.D. and Crosby, E.J., eds., The Mississippian and Pennsylvanian (Carboniferous) Systems in the United States: U.S. Geological Survey Professional Paper 1110-X, p. X1–X32.

Sturm, S.D., 1983, Depositional environments and sandstone diagenesis in the Tyler Formation (Pennsylvanian), southwestern North Dakota: North Dakota Geological Survey, Report of Investigations 76, 48 p.

Sturm, S.D., 1987, Depositional history and cyclicity in the Tyler Formation (Pennsylvanian), Southwestern North Dakota, *in* Longman, M. W., ed., Williston Basin: Anatomy of a Cratonic Oil Province: Rocky Mountain Association of Geologists, p. 209–221.

Thomas, G.E., 1974, Lineament-block tectonics: Williston-Blood Creek Basin: American Association of Petroleum Geologists Bulletin, v. 58, p. 1305–1322.

Williams, J.A., 1974, Characterization of oil types in Williston Basin: American Association of Petroleum Geologists Bulletin, v. 58, p. 1243–1252.

Williams, L.A., 1983, Deposition of the Bear Gulch Limestone: A Carboniferous plattenkalk from central Montana: Sedimentology, v. 30, p. 843–860.

Willis, R.P., 1959, Upper Mississippian-Lower Pennsylvanian stratigraphy of central Montana and Williston Basin: American Association of Petroleum Geologists Bulletin, v. 43, p. 1949–1966.

Ziebarth, H.C., 1962, The Micropaleontology and Stratigraphy of the Subsurface Heath Formation (Mississippian-Pennsylvanian) of Western North Dakota [M.S. thesis]: Grand Forks, North Dakota, University of North Dakota, 145 p.

Ziebarth, H.C., 1972, The Stratigraphy and Economic Potential of Permo-Pennsylvanian Strata in Southwestern North Dakota [Ph.D. thesis]: Grand Forks, North Dakota, University of North Dakota, 414 p.

Key Words

Medora Field, North Dakota, Williston basin, Tyler Formation, Pennsylvanian Bashkirian, barrier island-lagoon, back-barrier tidal channel deposits, flood-tidal-delta deposits, transgressive stratigraphy, sea levels, Carboniferous.

Shelf Environments

17

Shelf and Shoreface Reservoirs, Tom Walsh-Owen Field, Texas

John W. Snedden and R. Stan Jumper

Mobil Exploration Norway, Inc., 4001 Stavanger, Norway; HarCor Exploration, Inc., Dallas, Texas 75251

Introduction

In past years, as explorationists concentrated on better-known marginal marine and deepwater siliciclastic reservoirs, the economic potential of stratigraphic traps in marine-shelf sand bodies has been overlooked (Tillman, 1985). However, major petroleum discoveries in shelf sandstones and recent advances in the understanding of modern shelf processes have stimulated interest in these types of exploration targets. Unfortunately, the paucity of published information on these reservoirs limits our ability to predict production performance. The Tom Walsh and Owen fields of south Texas, the subject of this chapter, have been producing gas from shelf and shoreface sandstones of the Cretaceous Olmos Formation for nearly twenty years. Study of these fields provides a rare look at how depositional processes impact many aspects of reservoir quality and production in sandstones of shelf origin.

Tom Walsh and Owen fields are among the large number of fields which produce hydrocarbons from the Olmos Formation throughout south Texas (Tyler and Ambrose, 1986). Located in western Webb County about 100 miles (161 km) south of San Antonio, the two fields are part of a larger belt of gas production which is oriented parallel to depositional and structural strike (Fig. 17-1). This belt of eight fields is known as the "downdip trend" and is separated spatially and structurally from the "updip trend" of Olmos fields situated to the west. The updip trend contains Olmos sandstones in a 50 to 250 feet (15–76 m) thick, multistoried sequence. Examination of cores indicates that the updip Olmos comprises fluviodeltaic deposits (Snedden and Kersey, 1982; Tyler and Ambrose, 1986).

In contrast, the Olmos of the downdip trend exhibits a more sheet-like geometry and net thicknesses of less than 50 feet (15 m). Initial work on these downdip facies revealed that the sandstones were formed in an open marine setting (Snedden and Kersey, 1982). Their shelfal genesis was proposed as an explanation for their thin-bedded, blanket geometry.

Recent acquisition of additional cores from Tom Walsh and Owen fields offered the opportunity to study these downdip sandstones further. The two fields are significant in that they contain some of the most productive wells in the downdip trend. Gas entrapment here is stratigraphically controlled in contrast to the other downdip fields which are primarily structural traps.

Development History

Eight fields (Tom Walsh, Owen, Booth Range, Las Tiendas, Garner Ranch, H. E. Clark, La Cruz, and Segundo) comprise the 35-mile (56-km) long downdip gas-production belt of Webb County. Garner Ranch, the smallest of these fields, was discovered first when in 1970 gas shows in the Olmos were

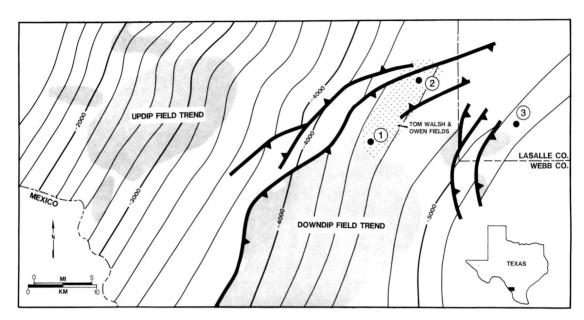

Fig. 17-1. Structure contour map on top of Cretaceous strata in Webb County, Texas. Contour interval is 200 feet (61 m). Locations of updip and downdip field trends and Tom Walsh-Owen Field are also shown. Dots indicate locations of the three cored wells discussed in the text: (1) Tesoro 3 Webb CSL; (2) Murexco 1 Leyendecker; (3) Drexel 70-4 Fansler.

noticed in a test of the deeper Edwards/Stuart City reef trend. Field development progressed rapidly through the early 1970s as independent operators drilled along strike to offset adjacent production. Drilling activity peaked in 1975, but small companies continue to explore, motivated by the relatively shallow drilling depths, perceptions of generally low risk (due to the apparent sheet-like nature of the sandstone distribution), and potential for uphole reserves in the overlying Cretaceous and Tertiary sandstones. As of December 1987, approximately 170 BCFG (4.8×10^9 m³) and 0.7 MMBO (1.0×10^5 m³) had been produced from 465 wells in eight fields.

Approximately 41% of the cumulative production is from Tom Walsh and Owen fields, which include only 28% of producing wells in the trend. The two fields together cover about 21,230 acres (8,600 ha.) or 11 miles (18 km) along strike and 3 miles (5 km) along the dip direction. Although they have been designated formally as separate fields, no evidence exists for physical separation of hydrocarbons between the two (Govett, 1983). Most operators agree that the two fields, referred to as "Tom Walsh-Owen Field," are the most economically viable portion of the downdip Olmos trend.

Regional Data

Stratigraphy

Upper Cretaceous strata of south Texas contain three siliciclastic intervals, the Escondido, the Olmos, and the San Miguel formations (Fig. 17-2). Each unit was originally defined from exposures in Maverick County (Dumble, 1892), located approximately 62 miles (100 km) northwest of Tom Walsh-Owen Field (Fig. 17-3, inset). Regional biostratigraphic work undertaken as part of this study indicates that all three sandstones are confined to the Maastrichtian and the Danian, rather than the Campanian age previously invoked (Weise, 1980; Tyler and Ambrose, 1986).

Figure 17-3 illustrates the relationship between the informal subsurface units of the study area and the formal surface-defined rock units of the Olmos and the lower portion of the Escondido Formation (Cooper, 1970). Correlations are based on examination of cores and wireline logs, surface exposures, and biostratigraphic analyses of outcrops and well cuttings. The cross-sectional datum is the lowermost transgressive surface (TS) present at the top of the updip Olmos section, which is recognizable in cores

Fig. 17-2. Stratigraphy of the Maastrichtian and Danian intervals of south Texas. Numerical zonation terminology from van Hinte (1976).

EPOCH	AGE	NUMERICAL ZONE	LITHOSTRATIGRAPHY
PALEOCENE	DANIAN	P 1	
LATE CRETACEOUS	MAASTRICHTIAN	UC 17	ESCONDIDO FORMATION
		UC 16	
		UC 15	
		UC 14	OLMOS FORMATION
		UC 13	SAN MIGUEL FORMATION

as a concentration of mechanically sorted, disarticulated bivalve shells. This surface is believed to correlate with a similar surface at the top of the Seven Points submember of the Escondido Formation. Although it is probably diachronous on a fine scale, biostratigraphic data suggest that it approximates a general chronostratigraphic boundary separating sediments of faunal zones UC 14 and 15.

The cross section (Fig. 17-3) illustrates the complexities generated when carrying surface-defined units into the subsurface and demonstrates the importance of differentiating between lithostratigraphic and chronostratigraphic units. The "updip" Olmos is an informal rock-stratigraphic unit which encompasses the coal-bearing sandstone and shale sequence of coastal-plain/deltaic origin, as characterized by Snedden and Kersey (1982). Although specific stratal geometries are difficult to define, the updip Olmos is considered to be a deltaic "highstand systems tract" (Posamentier and Vail, 1988).

Downdip, however, the sequence is substantially thinner. Operators refer to the entire section as the "downdip" Olmos, although both biostratigraphic and sedimentologic data support a greater affinity to the Escondido Formation. The Escondido Formation is largely a shale section, but it contains thick sandstone in the updip areas (e.g., the Fortress Bluff submember of the Escondido Formation). The stratigraphy and facies architecture of the downdip Olmos and Escondido formations as a whole are more comparable with a transgressive systems tract. Thus, the lowermost transgressive surface (TS) of Figure 17-3 is an appropriate boundary separating the two systems tracts of the Olmos-Escondido interval.

The transgressive surfaces separate fluviodeltaic deposits from overlying marine sediments and thus correspond to the coastal "ravinement" surface described by Schwartz (1967) and Swift (1968). These stratigraphic boundaries are produced by the landward migration of the high-energy shoreface during transgressive coastal onlap. Such surfaces can be coincident with sequence-bounding unconformities by virtue of this coastal erosion (Demarest and Kraft, 1987). Because of their distinctive appearance on wireline logs (e.g., sharp sandstone/shale contacts) and association with increases in marine faunal occurrences, these transgressive surfaces are often more easily distinguished in the subsurface than sequence-bounding unconformities (Galloway, 1989).

The presence of a transgressive surface passing downdip beyond the Tom Walsh-Owen Field implies the possible existence of a lowstand shoreline basinward of this location. Indeed, sandstones are positioned further downdip at Encinal, Tri-Bar, AWP, and Bobcat fields (Tyler and Ambrose, 1986). The Olmos B zone could therefore represent a portion of a stillstand shoreline developed during a relative sea-level rise from an earlier position near these more basinward fields and may have originated as a transgressive barrier island flanking a wave-dominated, erosional deltaic headland (Fig. 17-4A). During the ensuing rapid relative sea-level rise, the shoreface would have moved upward and landward, truncating the uppermost portions of the barrier island. Submerged and detached from the shoreline, this sand body developed into a shelf sand shoal (Fig. 17-4B). With less than 3 miles (5 km) of landward migration, the B zone sand body was deposited on the lowermost transgressive surface shown in Figure 17-3.

This proposed depositional history is comparable with the evolutionary sequence proposed for trans-

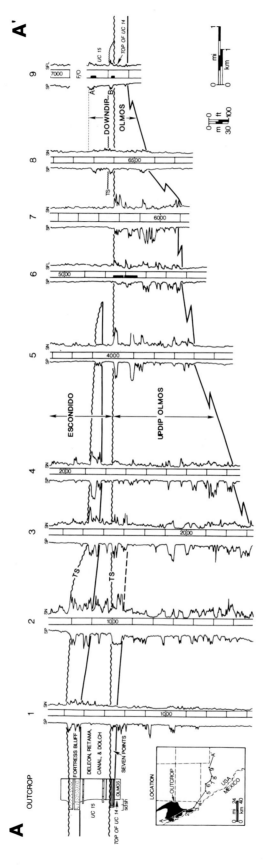

Fig. 17-3. Dip-oriented cross section from the Maastrichtian outcrop belt to Tom Walsh-Owen Field (well No. 9, Tesoro 3 Webb CSL; see Fig. 17-10 for location). Biostratigraphic information is shown in italics. Inset shows cross section location. Locations of cores are shown by black rectangles in depth track of log. Abbrevia- tions: TS, Transgressive surface; F/O, section faulted out; A, Olmos zone A; B, Olmos zone B; SP, spontaneous potential; SN, short normal; SFL, shallow-focus log. Depths are in feet.

NORTHWEST **SOUTHEAST**

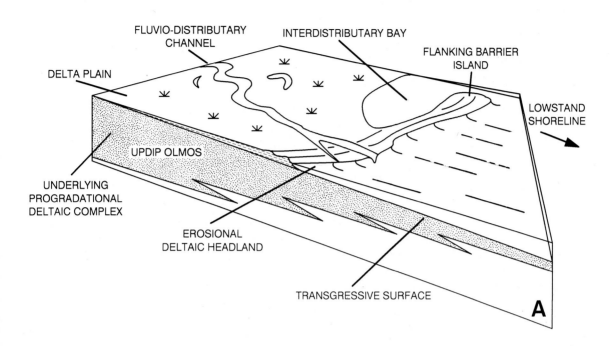

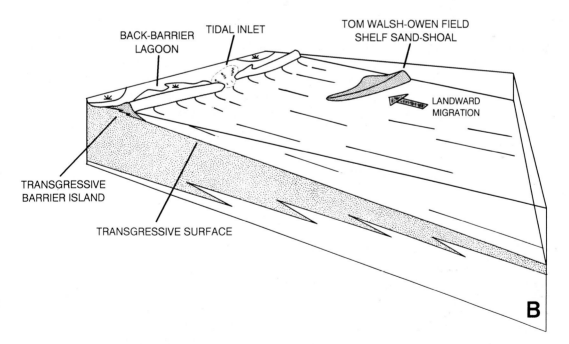

Fig. 17-4. Depositional model for the Olmos B zone sand body: (A) Formation of an erosional deltaic headland and flanking barrier island during a stillstand in the long-term retreat from the lowstand shoreline position. (B) Barrier island is later transgressed and submerged and develops into a shoreline-detached shelf sand shoal.

gressive sand shoals along the modern Louisiana coast (Penland et al., 1988). Their model requires that a transgressive erosion surface also develop on top of the sand body. Unfortunately, cores from the B zone of the Tom Walsh-Owen Field do not extend across this upper contact. A distinctive burrowed and highly glauconitic contact zone does occur at the top of the Olmos in other downdip Olmos fields such as AWP Field in McMullen County.

Sandstones of the A or upper zone of the downdip Olmos, in contrast, are underlain by marine shale and appear to be separated from any transgressive surface. The A zone sandstones are part of a thin but widespread sand sheet developed on the marine shelf, possibly during a minor progradational episode. This episode possibly included deposition of the marginal marine and marine sandstones of the Fortress Bluff submember of the Escondido Formation (Fig. 17-3), but such a correlation cannot be established with the present biostratigraphic control.

Structure

The major structural elements of the Tom Walsh-Owen Field area are a series of normal faults which bound many of the downdip gas fields (Fig. 17-1). These faults interrupt the gentle homoclinal dip of western Webb County and were developed as the result of latest Cretaceous to Early Tertiary displacement along the Albian Stuart City shelf margin which directly underlies the downdip field trend (Snedden and Kersey, 1982). The maximum displacement observed on any of these faults is approximately 100 feet (30 m), with little or no expanded section on the downthrown fault blocks. The limited growth across these faults suggests that the area was tectonically stable during Olmos and Escondido

deposition, in contrast to the extensive growth faulting which characterizes the early Tertiary section.

Entrapment of gas in many of the downdip fields appears to be related to postdepositional normal faulting. Tom Walsh-Owen Field differs, however, because gas production is centered on an oblique trending belt of improved reservoir quality which diverges away from the faults (Fig. 17-1). Hydrocarbon accumulations cross some of the faults, suggesting that the faults are nonsealing and that traps are stratigraphic.

Hydrocarbon Origin and Migration

Little information is available on the origin of hydrocarbons in Webb County. The two most likely sources are the overlying Maastrichtian shale and the underlying basinal shale of the Del Rio Formation (Cenomanian). Simple hydrocarbon analysis of the "lit shale" in adjacent counties suggests that the Maastrichtian has little generative capacity due to low organic carbon concentrations. The underlying black shale of the Del Rio Formation is richer and is therefore a strong candidate for sourcing the gas in this area. Unfortunately, no kerogen analyses of this unit have been published. Uncertainty exists concerning whether the dry gas (GOR of 250,000:1) in the downdip fields is a function of the Del Rio shale kerogen type or is related to the relatively high geothermal gradient in this area ($>2.5°F/100$ ft or $>4.5°C/100$ m).

Information on the timing of hydrocarbon migration is also lacking. Most operators think that migration occurred following early Tertiary burial of the area under several thousand feet ($>1,000$ m) of sediment. Normal faults, which bound many of the downdip fields, could have acted as conduits for funneling hydrocarbons into the Olmos. Leakage of

---

Fig. 17-5. Cores of Olmos Formation shelf deposits: (A) Discrete sandstone beds (DSB lithofacies) and intervening siltstone lithofacies of the Olmos A zone in the Tesoro 3 Webb CSL, 7077 to 7083 feet (2157.1–2158.9 m) core depths. (B) Discrete sandstone beds (DSB lithofacies) and intervening siltstone lithofacies of the Olmos B zone in the Tesoro 3 Webb CSL, 7156 to 7162 feet (2181.1–2183.0 m) core depths. Note improvement in reservoir quality of the intervening siltstone lithofacies, especially from 7156 to 7157 feet (2181.1–2181.5 m). (C) Discrete sandstone bed

(DSB lithofacies) showing a sharp basal contact, low-angle cross-stratification grading upwards into wave ripple bedding (arrow), and a burrowed upper surface, Olmos B zone, Drexel 70-4 Fansler, 8167 feet (2489.3 m) core depth. (D) Discrete sandstone bed (DSB lithofacies) displaying numerous internal truncation surfaces, where laminasets terminate against other sets, Olmos B zone, Tesoro 3 Webb CSL, 7164 feet (2183.6 m) core depth. Arrows show truncation surfaces.

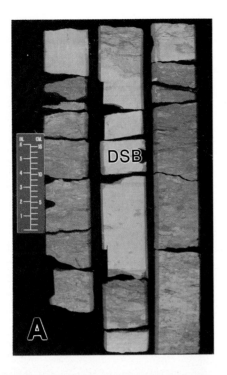

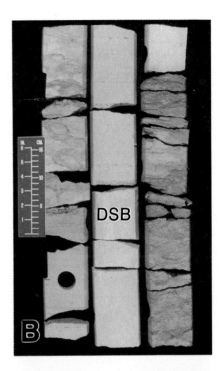

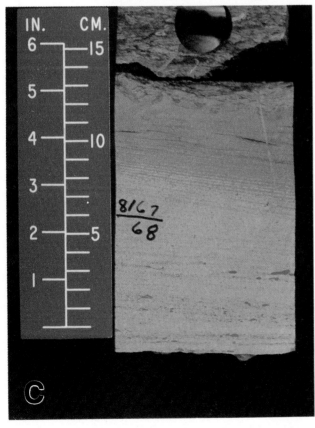

hydrocarbons up fault planes has also been suggested as one explanation for the paucity of gas in the underlying Stuart City reef carbonates (Snedden and Kersey, 1982). Continued burial of the area during the Late Tertiary may have inhibited later fault leakage out of the Olmos, Escondido, and Wilcox reservoirs which contain gas in this area.

Reservoir Characterization

Interpretation of the Olmos A and B zones as shelf and shoreface deposits is based on study of lithofacies characteristics, sand-body geometries, and production behavior. The depositional model developed from these features not only provides an understanding of basin-scale stratigraphic architecture but also helps to explain the differing economic contributions of the two pay horizons, the Olmos A and B zones.

Lithofacies

Cores of the Olmos A and B zones in Tom Walsh-Owen Field were available in the Tesoro No. 3 Webb CSL and Murexco No. 1 Leyendecker wells (Fig. 17-1). In addition, a core of the Olmos B zone was available for comparison purposes from the Drexel 70-4 Fansler well, located 12 miles (19 km) east-southeast of Tom Walsh-Owen Field.

Examination of the cored sequences of both the A and the B zones in all wells indicates that the Olmos comprises two lithofacies: (1) a series of thin (0.07–2.3 ft; 0.02–0.70 m) beds of very fine-grained sandstone termed the "discrete sandstone bed" lithofacies, and (2) siltstone, sandy shale, or, rarely, very fine-grained sandstones which separate the sandstone beds, called the "intervening siltstone" lithofacies (Fig. 17-5A, 17-5B).

Discrete Sandstone Bed Lithofacies. The discrete sandstone beds typically display sharp to clearly erosional basal contacts and either sharp, burrowed, or gradational upper contacts (Fig. 17-5A-C). Individual beds may contain an upward-increasing number of shale laminae, mimicking an upward-fining textural trend, but grading within the coarse fraction is minimal. Petrographic analyses of the Olmos A and B in the Tesoro 3 Webb CSL well indicate that any upward fining is usually limited to a 20-micron average quartz grain-size range (Fig. 17-6). Physical sedimentary structures are primarily planar-horizontal

Fig. 17-6. Summary of petrographic, petrophysical, and sedimentological features in the Olmos A zone (A) and B zones (B) in the Tesoro 3 Webb CSL well. Well location is shown in Figure 17-1. Abbreviations: SD, sand size; FG, fine- grained; VFG, very fine-grained; S, siltstone; SH, shale. Legend of symbols shown below.

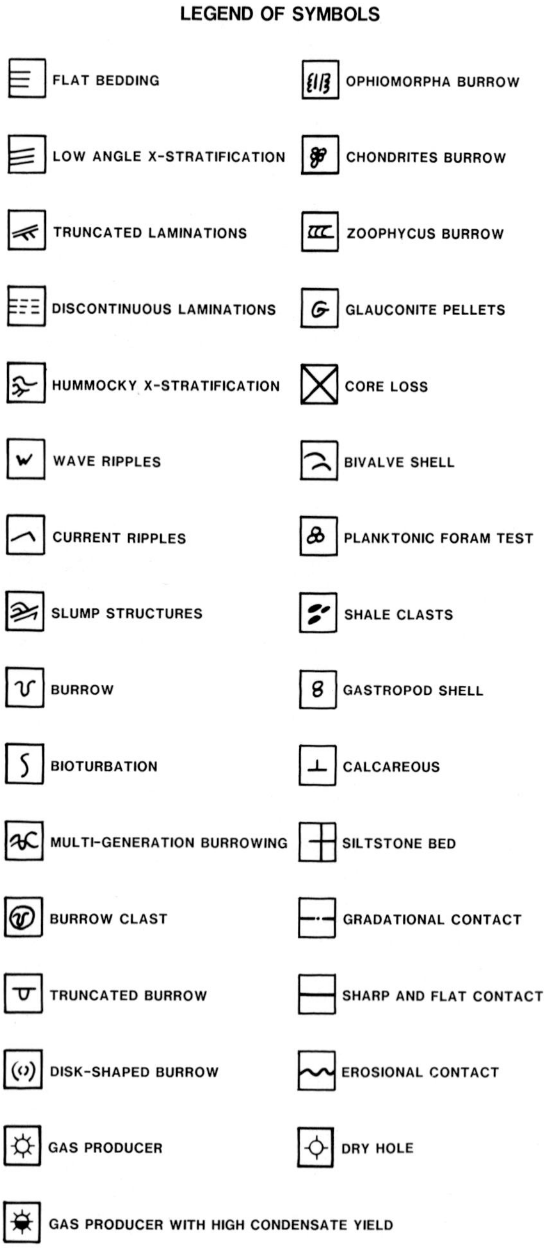

LEGEND OF SYMBOLS

☰	FLAT BEDDING	[I/3]	OPHIOMORPHA BURROW
☰	LOW ANGLE X-STRATIFICATION	8P	CHONDRITES BURROW
⇗	TRUNCATED LAMINATIONS	ccc	ZOOPHYCUS BURROW
≡≡≡	DISCONTINUOUS LAMINATIONS	G	GLAUCONITE PELLETS
⌇	HUMMOCKY X-STRATIFICATION	⊠	CORE LOSS
ᴡ	WAVE RIPPLES	⌒	BIVALVE SHELL
⌢	CURRENT RIPPLES	&	PLANKTONIC FORAM TEST
⩘	SLUMP STRUCTURES	⸮	SHALE CLASTS
℧	BURROW	8	GASTROPOD SHELL
⌠	BIOTURBATION	⊥	CALCAREOUS
ᔆᑕ	MULTI-GENERATION BURROWING	⊞	SILTSTONE BED
ⓥ	BURROW CLAST	⊟	GRADATIONAL CONTACT
⊔	TRUNCATED BURROW	⊟	SHARP AND FLAT CONTACT
(◦)	DISK-SHAPED BURROW	⌇	EROSIONAL CONTACT
☼	GAS PRODUCER	◇	DRY HOLE
☀	GAS PRODUCER WITH HIGH CONDENSATE YIELD		

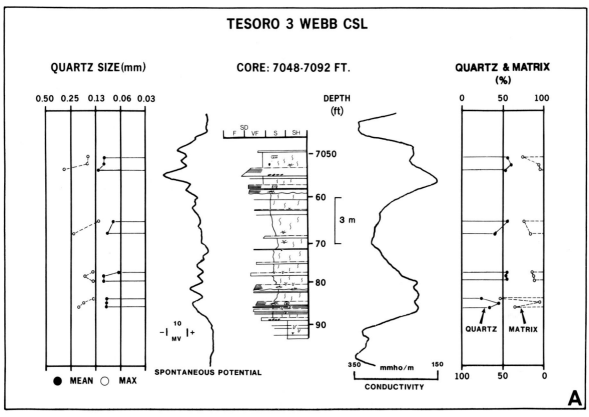

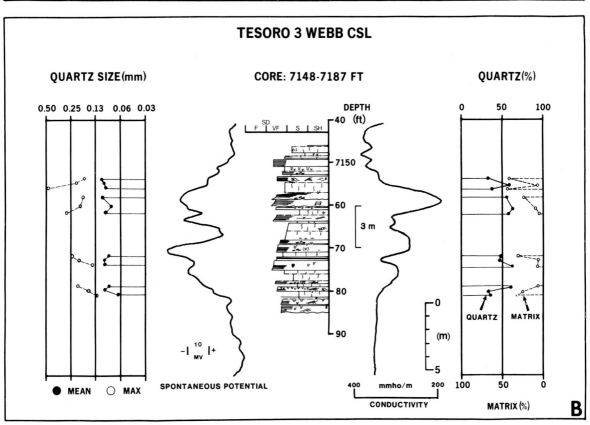

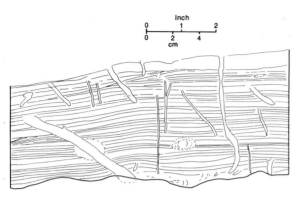

Fig. 17-7. Tracing of an X-ray radiograph of a sand bed in boxcore CB-3, taken in the near-surface modern sediments of the Texas shelf. Sand bed is at 7.1 inches (18.0 cm) depth in a core taken in 85 feet (26 m) of water, about 20 miles (32 km) offshore of Mustang Island, Texas. Unpublished data from Snedden (1985).

(flat) and low-angle cross-stratification (Fig. 17-5C), which locally grades upward into wave and combined-flow (wave and current) cross-lamination (terminology of Harms, 1969). Subtle internal truncation surfaces, where laminasets terminate against other sets dipping at different angles, are common (Fig. 17-5D). Truncation surfaces are often mantled by overlying concordant laminae which display a progressive lateral and downward thickening. Downward fanning of the laminae and truncation surfaces resembles features associated with hummocky cross-stratification as it appears in thin, discrete beds (the "micro-hummocky lenses" of Dott and Bourgeois, 1982). Truncation surfaces probably correspond to the second-order bounding surfaces which separate laminasets in hummocky cross-stratification.

Stratification within the sandstones is remarkably well preserved considering the extensive bioturbation of the siltstones and shales separating these beds. Trace fossils such as small forms of *Ophiomorpha nodosa* are present and may represent a vertically limited but areally extensive invasion by an infaunal organism which is not normally resident in a particular substrate (Pemberton and Frey, 1984).

Flat, rounded shale clasts usually less than two inches (5 cm) in diameter are abundant at the bases of these sandstone beds. Their lithology and internal structure imply a local origin, probably from erosion of the intervening lithofacies. Also common are semilithified burrow casts, probably derived from the same substrate as the shale clasts.

Sedimentary features displayed by individual sandstone beds, including their stratification, verti-

cal textural trends, and associated trace fossils, are consistent with a marine-shelf to shoreface origin. Discrete sandstone beds of the Olmos are remarkably similar to modern sediments of the Texas continental shelf and shoreface (Fig. 17-7). X-ray radiograph tracings from these modern sediments show sharp lower contacts, flat to slightly inclined bedding with numerous internal truncation surfaces, gradational (often burrowed) upper contacts, and an abundant and diverse series of infaunal traces. The discrete nature of the Olmos sandstone beds, particularly their regular interbedding with shalier lithologies, is also comparable with modern Texas shelf sediments. Storm-induced combined flows (combined waves and currents), which were identified as the dominant transport agent on the modern Texas shelf (Snedden et al., 1988), are important in producing hummocky cross-stratification and other storm-generated bedding (Swift et al., 1983; Nøttvedt and Kreisa, 1987) and are probably responsible for the formation of the individual sandstone beds of the downdip Olmos.

Intervening Siltstone Lithofacies. Siltstone, sandy shale, and rare very fine-grained sandstone forming the intervals between the discrete sandstone beds are referred to as the intervening siltstone lithofacies (Fig. 17-5A-B). These are frequently bioturbated (greater than 75% burrowed) and display a diverse assemblage of marine trace fossils, including *Chondrites, Teichichnus, Terebellina,* and *Rhizocorallium.* Intimate interbedding of this lithofacies with the discrete sandstone beds suggests a common paleoenvironment, a marine shelf and shoreface. The presence of an abundant and diverse trace-fossil assemblage indicates that normal marine conditions prevailed here.

However, the high detrital matrix content, lack of physical sedimentary structures, and abundant burrow traces indicate that this lithofacies formed during periods of reduced energy and sediment accumulation. These deposits are similar to the "fairweather" muds of the Texas shelf, which accumulate in the intervening periods between major tropical storms (Snedden, 1985; Snedden and Nummedal, in press). The lack of body fossils in this lithofacies suggests that the infauna was dominated by soft-bodied organisms.

Petrography and Reservoir Properties

Olmos sandstone is very fine grained, with average quartz grain size in the A zone of 107 microns and

Table 17-1. Petrographic analyses of Olmos Sandstone, Tom Walsh-Owen Field.

Unit[1]		Composition [2] (%)					Grain size[3] (microns)	
Zone	Lithofacies	Q	F	R	MX	Other	Mean	Maximum
A	DSB	57	11	6	11	15	107	201
	IV	44	12	7	29	8	89	176
B	DSB	55	12	7	15	11	101	208
	IV	44	12	7	19	18	91	181

[1]Abbreviations: DSB, discrete sandstone beds; IV, intervening siltstone lithofacies.
[2]Normalized volume percentage of Q, monocrystalline quartz; F, feldspar; R, rock fragments (includes polycrystalline quartz and chert); MX, detrital clay matrix; Other, calcite and dolomite cement, glaucony, and other accessory minerals.
[3]Petrographic measurement of the long axes of 300 detrital quartz grains. Abbreviations: Mean, arithmetic mean; Maximum, maximum grain size observed.

101 microns in the B zone (Table 17-1). This is much coarser, on average, than the measured grain size of the intervening siltstone lithofacies (Fig. 17-8). Both the A and B zones show an overall upward coarsening, particularly in the maximum measured grain sizes (Fig. 17-6). Individual beds often show upward-fining, although reverse grading is not unusual in the B zone.

Compositionally, the sandstones are dominated by quartz and detrital clay matrix (Table 17-1). Feldspar and rock fragments typically comprise less than 25% of the whole-rock composition. In the classification of Folk and others (1970), the sandstones typically plot in a tight cluster at the border of the subfeld-sarenite and lithic feldsarenite compartments. Glaucony (glauconitized pellet material) is one of the more important accessory minerals, averaging 2 to 3% of the normalized whole-rock volume. X-ray diffraction and scanning electron microscope observations suggest that the clay minerals are largely detrital in origin and include, in subequal amounts, illite/smectite mixed-layer clays and chlorite.

Detrital clay matrix content of the intervening siltstone lithofacies is high, averaging 29% by volume in the A zone and 19% in the B zone (Table 17-1, Fig. 17-8). Detrital quartz content averages 44% in both zones. Clay minerals are mixed-layer illite/smectite (75% of the < 2 micron fraction), chlorite (17%), and illite (8%). All are detrital in origin.

SEM and petrographic analyses also indicate that observable porosity encompasses 80% primary macroporosity, 10% secondary macroporosity, and 10% primary microporosity (pores less than 1.0 micron). Secondary pores result from dissolution of feldspars, calcite cement, and shale clasts. Microporosity is associated with the detrital clays and glaucony.

The very fine texture and high clay content of these beds have a large impact on their reservoir

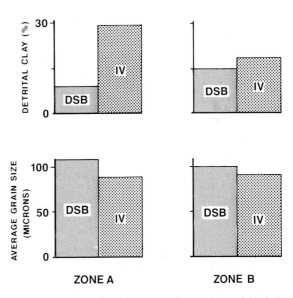

Fig. 17-8. Comparison of petrographic analysis of detrital clay content and average quartz grain size between zones (A and B) and lithofacies (DSB and IV) in the Tesoro 3 Webb CSL cores. Abbreviations: DSB, discrete sand bed lithofacies; IV, intervening siltstone lithofacies.

quality, particularly their permeability (Fig. 17-9). Measured unstressed permeabilities are low, ranging from 0.01 to 8.0 md, with a geometric average of 0.09 md in the A zone and 0.56 md in the B zone of the Murexco well, for example. In situ permeabilities, however, are undoubtedly much lower. Core porosities in the sandstones range from 6 to 16% with a mean of just under 10% in both the A zone and the B zone. Log measurements indicate higher porosities, with typical arithmetic mean porosities of 15% in the better wells. The apparent discrepancy between core and log porosities reported here is due to the fact that core data are not available from wells with better reservoirs.

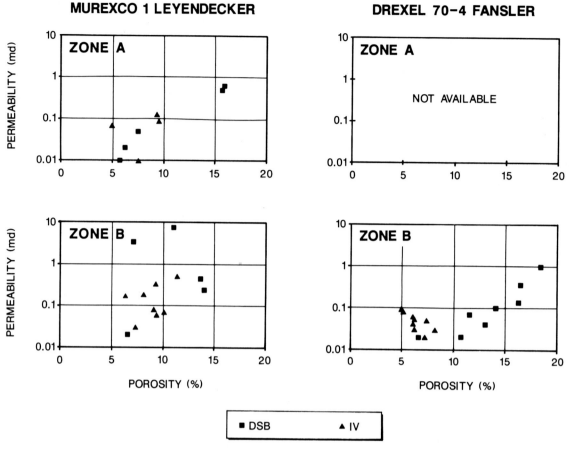

Fig. 17-9. Comparison of measured core permeability and porosity between zones (A and B) and lithofacies (DSB and IV) in the Murexco 1 Leyendecker and Drexel 70-4 Fansler cores. Permeability is horizontal unstressed air permeability measurement in millidarcies. Porosity (percent) is from the summation of fluids method. Abbreviations: DSB, discrete sand bed lithofacies; IV, intervening siltstone lithofacies.

Measured core permeabilities of the intervening siltstone lithofacies in the A zone are low, typically 0.1 md or less. In the B zone, where this lithofacies' texture and clay content approach those of the discrete sandstone beds, the measured permeability can exceed 0.2 md (Fig. 17-9).

Sand-Body Geometry

Discrete sandstone beds and the intervening siltstone lithofacies are both present in the A and B producing horizons of the Olmos at the Tom Walsh-Owen Field. However, the two zones differ considerably in the areal and vertical development of these lithofacies. This provides a significant clue to understanding the differing genesis and productivity of the two zones.

Sandstone thickness maps of the two zones demonstrate significant differences. The Olmos A zone displays a sheet-like geometry, with no well-defined thick or thin trends (Fig. 17-10A). Production trends in the field do not follow the sandstone distribution, as several dry holes are present in the thickest portion of the A zone. This lack of conformity between production patterns and sandstone thickness may be related to the difficulty of mapping net sandstone in a unit with poor wireline-log character and the different fracture treatments in use by the operators. A factor which probably outweighs these pertains to the likelihood that the B zone contributes the majority of the gas recovered in these fields. This view is supported by all the sedimentological

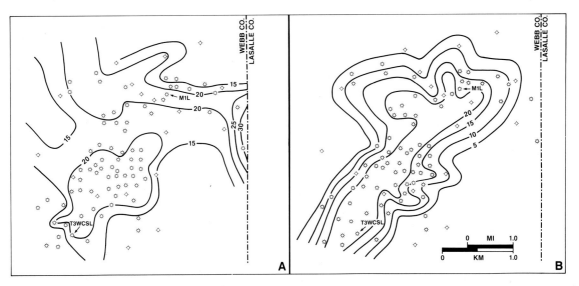

Fig. 17-10. Sandstone isopach maps. (A) Olmos zone A. (B) Olmos zone B. Contour interval is 5 feet (1.5 m). Locations of cored wells are also shown. Abbreviations: T3WCSL, Tesoro 3 Webb CSL; M1L, Murexco 1 Leyendecker.

and petrological indications as well as reservoir quality trends. Unfortunately, gas production from the A and B zones is commingled, which prevents comparison of their individual zone production histories.

The Olmos B zone sandstone isopach map, in contrast, shows the unit as a more distinct sand body, with a well-defined thick area centered on the Tom Walsh-Owen Field (Fig. 17-10B). The sand body trends parallel to structural and depositional strike, with a superimposed dip-trending feature in the northern portion of the map area. Production rates conform well to sandstone thickness.

Facies characteristics of the two zones are also dissimilar (Fig. 17-6). In the A zone, sandstone beds are thin, often widely separated, and rarely incised into other discrete beds (i.e., the degree of amalgamation is low). In the B zone, by contrast, these beds are thicker and more amalgamated (Fig. 17-11, Table 17-2). In addition, the intervening siltstone lithofacies is consistently coarser grained and has less detrital matrix in the B zone than in the A zone (Fig. 17-8). This is evidenced by relatively higher B-zone permeabilities (Fig. 17-9).

The isopach map pattern of the A zone (Fig. 17-10A) suggests that the sandstone beds in this unit are part of a widespread sand sheet that developed on the storm-dominated marine shelf of the downdip Olmos (Fig. 17-4). The wide lateral extent of the A zone throughout the downdip trend is consistent

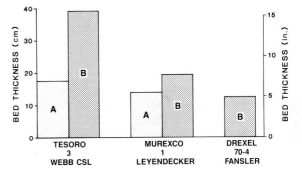

Fig. 17-11. Measured bed-thickness trends in the discrete sandstone bed lithofacies of the three cored wells. Abbreviations: A, Olmos zone A; B, Olmos zone B.

with this interpretation. Similar fine-grained sheet sandstones have been recognized in many other ancient shelf settings (Goldring and Bridges, 1973; Brenchley et al., 1979; Atkinson et al., 1986).

Substantial lateral continuity of individual sandstone beds is not unusual, given the nature of "event" bed stratification and the kinematics of storm flows that have been observed on modern shelves. On the modern Texas shelf and shoreface, tropical storms are known to produce a near-bottom current flow oriented along-shelf and obliquely offshore over large portions of the shelf (Snedden et al., 1988). Evidence suggests that one sand bed, deposited during Hurricane Carla in 1961, was continuous over a

Table 17-2. Thicknesses of Olmos sandstone beds measured in three cored wells, Tom Walsh-Owen Field.

Well	Zone	No. beds measured	Range of thicknesses	Average thickness	Standard deviation
Tesoro 3 Webb CSL	A	19	0.1–2.3 feet (2-71 cm)	0.6 feet (18.1 cm)	0.5 feet (15.5 cm)
	B	15	0.1–2.2 feet (3-66 cm)	1.3 feet (39.2 cm)	0.6 feet (19.0 cm)
Murexco 1 Leyendecker	A	19	0.1–2.0 feet (3-62 cm)	0.5 feet (13.8 cm)	0.5 feet (14.3 cm)
	B	12	0.1–2.0 feet (4-62 cm)	0.6 feet (19.3 cm)	0.6 feet (17.5 cm)
Drexel 70-4 Fansler	B	16	0.1–1.0 feet (4-29 cm)	0.4 feet (12.2 cm)	0.3 feet (8.3 cm)

distance of more than 100 miles (161 km) along depositional strike (Hayes, 1967; Snedden and Nummedal, 1989; Snedden and Nummedal, in press).

Lithology and sand-body geometry of the B zone, in contrast, indicate that the discrete sandstone beds of this unit are components of a distinct morphological element developed on a marine shelf: a shelf sand ridge or sand shoal (Fig. 17-4). Shelf sand ridges/shoals have been recognized in both storm-dominated and tide-dominated shelf settings (Stride et al., 1982; Swift et al., 1986). The dominance of storm-generated stratification in the Olmos B zone points to the general applicability of ridge/shoal models derived from work in modern and ancient storm-dominated shelves (Figueiredo, 1980; Boyles and Scott, 1982; Parker et al., 1982; Tillman and Martinsen, 1984; Swift et al., 1986; Penland et al., 1988).

Work on modern shelf environments and interpreted rock-record equivalents have demonstrated that shelf sand ridges/shoals are common in occurrence but diverse in internal facies architecture and external form. This is partly a result of the degree of posttransgressive reworking. For example, the Holocene midshelf ridges of the New Jersey (USA) shelf are thought to have built upward from a transgressive-shelf sand sheet and to have formed entirely in midshelf water depths by storm-related processes (Rine et al., 1986). Unlike the Olmos B sand body, these ridges occur as multiple topographic features in distinct ridge "fields."

Other ridge/shoal sand bodies are believed to represent transgressed barrier islands, portions of lowstand or stillstand shorelines which were submerged during rapid relative sea-level rise. Ship Shoal, an elongate, coast-parallel, relatively iso-lated feature on the Louisiana continental shelf, is derived from transgressive submergence of a former barrier island arc and deltaic headland (Penland et al., 1988). Evidence of the solitary nature of the Olmos B zone sand body in the Tom Walsh-Owen Field argues for its formation under similar circumstances.

However, disagreement exists concerning how lowstand/stillstand shorelines are transgressed and incorporated into the coastal sedimentary record (Rampino and Sanders, 1982; Swift and Moslow, 1982; Leatherman, 1983). A recent study proposed that preservation of transgressive barriers is facilitated by an alternation of shoreline progradation and retreat which causes the shoreface erosional surface to "step up" and translate landward across the sand bodies (Donselaar, 1989).

The stratigraphic sequence and facies characteristics of a modern shoreface developed under similar conditions are illustrated by the work of Schwartz and others (1981), who cored the seaward side of Topsail Island, North Carolina (USA) (Fig. 17-12). Shoreface deposits are developed in a regressive unit embedded within a larger transgressive sequence. The medium- to fine-grained sands are separated from underlying muddy lagoonal sediments by a sharp but thin erosional surface. These shoreface deposits display numerous similarities to the Olmos B zone of Tom Walsh-Owen Field: thin (0.13-3.9 feet; 0.04–1.2 m) sharp-based beds, common burrow mottling, and an abundance of horizontal (flat) stratification. The paucity of megaripple-scale (dune) cross-stratification, in spite of the availability of sufficiently coarse bedload material, indicates the extremely high velocities of the storm flows which deposited these beds. Like the Olmos B zone, the Topsail Island shoreface sequence shows an overall

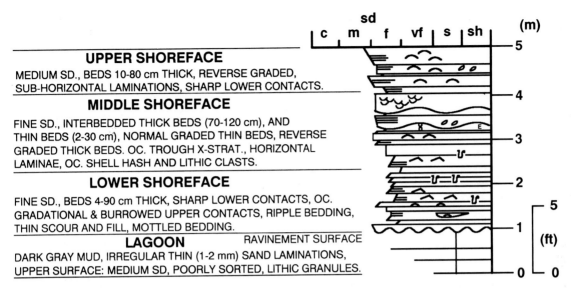

UPPER SHOREFACE
MEDIUM SD., BEDS 10-80 cm THICK, REVERSE GRADED,
SUB-HORIZONTAL LAMINATIONS, SHARP LOWER CONTACTS.

MIDDLE SHOREFACE
FINE SD., INTERBEDDED THICK BEDS (70-120 cm), AND
THIN BEDS (2-30 cm), NORMAL GRADED THIN BEDS, REVERSE
GRADED THICK BEDS. OC. TROUGH X-STRAT., HORIZONTAL
LAMINAE, OC. SHELL HASH AND LITHIC CLASTS.

LOWER SHOREFACE
FINE SD., BEDS 4-90 cm THICK, SHARP LOWER CONTACTS, OC.
GRADATIONAL & BURROWED UPPER CONTACTS, RIPPLE BEDDING,
THIN SCOUR AND FILL, MOTTLED BEDDING.

LAGOON   RAVINEMENT SURFACE
DARK GRAY MUD, IRREGULAR THIN (1-2 mm) SAND LAMINATIONS,
UPPER SURFACE: MEDIUM SD, POORLY SORTED, LITHIC GRANULES.

Fig. 17-12. Sedimentary facies sequence at Topsail Island, North Carolina, USA. Based upon data presented in Schwartz and others (1981). See Figure 17-6 for legend of symbols.

upward coarsening but includes both normally and inversely graded beds. Although questions exist concerning the preservation potential of the shoreface sediments here (Hine and Snyder, 1985), the Topsail Island model provides an example of a sequence formed by a wave-dominated, retreating coastline such as proposed for the Olmos B zone.

Production Characteristics

Determination of sandstone origin is of great importance, but it is just the first step in the reservoir characterization process. Establishing the correct sandstone paleoenvironment can provide a better understanding of the production data. The reverse is also true: analysis of production information also facilitates independent testing of geological predictions based on a given facies model.

Reservoir Heterogeneity

Bed thickness is of particular importance in the producibility of the Olmos at Tom Walsh-Owen Field. However, because of inherent problems with wireline-log measurements (Snedden, 1984) and because modern high-resolution dipmeter logs like SHDT* and FMS* are not available on existing

_____
*Trademarks of Schlumberger.

Olmos wells, reliable measurement of sandstone bed thicknesses can only be made with cored sequences. Measured sandstone bed thicknesses show the differences between the Olmos A and B zones within wells and between wells (Table 17-2, Fig. 17-11). Average bed thickness in the Olmos B zone decreases from the Tesoro to the Murexco well in the Tom Walsh Field and decreases even more in the Drexel well, located 12 miles (19 km) east-southeast of the field. The isopach map patterns show these thickness differences, as the Tesoro well is located in the thickest portion of the B zone, while the Murexco well is on the margin of that zone (arrows, Fig. 17-10B).

The A zone, as expected, is consistently thinner than the B zone (Fig. 17-11). This is true in both lithofacies, in wells which penetrate both thick and thin areas of the reservoir. The A zone is a more "layered" reservoir than the B zone. This contrast results from the different origins of the two units, with the A zone representing a shelf sheet sand and the B zone a sand shoal (transgressed barrier).

Differences in sandstone bed thickness parallel variations in petrophysical characteristics, reservoir quality, and gas production. The spontaneous potential (SP) response of the Olmos is variable but displays the largest deflection in zones with the thickest discrete sandstone beds, the most amalgamation, and the fewest intervening shale and siltstone beds. Compare, for example, the difference in SP log

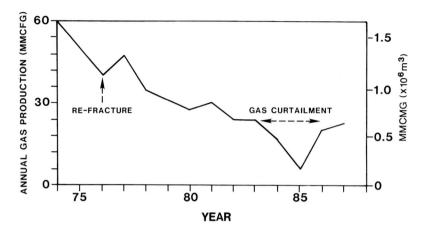

Fig. 17-13. Annual gas production in the Tesoro 3 Webb CSL well, 1974 to 1987.

response between the A and B zones in the Tesoro well (Fig. 17-6). The trend in log-derived measurements of porosity parallels the SP response, implying that the SP can be used as a proxy for porosity. Since the areal differences in gas production rates reflect these porosity differences, it is proposed that bed thickness and degree of bedding amalgamation are the best indicators of gas production rates. For example, the Tesoro well, with the thickest beds, produced 100 MMCFG (2.8×10^6 m³) in its first two full years of production (prior to restimulation). The Murexco well, with thinner beds, produced only 14 MMCFG (4.0×10^5 m³) in its first two full years. The Drexel well, with the lowest bed thickness, was a dry hole.

Reservoir Continuity

Although the discrete sandstone beds are quite thin, data imply that individual beds may have substantial lateral continuity. This lateral continuity is a function of the dynamics and kinematics of sediment transport in the marine environment.

Early in the development of the field, operators recognized similar reservoir pressures in wells on 40-acre (16.2-ha.) spacing, particularly in the B zone. It was concluded that this was due to the reservoir being a continuous horizontal "layer." This conclusion was later substantiated by production-decline curve analysis, which showed a distinct flattening-out of the production rates for up to 10 years (Fig. 17-13). Most workers interpret this to mean that the wells are not draining a limited radius created by artificial fracturing but a well-developed, laterally continuous reservoir layer or series of layers. The poor vertical continuity of the beds as

revealed in cores is indisputable. Thus, the decline curves and similar reservoir pressures can only be explained by the presence of a reservoir with considerable lateral continuity.

Detailed correlation of the cored wells also supports this inference (Fig. 17-14). Using a shaly horizon in the middle of the B zone as a datum, it is possible to trace several of the discrete sandstones from the Tesoro well to the Murexco well, a distance of more than 4 miles (6.4 km), with some confidence. The slight inclination of the dashed lines can be attributed to minor differences in sediment accumulation rates over this area. This correlation demonstrates that the shelf-sand shoal proposed for the Olmos B zone is composed of a series of storm "event" beds with considerable lateral continuity. This would be expected if this shoal originated as a barrier-island shoreface.

Although the unit well spacings in the field have been designated at 40 acres (16.2 ha.), many operators think that the wells have at least 600 to 800 feet (180–240 m) or 80 acres (32.4 ha.) of drainage. This may be pessimistic, if one accepts the interpretation of thin beds which are continuous over a distance of 4 miles (6.4 km).

Well Completion Procedures

The poor vertical flow continuity between discrete sandstone beds is an important consideration with regard to the techniques used to establish production. Completion procedures were not developed by the detailed study of the cores, as few wells in this area are cored, but were formulated in an iterative fashion. Most wells are stimulated by fracture treatment prior to initial production and often later in the

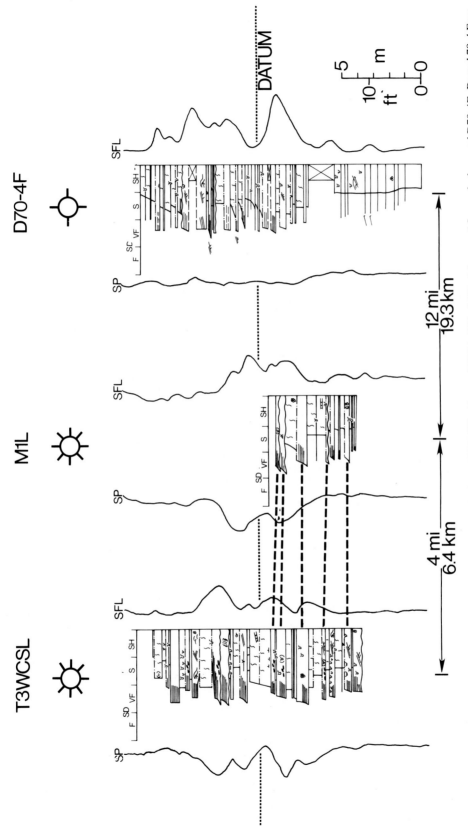

Fig. 17-14. Correlation of Olmos B zone in the cored wells. Abbreviations: T3WCSL, Tesoro 3 Webb CSL; M1L, Murexco 1 Leyendecker; and D70-4F, Drexel 70-4 Fansler. Well locations are shown in Figure 17-1. Abbreviations and symbols are the same as in Figure 17-6.

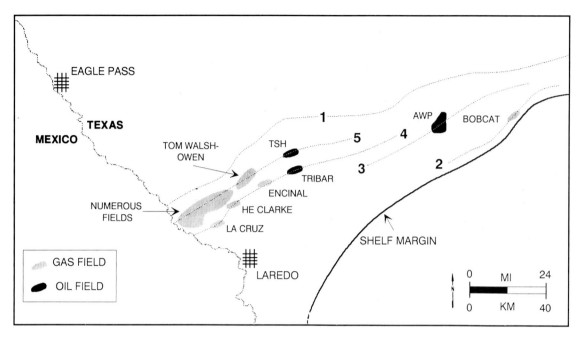

Fig. 17-15. Regional distribution of downdip Olmos fields and possible shoreline positions. Rapid basinward shift from updip Olmos shoreline (1) to lowstand position (2) is followed by successively more landward stillstand positions at 3, 4, and 5. Only portions of each shoreline position are shown due to uncertainties in locating these outside of the areas of present well control. Shelf margin location is based on observed change in seismic character and, in some areas, coincidence with underlying carbonate shelf-margin trend.

well's lifespan. Typical hydrofracture treatments involve 350,000 pounds (1.6 × 10⁵ kg) of 20 to 40 Ottawa sand and 65,000 gallons (2.5 × 10⁵ L) of gel water. In most cases, gas flow rates increase substantially after fracturing (Fig. 17-13). The relative success of these treatments results from fractures allowing greater communication between the various reservoir "layers" (the thin discrete sandstone beds). Operators have also found that, following casing and stimulation of the well, reperforation (>4 shots/ft; >13 shots/m) increases the production rate by allowing more of the thin discrete sandstone beds to flow gas into the wellbore.

Well Performance

Like the Olmos in most of Webb County, the reservoirs at Tom Walsh-Owen Field produce dry gas with relatively little condensate (specific gravity is 0.7). Gas-oil ratios average 250,000:1, and recovery efficiencies are 50%. A typical well produces 690 MMCFG (2.0 × 10⁷ m³) at an average sustained rate of 100 MCFGPD (2.8 × 10³ m³/D).

Exploration and Production Strategy

A model for finding other gas accumulations in this trend, and in other trends with similar depositional characteristics, can be developed from this case study of the Tom Walsh-Owen Field. A key factor is establishing the proper stratigraphic correlations and recognizing the existence of separate depositional systems tracts. The architecture of the Olmos interval was discovered only after detailed biostratigraphic and sedimentological analysis of cuttings, cores, and outcrops (Fig. 17-3). Simple well-log correlations did not indicate that the downdip Olmos is younger than the updip Olmos and that the two are separated by a regional transgressive surface. The retrogradational nature of the Upper Olmos and Escondido interval also was not apparent. The presence of marine shale in outcrops of the lower Escondido Formation confirms the transgressive stratigraphy.

With the recognition that the transgressive surface at the base of the Olmos B zone continues beyond the Tom Walsh-Owen Field area, the possibility of reservoir sand bodies being located in downdip shoreline positions has to be considered (Fig. 17-15). Indeed, drilling in a basinward direction has resulted in field discoveries such as at Tribar, AWP, and Bobcat fields (Tyler and Ambrose, 1986). The sand bodies here are elongate, roughly shore-parallel, and detached from any associated coastal sedimentary facies (Travis, 1985). Many of the fields show a preferred alignment in what is interpreted as a series of successive stillstand shoreline positions located landward of the lowstand shoreline near the contemporaneous shelf margin. Drilling along depositional strike with these discoveries holds some promise, particularly when there are additional indications of sand body occurrence (e.g., hydrocarbon shows in deeper but abandoned tests, as with the Tom Walsh-Owen and AWP fields).

Further exploration in the Olmos play as well as in other similar plays should recognize the potential for stratigraphic traps developed seaward of the traditional targets in deltaic/shoreline clastics. This is particularly important where there is evidence of a well-developed transgressive stratigraphy.

Production data indicate that good lateral continuity of the reservoir sandstone allows use of larger well spacings (80 ac; 32.4 ha.) than originally used. Careful production testing should be conducted to determine the minimum number of wells needed to drain the reservoirs efficiently. Artificial stimulation and high-density perforation of pay zones help to increase wellbore access to these thin-bedded sandstones. Isopach mapping at smaller contour intervals (<10 ft; 3m) allows distinction between shelf-sand-shoal deposits, where thin beds are often amalgamated into thicker bedsets with better reservoir properties, and shelf-sand-sheet deposits. Efficient field development always begins with detailed reservoir description and accurate paleoenvironmental determination.

Conclusions

Gas reservoirs of the Olmos Formation at relatively shallow depths at Tom Walsh-Owen Field are developed in thin, storm-deposited shelf and shoreface sandstones with individual thicknesses averaging less than 3.2 ft (1.0 m). The discrete sandstones are components of a shelf-sand sheet in the upper producing horizon, A zone, and a transgressed barrier/sand shoal in the lower horizon, B zone. Thicker, amalgamated beds in the B zone and their better reservoir properties make them the most prolific gas producers in this field and in the entire downdip Olmos trend of Webb County, Texas.

The Olmos A and B zone reservoirs are complex in that they are composites of many thin layers (as many as 20 in any one zone). Even though the vertical continuity of these beds is quite limited, production trends indicate that these layers have good lateral continuity. Individual sandstone beds may have a lateral continuity of up to 4 miles (6.4 km), far exceeding the spacing of individual wells. Well spacings of 80 acres (32.4 ha.) or more are thus quite reasonable in the center of the field, although 40-acre (16.2-ha.) spacings are employed elsewhere in the downdip trend of fields. Artificial stimulation by hydrofracture and high-density perforation (>4 shots/ft; >13 shots/m) of the wellbores have facilitated the drainage of the thin sandstone beds and increased gas production.

The trend is an active exploratory area. In Tom Walsh-Owen Field, wells are economic although even after fracture treatment they produce at very low rates. The trend is economic due to shallow drilling depths and long production periods. The latter result from the good lateral continuity and, thus, wide drainage of thin shelf and shoreface sandstones.

Acknowledgements. The authors appreciate the technical assistance of J.C. Cooke and D.G. Kersey. We are grateful to D.C. Caraway, H.W. Posamentier, and the editors of this volume for manuscript review. We also benefitted from discussions with J.S. Janks, R.D. Kreisa, and D. Imperato. Thanks to Mobil Oil Corporation and HarCor Exploration for permission to publish this chapter.

Reservoir Summary

Field : Tom Walsh-Owen
Location: Webb County, Texas
Operators: HarCor, Tesoro, Coastal, H.N.G., Shar-Alan, Stampede, Peninsula
Discovery: 1970
Basin: Rio Grande embayment, Gulf of Mexico
Tectonic/Regional Paleosetting: Subsiding aulacogen/passive margin
Geologic Structure: Homoclinal dip
Trap Type: Stratigraphic pinch-out
Reservoir Drive Mechanism: Depletion
  • **Original Reservoir Pressure:** 3,000 psi (2.1×10^4 kPa) at 6,350 feet (1,935 m) subsea
Reservoir Rocks
  • **Age:** Late Cretaceous, Maastrichtian
  • **Stratigraphic Unit:** Olmos Formation
  • **Lithology:** Very fine-grained sandstone, lithic feldsarenites
  • **Depositional Environments:** Marine shelf
  • **Productive Facies:** Shelf sandstones (shoal and sand sheet)
  • **Petrophysics**
    • **Porosity Type:** Total = 10% (9% primary and 1% secondary inter- and intragranular) All values from thin-
                section measurements
    • **ϕ:** Average 15%, range 8 to 23%, cutoff 12% (logs)
    • **k:** Average 0.4 md, range 0.01 to 8.0 md, cutoff 0.1 md (unstressed horizontal air permeabilities measured on
         conventional cores)
    • **S_w:** Average 50%, range 0 to 100%, cutoff 72% (logs)
Reservoir Dimensions
  • **Depth:** 7,200 feet (2,195 m)
  • **Areal Dimensions:** 11 miles (18 km) along strike and 3 miles (5 km) along dip direction
  • **Productive Area:** 21,120 acres (8,555 ha.)
  • **Number of Reservoirs:** 2 (A,B)
  • **Hydrocarbon Column Height:** 35 feet (10.7 m)
  • **Fluid Contacts:** No apparent water level present
  • **Number of Pay Zones:** 2 (A,B), with majority of pay in lower zone (B)
  • **Gross Sandstone Thickness:** 35 feet (10.7 m)
  • **Net Sandstone Thickness:** 20 feet (6.1 m)
  • **Net/Gross:** 0.57
Source Rocks:
  • **Lithology & Stratigraphic Unit:** Basinal shale, Del Rio Formation (Cenomanian)
  • **Time of Hydrocarbon Maturation:** Tertiary
  • **Time of Trap Formation:** Late Cretaceous (at deposition)
  • **Time of Migration:** Tertiary
Hydrocarbons
  • **Type:** Gas, with very small amounts of condensate
  • **GOR:** 250,000:1
Volumetrics
  • **Specific Gravity:** 0.7
  • **In-Place:** 228 BCFG (6.5×10^9 m³)
  • **Cumulative Production:** 68.8 BCFG (1.9×10^9 m³)
  • **Ultimate Recovery:** 115 BCFG (3.3×10^9 m³)
  • **Recovery Efficiency:** 51%
Wells
  • **Spacing:** 1,100 feet (335 m); 40 acre (16.2 ha.) Drainage may exceed this in some wells.
  • **Total:** 161
  • **Dry Holes:** 30

Typical Well Production:
- **Average Daily:** 96 MCFG (2.7 $\times$ 10^3 m^3)
- **Cumulative:** 690 MMCFG (2.0 $\times$ 10^7 m^3)

Other: Wells are artificially fractured with 350,000 pounds (1.6 $\times$ 10^5 kg) of 20 to 40 Ottawa sand, and 65,000 gallons (2.5 $\times$ 10^5 L) of gel water. Reperforated with 4 shots per foot (13 shots/m).

References

Atkinson, C.D., Goeston, M.J. B.G., Speksnijder, A., and van der Vlugt, W., 1986, Storm-generated sandstone in the Miocene Miri Formation, Seria Field, Brunei (N.W. Borneo), *in* Knight, R.J. and McLean, J.R., eds., Shelf Sands and Sandstones: Canadian Society of Petroleum Geologists Memoir 11, p. 213–240.

Boyles, J.M., and Scott, A.J., 1982, A model for migrating shelf bar sandstones in Upper Mancos Shale (Campanian), northwest Colorado: American Association of Petroleum Geologists Bulletin, v. 66, p. 491–508.

Brenchley, P.J., Romano, M., and Stanistreet, I.G., 1979, A storm surge origin for sandstone beds in an epicontinental platform sequence, Ordovician, Norway: Sedimentary Geology, v. 33, p. 783–815.

Cooper, J.D., 1970, Stratigraphy and paleontology of Escondido Formation (Upper Cretaceous), Maverick County, Texas and northern Mexico [Ph.D. thesis]: Austin, Texas, University of Texas, 288 p.

Demarest, J.M., and Kraft, J.C., 1987, Stratigraphic record of Quaternary sea-levels: Implications for more ancient strata, *in* Nummedal, D., Pilkey, O.H., and Howard, J.D., eds., Sea-Level Fluctuation and Coastal Evolution: Society of Economic Paleontologists and Mineralogists Special Publication 41, p. 223–240.

Donselaar, M.E., 1989, The Cliff House Sandstone, San Juan Basin, New Mexico: Model for the stacking of "transgressive" barrier complexes: Journal of Sedimentary Petrology, v. 59, p. 13–28.

Dott, R.H., and Bourgeois, J.D., 1982, Hummocky stratification: Significance of its variable bedding sequences: Geological Society of America Bulletin, v. 93, p. 663–680.

Dumble, E.T., 1892, Notes on the geology of the middle Rio Grande: Geological Society of America Bulletin, v. 3, p. 219–230.

Figueiredo, A.G., 1980, Response of water column to strong wind-forcing, southern Brazilian inner shelf—implications for sand ridge formation: Marine Geology, v. 35, p. 367–376.

Folk, R.L., Andrews, P.B., and Lewis, D.W., 1970, Detrital sedimentary rock classification and nomenclature for use in New Zealand: New Zealand Journal of Geology and Geophysics, v. 13, p. 937–68.

Galloway, W.E., 1989, Genetic stratigraphic sequences in basin analysis, I: Architecture and genesis of flooding-surface bounded depositional units: American Association of Petroleum Geologists Bulletin, v. 73, p. 125–143.

Goldring, R. and Bridges, P., 1973, Sublittoral sheet sandstones: Journal of Sedimentary Petrology, v. 43, p. 736–747.

Govett, R., 1983, Tom Walsh Field, Webb County, Texas: Corpus Christi Geological Society Bulletin, v. 3, p. 19–21.

Harms, J.C., 1969, Hydraulic significance of some sand ripples: Geological Society of America Bulletin, v. 80, p. 363–396.

Hayes, M.O., 1967, Hurricanes as geological agents: Case studies of Hurricane Carla, 1961 and Cindy, 1963: University of Texas Bureau of Economic Geology Report of Investigations 61, 56 p.

Hine, A.C., and Snyder, S.W., 1985, Coastal lithosome preservation: Evidence from the shoreface and inner continental shelf off Bogue Banks, North Carolina: Marine Geology, v. 63, p. 307–330.

Hinte, J.E. van, 1976, A Cretaceous time scale: American Association of Petroleum Geologists Bulletin, v. 60, p. 498–516.

Leatherman, S.P., 1983, Barrier island evolution in response to sea level rise—Discussion: Journal of Sedimentary Petrology, v. 53, p. 1026–1031.

Nøttvedt, A. and Kreisa, R.D., 1987, Model for the combined-flow origin of hummocky cross-stratification: Geology, v. 15, p. 357–361.

Parker, G., Langfredi, N.W., and Swift, D.J.P., 1982, Seafloor response to flow in a southern hemisphere sand-ridge field—Argentine inner shelf: Sedimentary Geology, v. 33, p. 195–216.

Pemberton, S.G., and Frey, R.W., 1984, Ichnology of a storm-influenced shallow marine sequence, Cardium Formation (Upper Cretaceous) at Seebe, Alberta, *in* Stott, D.F. and Glass, D.J., eds., The Mesozoic of Middle North America: Canadian Society of Petroleum Geologists Memoir 9, p. 281–304.

Penland, S., Boyd, R., and Suter, J.R., 1988, Transgressive depositional systems of the Mississippi delta plain: A model for barrier shoreline and shelf sand development: Journal of Sedimentary Petrology, v. 58, p. 932–949.

Posamentier, H., and Vail, P.R., 1988, Eustatic controls on clastic deposition, II—Sequence and systems tract models, *in* Wilgus, C.K., Hastings, B.S., Kendall, C.G.St.C., Posamentier, H.W., Ross, C.A., and Van

Wagoner, J.C., eds., Sea Level Changes – An Integrated Approach: Society of Economic Paleontologists and Mineralogists Special Publication 42, p. 125–154.

Rampino, M.R., and Sanders, J.E., 1982, Barrier island evolution in response to sea level rise – Reply: Journal of Sedimentary Petrology, v. 53, p. 1031–1034.

Rine, J.M., Tillman, R.W., Stubblefield, W.L., and Swift, D.J.P., 1986, Lithostratigraphy of Holocene sand ridges from the nearshore and middle continental shelf of New Jersey, U.S.A., in Moslow, T.F. and Rhodes, E.G., eds., Modern and Ancient Shelf Clastics – A Core Workshop: Society of Economic Paleontologists and Mineralogists Core Workshop 9, p. 1–72.

Schwartz, M.L., 1967, The Bruun theory of sea-level rise as a cause of shore erosion: Journal of Geology, v. 75, p. 76–91.

Schwartz, R.K., Hobson, R., and Musialowski, F.R., 1981, Subsurface facies of a modern barrier island and relationship to the active nearshore profile: Northeastern Geology, v. 3, p. 283–296.

Snedden, J.W., 1984, Validity of the use of the spontaneous potential curve shape in the interpretation of sandstone depositional environments: Gulf Coast Association of Geological Societies Transactions, v. 34, p. 255–264.

Snedden, J.W., 1985, Origin and sedimentary characteristics of discrete sand beds in modern sediments of the Central Texas continental shelf [Ph.D. thesis]: Baton Rouge, Louisiana, Louisiana State University, 247 p.

Snedden, J.W., and Kersey, D.G., 1982, Depositional environments and gas production trends, Olmos Sandstone, Upper Cretaceous, Webb County, Texas: Gulf Coast Association of Geological Societies Transactions, v. 32, p. 497–518.

Snedden, J.W., Nummedal, D., and Amos, A.F., 1988, Storm and fairweather combined-flow on the Central Texas continental shelf: Journal of Sedimentary Petrology, v. 58, p. 580–595.

Snedden, J.W., and Nummedal, D., 1989, Sand transport kinematics on the Texas shelf during Hurricane Carla, September 1961, in Morton, R.A. and Nummedal, D., eds., Shelf Sedimentation, Shelf Sequences, and Related Hydrocarbon Accumulation: Proceedings of the Seventh Annual Research Conference of the Gulf Coast Section of the Society Economic Paleontologists and Mineralogists, p. 63–76.

Snedden, J.W., and Nummedal, D., in press, Origin and geometry of storm deposited sand beds in modern sediments of the Texas continental shelf, in Swift, D.J.P., Tillman, R.W., and Oertel, G.F., eds., Geometry of Shelf Sands and Sandstones: International Association of Sedimentologists Special Publication, 25 p.

Stride, A.H., Belderson, R.H., Kenyon, N.H., and Johnson, M.A., 1982, Offshore tidal deposits – Sand sheet and sand bank facies, in Stride, A.H., ed., Offshore Tidal Sands – Processes and Deposits: New York, Chapman and Hall, p. 95–125.

Swift, D.J.P., 1968, Coastal erosion and transgressive stratigraphy: Journal of Geology, v. 76, p. 444–456.

Swift, D.J.P., and Moslow, T.F., 1982, Holocene transgression in south-central Long Island, New York – Discussion: Journal of Sedimentary Petrology, v. 52, p. 1014–1019.

Swift, D.J.P., Figueiredo, A.G., Freeman, G. L., and Oertel, G.F., 1983, Hummocky cross-stratification and mega- ripples A geological double standard?: Journal of Sedimentary Petrology, v. 53, p. 1295–1318.

Swift, D.J.P., Thorne, J., and Oertel, G.F., 1986, Fluid processes and sea-floor response on a modern storm-dominated shelf – Middle Atlantic shelf of north America – Part II. Response of the shelf floor, in Knight, R.J. and McLean, J.R., eds., Shelf Sands and Sandstones: Canadian Society of Petroleum Geologists Memoir 11, p. 191–211.

Tillman, R.W., 1985, A spectrum of shelf sands and sandstones, in Tillman, R.W., Swift, D.J.P., and Walker, R.G., eds., Shelf Sands and Sandstone Reservoirs: Society of Economic Paleontologists and Mineralogists Short Course Notes 13, p. 1–46.

Tillman, R.W., and Martinsen, R.S., 1984, The Shannon shelf-ridge sandstone complex, Salt Creek Anticline area, Powder River basin, Wyoming, in Tillman, R.W. and Siemers, C.T., eds., Siliciclastic Shelf Sediments: Society of Economic Paleontologists and Mineralogists Special Publication 34, p. 85–142.

Travis, R.P., 1985, A.W.P. Field, McMullen County, Texas: South Texas Geological Society Bulletin, v. 25, p. 37–39.

Tyler, N., and Ambrose, W.A., 1986, Depositional systems and oil and gas plays in the Cretaceous Olmos Formation, South Texas: University of Texas Bureau of Economic Geology Report of Investigations 152, 42 p.

Weise, B.R., 1980, Wave-dominated delta systems of the Upper Cretaceous San Miguel Formation, Maverick Basin, Texas: University of Texas Bureau of Economic Geology Report of Investigations 107, 42 p.

Key Words

Tom Walsh-Owen Field, Texas, Rio Grande embayment, Olmos Formation, Late Cretaceous, Maastrichtian, shelf, stratigraphic trap, shallow-marine sandstone, shelf sandstone, shoreface sandstone, reservoir continuity.

18

Reservoir Characteristics of Nearshore and Shelf Sandstones in the Jurassic Smackover Formation, Thomasville Field, Mississippi

Roger D. Shew and Mark M. Garner

Shell Development Company, Houston, Texas 77001; Shell Offshore Inc., New Orleans, Louisiana 70161

Introduction

General

The Jurassic Smackover trend throughout the U. S. Gulf Coast is typified by production from a high-energy carbonate facies, most commonly at depths of less than 15,000 feet (4,575 m). However, Thomasville Field and four other nearby fields in the trend produce sour gas (containing hydrogen sulfide) with sustained high production rates from low porosity and permeability sandstones that are interbedded with tight nonreservoir carbonates. These fields are located in central Mississippi in the informally designated sour gas trend. Thomasville Field was the first discovered and largest of these fields, containing more than one-half of the more than one trillion cubic feet (2.8 × 10¹⁰ m³) of gas in the sour gas trend. Thomasville was developed over a span of more than 10 years, requiring both detailed geological interpretations and the development of new engineering technology.

New engineering technology was needed to drill, evaluate, and produce the sour gas from a hostile and very variable subsurface environment. The main problems were the great depths (> 19,300 ft; > 5,880 m), high geopressures (> 0.88 psi/ft; 19.9 kPa/m), high temperatures (> 350°F, > 177°C), and unusual gas mixtures that include large amounts of hydrogen sulfide (H_2S > 25%) and lesser amounts of carbon dioxide ($CO_2 \leq$ 9%). Structural, sedimentologic, and diagenetic studies were neces-

sary to define the distribution of reservoir-quality sandstones. Field and well productivities are governed by the sandstone distribution and environments of deposition (upward-shoaling shelf to nearshore sediments) within a mixed carbonate/siliciclastic setting, and by a complex diagenetic sequence which strongly overprints but does not obscure the original depositional control. This chapter will discuss two areas of study which have been essential to the discovery, delineation, and production of Thomasville Field. These include (1) a brief review of the seismic data which were used to define the structural and stratigraphic characteristics, and (2) geological and petrophysical studies used to determine the depositional and diagenetic controls on reservoir quality and distribution. The subsurface and surface engineering techniques developed for safe drilling and production from these hostile environments are beyond the scope of this chapter.

Development and Production History

The discovery of Thomasville Field in 1969 followed two other significant updip discoveries: the Pisgah anticline (now referred to as Goshen Springs and South Pisgah fields; see Studlick et al., this volume) by Chevron, Inc. in 1967 and Pelahatchie Field by Shell Oil Company in 1968 (Fig. 18-1). In addition to finding major gas reserves, these discoveries were significant in their use of improved reflection-seismic data to recognize previously unresolved deep (> 15,000 ft; > 4,570 m) structures.

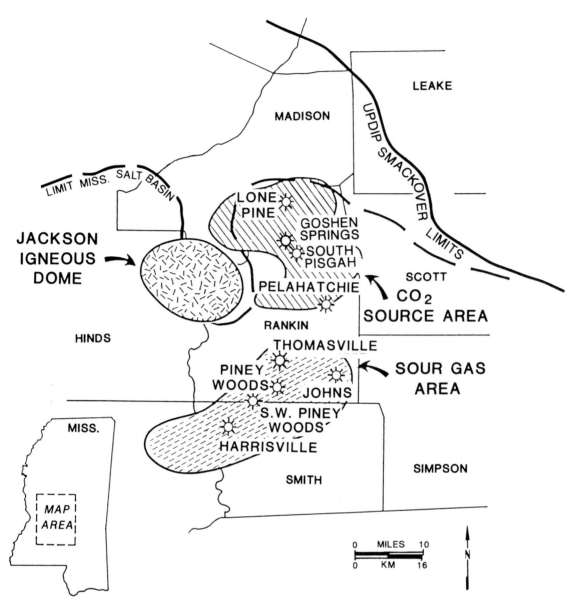

Fig. 18-1. Thomasville Field is located within the sour gas trend in central Mississippi and is one of five fields that are sour-gas productive. Other structural features in the area include the carbon dioxide source area (comprising numerous salt-generated domes and anticlines) and the Jackson Igneous Dome.

The Shell Garrett No. 1, the discovery well for Thomasville Field, was drilled based on the seismic interpretation of closure at two Jurassic levels, the Oxfordian Smackover and Norphlet intervals. Although the well was planned to test both of these objectives, it could be drilled only to a total depth of 20,451 feet (6,233 m), still within the Smackover Formation, the shallower of the two. Two other Thomasville Field wells have since penetrated the Norphlet and found it to be water-bearing. Within the Smackover, fifteen sour gas-bearing ($CH_4-55\%$, $H_2S-35\%$, $CO_2-9\%$) sandstone layers interbedded with tight carbonates were penetrated in the Garrett No. 1. A total of 203 NFG (62 m) was evaluated in beds ranging in thickness from 4 to 37 feet (1.2–11.3 m). The productive interval is from 19,300 feet (5,880 m) subsea to a water level at approximately 20,000 feet

(6,100 m) subsea, a hydrocarbon column of 700 feet (210 m).

In addition to the corrosive gas mixture and great depths, the reservoir is characterized by high temperatures ($>350°F$; $>177°C$) and hard geopressures (>0.88 psi/ft; 19.9 kPa/m). The geopressures, although problematic to drill, are beneficial to production; high flow rates are possible even in these poor reservoirs, where porosity and permeability average 7.0% and 0.35 md unstressed, respectively. For example, the Shell Burch No. 1, although not representative of the typical field well (5–10 MMCFGPD; 1.4–2.8×10^5 m³/D), has been producing at a rate of approximately 20 MMCFGPD (5.7×10^5 m³/D) since 1972 and has produced more than 100 BCFG (2.8×10^9 m³).

It took more than a decade to develop the field, primarily because of the engineering challenge of safely and economically exploiting these deep reservoirs in a high-pressure, high-temperature environment with corrosive gas mixtures. Field limits have been defined by 14 wells, including 4 dry holes and two redrills. The producing wells (four Shell Oil and four Pursue Energy) are located on 1,280-acre (518-ha.) spacing.

Regional Data

Regional and Tectonic Setting

Thomasville Field is located in the Mississippi interior salt basin (Fig. 18-2), which was formed during the initial rifting of the Gulf of Mexico (Pindell and Dewey, 1982). The Pickens-Gilbertown fault system on the north and the South Mississippi and Wiggins uplifts on the south bound the basin (Dinkins et al., 1968). The Gulf of Mexico opened south of these uplifts, but influx of oceanic waters into the interior basin and rapid evaporation in an arid setting led to the accumulation of thick deposits of salt. These extensive salt deposits are the Middle to Upper Jurassic (Callovian) Louann Salt (Fig. 18-3), which is the effective sedimentary "base-

Fig. 18-2. Thomasville Field is located in the north-central portion of the Mississippi interior salt basin. The salt basin was separated from the main Gulf of Mexico opening by the Wiggins and South Mississippi uplifts.

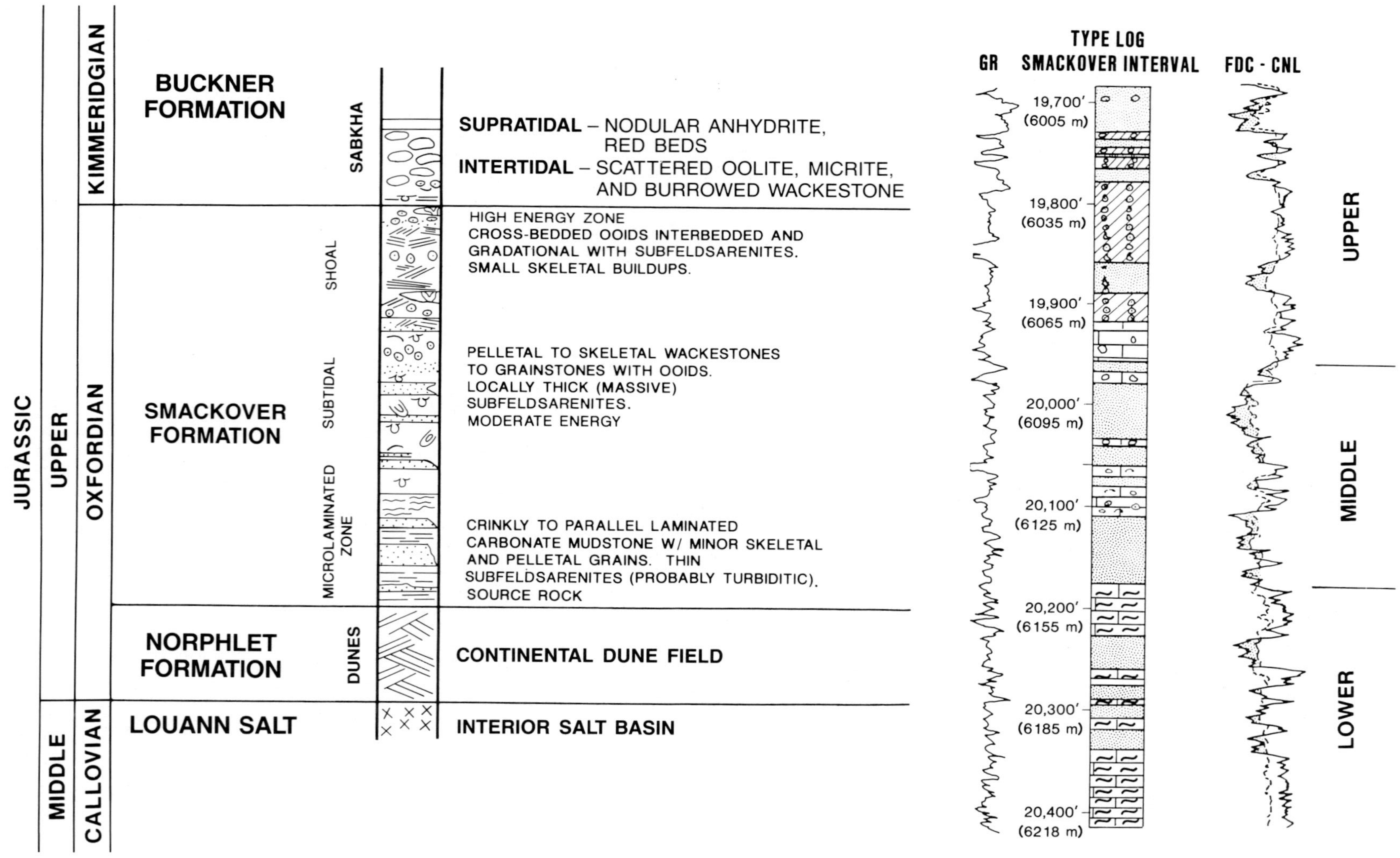

Fig. 18-3. Simplified stratigraphic column for the Jurassic section in the sour gas trend. The Smackover and the Buckner represent an overall upward-shoaling sequence. Also shown is a type log for Thomasville Field; reservoirs may be divided into lower, middle, and upper zones but do not include the very basal microlaminated zone. The shaded zones on the FDC/CNL log indicate the presence of porous sandstones. FDC/CNL is a Schlumberger trademark.

SOUTH   **NORTH**

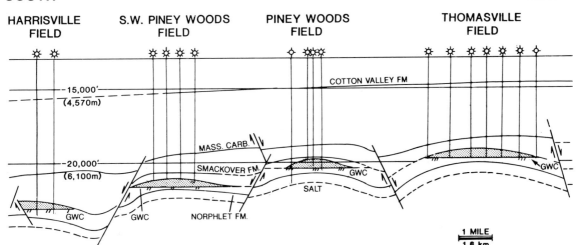

Fig. 18-4. Schematic diagram of the structures and hydrocarbon accumulations of four sour gas fields. Traps are both four-way and fault closures. Depths shown are subsea. Thomasville is the shallowest of the structures which have been generated by salt movement.

ment" for Jurassic hydrocarbon-bearing rocks in central Mississippi. Early loading of the salt by the Upper Jurassic (Oxfordian) Norphlet and Smackover sediments created faulted anticlinal and domal structures (Hughes, 1968; Parker, 1974), which became traps for later hydrocarbon migration. These salt-generated highs are often termed the salt roller belt, which roughly parallels but is downdip of the Jurassic paleoshoreline. Seismic data and well control indicate that the structural development of many fields in this area of central Mississippi was essentially complete by the middle to end of Upper Jurassic (Kimmeridgian) Buckner Formation deposition. Figure 18-4 illustrates the structural style of the four fields in the sour gas complex. Four-way and fault closure are both important in forming structural traps.

The structure at Thomasville Field is a faulted anticline (Fig. 18-5) generated by a nonpiercement salt pillow. A major sealing fault with displacement greater than 500 feet (150 m) bounds the structure on the west. Additional smaller faults are present within the field, but pressure data suggest they are not major barriers to lateral continuity and fluid flow. One such fault (Fault 2 on Fig. 18-5) has additional offset with depth and partially separates the field into eastern and western highs. The eastern area is informally referred to as "East Thomas" and the western area as "Thomasville." Pressure data in wells on either side of Fault 2 indicate fluid communication across the fault.

At least two periods of structural development were important in localizing the hydrocarbon accumulation in Thomasville Field. First, early salt movement during Smackover deposition led to a Middle Smackover unconformity, where sediments were removed on the paleohigh located near the Shell Crain well locations (Fig. 18-5). Stratigraphically equivalent high-energy sediments to those removed are present elsewhere in the field but are generally below the field water level. Second, uplift near the conclusion of and following Smackover deposition form the faulted and four-way closure that created the structural trap.

Stratigraphy

A generalized vertical sequence for the Jurassic interval in the sour gas trend is shown in Figure 18-3. The deepest exploration objective in central Mississippi is the Norphlet Sandstone, which overlies the Louann Salt and consists of a sequence of eolian dune and interdune sediments ranging in thickness from 200 to 1,200 feet (61–365 m). The Norphlet is locally productive for both hydrocarbons and carbon dioxide (Hartman, 1968; Studlick et al., this volume).

Unconformably overlying the Norphlet is the Smackover Formation, which throughout most of the Gulf Coast rim is a ramp-deposited, upward-

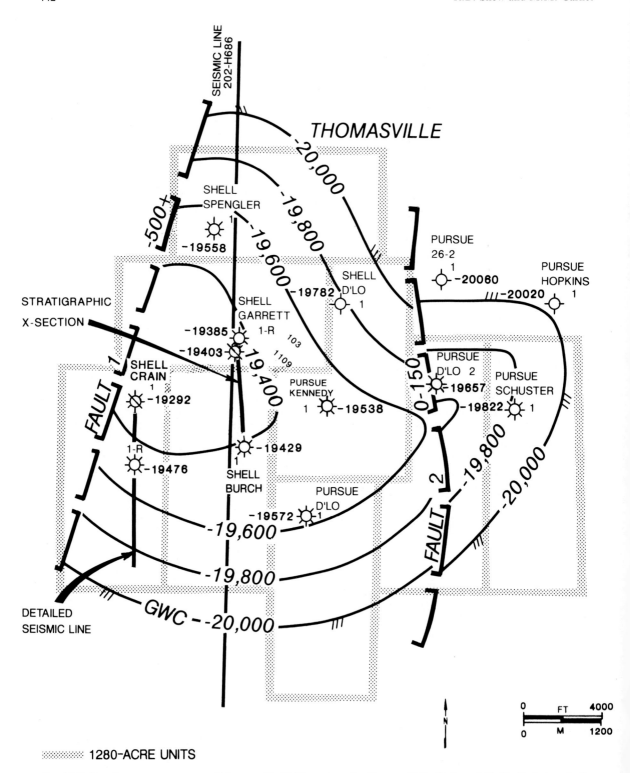

Fig. 18-5. Smackover structure map of Thomasville Field, contoured in feet, illustrating the faulted anticline, well locations, units, and cross section and seismic line loca-tions for Figures 18-7, 18-8, and 18-11. Fault 1 seals the structure, but Fault 2 does not isolate the two highs based on production and pressure data.

shoaling carbonate sequence (Ahr, 1973; Budd and Loucks, 1981). The onset of Smackover carbonate deposition corresponds to a rapid rise in relative sea level (Budd and Loucks, 1981; Vail et al., 1984). During this rise, the Norphlet was locally transgressively reworked, and a thin basal Smackover unit was deposited. The remainder of Smackover deposition in the Gulf Coast, although complicated by salt tectonics, generally represents deposition of a carbonate regressive wedge during a slow relative sea level rise. The typical Smackover sequence, from base to top, consists of (1) a thin transgressive deposit, (2) a basal low-energy zone composed of laminated, organic-rich shaly carbonates and carbonate mudstones, which includes a microlaminated zone (MLZ) evaluated as source rock, (3) a middle ramp zone, termed the brown dense zone (BDZ), which is composed of wackestones and packstones with local grainstone lenses, and (4) an uppermost high-energy zone (HEZ) with thick-bedded grainstones deposited as ooid shoal and beach deposits. The HEZ is the most distinctive facies of the Smackover and is the primary exploration target in most of the U.S. Gulf Coast. Scattered skeletal buildups are present in the HEZ and the BDZ in Thomasville Field and elsewhere (Baria et al., 1982).

Variations of this typical upward-shoaling carbonate sequence are most often associated with syndepositional salt highs. However, in the areas of central and western Mississippi and eastern Louisiana, the influx of abundant siliciclastics (Fig. 18-6) and salt tectonism both complicate the vertical sequence. The variations include (1) abrupt lateral and some vertical changes from low- to high-energy facies around the salt highs and (2) the siliciclastics both interfinger with and abruptly replace carbonate deposition. Although siliciclastics have been recognized elsewhere in the Smackover of the Gulf Coast (Chimene, 1976; Ahr and Palko, 1981; Budd and Loucks, 1981; Judice and Mazzullo, 1982), they predominate only in portions of Mississippi and Louisiana as previously mentioned. The siliciclastics are the objectives in Thomasville.

Overlying the Smackover and capping this Jurassic regressive wedge is the Buckner Formation, a sequence of nodular and disseminated anhydrite with minor supratidal and intertidal carbonates. In Thomasville Field, structural growth of the anticline occurred prior to and during early Buckner deposition (Shew and Garner, 1986), and as a result, the Buckner is locally unconformable with the Smackover on the crest. However, the Buckner in many

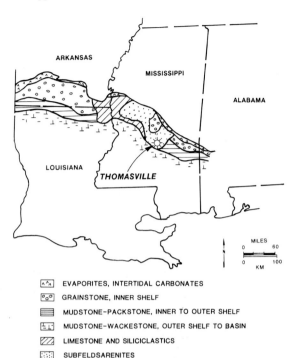

EVAPORITES, INTERTIDAL CARBONATES

GRAINSTONE, INNER SHELF

MUDSTONE-PACKSTONE, INNER TO OUTER SHELF

MUDSTONE-WACKESTONE, OUTER SHELF TO BASIN

LIMESTONE AND SILICICLASTICS

SUBFELDSARENITES

Fig. 18-6. Smackover facies distribution map, Arkansas, Louisiana, and Mississippi. Note the large area of siliciclastics in central and western Mississippi and the location of Thomasville Field within this belt (modified from Bishop, 1968; reprinted by permission of the American Association of Petroleum Geologists).

parts of the Gulf Coast (Ahr, 1973; Budd and Loucks, 1981), including most of central Mississippi, is in part laterally equivalent to the Smackover. It is the most proximal facies (sabkha) of a regressive, upward-shoaling sequence.

Reservoir Characterization

Seismic Data

Seismic reflection data were the basis for drilling the Thomasville structure in 1969 when simple and fault closures (Fig. 18-7) were recognized at the Smackover and Norphlet horizons. Larger faults and closure are clearly evident and easily mapped using newer seismic data. Recent seismic data, using improved acquisition and processing techniques, have also provided stratigraphic mapping capabilities. Although individual sandstone beds still cannot be seismically resolved, zones of porous sandstone

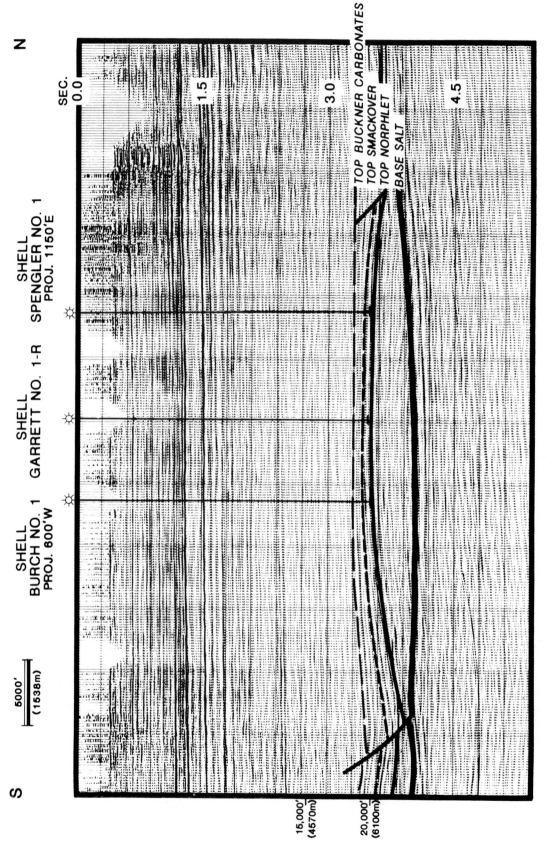

Fig. 18-7. Seismic line (202-H686; location shown in Figure 18-5) showing structure and closure at the Smackover and Norphlet horizons. Structural closure is diminished at the overlying Buckner horizon. Note lenticular shape of the salt.

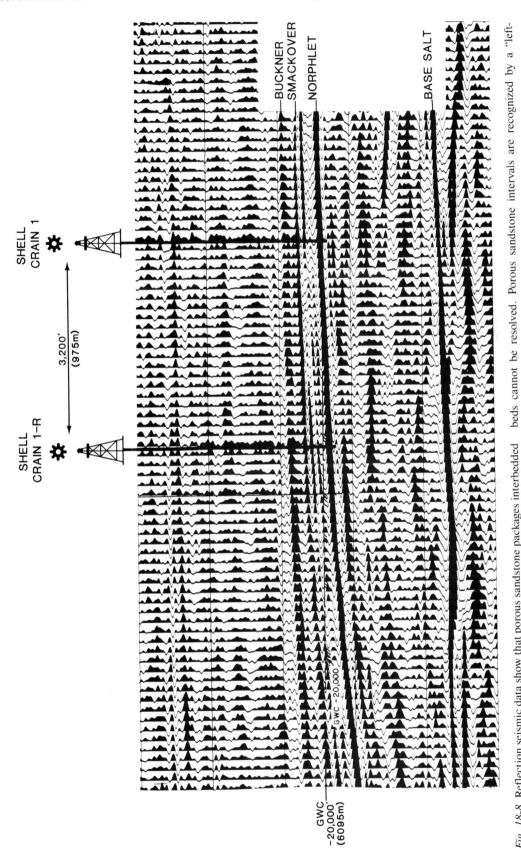

Fig. 18-8. Reflection seismic data show that porous sandstone packages interbedded with thicker tight carbonates can be recognized and sometimes correlated with the seismic data (location of line shown in Figure 18-5). However, individual sandstone beds cannot be resolved. Porous sandstone intervals are recognized by a "left-kicking," relatively low-velocity zone.

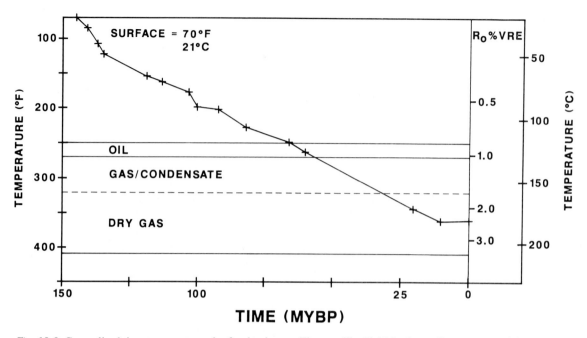

Fig. 18-9. Generalized time-temperature plot for the thermal history of Thomasville Field. Also shown are the approximate hydrocarbon type formation windows. Thomasville Field is thermally mature and is currently within the dry-gas window at a vitrinite reflectance (VRE) of approximately 2.3%.

can be delineated and often traced between wells (Fig. 18-8). The acoustic impedance contrast of the porous sandstones and tight carbonates is the reason for the "soft" seismic expression of the sandstones.

Hydrocarbon Generation and Trap Formation

The hydrocarbon source rock for the Smackover and the Norphlet reservoirs throughout the Gulf is the basal Smackover microlaminated zone (MLZ). Hydrocarbon generation and migration probably began during Late Cretaceous or very early Tertiary at burial depths of more than 10,000 feet (3,050 m) and at approximately 250°F (120°C). Sassen and others (1987) and Sassen and Moore (1988) have shown that hydrocarbon migration and entrapment in Smackover reservoirs occurred elsewhere in the Mississippi interior salt basin at similar depths. The thermal maturity level of the Smackover in Thomasville Field is extremely high, with a vitrinite reflectance of approximately 2.3%. This is within the dry gas window as shown in Figure 18-9. Destruction of the oil is evidenced by a large amount of bitumen remaining in the pore space. Thermochemical sulfate reduction (see discussion in Origin of Sour Gas

section), the reduction of sulfate by hydrocarbons at high temperature, is responsible for the formation of large amounts of hydrogen sulfide and minor amounts of carbon dioxide.

All of the structures in central Mississippi, including Thomasville, were formed early, before the end of Buckner deposition (Late Jurassic) and before hydrocarbon generation and migration. In addition to the sealing western fault at Thomasville, impermeable carbonates and evaporites in the Buckner and the Smackover form an effective local and even regional top seal to the hydrocarbon-bearing sandstones.

Depositional Environment

Smackover deposition in most areas of the Gulf Coast is best described as a regressive wedge of carbonates deposited on a ramp (Ahr, 1973; Budd and Loucks, 1981; Moore, 1984). The ramp model consists of coast-parallel belts of sediments with high-energy deposits nearshore, grading seaward into upper- to middle-ramp facies and lower energy, outer-ramp to basinal deposits (Fig. 18-6 and uppermost part of block diagram in Fig. 18-10). Figure 18-10 illustrates the depositional setting envisioned for the sour gas trend and Thomasville in particular,

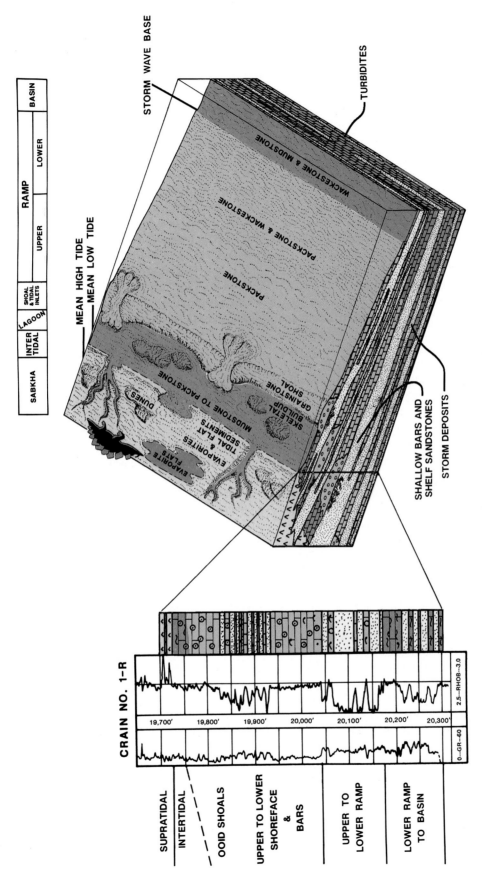

SABKHA	INTER TIDAL	LAGOON	SHOAL & TIDAL INLETS	RAMP		BASIN
				UPPER	LOWER	

MEAN HIGH TIDE
MEAN LOW TIDE

STORM WAVE BASE

TURBIDITES

WACKESTONE & MUDSTONE

PACKSTONE & WACKESTONE

PACKSTONE

SKELETAL BUILDUP
GRAINSTONE SHOAL

DUNES
MUDSTONE TO PACKSTONE
TIDAL FLAT SEDIMENTS
EVAPORITE FLATS
EVAPORITES

SHALLOW BARS AND SHELF SANDSTONES

STORM DEPOSITS

CRAIN NO. 1-R

| 19,700' | 19,800' | 19,900' | 20,000' | 20,100' | 20,200' | 20,300' |

2.5—RHOB—3.0

0—GR—60

SUPRATIDAL

INTERTIDAL

OOID SHOALS

UPPER TO LOWER SHOREFACE & BARS

UPPER TO LOWER RAMP

LOWER RAMP TO BASIN

Fig. 18-10. Depositional ramp model, log response, and lithology log for Thomasville Field. Many of the upper and middle siliciclastics and carbonates have gradational contacts. Vertical and lateral facies changes correspond to a general progradation and upward-shoaling sequence. Depths are in feet.

447

a modified ramp with mixed carbonate/siliciclastic sediments deposited in an upward-shoaling sequence. Smaller-scale facies changes in the sour gas trend result from local depositional controls and salt tectonics.

The depositional model for the sour gas trend, and for Thomasville Field in particular, must account for the type, abundance, and distribution of siliciclastics as well as their relationship to the interbedded carbonates. The depositional interpretation of the objective sandstone beds is based on physical sedimentary structures, macrofossils, trace fossils, vertical sequences, and on environmental information from the interbedded nonreservoir carbonates.

From the base to the top of the productive Smackover interval, the sandstones consist of (1) a lower zone of outer-ramp storm deposits which consists of numerous thin sandstones with sharp bases and bioturbated tops, (2) a middle zone of shallower inner ramp and bar sandstones (massive, amalgamated thick sandstones that sometimes show low-angle cross-bedding), and (3) an upper zone of thinner, higher energy nearshore sandstones (low-angle and planar stratification). The sandstone-carbonate contacts are mostly gradational in the middle and upper zones ("facies mixing" model of Mount, 1984) and mostly sharp ("punctuated" model of Mount, 1984) in the lower zone. Parker (1974) referred to the mixing as "competitive sedimentation," where the siliciclastics slowly overwhelmed carbonate deposition. Trace fossils (*Ophiomorpha, Terebellina, Asterosoma*, and *Teichichnus*) and macrofossils (coral, echinoderm, bryozoa, bivalves, tubiphytes, etc.) also support a ramp (shelf) and nearshore depositional environment interpretation. These faunal insights and vertical sequences are important to the Smackover environmental reconstruction. Using primary sedimentary structures alone, Olsen (1982) concluded that Smackover and Buckner sediments at Thomasville Field were all turbidites. Although we have interpreted turbidite deposits to be present downdip and perhaps locally in the deeper section at Thomasville Field, the combined evidence of physical and faunal data suggests a well-defined upward-shoaling sequence of in situ sandstone and carbonate deposits.

A large source of sand must have been present in central Mississippi during the Jurassic to result in the deposition of several hundred feet (>60 m) of sandstone within the Smackover Formation. Deltas (Dinkins et al., 1968) and ancestral river drainage systems (Mann and Thomas, 1968) have been suggested as the source of these sediments. Although fluviodeltaic deposits are not present near Thomasville and deep drilling and seismic data are too sparse to determine the sandstone source, the abundant nearshore shallow marine sandstones, shelf deposits, and "basinal" turbidites downdip support the interpretation of a proximal sediment source.

The theory of a nearby major river system and other minor ones supplying the sediments to the basin is probably correct. Large river systems often form in the failed arms of triple junctions (Burke and Dewey, 1973). The ancestral Mississippi River is postulated to have formed at the time of initial Mesozoic Gulf rifting in a failed arm (Ervin and McGinnis, 1975). This ancestral river system could have delivered a large amount of terrigenous sediment which was subsequently reworked by longshore currents, and spread downshelf by storm processes. Sea-level lowstands may also be important in the shelf and turbidite sandstone distribution, but the effect cannot be documented because of the lack of well data, inadequate seismic resolution, and structural complexities associated with salt tectonics.

After the initial rapid sea-level rise, at the beginning of Smackover deposition, a slow relative rise was accompanied by aggradation and slow progradation of the lower to the upper Smackover sediments. In Thomasville Field, progradation is evidenced by the upward-shoaling sequence capped by Buckner evaporites. The final shoreline position occurred just south of Thomasville Field as evidenced by the absence of massive evaporites in downdip fields. Small, higher frequency fluctuations in sea level may have led to the cyclic interbedding of clastics and carbonates, with the sandstone intervals representing the higher-order relative lowstands. Variable sediment input rates, alone or coupled with sea-level lowstands, may be important also. Similar stacked deposits have been described in mixed systems elsewhere (Mack and James, 1986; Horne et al., 1989).

Subdivisions and Characteristics of the Productive Interval

The productive interval in Thomasville Field has been divided into lower, middle, and upper zones based on lithofacies and depositional environment (see Figures 18-3 and 18-10 for a summary of these zones). Figure 18-11 illustrates two of the primary distinctions among these intervals: variable thick-

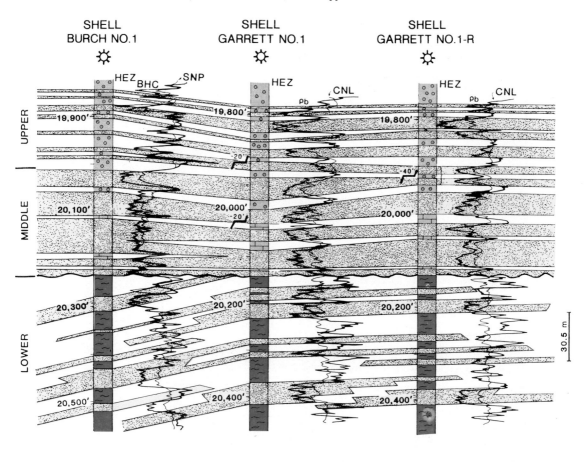

Fig. 18-11. Stratigraphic cross section of Thomasville Field. The middle interval contains the thickest, most laterally continuous, and best-quality sandstones. The lower interval contains thin but amalgamated and biotur- bated sandstones, which are generally poorer reservoirs. The upper zone is continuous, but sandstones are thin and highly cemented. Thin carbonate layers may be present within the thick sandstones. Scale bar is 1 foot (0.3 m).

nesses and continuity. The third variable is the reservoir quality. Each zone's reservoir properties, depositional environment, and productive characteristics are discussed below.

Lower Zone. The lower zone contains the smallest reservoir volume in the field, approximately 10% of the total, because most of these sandstones are below the field's gas-water contact and/or reservoir quality is poor. The lower zone is separated from the middle zone by an erosional unconformity, which represents removal of more than 180 feet (>60 m)

of sediment in the western portion of the field. Recognition of this unconformity is important because it explains the distribution of facies in the lower interval. Clean, well-sorted sandstones were recognized locally from cores and logs in the eastern part of the field (off the paleohigh). These are generally below the gas-water contact in the east and have been removed on the paleohigh in the western half of the field (near the Shell Crain well locations; Fig. 18-5). In other words, syndepositional structural growth led to the removal of the upper part of an early upward-shoaling facies in the west.

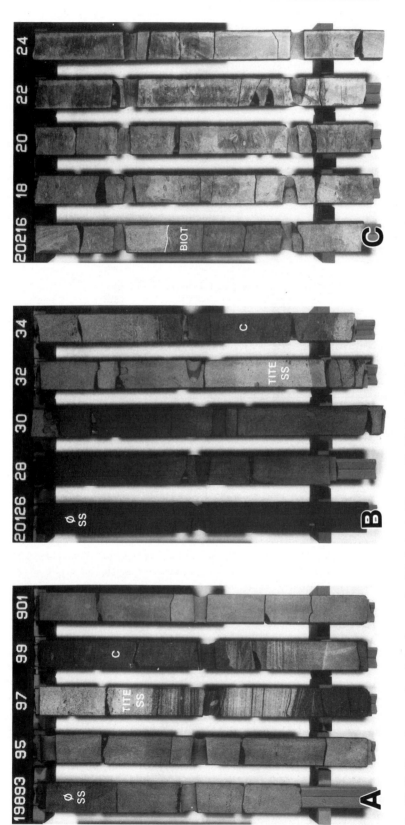

Fig. 18-12. Core photographs of the three major lithofacies in Thomasville Field from Shell Crain No. 1-R. Scale bar at right is 1 foot (30 cm). (A) The upper high-energy zone is composed of carbonates (c), gradational contacts with ooid laminae, and thinner sandstones. Sandstones (ss) grade from low porosity close to the carbonate zone to higher porosity in the middle of the sandstone bed. (B) The massive middle interval is composed of higher quality sandstone and tight interbedded carbonate (c). The black core color is related to bitumen within the porous sandstone. The sandstones once again have greater porosity away from the carbonate interbeds. (C) The lower section consists of storm deposits, the tops of which are intensely bioturbated (BIOT).

The only productive sands remaining in the western area are generally poor reservoir quality, bioturbated storm deposits (Fig. 18-12C). These storm deposits have sharp bases and intensely bioturbated tops leading to extreme reductions in horizontal and vertical permeability. Multiple storm deposits may be stacked to form a composite sandstone layer. These deposits are similar to those described for the Olmos Formation by Snedden and Jumper (this volume). The thin, laterally continuous lower zone sandstones in Thomasville are conspicuously devoid of clay (carbonate mud instead) in contrast to the Olmos. The storm deposits generally have average porosities and permeabilities below the average field pay cutoffs of 6.0% porosity and 0.05 md permeability (unstressed). The reduction in permeability is associated with the bioturbated layers. The porosity is similar to that of the nonbioturbated layers, but permeability averages only 0.03 md (unstressed).

Middle Zone. Above the unconformity and below the upper zone is the middle reservoir interval. Thin to thick, laterally continuous sandstone beds interbedded with packstones characterize this zone. Most of the sandstone/carbonate contacts are gradational (Fig. 18-12B). The sandstones are interpreted as having been deposited in moderate- to locally high-energy environments as evidenced by the absence of bioturbation, coarser grain size, amalgamated beds, local scour, undulatory bedding surfaces, and low-angle cross-bedding. The depositional environment is interpreted to be lower shoreface shallow bars and shallow shelf sandstones. These stacked, thick amalgamated sandstones must have had a proximal source and repeated storm conditions. There is no indication of subaerial exposure in these sandstones. The sandstones have characteristics of ridge/shoal sand bodies (described by Tillman and Martinsen, 1984, and Penland et al., 1988). These inner shelf/shoal sequences have been recognized in several settings (tide and storm-dominated shelves; Swift et al., 1986), but the storm-dominated shelf environmental interpretation is most likely for Thomasville Field. Sand bypass of the nearshore or sand deposition on the shallow shelf during minor relative sea level lowering may be responsible for the large amount of sandstone present in the middle zone. Sand-body geometries and origins may be similar to those described by Brenner (1978) and Tillman and Martinsen (1984). Amalgamation of these types of sandstones would result in the observed thick, middle-interval sandstones.

More than 75% of the field's reserves occur in the middle zone. This high volume is the result of a higher proportion of sandstone (40–80%) and increased reservoir quality associated with thicker individual beds (Figs. 18-11, 18-13), which contain smaller proportions of carbonate cement (see Petrography and Diagenesis discussion). In addition, the thicker sandstone beds are probably more laterally continuous.

Figure 18-14 is a net pay (using previously defined porosity and permeability cutoffs) isopach map of the middle zone. It is evident from this map and that of Figure 18-13 that the presence and distribution of reservoir-quality sandstones are a function of both depositional and diagenetic processes. For example, the Shell D'Lo No. 1 contains abundant sandstone but no net pay. This is a result of extreme porosity reduction associated with increased carbonate cementation, primarily dolomite, and compaction. One possible explanation for the increased cementation at this location is the increased thickness and proximity of the Buckner evaporites to the reservoir sandstones (possible reflux or mixing model to dolomitize the carbonates and precipitate dolomite in the sandstones).

Upper Zone. Gradational with the middle interval is the overlying upper zone (HEZ), composed primarily of carbonate grainstones and subfeldsarenites. Carbonates consist of cross-bedded ooid grainstones with thin interbeds of pelletal, skeletal, and oncolite grainstones. Rubble zones of coral and other skeletal material zones are present as well as mudstones and packstones. Sandstones have gradational contacts with the carbonates (Fig. 18-12A), ooid laminae occur within the sandstones, and sandstone laminae occur within the ooid grainstones. Low angle cross-stratification and planar stratification are present in the sandstones, which are interpreted as shoal and shoreface deposits intercalated with high-energy carbonate shoals and lower-energy back-shoal carbonate lagoons.

Based on log correlation, the sandstones are thin but continuous in the upper zone (Fig. 18-11). They generally have moderate to low reservoir quality, because porosity and permeability have been reduced by depositional processes and diagenesis. Alternating fine and coarse laminae result in decreased vertical permeability, and many of the very thin sandstones are tightly cemented. The thicker sandstones have higher reservoir quality. Minor faults, recognized on logs and in cores, are suspected to have isolated some of the sandstones

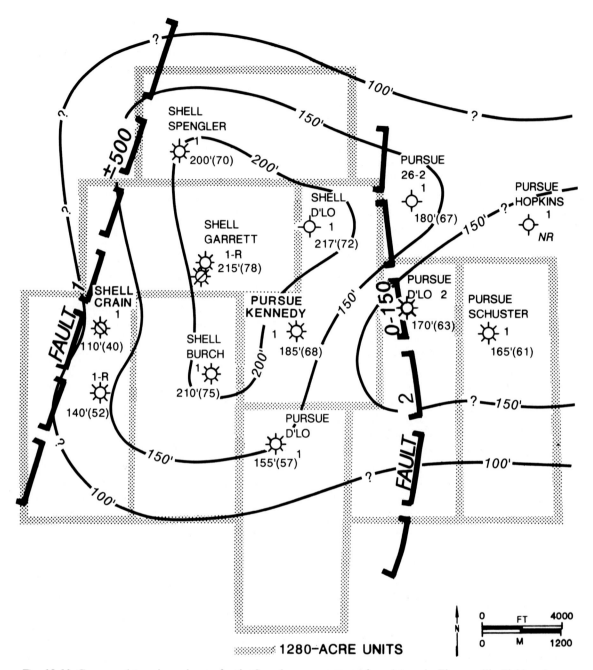

Fig. 18-13. Gross-sandstone isopach map for the Smackover middle zone. Numbers in parentheses are percent
sandstone. This zone contains the thickest and highest
percentage of sandstone in Thomasville Field and more
than 75% of the hydrocarbon reserves. (NR = not
reached.) Contour interval is 50 feet (15.2 m).

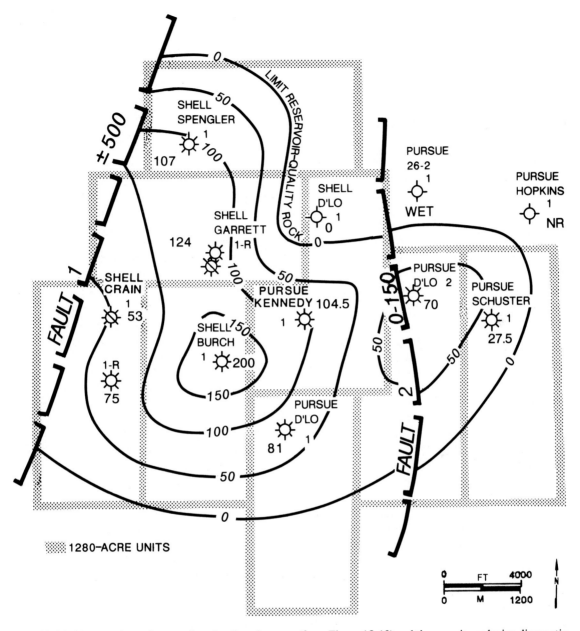

Fig. 18-14. Net-pay isopach map for the Smackover middle zone. The map is based on cutoff porosities and permeabilities (unstressed) of 6.0% and 0.05 md, respectively. Net pay is related to both sandstone thickness (from Figure 18-13) and the porosity-reducing diagenetic overprint. Contour interval is 50 feet (15.2 m). NR = not reached. WET = water bearing.

by juxtaposing them against impermeable carbonates. The upper zone comprises a volumetrically small amount of the net sandstone and gas reserves, approximately 15%.

Petrography and Diagenesis

Carbonate interbeds in the sandstone reservoirs are generally nonporous and impermeable and contain only a very minor amount of the field's reserves. Even the ooid grainstones, the primary objective throughout most of the Smackover trend, have less than 1% porosity as a result of early cementation and/or severe compaction. Dolomitized intervals (upper zone) are the only hydrocarbon-bearing carbonates and they account for less than 2% of the hydrocarbon reserves of the field.

The sandstones are fine- to medium-grained, poorly to moderately well-sorted subfeldsarenites (Fig. 18-15A). The average framework composition (bulk volume excluding porosity) for a reservoir quality sandstone is 61.5% quartz, 6.1% potassium feldspar, and less than 2% each of rock fragments, chert, plagioclase, and carbonate allochems. The cements and matrix (Fig. 18-15B) include subhedral to euhedral dolomite rhombs (16.3%), bitumen (8.1%), calcite (2.7%), and quartz overgrowths, potassium feldspar overgrowths, and pyrite (<2.0%). The amounts of carbonate cement and bitumen display the greatest variation throughout the sandstone layers. Patchy calcite cement occurs locally within the sandstones.

Detrital clays are noticeably absent in the sandstones. Minor amounts are present in the lower zone, but these are volumetrically insignificant. This is probably due to fine sediment bypass to a more basinal setting from the higher energy nearshore and storm-dominated shelf settings. Increased amounts of clay are present downdip. A minor amount of authigenic fibrous illite is locally abundant and important in reducing permeability. Minor mixed-layer clays and zeolites are also present but have little effect on reservoir quality. Despite the absence of detrital clays, reservoir quality is generally low and difficult to predict as the result of a complex diagenetic history (Fig. 18-16) that overprinted the original depositional fabric. Compaction, numerous cementation and dissolution events, hydrocarbon migration and alteration, thermal sulfate reduction, and the formation of authigenic clay minerals were all important in determining the final reservoir quality of the sandstones.

Porosity occurs as both secondary and altered primary types (Figs. 18-15A, 18-17). Most of the secondary porosity is a result of calcite dissolution, but chert, plagioclase, and potassium feldspar also exhibit dissolution textures. Zones with only secondary porosity have a vuggy texture and exhibit lower permeabilities than do those with both pore types. The primary porosity is considered altered because of the presence of dolomite and other cements and bitumen which occlude part of the original pore space. There is no evidence of complete framework grain and/or cement dissolution to suggest secondary porosity.

Although rock properties of Smackover sandstones are controlled by numerous diagenetic processes, one of the major factors in determining whether the sandstones are of reservoir quality is the relative abundance of carbonate cement. Sandstones with high proportions of carbonate content (>25%) generally have porosities less than 5% (Fig. 18-17). Similarly, sandstones with very small amounts of carbonate cement also have low porosities, but these are commonly highly compacted and quartz-cemented. The local occurrence of high minus-carbonate-cement porosity, particularly adjacent to carbonate interbeds, indicates probable early cementation. Reservoir quality is often best where an "optimum" amount, 6 to 22%, of carbonate cement (primarily rhombic dolomite) occurs. These early-formed rhombs appear to have been important in partially preventing compaction by "propping" the framework grains (Figs. 18-15A, 18-17). Sandstones with a combination of primary and secondary porosity have the highest reservoir quality.

Reservoir deterioration is commonly caused by increased cementation in sandstones proximal to carbonate interbeds (Fig. 18-17). Higher porosities and permeabilities therefore occur in the central portions of the sandstones, which is another reason why the thicker sandstones are the best reservoirs (e.g., the thick sandstones in the middle zone). However, most sandstone beds are variable in thickness, particularly in the upper and lower zones, and display variable degrees of cementation. Thus, it is difficult to predict reservoir quality laterally even where the sandstone beds are interpreted to be continuous.

In thin sections, sandstones contain 5 to 15% bitumen (bulk volume), a constituent which is important in reducing porosity and permeability. Leaching

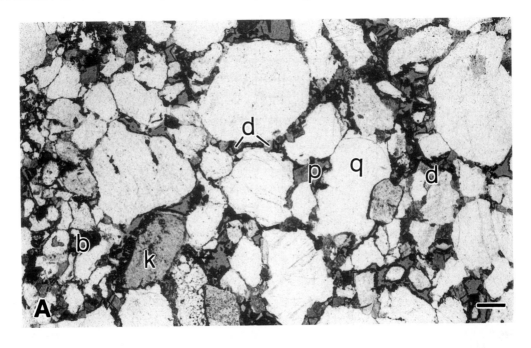

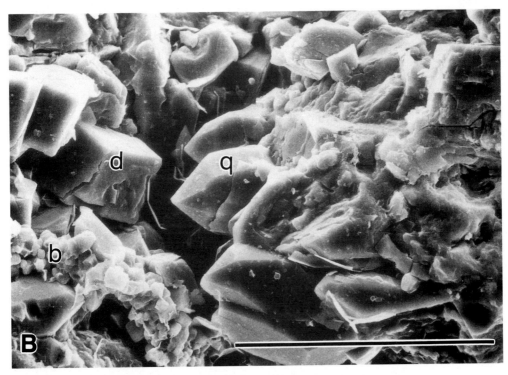

Fig. 18-15. (A and B) Photomicrographs of typical reservoir sandstones which are subfeldsarenites composed of predominantly quartz (q) and potassium feldspar (k). Both secondary and altered primary porosity (p) are present. Dolomite (d) is the dominant cement in these sandstones. Bitumen (b) fills many of the pores and reduces both porosity and permeability. Other cements include calcite and quartz and k-feldspar overgrowths. Scale bars are 0.1 mm.

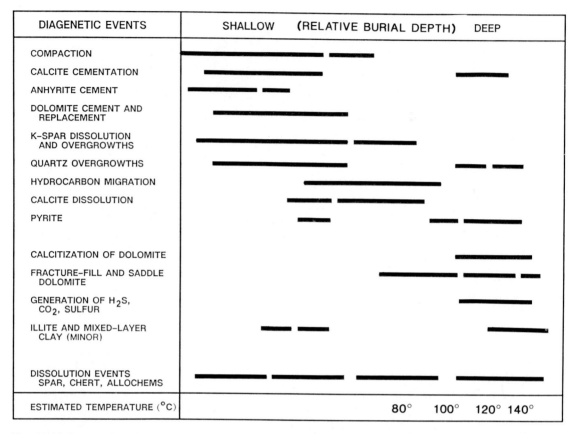

DIAGENETIC EVENTS	SHALLOW (RELATIVE BURIAL DEPTH) DEEP
COMPACTION	
CALCITE CEMENTATION	
ANHYRITE CEMENT	
DOLOMITE CEMENT AND REPLACEMENT	
K-SPAR DISSOLUTION AND OVERGROWTHS	
QUARTZ OVERGROWTHS	
HYDROCARBON MIGRATION	
CALCITE DISSOLUTION	
PYRITE	
CALCITIZATION OF DOLOMITE	
FRACTURE-FILL AND SADDLE DOLOMITE	
GENERATION OF H_2S, CO_2, SULFUR	
ILLITE AND MIXED-LAYER CLAY (MINOR)	
DISSOLUTION EVENTS SPAR, CHERT, ALLOCHEMS	
ESTIMATED TEMPERATURE (°C)	80° 100° 120° 140°

Fig. 18-16. Paragenetic sequence for the sandstones at Thomasville Field. Fourteen separate events led to a reduction in reservoir quality. Compaction, calcite and quartz cement, formation of bitumen, calcite dissolution, and thermal sulfate reduction are the most important diagenetic events.

experiments to remove both framework grains and cements revealed a fine network of bitumen partially filling the pores and pore throats. It is significant that most reservoir-quality sandstones have bitumen in the pore space, which indicates the presence of a former oil pool. Calcite dissolution is often observed where bitumen is present, probably resulting from fluids associated with hydrocarbon migration. Reduced cementation occurred after oil migration except for calcite, which possibly formed during thermal sulfate reduction. The bitumen was formed from thermal maturation of the original oil pool and thermal sulfate reduction.

Petrophysics

Average (arithmetic, unstressed) porosities and permeabilities for the sandstones are 7.0% and 0.35 md,

respectively. These low values are due to the highly compacted and diagenetically altered nature of the sandstones. Data from the Shell Crain No. 1-R, the only Shell well in the field with continuous cores, illustrate the range of porosity and permeability values within the Smackover interval. Figure 18-18 shows all of the measured values (unstressed) for the sandstones previously discussed in the lower, middle, and upper zones from the Crain No. 1-R. Two trend lines may be drawn through these data. The lower trend includes most of the lower zone which is bioturbated, where microporosity accounts for part of the porosity. The lower trend also includes sandstones from the middle and upper zones with dominantly secondary "vuggy" porosity. These are mostly nonreservoir sandstones. The upper trend is related to those sandstones with a combination of altered primary and secondary porosity, which is typical of the reservoir quality sandstones. The large

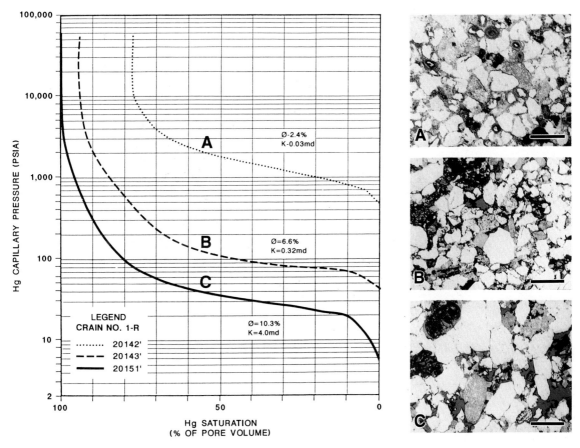

Fig. 18-17. Lithologic and petrophysical properties of the sandstones are strongly affected by grain size, cements, and dissolution fabrics. Reservoir quality also is strongly influenced by proximity to carbonate interbeds; carbonate cement occludes porosity at these gradational contacts. This is illustrated by the capillary pressure curves and thin-section photos. Sample A is closest to the carbonate interbed and C is from the center of a thick sandstone.

scatter is related to varying degrees of cementation and compaction. It is evident from this plot that using just a porosity cutoff is inadequate to portray reservoir quality accurately.

Capillary pressure curves indicate a relatively high but uniform entry pressure into the reservoir sandstones, but the highly cemented or vuggy zones have much higher entry pressures (Fig. 18-17).

Evaluation of the interbedded carbonate/sandstone layers is best accomplished using the FDC/CNL* logs. Using a limestone matrix density value of 2.72 gm/cc, the less dense, porous sandstones are clearly shown by a lower bulk density (FDC-shift to the left in Fig. 18-3).

---

*Schlumberger trademark.

Origin of Sour Gas

Smackover sandstone reservoirs in the sour gas trend contain gases with high percentages of hydrogen sulfide (>25–45%). In Thomasville Field, although slight variations are observed between wells, the average gas composition is 55% methane, 35% hydrogen sulfide, and 9% carbon dioxide. The origin of high H_2S concentrations in reservoirs has commonly been explained by either thermal or bacterial sulfate reduction (Orr, 1974, 1977). Sulfur isotope and petrographic data support a thermal sulfate reduction origin of the H_2S in Thomasville Field, that is, the reduction of sulfate by hydrocarbons at high temperatures and the resultant generation of H_2S and other products. The reaction and

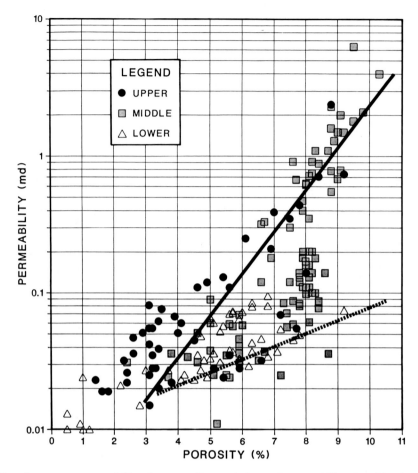

Fig. 18-18. Porosity versus permeability (unstressed) from the Shell Crain No. 1-R. Data have been separated into the upper, middle, and lower zones to illustrate the distribution of reservoir quality. Although data are scattered, two general trends are evident. The upper trend (solid line) is sandstones with better-connected porosity and permeability (combination of altered primary and secondary porosity), whereas the lower trend (dashed line) has mainly disconnected secondary dissolution porosity and microporosity within the bioturbated interval. Both of these latter porosity types are associated with low permeabilities.

products depend on the reservoir's thermal history and on the availability of sulfate to react with the hydrocarbons. A small amount of H_2S, perhaps originating from the thermal maturation of the original oil pool, must be present initially to catalyze the reaction.

Sulfur isotope data are the primary evidence for thermal sulfate reduction. The sulfur isotopes should be similar for the reactants and the products in thermal sulfate reduction because little or no fractionation occurs at high temperatures. The isotopic values of the reactants (anhydrite = $+16.3$ to $+17.3\permil$ CDT) and the products (sulfur = $+16.1\permil$ CDT and H_2S = $+16.1$ to $+17.3\permil$ CDT) are similar in Thomasville Field (Shew,

1987). Secondary evidence for thermal sulfate reduction includes the presence of calcite-replacing anhydrite, the local presence of elemental sulfur, and high temperatures. The current reservoir temperature is 375°F (190°C), considerably higher than the temperatures (175–285°F; 80–140°C) suggested for the reaction to proceed (Orr, 1977; Siebert, 1985). Siebert (1985) and Sassen and others (1987) have made similar observations on the formation of sour gas in other Smackover carbonate reservoirs in the Gulf Coast. In addition to altering the reservoir fluids and gases (generation of H_2S and CO_2), minor redistribution of porosity may occur by the dissolution of anhydrite and the formation of calcite cement.

Thermal sulfate reduction is one of the last diagenetic processes to have occurred in Thomasville Field but is one of the most important in determining field economics. The high H_2S content required special drilling and completion equipment, and the gas must be treated to separate the methane and H_2S. The H_2S is processed into sulfur, which is an important product in the overall economics of the sour gas trend.

Siliciclastic reservoirs are rarely the sites of high H_2S concentrations because of the usual presence of iron that may combine with and remove the H_2S and/or the absence of a proximal sulfate source. However, Thomasville Field sandstones are the exception, because they contain minimal amounts of clay and other iron-bearing minerals and, more importantly, are close to an abundant source of sulfate. Anhydrite occurs both as a secondary pore-filling patchy cement within the sandstones and as massive nodular beds in the overlying Buckner Formation. This pore-filling anhydrite is probably the most important source of sulfate because it is more accessible to the hydrocarbons. Core data indicate that very little anhydrite is present within the hydrocarbon column but that pore-filling anhydrite is more abundant below the gas-water contact.

Production Characteristics

Primary development in Thomasville Field occurred from 1969 to 1980 by Shell Oil Company and Pursue Energy. Several reasons for this long development history include construction of a sour-gas processing plant, a Shell Oil-imposed drilling moratorium following a blowout in Piney Woods Field to the south, and field delineation by Pursue Energy to the east. Another dry hole was drilled in 1981 and a replacement well, the Shell Crain No. 1-R (Fig. 18-5), was drilled in 1985. Development drilling by Shell was cautious, with outsteps to define the reservoir quality on the flanks of the structure. Pursue Energy established the field extension to the east.

Engineering considerations have been a major control on the success and timing of the drilling and completion of wells in Thomasville Field. The hostile environment that includes a complex gas mixture, high temperatures and pressures, and great depths required special safety and design efforts. For example, wellheads and tubulars required special pressure and corrosion resistance. Drilling of these wells requires that a liner be set in the tight Buckner carbonates above the first porous sandstones of the Smackover. Mud weights are then increased, and drilling continues into the highly geopressured Smackover interval.

Production is currently from 8 wells on 1,280-acre (518-ha.) units. With the low porosities and permeabilities previously discussed, these large units would seem inadequate to produce the majority of reserves. However, pressure data indicate that many of the sandstones are in flow communication across large areas of the field, supporting the interpretation of generally continuous sandstones in this "shelf" setting. Additionally, high production rates and large recoveries per well are possible because of the high reservoir pressures and large drainage areas, particularly in the middle zone. Original bottom-hole pressures were 17,500 psi (1.2 $\times 10^5$ kPa). The average per-well production rate is approximately 5 MMCFGPD (1.4 $\times 10^5$ m³/D), although rates of 10 to 20 MMCFGPD (2.8–5.7 $\times 10^5$ m³/D) have been sustained in some wells. More than 275 BCFG (7.8 $\times 10^9$ m³) have been produced of the more than 600 BCFG (1.7 $\times 10^{10}$ m³) originally in place. The reservoir drive mechanism is depletion gas drive. A field-decline curve for the Shell-operated wells is shown in Figure 18-19.

The ultimate top seal for the productive sandstones is the supratidal and intertidal carbonates and evaporites of the Upper Smackover and Buckner. The interbedded carbonates also provide intrareservoir barriers to fluid flow, particularly between the middle and the upper zones. However, pressure data and the fact that all sandstone beds above the water level are hydrocarbon-charged indicate some current or former vertical communication existed between sandstone layers within zones. This charging may have originated from small fractures and faults either formed by structural growth and/or by overpressuring. Although the wells are relatively far apart, some of these fracture/fault zones were identified from detailed well-log correlations and from observations in cores and cuttings. Small fault cutouts picked from logs could sometimes be correlated to mineralized zones such as calcite slickensides. These mineralized zones are often associated with faults and fractures. Minor faulting is more significant in the thinner sandstones, i.e., upper zone, than in the middle and lower zones. Minor faults in the thin sandstones may completely isolate parts of an originally continuous sandstone bed by juxtaposing porous sandstones with tight carbonates.

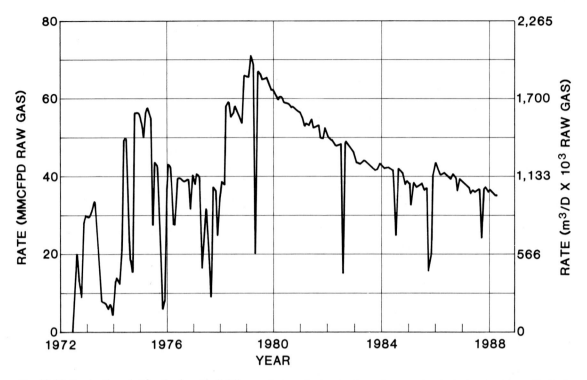

Fig. 18-19. Production plot for the four Shell Oil-operated wells in Thomasville Field. The many downward spikes result from required maintenance shutdowns of the sour-gas processing plant. The rates had been constrained by plant capacity and gas contracts, but are no longer limited. The overall increase in production evident through 1979 reflects the development phase of the field. Raw gas refers to total volume of produced, untreated gas.

Of the three productive zones within the Smackover, the middle zone (inner to middle ramp setting with thick, massive sandstone beds) is estimated to contain more than 75% of the reserves, based on mapping sandstone quality and distribution. However, the entire interval (upper to lower zones) in a well is open to production either through slotted liners or open-hole completions. Thus, gas compositions and production data by individual zone or sandstone layer are not available. The small differences in gas composition and local pressures observed in field wells may be related to the presence of locally isolated sandstones.

Exploration and Production Strategy

Jurassic prospects in central Mississippi and in the sour gas trend in particular occur in four-way and/or fault closures. Thomasville Field was discovered based on the seismic interpretation of closure at the Smackover and the Norphlet horizons. Early drilling in the field was based solely on structural closure.

However, core and log data soon indicated that reservoir quality varied laterally and vertically both within Thomasville Field and in nearby structures. An understanding of the structural, sedimentologic, and diagenetic properties is essential in exploring for and developing fields in this area and in similar depositional settings.

The structures in central Mississippi are salt-generated nonpiercement domes and anticlines. These are shallow to intermediate salt structures, formed in the salt roller belt. Early loading of the relatively thin salt in this area by Norphlet and Smackover sediments led to localization of trap configurations associated with faults and four-way closure. Seismic reflection data have been used to map these gross-scale structural features accurately. However, smaller faults, recognized on logs and in cores, cannot be seismically resolved. Improved acquisition and processing techniques have resulted in some seismic stratigraphic mapping capabilities even at these great depths. The large acoustic impedance contrasts of the porous sandstones and dense, nonporous carbonates allow the sandstone-rich zones, not individual beds, to be locally traced between wells in Thomasville Field.

Because of the high drilling and production costs and large gas volumes needed for economic discoveries, determination of reservoir potential requires more than just structural information. Evidence of this fact is shown by the industry drilling history in the central Mississippi sour gas trend. Of the 17 prospects drilled, only five have been gas discoveries in the Smackover (one encountered an accumulation at shallower depths). Although this is a higher than average discovery rate (29%), it should be noted that the Mississippi interior salt basin is a hydrocarbon-rich province and that all of these prospects were defined by seismically interpreted structural closure (charge and structure have high probabilities of success). The real uncertainties to consider are trap integrity, reservoir size, and reservoir quality. These are determined by depositional facies, preservation and/or formation of later porosity, and occurrence of economic gas mixtures.

The evolution of trap integrity through time in Smackover reservoirs is difficult to assess. Evidence of former oil accumulations (presence of bitumen) in some structures with no or minimal hydrocarbons remaining obviously attest to these as "leaky" systems. Several points to consider in seismic prospecting and interpretation are (a) fault-sealing capacity and the potential for sand-on-sand contacts across faults and (b) continuity and thickness of the top seal (Buckner carbonate and anhydrite). Originally, fault traps were considered to have a low potential for trapping hydrocarbons. However, several fields with sandstone on sandstone contacts across faults are hydrocarbon-bearing in the area (see Studlick et al., this volume). A continuous, non-faulted evaporite interval will serve as an effective top seal.

Depositional facies and diagenetic controls, as discussed in this chapter, are also important in determining the presence and quality of reservoir sandstones. Deep drilling, although sparse, indicates that a corridor of sandstone extends from the carbon dioxide source area (see Studlick et al., this volume) southward past Thomasville Field and downdip of Harrisville Field, the deepest of the productive sour gas accumulations. A progressive change occurs from higher energy sabkha and nearshore sandstones in the north to nearshore and shelfal (ramp) sandstones at Thomasville Field, to outer ramp and "basinal" turbidite sandstones at Harrisville Field. Thomasville Field is highly productive because of the large area of structural closure but mainly because of the presence of thick, lower-shoreface and shelf sandstones, resulting from its position on the ramp. Harrisville Field to the south contains continuous but fewer and thinner sandstones that have somewhat lower reservoir quality and thus smaller reserves. Other structures outside of this area contain thinner sandstones and lower sandstone percentages, which result in generally noneconomic accumulations. However, the Smackover is locally productive from carbonate facies outside of the sandstone corridor. For instance, there are carbon dioxide-bearing reservoirs in Smackover carbonates to the north and west of Thomasville Field in the carbon dioxide source area.

The final, but very important control on the occurrence of economic accumulations is diagenesis. Diagenesis controls both reservoir quality and the resulting gas composition. As previously discussed, porosity and permeability, although low, are high enough for high flow rates and volumes in Thomasville Field and in four other nearby fields. A major factor in preservation of reservoir quality is the distribution and amount of carbonate cement, particularly early dolomite cements that may have helped maintain an open porous framework. Thicker sandstone beds generally have better quality, illustrating the combined effects of depositional and diagenetic properties.

Another aspect of diagenesis important to the final determination of the Smackover reservoir's economic potential is that hydrocarbon accumulations have been modified by their thermal history. These fields are thermally mature and are in the lower part of the dry-gas generation window. The hydrocarbon gases have been altered by thermochemical sulfate reduction leading to gas mixtures rich in H_2S. The extent of this reaction is dependent on reservoir temperature (correlation of high temperature with increased H_2S), on the availability of a small amount of H_2S to act as a catalyst, and on the presence of sulfate. All of these properties are present in the sour-gas trend. Therefore, drilling in this type of setting with similar reservoir properties is likely to encounter variable gas mixtures. There is probably an upper limit of H_2S concentration in the gas because of a finite source of sulfate, either from pore fill or from the absence further downdip of the overlying massive Buckner evaporites. Additional drilling downdip and laterally should consider the deterioration of sandstone quality and quantity, variable gas mixtures, structural integrity, and diagenetic alteration.

Conclusions

Variations in lithology and reservoir quality occur throughout Thomasville Field, but the reservoir interval may be divided into three lithofacies. The lower zone comprises storm and middle- to outer-ramp sandstone and carbonate deposits. This zone has a small volume (10%) of the field's reserves for two reasons. First, most of the higher quality sandstones are below the fieldwide water level or have been removed on the structural crest of the field. Second, the more distal, bioturbated storm deposits are relatively poor reservoirs. The middle zone consists of thick, amalgamated shelf ridge/shoal and lower-shoreface sandstones that have the best reservoir quality and continuity. They contain more than 75% of the field's reserves because of greater sandstone thicknesses and high lateral sandstone continuity. Individual bed thickness is also important because the proportion of carbonate cement decreases away from contacts with carbonate interbeds. The upper zone consists of high-energy shoreface and shoal deposits. These sandstones have moderate reservoir quality and were originally very continuous. However, minor faults have isolated some of the sandstones, and most thin sandstones are extensively cemented. This interval contains approximately 15% of the reserves.

The productive potential of the field is high even though reservoir quality is generally low. This results from high pressures, large well-drainage areas, and high gross-sandstone percentages, particularly in the middle zone. Reservoir variability within each zone is a result of diagenetic alteration. A complex diagenetic history has generally reduced the reservoir quality, but enough altered primary and secondary porosity remains to allow economic prospects even at these great depths. The variable gas mixture is also a product of diagenetic alteration. Thermal sulfate reduction has led to the formation of high H_2S concentrations in the gas, which is a significant source of sulfur.

Acknowledgments. The authors thank Shell Offshore Inc. and Shell Development Company for permission to publish this paper. We also acknowledge the many previous Shell workers in both Exploration and Production who contributed to an improved knowledge of this area and Thomasville Field in particular.

Reservoir Summary

Field: Thomasville
Location: Central Mississippi
Operators: Shell Oil, Pursue Energy Corp.
Discovery: 1969
Basin: Mississippi interior salt basin, Gulf of Mexico
Tectonic/Regional Paleosetting: Salt tectonics and associated faulting along a rifted margin
Geologic Structure: Faulted domal to anticlinal feature
Trap Type: Four-way and fault closure
Reservoir Drive Mechanism: Depletion
 • **Original Reservoir Pressure:** 17,500 psi (1.2 × 10⁵ kPa) at 19,300 feet (5,880 m) subsea
Reservoir Rocks
 • **Age:** Late Jurassic, Oxfordian
 • **Stratigraphic Unit:** Smackover Formation
 • **Lithology:** Fine to medium-grained, subfeldsarenites
 • **Depositional Environment:** Nearshore to mid-ramp facies
 • **Productive Facies:** Shoreface to shelf sandstones
 • **Petrophysics**
  • **Porosity Types:** Altered primary and secondary porosity, approximately equal amounts
  • **ϕ:** Average 7.0%, range <5 to 10%, cutoff 6.0% (unstressed cores)
  • **k:** Average 0.35 md, range <0.001 to 6.0 md, cutoff 0.05 md (unstressed cores)
  • **S_w:** Average 35%, range <10 to 60% (logs)
Reservoir Geometry
 • **Depth:** 19,750 feet (6,075 m)

- **Areal Dimensions:** 4.6 by 4.2 miles (7.4 × 6.8 km)
- **Productive Area:** 10,250 acres (4,150 ha.)
- **Number of Reservoirs:** One unitized reservoir with multiple sandstone layers
- **Hydrocarbon Column Height:** 700 feet (210 m)
- **Fluid Contacts:** Gas-water at 20,000 feet (6,100 m) subsea
- **Number of Pay Zones:** One unitized pay zone from 19,300 feet to 20,000 feet (5,880–6,100 m) subsea consisting of up to 15 sandstone beds.
- **Gross Sandstone Thickness:** 700 feet (210 m)
- **Net Sandstone Thickness:** 300 feet (90 m)
- **Net/Gross:** 0.43 (variable depending on location)

Hydrocarbon Source, Migration
- **Lithology and Stratigraphic Unit:** Marine carbonate, Smackover Microlaminated Zone
- **Time of Hydrocarbon Maturation:** Cretaceous
- **Time Of Trap Formation:** Late Jurassic
- **Time of Migration:** Cretaceous

Hydrocarbons
- **Type:** Sour gas (CH_4 - 55%, H_2S - 35%, CO_2 - 9%)

Volumetrics
- **In-Place:** >600 BCFG ($>1.7 \times 10^1$) m³)
- **Cumulative Production:** > 275 BCFG ($>7.8 \times 10^9$ m³)
- **Ultimate Recovery:** NA
- **Recovery Efficiency:** NA

Wells
- **Spacing:** 1,280 acres (518 ha.)
- **Total:** 14 (includes 2 redrills)
- **Dry Holes:** 4

Typical Well Production
- **Average Daily:** 5 MMCFG (1.4×10^5 m³)
- **Cumulative:** 30 to 60 BCFG ($0.9–1.7 \times 10^9$ m³)

References

Ahr, W.M., 1973, The carbonate ramp—An alternative to the shelf model: Gulf Coast Association of Geological Societies Transactions, v. 23, p. 221–225.

Ahr, W.M., and Palko, G.J., 1981, Depositional and diagenetic cycles in Smackover limestone-sandstone sequences, Lincoln Parish, Louisiana: Gulf Coast Association of Geological Societies Transactions., v. 31, p. 7–17.

Baria, L.R., Stoudt, D.L., Harris, P.M., and Crevello, P.D., 1982, Upper Jurassic reefs of the Smackover Formation, U.S. Gulf Coast: American Association of Petroleum Geologists Bulletin., v. 66, p. 1449–1482.

Bishop, W.F., 1968, Petrology of Upper Smackover limestone in North Haynesville Field, Claiborne Parish, Louisiana: American Association of Petroleum Geologists Bulletin, v. 52, p. 92–128.

Brenner, R.L., 1978, Sussex sandstone of Wyoming: An example of Cretaceous offshore sedimentation: American Association of Petroleum Geologists Bulletin, v. 62, p. 181–200.

Budd, D.A., and Loucks, R.G., 1981, Smackover and Lower Buckner formations, South Texas: Depositional systems on a Jurassic carbonate ramp: Bureau of Economic Geology, the University of Texas, Austin, Report of Investigations 112, 38 p.

Burke, K., and Dewey, J.F., 1973, Plume-generated triple junctions: Key indicators in applying plate tectonics to old rocks: Journal of Geology, v. 81, p. 406–433.

Chimene, C.A., 1976, Upper Jurassic reservoirs, Walker Creek Field area, Lafayette and Columbia counties, Arkansas, in North American Oil and Gas Fields: American Association of Petroleum Geologists Memoir 24, p. 177–204.

Dinkins, T.H., Oxley, M.L, Minihan, E., and Ridgeway, J.M., 1968, Jurassic Stratigraphy of Mississippi: Mississippi Geological, Economic and Topographical Survey Bulletin 109, 77 p.

Ervin, C.P., and McGinnis, L.D., 1975, Reelfoot rift: Reactivated precursor to the Mississippi Embayment: Geological Society of America Bulletin, v. 86, p. 1287–1295.

Hartman, J.A., 1968, The Norphlet Sandstone, Pelahatchie Field, Rankin County, Mississippi: Gulf Coast Association of Geological Societies Transactions, v. 18, p. 2–11.

Horne, J.C., Reel, C.L., and Cummins, G.D., 1989, Effects of sequence stratigraphy on distribution of Cambro-Ordovician siliciclastic hydrocarbon reservoirs in Michigan Basin [abst.]: American Association of Petroleum Geologists Bulletin, v. 75, p. 1034.

Hughes, D.J., 1968, Salt tectonics as related to several Smackover fields along the northeast rim of the Gulf of Mexico basin: Gulf Coast Association of Geological Societies Transactions, v. 18, p. 320–330.

Judice, P.C., and Mazzullo, S.J., 1982, The gray sandstones (Jurassic) in Terryville Field, Louisiana: Basinal deposition and exploration model: Gulf Coast Association of Geological Societies Transactions, v. 32, p. 23–43.

Mack, G.H., and James, W.C., 1986, Cyclic sedimentation in the mixed siliciclastic-carbonate Abo-Hueco transitional zone (Lower Permian), southwestern New Mexico: Journal of Sedimentary Petrology, v. 56, p. 635–647.

Mann, M.C., and Thomas, W.A., 1968, The ancient Mississippi River: Gulf Coast Association of Geological Societies Transactions, v. 18, p. 187–197.

Moore, C.H., 1984, The Upper Smackover of the Gulf Rim: Depositional systems, diagenesis, porosity evolution and hydrocarbon production, in Ventress, W.P.S., Bebout, D.G., Perkins, B.F., and Moore, C.H., eds., The Jurassic of the Gulf Rim: Third Annual Research Conference, Gulf Coast Section, Society of Economic Paleontologists and Mineralogists, p. 283–307.

Mount, J.F., 1984, Mixing of siliciclastic and carbonate sediments in shallow shelf environments: Geology, v. 12, 432–435.

Olsen, R.S., 1982, Depositional environment of Jurassic Smackover sandstones: Gulf Coast Association of Geological Societies Transactions, v. 32, p. 59–65.

Orr, W.L., 1974, Changes in sulfur content and isotopic ratios of sulfur during petroleum maturation—Study of Big Horn Basin Paleozoic oils: American Association of Petroleum Geologists Bulletin, v. 58, No. 11, p. 2295–2318.

Orr, W.L., 1977, Geologic and geochemical controls on the distribution of hydrogen sulfide in natural gas, in Advances in Organic Geochemistry: Proceedings of 7th International Meeting on Organic Geochemistry, Madrid, Spain, p. 571–597.

Parker, C.A., 1974, Geopressures and secondary porosity in the deep Jurassic of Mississippi: Gulf Coast Association of Geological Societies Transactions, v. 24, p. 69–80.

Penland, S., Boyd, R., and Suter, J.R., 1988, Transgressive depositional systems of the Mississippi delta plain: A model for barrier shoreline and shelf sand development: Journal of Sedimentary Petrology, v. 58, p. 932–949.

Pindell, J., and Dewey, J.F., 1982, Permo-Triassic reconstruction of western Pangea and the evolution of the Gulf of Mexico/Caribbean region: Tectonics, v. 1, no. 2, p. 179–211.

Sassen, R., Moore, C.H., Nunn, J.A., Meendsen, F.C., and Heydari, E., 1987, Geochemical studies of crude oil generation, migration, and destruction in the Mississippi Salt Basin: Gulf Coast Association of Geological Societies Transactions, v. 37, p. 217–224.

Sassen, R., and Moore, C.H., 1988, Framework of hydrocarbon generation and destruction in eastern Smackover Trend: American Association of Petroleum Geologists Bulletin, v. 72, p. 649–663.

Shew, R.D., 1987, Isotopic and petrographic evidence for a thermochemical sulfate reduction origin of high H2S concentrations in Central Mississippi: Society of Economic Paleontologists and Mineralogists, Annual Mid-year Meeting, Abstracts with programs, August, p. 77.

Shew, R.D., and Garner, M.M., 1986, Geologic study and engineering review of the Jurassic Smackover Formation of Thomasville Field, Rankin County, Mississippi: Gulf Coast Association of Geological Societies Transactions, v. 36, p. 283–296.

Siebert, R.M., 1985, The origin of hydrogen sulfide, elemental sulfur, carbon dioxide, and nitrogen in reservoirs: Sixth Annual Research Conference, Gulf Coast Section, Society of Economic Paleontologists and Mineralogists, p. 30–31.

Swift, D.J.P., Thorne, J., and Oertel, G.F., 1986, Fluid Processes and sea-floor response on a modern storm-dominated shelf—Middle Atlantic shelf of North America—Part II. Response of the shelf floor, in Knight, R.J. and McLean, J.R., eds., Shelf sands and sandstones: Canadian Society of Petroleum Geologists Memoir 11, p. 191–211.

Tillman, R.W., and Martinsen, R.S., 1984, The Shannon shelf-ridge sandstone complex, Salt Creek Anticline area, Powder River Basin, Wyoming, in Tillman, R.W. and Siemers, C.T., eds., Siliciclastic shelf sediments: Society of Economic Paleontologists and Mineralogists Special Publication 34, p. 85–142.

Vail, P.R., Hardenbol, J., and Todd, R.G., 1984, Jurassic unconformities, chronostratigraphy and sea-level changes from seismic stratigraphy and biostratigraphy, in Ventress, W.P.S., Bebout, D.G., Perkins, B.F., and Moore, C.H., eds., The Jurassic of the Gulf Rim: Third Annual Research Conference, Gulf Coast Section, Society of Economic Paleontologists and Mineralogists, p. 283–307.

Key Words

Thomasville Field, Mississippi, Mississippi Interior Salt basin, Smackover Formation, Jurassic, Oxfordian, nearshore/shoreface sandstones, shelf sandstones, storm deposits, ramp model, mixed siliciclastics/carbonates, sour gas, diagenesis, thermal sulfate reduction, sulfur isotopes, bitumen, carbonate cements, salt-generated structures, seismic, capillary-pressure curves, geopressures, upward-shoaling sequence, Louann Salt.

Turbidite Environments

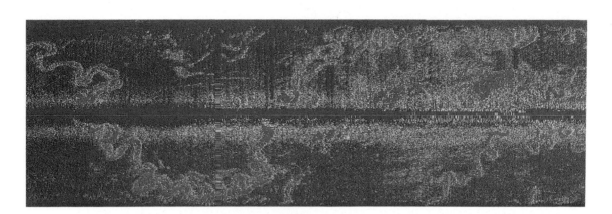

19

Channel-Fill Turbidite Reservoir, Yowlumne Field, California

Robert R. Berg and Gilberto R. Royo

Texas A&M University, College Station, Texas 77843; Consultant, College Station, Texas 77840

Introduction

The Yowlumne reservoir has internal structures that have an important effect on oil recovery. The original depositional conditions produced heterogeneity in the distribution of textural, compositional, and structural elements that result in flow anisotropy. However, such heterogeneities are commonly not detected until problems of fluid injection or hydrocarbon recovery are encountered. Therefore, it is important to predict the internal structures at an early stage of field development. Well logs provide the main source of this reservoir data, and dipmeter logs can be especially helpful in the interpretation.

The Yowlumne sandstone was interpreted to be composed of thin lenses that result in a complexly layered reservoir (Metz and Whitworth, 1983). This original interpretation was based largely on the distribution of recorded pressures in the reservoir during oil production. However, the same interpretation could have been made from dipmeter logs early in the development history. The logs show consistent dip patterns that reflect the internal reservoir structure which has a pronounced effect on oil

recovery during waterflooding. Similar interpretations might be used to predict the performance of other layered reservoirs which contain lensing units that control the production and injection of fluids.

Discovery and Development

The Yowlumne reservoir was discovered in 1974 at a depth of 11,300 feet (3,445 m) on a structural closure in the southwest part of the San Joaquin basin (Fig. 19-1). The discovery was preceded by 10 dry holes that had been drilled on the eastward-plunging structure called the San Emidio Nose beginning in 1938 (Taylor, 1978). Subsequent development showed a channel-form reservoir with a maximum thickness of 300 feet (91 m) which was 1 mile (1.6 km) wide at the crest of the structure and increased to 2.5 miles (4.0 km) wide downdip and 4.5 miles (7.2 km) to the northwest.

An oil-water contact at a depth of 10,700 feet (3,261 m) subsea occurs at the top of the structure, but production extends to 12,700 feet (3,871 m) subsea on the north flank of the structure. Therefore, oil production is confined stratigraphically to the

---

Editor's note: The internal structure of many types of sandstone reservoirs is understood largely as a result of detailed studies of modern depositional analogues. For logistical reasons, there are few detailed analog studies of modern submarine fans. Consequently, the shapes of flow units and reservoir heterogeneities must be inferred. However, it is possible to establish the inter-

nal structure of turbidites from data available within a reservoir itself. Detailed wireline-log correlations, combined with core analyses and reservoir performance, can reveal the geometries of reservoir components and provide analogues for other fields. This chapter explains such a technique applied to a turbidite reservoir in California.

Casebooks in Earth Science
Sandstone Petroleum Reservoirs
Eds.: Barwis/McPherson/Studlick
© 1990 by Springer-Verlag New York, Inc.

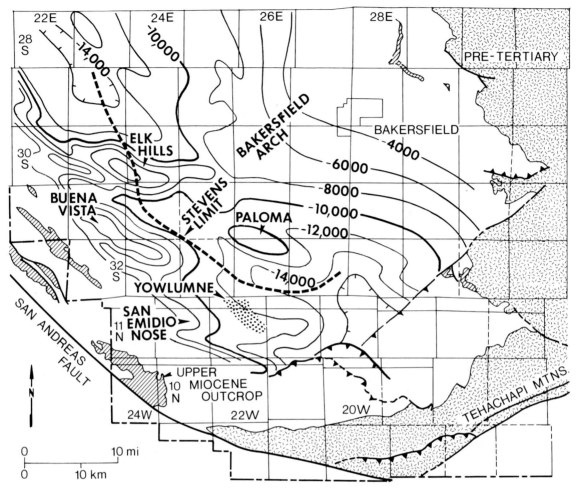

Fig. 19-1. Generalized structure map on the "N" marker bed near the top of the Stevens sandstone, southern San Joaquin basin, California. Contour interval is 2,000 feet (610 m). Dashed line marks the western limit of the main body of the Stevens sandstone. Structure adapted from Webb (1981) and published by permission of the American Association of Petroleum Geologists.

broad, channel-form reservoir, but the oil accumulation is tilted northward, either by tectonic tilting after accumulation or by hydrodynamic flow.

Two facts tend to support a hydrodynamic influence on the oil accumulation. First, the formation water has a relatively low salinity of 15,000 to 20,000 ppm (NaCl) which suggests dilution by meteoric water. Second, the wet sandstone at the top of the structure is porous and permeable, and no evidence exists for a permeability barrier which would be required to maintain an oil column of 2,000 feet (610 m) down the north flank of the structure. Therefore, the primary trapping mechanism is stratigraphic, although a hydrodynamic effect is possible.

A peak production rate of 26,400 BOPD (4.2 × 10³ m³/D) was reached in 1978, and the field had 87 wells by 1983 that had produced a total of 42 million barrels (6.7 × 10⁶ m³) of oil (Burzlaff, 1983). Waterflooding was initiated in 1978, first in the area designated Unit A and then in Unit B (Fig. 19-2). Unit A was estimated to contain 72 million barrels (1.1 × 10⁷ m³) of oil in place or 26% of the field total (Clark, 1987). Unit B contains 205 million barrels (3.3 × 10⁷ m³) of oil in place or 74% of the total. The waterflood was expected to add significantly to the estimated ultimate recovery of 78 million barrels (1.2 × 10⁷ m³).

The reservoir sandstones were interpreted as channel-fill turbidites from cores in four wells dur-

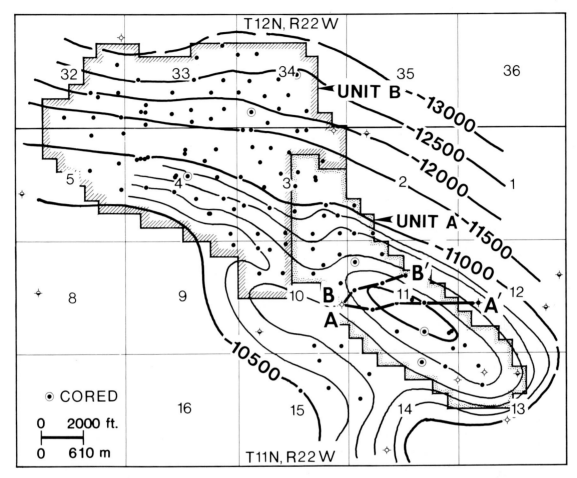

Fig. 19-2. Structure on top of the Yowlumne reservoir sandstone and equivalent beds in the Yowlumne Field area. Contour interval is 100 feet (30.5 m). Cores were examined from the circled wells, and the reservoir study was conducted in the producing unit "A" shown by the outlined area. Correlation section A-A' is given in Figure 19-3 and section BB' is given in Figure 19-10.

ing development of the field (Berg, 1986). Diagenesis and the development of secondary porosity were described by Tieh and others (1986). This study was undertaken to relate the internal structure of the reservoir to the production history (Royo, 1986) and was focused on Unit A.

Regional Setting

The San Joaquin basin is a narrow and deep syncline that underlies the Great Valley of central California (Fig. 19-1). The south part of the basin is bounded on the northeast by the Sierra Nevada uplift, on the southwest by the San Andreas fault, and on the southeast by the Tehachapi Mountains uplift and associated thrust faults. Large folds are present on the southwest, the Elk Hills, Buena Vista, and Midway anticlines, and, in the central basin, the Paloma anticline. A broad nose, the Bakersfield arch, plunges southwestward and separates the Buttonwillow depocenter to the north from the Tejon depocenter to the south (Zieglar and Spotts, 1978). At the southwest corner of the basin is the San Emidio nose on which the Yowlumne Field is located.

The Tejon depocenter contains more than 30,000 feet (9,150 m) of sediments that range in age from Eocene to Pliocene. The major reservoirs are sandstones of upper Miocene age that are encased in the basinal Monterey Shale. The reservoirs are called

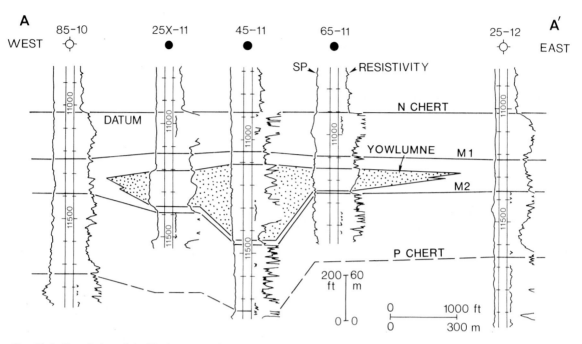

Fig. 19-3. Correlation of the Yowlumne sandstone across Unit A showing the lenticular nature of the reservoir. Datum is the N-chert marker bed; location of the section is shown in Figure 19-2.

Stevens sandstones and have a net thickness of 2,000 feet (610 m) in the east-central part of the basin (Sullwold, 1961; MacPherson, 1978; Webb, 1981). The main body of Stevens sandstone was derived from an eastern source area, and the sandstones grade abruptly into Monterey Shale to the southwest.

During the Miocene, water depths in the basin were on the order of 2,000 to 4,000 feet (610–1,220 m) (Bandy and Arnal, 1969). Sands were transported from adjacent highlands and delivered to the shoreline and shallow shelf and, subsequently, to the deep basin by turbidity currents through major submarine canyons which were cut into the continental slope. The main body of the Stevens was deposited as prograding, submarine fans that thicken basinward and spread across the basin floor (MacPherson, 1978). The fan-like bodies are composed predominantly of channel-fill deposits which form a complex of interfingering, braided-channel sequences.

Beyond the Stevens sandstone limit, other sandstones are found within the Monterey, and these thinner, lenticular sandstones are also channel-fill turbidites that were derived from a southern source

(Tieh et al., 1986; Berg, 1986). The main body of Stevens sandstones to the northeast is rich in volcanic rock fragments, whereas the lenticular sandstones to the southwest contain large amounts of feldspar.

The main body of the Stevens contains the principal oil reservoirs in numerous fields which extend across the Bakersfield arch to the east half of Elk Hills anticline. Beyond the Stevens limits, the reservoirs are in fractured Monterey Shale or in the thinner, lenticular sandstones within the Monterey. It has been estimated that the Stevens and equivalent rocks contain reserves of more than 4 billion barrels (6.4×10^8 m^3) and that more than 2 billion barrels (3.2×10^8 m^3) have been produced (Callaway, 1971).

The Yowlumne reservoir is one of the thinner sandstones derived from a granitic terrane to the south. The sandstone extends northwestward across the San Emidio Nose in a relatively narrow, channel-like body, and the reservoir has a maximum thickness of about 300 feet (91 m) at the top of the structure (Fig. 19-3). The reservoir is lenticular in cross section, and the correlation of marker beds in the enclosing shales indicates that intervals have expanded to accommodate the thickness of sand-

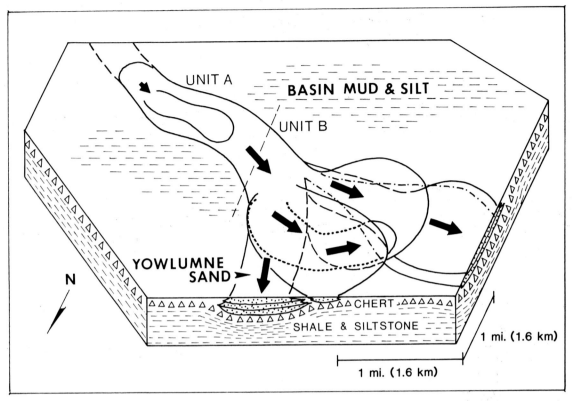

Fig. 19-4. Conceptual diagram showing deposition of Yowlumne sands within a relatively narrow channel in Unit A and spreading out to form a broad channel complex to the north in Unit B. Arrows represent directions of paleocurrent flow. Diagram adapted from Metz and Whitworth (1983) and reprinted by permission of the San Joaquin Geological Society.

stone. There is no evidence for erosion at the base of the channel, and it has been concluded that deposition of sand was confined to a topographic low on the seafloor (Dorman, 1980).

During deposition of the Yowlumne sandstone, the channel was filled from south to north by sediments of successive turbidity flows (Fig. 19-4). Early channel deposits were more narrowly confined in what is now the field Unit A, and later deposits were spread more widely in a broad, fanlike pattern across the field Unit B. The spreading of the sediments could be explained by breaching of a levee that confined earlier flows to a main, eastern channel (Metz and Whitworth, 1983). Filling of the main channel may have forced the later flows to extend westward in a series of splays that deposited sands more widely in the northwestern part of the field. Whatever the explanation, the sandstones are similar in character throughout the field, and the widening of the reservoir to the north suggests that sand transport was also in the same direction.

Source Rocks and Migration History

The Yowlumne sandstone is enclosed in basinal Monterey Shale, the principal source rock for oil in the Miocene and younger sandstones in the San Joaquin basin as well as in other California basins (Zieglar and Spotts, 1978; Isaacs and Petersen, 1987; Petersen and Hickey, 1987). The Monterey is a dark, siliceous to dolomitic shale which contains a total organic content ranging from 0.4 to 9.0% and averaging 3.4% (Graham and Williams, 1985). The organic material has a relatively large amount of extractable hydrocarbons where the Monterey is deeply buried. Monterey kerogen is typically type II, with minor contributions of type III. The type II

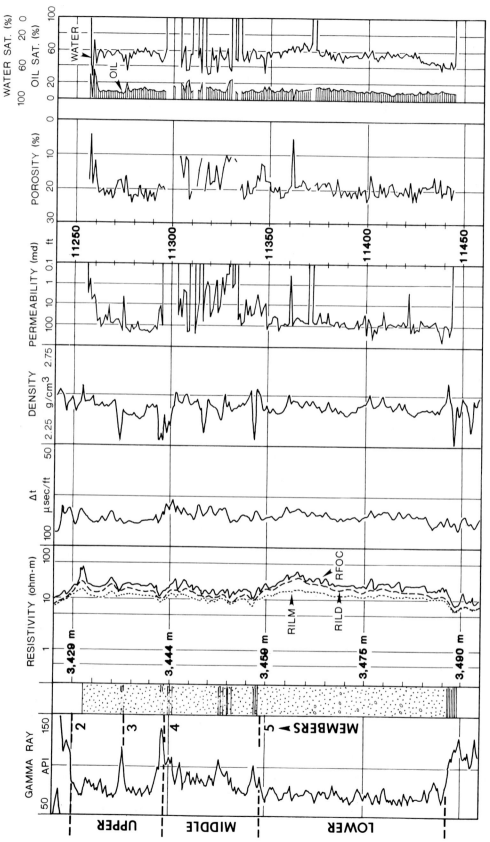

Fig. 19-5. Well logs and core analysis for the Yowlumne reservoir in Tenneco 12X-11 well showing the general character of the lower and upper units that represent channel facies and the middle unit that represents a channel-margin facies. Member numbers refer to mapped reservoir divisions. Location of the well is shown in Figure 19-11.

kerogen is dominantly amorphous, marine organic matter (Isaacs and Petersen, 1987).

The Monterey Shale entered the stage of oil generation about 6 million years ago, or in latest Miocene time, at depths of 13,000 feet (3,960 m) as based on its burial and thermal history (Zieglar and Spotts, 1978). Much of the organic material is immature at shallow depths, but the Monterey is presently generating oil in the deeper parts of the basin (Zieglar and Spotts, 1978).

Oil migration into the Yowlumne sandstone is a relatively recent event. Folding in the southern San Joaquin basin occurred during the middle Pleistocene and is responsible for the present configuration of the basin (Hackel, 1966). Previous to folding, the Monterey was deeply buried and began generating oil and, after uplift of the channel sandstones along the San Emidio Nose, oil generation continued and migration began.

Erosion and exposure of the Miocene section around the flanks of the basin occurred in the latest Pleistocene or Holocene and permitted the influx of meteoric waters, presumably along fractures in the Monterey Shale as well as through permeable sandstones. The northward tilt of the oil-water contact at Yowlumne Field may be the result of hydrodynamic flow of formation waters from the outcrops located a short distance south of the field.

Reservoir Characterization

The reservoir has a gross thickness of from 50 to 300 feet (15–91 m) and is composed of thin beds separated by shales (Fig. 19-5). Single beds are 1 to 4 feet (0.3 to 1.2 m) thick and average 2 feet (0.6 m). Each bed shows an ordered sequence of sedimentary structures from massive in the lower part to laminated at the top. These divisions represent part of the turbidite sequence (Bouma, 1962) that consists, from the base upward, of a structureless sand unit (T_a), an overlying laminated sand unit (T_b), a ripple cross-laminated sand unit (T_c), an upper laminated silt (T_d), and an overlying pelagic shale (T_e). The ideal sequence (T_{abcde}) is not common, and typical turbidites contain only a partial sequence. Thicker beds of the T_a or T_{ab} type represent channel-fill deposits; thinner, more complete beds of the T_{abcd} type represent proximal overbank deposits; and very thin, less complete beds of the T_{bcd} type represent distal overbank deposits (Berg, 1986).

The Yowlumne sandstones are dominantly massive to laminated beds (Type T_{ab}) that represent channel-fill turbidites (Fig. 19-6). Rarely, a ripple cross-laminated division is present at the top of beds (Type T_{abc}) to form a somewhat more complete turbidite sequence. Nowhere are the complete, proximal to distal overbank sequences present within the reservoir. The beds also show other common features of turbidites such as sharp basal contacts on underlying shales, basal load features, and some contorted bedding.

The reservoir section contains variable amounts of interbedded shale that, together with bed thickness, denotes differences in position within the depositional channel. These differences are reflected by the gamma-ray log in a typical section (Fig. 19-5). A lower section contains thicker sandstones and the interbedded shales are thin; a middle section consists of thinner sandstone and more numerous interbedded shale; and an upper section has thicker sandstones and a greater number of shales. The facies containing thicker beds with less shale are deposits of the central channel; the thinner beds with more interbedded shales are the channel-margin deposits.

The total section has been divided into six informal members for reservoir mapping. Members 2, 3, 4, and 5 are more widely present throughout the unit, whereas members 1 and 6 have a limited extent. Each of the members is bounded by persistent shale beds that can be correlated throughout the reservoir, and each member shows lateral changes from the thicker, central-channel facies to the thinner, shaly channel-margin facies.

Petrography and Diagenesis

The mean grain size of the sandstones is 0.3 mm (medium grained), and each bed shows an upward-fining grain size from the massive to the laminated or rippled units (Fig. 19-7). The sandstones are feldsarenites (arkoses) that on average contain 53% quartz and 35% feldspar, largely orthoclase, as detrital grains and in granitic rock fragments. The average matrix content is 9% and is composed largely of authigenic kaolinite (Tieh et al., 1986). Carbonate cement averages 5% of the bulk volume.

Burial diagenesis has altered the original composition and resulted in abundant secondary pores (Tieh et al., 1986). Carbonate cements were partly removed by dissolution, and feldspars were altered by reaction with pore fluids. Some feldspar grains

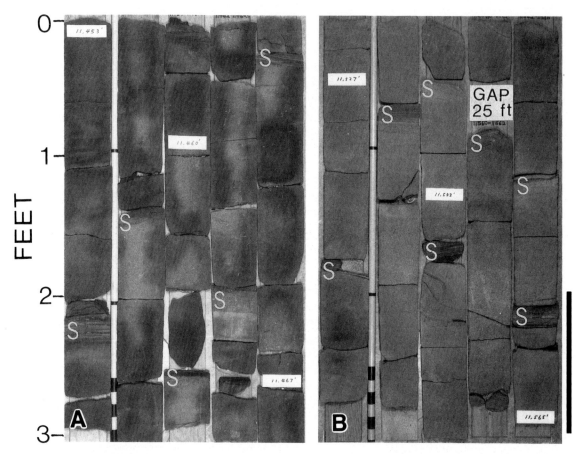

Fig. 19-6. Representative core slabs of the Yowlumne sandstone showing bedding characteristics. Massive sandstones are 1 to 3.5 feet (0.3–1.1 m) thick and separated by thin shales (labeled S). The core is from well 54X-4 in Unit B, Yowlumne Field (Fig. 19-2). (A) Section from 11,453 to 11,467.5 feet (3,491–3,495 m). (B) Section from 11,526.5 to 11,565 feet (3,513–3,525 m). Gap denotes missing section of 25 feet (7.6 m).

were almost completely dissolved, leaving oversize pores. Relict feldspar grains show that alteration took place by dissolution along compositional and cleavage planes (Fig. 19-8A). Kaolinite was the principal product of feldspar alteration and constitutes approximately 66% of the clay minerals (Tieh et al., 1986). Kaolinite crystals are located adjacent to altered feldspar grains or occupy pores (Fig. 19-8B). Other clays are present in minor amounts: illite (15%), vermiculite and chlorite (16%), and montmorillonite (3%).

The net result of diagenesis is that the Yowlumne sandstone has higher porosities and permeabilities than do the volcanic-rich sandstones of the main body of the Stevens sandstone at the same depth (Berg, 1986). Diagenesis of the volcanic rock fragments altered and recrystallized the groundmass to a labile matrix that reduced permeabilities to low values at depths below 11,000 feet (3,350 m) (Berg, 1986). Although the authigenic kaolinite is relatively common in the Yowlumne Field, it occurs in amounts too small to have a significant effect on permeability.

Reservoir Properties

Porosity and permeability of the sandstones can be related to the depositional facies. The central-channel sandstones have higher porosities in the range of 17 to 20% and permeabilities of 50 to 200 md (Fig. 19-5). The channel-margin sandstones have lower and variable porosities of 10 to 20% and permeabilities of 10 to 100 md. Consequently, the central-channel facies has produced most of the oil

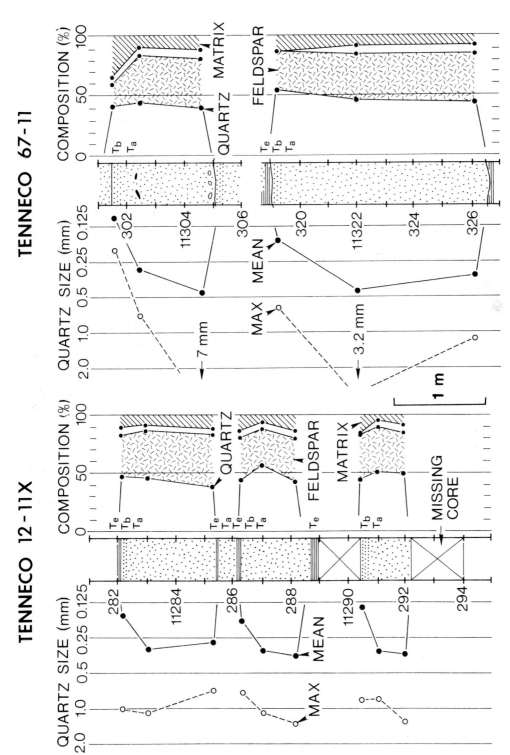

Fig. 19-7. Texture, composition, and sedimentary structures of typical beds in the Yowlumne reservoir. Letters at the right of the depth columns (in feet) represent turbidite divisions of Bouma (1962) (T_a = massive, T_b = laminated, T_e = shale).

Fig. 19-8. Petrographic characteristics of the Yowlumne sandstone. (A) Thin-section photomicrograph of relict plagioclase-feldspar grain (f) showing dissolution along compositional and cleavage planes. Scale bar is 0.05 mm long. (B) SEM photomicrograph showing kaolinite crystals (k) filling a pore adjacent to quartz grain with overgrowth (q_o). Scale bar is 0.02 mm long. Both photographs from T. Tieh (personal communication, 1989).

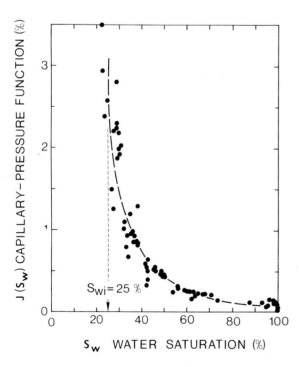

Fig. 19-9. Capillary displacement pressures in the Yowlumne reservoir normalized by the $J(S_w)$ function and showing irreducible water saturation (S_{wi}) of 25%.

and appears to have the greater connectivity for water injection.

Residual oil saturations measured in cores range from 10 to 15%, and residual water saturations range from 40 to 60%. Capillary displacement-pressure measurements (Fig. 19-9) indicate that the irreducible water saturation (S_{wi}) is 25% (Clark, 1987). The capillary pressures are expressed in terms of the nondimensional $J(S_w)$ function which normalizes the measurements on different samples for variations in porosity and permeability (Amyx et al., 1960). In well-log evaluation, the cutoff for water-free oil production is 55% of water saturation.

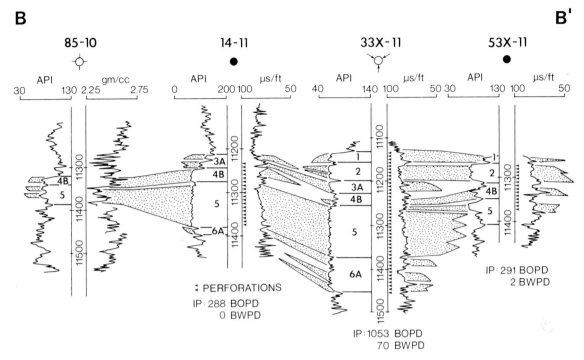

Fig. 19-10. Correlation of well logs showing the lenticular nature of members that are defined by persistent shale units. Perforations and initial production (IP) are also shown. Location of section is shown in Figure 19-2 and in Figures 19-11A and C.

Well-Log Response

Both composition and bedding play important roles in the response of well logs. The orthoclase content of the sandstones is reflected in the natural gamma-radiation, and the presence of interbedded shales also increases the gamma-ray response. The central-channel facies has gamma-ray values of 60 to 80 API, whereas the channel-margin facies has values of 70 to 90 API because of the greater number of interbedded shales (Fig. 19-5). Induction resistivity is also affected by shales and is generally higher in the central-channel facies than in the channel margin, although the fluid saturations may be essentially the same.

The porosity logs are similarly influenced by shales, especially the sonic logs (Fortner, 1988). Comparison of core porosities and interval-transit times from sonic logs indicates that velocities are decreased significantly in the shaly channel-margin facies. Consequently, the calculation of pore volumes for the reservoir from the sonic logs requires the use of different time-porosity functions for the two facies. The density-log response is less affected by the interbedded shales; however, density logs were run in fewer wells, whereas sonic logs were run in all wells.

Another useful tool for reservoir evaluation is the microlog, which delineates thin, permeable beds; however, this log also was run in only a few wells.

Reservoir Morphology

The size and shape of the reservoir were established by correlation of well logs and by defining and mapping the members which represent the major productive intervals. These members are a complex of interbedded sandstones and shales that are bounded by major shale beds.

Sandstone Members

The Yowlumne sandstone can be divided into six informal members (Fig. 19-10). Members 2, 3, 4, and 5 are present widely throughout the A unit, whereas members 1 and 6 are restricted in extent. The members are defined by thick beds of shale which are 20 feet (6.1 m) or more thick, persistent between wells, and recognized by high gamma-ray response. In addition, the member boundaries were defined so that dipmeter-log patterns, as described in a following section, were not divided between two members. Thicker shales are common in the upper part of the Yowlumne Field, whereas only thin shales are present in the lower part. Member 5 forms the thickest, more continuous section of sandstones and cannot be further subdivided.

The distribution of net porous sandstone is illustrated for members 5 and 6 (Fig. 19-11). Member 5 is present throughout the A unit and has a net thickness of 100 feet (30.5 m) at the south to nearly 200 feet (61 m) at the north. The member is composed of thick, channel-fill sandstones, and interbedded shales of the channel-margin facies are restricted to the channel edges. In contrast, the underlying member 6 is not continuous but is confined to two separate areas within the A unit and has a maximum net sand thickness of 80 feet (24.4 m). The channel-fill facies is more narrowly confined in member 6 by the interbedded sandstone and shale of the channel-margin facies.

Flow Units

The six members cannot be readily subdivided into mappable units, but each member is composed of numerous, thin turbidite sandstones separated by thin shales. Each bed represents a single depositional event and, where the separating shales are continuous, each bed also represents a *flow unit* that is homogeneous and can transmit fluid laterally through the extent of the bed (Ebanks, 1987).

Because the intervening shales have much lower permeability than do the sandstones, they can be expected to confine the flow of fluids to single beds. Therefore, the Yowlumne reservoir can be characterized as a thinly layered reservoir which lacks cross flow between the layers. Furthermore, the thin beds are probably lenses that have restricted lateral extent and may allow fluid communication through a limited area. Therefore, the reservoir is composed of layered and lensed flow units.

Internal Structures

The interpretation of dipmeter logs can aid in the definition of thin flow units in the Yowlumne reservoir. Distinctive patterns of dip magnitude and direction are observed in most sections, and these

patterns can define the general trends of sandstone beds and the directions of flow within them.

The interpretation of dipmeter logs in a thinly layered turbidite sequence must consider the source of dip. The interpretation of sandstone dips as representing transport directions, as in cross-bedded, fluvial sandstones (Selley, 1979), is not valid for turbidites. The turbidites are mostly massive and lack well-defined stratification, and no recorded dips can be expected within the sandstone beds. Rather, the recorded dips probably represent the attitudes of thin shales between the sandstones. In this case, the dip directions are most likely normal to the direction of current flow and not parallel to flow.

This interpretation of dipmeter logs in turbidites was first suggested for sandstones in the Delaware Mountain Group of southeastern New Mexico (Berg, 1979). Subsequently, the interpretation was documented on dipmeter logs within a single field, and consistent patterns of shale dips were found to reflect the extent of turbidite units (Phillips, 1987).

Dip Patterns

Dipmeter logs show characteristic patterns of increasing and then decreasing dip upwards through sections that range from 10 to 40 feet (3.0–12.2 m) in thickness (Fig. 19-12). These patterns are present in all wells for which dipmeter logs were run, but the patterns are especially well developed in thick sandstones, such as member 5, that cannot otherwise be subdivided by log response.

Typical dip patterns through member 5 are shown in well 81X-10 located in the center of Unit A (Fig. 19-12). Structural dip was removed from the recorded true dips such that the dips have been restored to reflect stratigraphic conditions within the section.

The dip patterns define 5 submembers, labeled *a* through *e* (Fig. 19-12). One submember (*d*) has horizontal dip, but the other submembers show dip magnitudes increasing upwards to maximum values of 22° to 32° and then decreasing to 10° or less at the top. Accompanying the dip change is a generally consistent change upward in dip azimuths in a clockwise direction. Azimuths migrate upward from east to south in submember *e*, from southeast to northwest in submembers *c* and *b*, and from southwest to northeast in submember *a*.

It is assumed that the dip patterns represent the attitudes of interbedded shales which were deposited

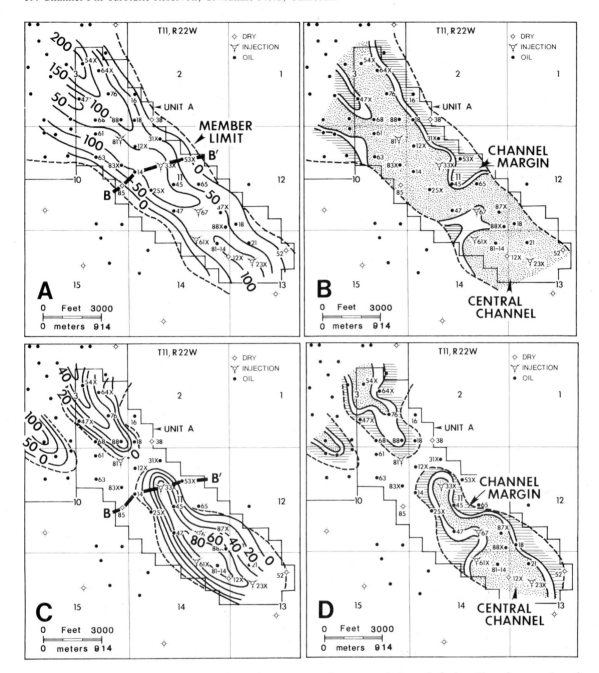

Fig. 19-11. Distribution of net porous sandstone in members 5 (A and B) and 6 (C and D) of the Yowlumne reservoir. Contoured in feet. Channel limits (dashed lines) enclose central-channel facies (dotted pattern) and channel-margin facies (horizontal lines). Correlation section B-B' is shown in Figure 19-10.

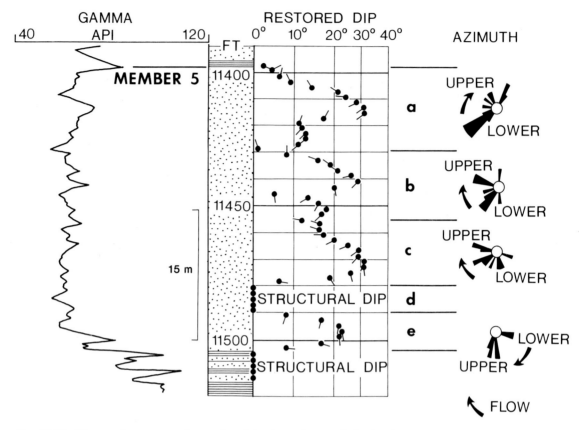

Fig. 19-12. Dipmeter-log patterns in member 5 in the 81X-10 injection well which define submembers **a** through **e** by sets of increasing, then decreasing, dip upward. Note clockwise rotation upward of dip azimuths. Structural dip has been removed to show stratigraphic (restored) dips. Location of well is shown in Figure 19-15 (NE NE section 10).

between the periods of sand deposition from the turbidity flows. Thus, the dips represent the drape of shales over the previously deposited sands, and the patterns can be interpreted in terms of the direction of flow.

It is well known that channelized flow has a sinuous pattern in subaerial streams as well as in deep-sea channels (Berg, 1986; Flood and Damuth, 1987). Deposition of the Yowlumne turbidites probably took place within low sinuousity channels, perhaps braided, because of relatively high discharge of the flows. It is expected that these sands were deposited within channel loops as thin lenses that were elongated in a downstream direction. Furthermore, the deposits of successive flows probably migrated downstream, similar to point bars deposited within the meander loops of modern streams.

The channel turbidites and interbedded shales probably formed composite units like those of the submembers defined by dipmeter patterns in the Yowlumne reservoir. Each submember represents the deposits within a single, laterally migrating channel, and each was succeeded by the deposits of another channel that migrated across the location.

The dipmeter pattern through a single series of migrating channel deposits can be visualized as follows (Fig. 19-13). A composite unit of three sandstone lenses and their bounding shales would show clay drapes with low dip at the base, increasing dip upward, and then decreasing dip at the top. The dip directions of the composite unit would be toward the channelway, and they should migrate upward in the direction of flow around the well location and should reflect the down-channel migration of sand lenses. In the diagram, the flow direction is interpreted to be clockwise around the location from southwest to northwest (Fig. 19-13).

Alternatively, the dipmeter patterns could be interpreted as a series of normal faults, but there is

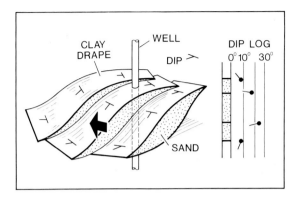

Fig. 19-13. Diagram of recorded dips through a section of overlapping sandstone lenses that were deposited on the inside of meandering channels. Dips are determined from clay drapes between the lenses, and the dips increase, then decrease, upward while dip azimuths rotate in a downstream direction as indicated by the large arrow.

no structural evidence for tectonic faulting. Rather, the close spacing of maximum dips (Fig. 19-13) suggests that they may represent small-scale slump faulting that occurred soon after deposition. Such faulting may have taken place periodically along shale beds after the accumulation of 20 to 30 feet (6.1–9.1 m) of newly deposited sediment. In this case, the dip patterns would indicate drag by soft-sediment deformation adjacent to the slump fault. The variations in dip directions could be the result of initial dips, as described above (Fig. 19-13), or by contortion of the sediments.

In the case of either depositional dip or post-depositional slumping, the dip from a dipmeter would point in a *cross-channel* direction or toward the open channelway and normal to the direction of turbidity-current flow. Clay drapes would inhibit the movement of reservoir fluids in this direction. On the other hand, the continuity of sandstone beds in an *along-channel* direction would promote the movement of fluids toward producing wells located near the channel axis. In other words, a strong flow anisotropy should be present within the reservoir.

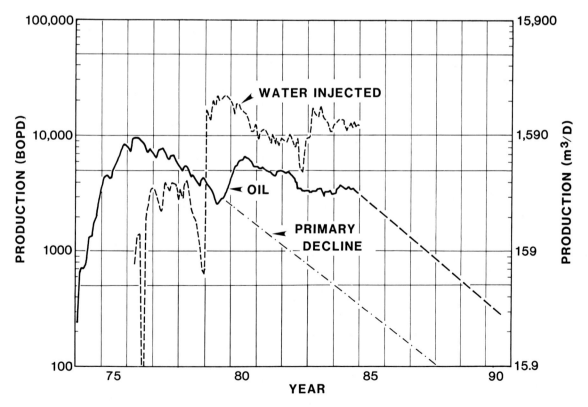

Fig. 19-14. Oil production in Unit A, Yowlumne Field, showing the response to water injection. Modified from Clark (1987).

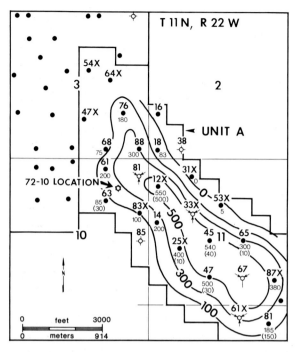

Fig. 19-15. Waterflood performance is illustrated by total fluid production in January, 1980, 12 months after water injection began. Note location of injection well 81X-10 (NE NE section 10) shown in Figure 19-10. Well 12X-11 had water breakthrough in 4 months. The 72-10 location refers to an infill well drilled after 42 months of water injection, which was completed as an oil producer (see discussion under Waterflood Response). Figures show total fluid rate in bbl/D and water rate in bbl/D in parentheses.

Reservoir Performance

Oil production reached nearly 10,000 BOPD (1.6×10^3 m³/D) in 1976, two years after discovery (Fig. 19-14). However, declining pressures soon indicated the need for water injection, and so the A Unit was formed while development continued along the north flank in the area that later became the B Unit. Full-scale water injection began in January, 1979 (Burzlaff, 1983). The internal structure of the reservoir influenced oil recovery. Little effect was noted during primary production, but a preferred direction of flow within the reservoir was seen during water-flooding and after infill drilling.

Waterflood Response

The waterflood was begun with one injection well (61X-14) in 1976 and, after unitization, three

more wells were converted to injectors along the crest of the structure in 1978. The additional injection wells were, from southeast to northwest, 67-11, 33X-11, and 81X-10 (Fig. 19-15). The response to water injection is shown by total fluid production at the end of 1979, 12 months after full-scale injection began. Oil production ranged from less than 100 BOPD (16 m³/D) to more than 500 BOPD (80 m³/D) with small amounts of water. The higher production rates were from wells located along the central part of the channel, and lower well production rates occurred at the margins. One exception was well 12X-11 in which water breakthrough occurred only 4 months after injection began in the nearby well 81X-10.

The response of producing wells to injection differed according to their locations with respect to the inferred dip of reservoir lenses. The differing response is best shown around injection well 81X-10, which has cross-channel dips toward the southwest (Fig. 19-12). Producing wells located to the east and west, generally along the channel trend, showed early response in oil production but earlier breakthrough of water (Fig. 19-16). Breakthrough occurred after 4 months in well 12X-11 to the east and after 38 months in well 61-10 to the west.

Producing wells located to the north and south, generally across the channel trend, showed a delayed response in oil production of 16 months in well 18-2 to the north of the injector and 15 months in well 83X-10 to the south (Fig. 19-16). There was no breakthrough of water, however, in the cross-channel direction, and the water-oil ratios showed a general decline in both producing wells.

Although greater oil recoveries were achieved in the cross-channel wells, there is evidence that the waterflood was not completely effective in that direction. An infill well (72X-10) was located about 1,000 feet (305 m) to the southwest of the 81X-10 injector and near well 63-10 (Fig. 19-16). The new well was completed in March, 1982, about 42 months after water injection began, but the well produced at an average rate of 130 BOPD (21 m³/D) of nearly water-free oil for 6 months and had a cumulative oil recovery of 55,000 barrels (8.7×10^3 m³) during a period of 33 months. Evidently, the produced oil came from reservoir lenses that were not drained in either the primary or the secondary stages. This conclusion suggests that single channel-fill lenses are less than 1,000 feet (305 m) wide in a cross-channel direction.

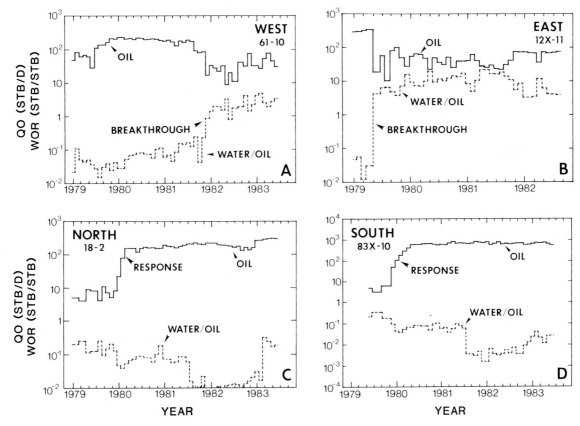

Fig. 19-16. Oil production rates (QO) and water-oil ratios (WOR) for producing wells located around well 81X-10 in which water injection began in October, 1978. (A, B) Early oil response and water breakthrough in along-channel wells to the east and west. (C, D) Delayed oil response and declining WOR in cross-channel wells to the north and south. Oil rates are in stock-tank barrels per day (STB/D).

Connectivity of Flow Units

The performance of wells during waterflooding can be explained in terms of the internal structure of the reservoir and the lack of complete connection of lenses between injection and producing wells. A low degree of connectivity is illustrated by a diagrammatic section in a cross-channel direction (Fig. 19-17). The reservoir is shown as two shale-bounded members, each of which consists of inclined lenses of sandstone that are also interbedded with thin shales.

During primary production, no barriers to oil flow are detected, and the producing wells drain parts of the different lenses. During waterflooding, however, the injected water invades only the connected lenses. Some lenses are continuous to a producing well in one direction, while others are connected in an opposite direction. Still other lenses remain

unflooded or have a connection to a producing well farther from the injector along the channel. The result is incomplete flooding of the reservoir.

The partly drained but unflooded lenses may represent a significant part of the total pore volume of the reservoir, perhaps 25% or more between wells located 1,000 feet (305 m) apart. These lenses probably contain a large amount of movable oil which can be recovered only by the drilling of additional, closely spaced infill wells. Prime locations for additional oil recovery can be detected by reservoir simulation of the produced and injected volumes of fluids.

The depositional patterns of the reservoir lenses are only indirectly confirmed by the production histories, and further details cannot be resolved throughout the extent of the reservoir. Most wells, both producers and injectors, were perforated through the entire reservoir section including parts

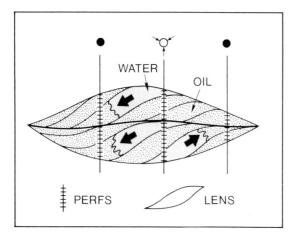

Fig. 19-17. Cross-channel diagram of a complexly layered and lensed reservoir like the Yowlumne sandstone showing shale-bounded flow units within two members. During water injection (middle well), the flood is effective only through a limited number of flow units that are intercepted by both the injection and production wells. Other lenses are not flooded in a cross-channel direction.

of two or more members so that fluid volumes cannot be assigned accurately to the members. In the case of the 81X-10 injector, however, most of the injected fluid can be attributed to flooding of member 5 as based on the interpretation of well spinner surveys (Clark, 1987). Furthermore, the connectivity of lenses cannot be established everywhere because dipmeter logs were run only in half of the wells.

Despite the lack of further detail, it seems likely that the internal structure of the reservoir exerts a strong control on oil recovery. In fact, a better reservoir description, and a prediction of performance, might have been made at an early stage of development had dipmeter logs been run in all wells.

Exploration and Production Strategy

Exploration for channel-fill turbidite reservoirs can be conducted by reconnaissance seismic surveys. Although the Yowlumne reservoir was discovered by drilling a structural closure, it became apparent during development that the closure was at least partly the result of differential compaction of shale over channel sandstone.

A seismic profile down the north flank of the San Emidio Nose and across the field shows that prominent reflections diverge and then converge to

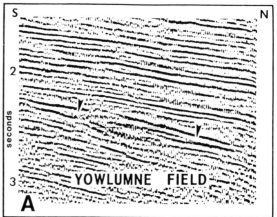

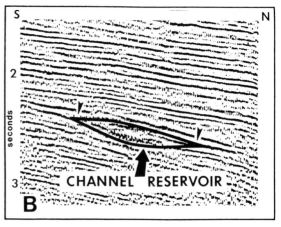

Fig. 19-18. Part of a seismic section across the Yowlumne Field showing the location of the channel-fill reservoir (arrows). (A) Migrated seismic section at a compressed horizontal scale. (B) Interpreted outline of the reservoir section. Width of channel reservoir is approximately 2 miles (3.2 km). Original section from Dorman (1980) and published by permission of San Joaquin Geological Society.

enclose the reservoir sandstones (Fig. 19-18). The same lensoidal configuration is also present at the top of the structure but is less distinct than on the north flank (Dorman, 1980). The lenticular outline of the seismic reflections encompasses about 110 milliseconds (two-way time) or a thickness of about 650 feet (200 m), considerably more than the reservoir interval of 300 feet (91 m), which suggests that a significant section of enclosing beds is incorporated within the lenticular configuration. Channel-fill sequences of a thickness similar to the Yowlumne channel should also be detectable by seismic methods.

Exploration must also consider the depth, composition, and source of the sands. The Yowlumne feldsarenites retain relatively high porosities and permeabilities at depths below 11,000 feet (3,350 m) because of the dissolution of grains and cements. In contrast, the volcanic-rich sandstones of the main body of the Stevens sandstone do not have secondary porosity at depths below 11,000 feet (3,350 m), and permeabilities are greatly reduced (Berg, 1986; Tieh et al., 1986).

The best prospect for improved oil recovery from the Yowlumne reservoir appears to be the drilling of additional infill wells. The production history should be simulated for both primary and secondary recoveries in order to detect locations that could contain undrained lenses within the reservoir.

Conclusions

The Yowlumne channel-fill, turbidite reservoir is composed of lenticular flow units that are narrow but elongated parallel to the channel trend. In terms of fluid production, the flow units are individual thin beds that are bounded by low permeability shale. Dips recorded by well logs reflect the attitudes of thin shales between flow units, and groups of flow units form submembers characterized by distinct dipmeter patterns in which dips are inclined in a cross-channel direction.

Well performance during water injection suggests that there is strong flow anisotropy within the reservoir. Water breakthrough occurred early in the along-channel direction, whereas there was good oil response but no water breakthrough in a cross-channel direction. Because of the layered and lensing nature of the reservoir, secondary recovery of oil cannot be entirely effective with existing wells. Infill drilling will be required to drain the reservoir more completely.

Evidence from infill drilling indicates that some reservoir flow units are 1,000 feet (305 m) or less in width in a cross-channel direction. These narrow lenses may contain original pressures and unproduced oil in fields that are drilled on wide well spacings.

The development of reservoirs similar to Yowlumne Field should include complete logging suites in every well. Dipmeter logs, especially, can prove to be most useful in defining reservoir flow units and in predicting internal barriers to flow of the produced or injected fluids.

Acknowledgments. This chapter presents a part of the geologic description of the Yowlumne reservoir undertaken by the Chrisman Institute for Reservoir Management in cooperation with the Department of Petroleum Engineering, Texas A&M University. Tenneco Oil Company provided all reservoir data as well as major financial support to the Chrisman Institute. We are indebted to John C. Calhoun, director of the Chrisman Institute, B. Desidier and M. Kudchadker, Tenneco Oil, and to faculty members of the Petroleum Engineering Department, Texas A&M, for their support of the study and review of the conclusions.

Reservoir Summary

Field: Yowlumne
Location: Kern County, central California
Operator: Tenneco Oil Company (now ARCO)
Discovery: 1974
Basin: San Joaquin basin
Tectonic/Regional Paleosetting: Remnant forearc/transform-margin basin
Geologic Structure: Anticlinal nose
Trap Type: Stratigraphic pinch-out across nose
Reservoir Drive Mechanism: Water drive and gas expansion
• **Original Reservoir Pressure:** 5600 psi (3.9 × 10⁴ kPa) at 10,500 feet (3,200 m) subsea
Reservoir Rocks
• **Age:** Late Miocene
• **Stratigraphic Unit:** Stevens sandstone
• **Lithology:** Medium-grained, feldsarenite (arkose)
• **Depositional Environment:** Submarine fan

- **Productive Facies:** Central channel-fill deposits
- **Petrophysics**
  - **Porosity Types:** Intergranular, dissolution
  - **ϕ:** Average 18%, range 5 to 23% (cores)
  - **k:** Average 100 md, range 1 to 700 md (cores)
  - **S_w:** Average irreducible 25%, average core 45%, range 25% to 50%, cutoff 55% (logs)
  - **S_{or}:** Average 15% (cores)

Reservoir Geometry
- **Depth:** 11,300 to 13,400 feet (3,445–4,085 m)
- **Areal Dimensions:** 1.5 by 3.5 miles (2.4 × 5.6 km)
- **Productive Area:** 3,100 acres (1,255 ha.)
- **Number of Reservoirs:** 1
- **Hydrocarbon Column Height:** 2,000 feet (610 m)
- **Fluid Contacts:** Oil-water contacts at 10,700 feet (3,261 m) subsea (south) to 12,700 feet (3871 m) subsea (north)
- **Number of Pay Zones:** 6
- **Gross Sandstone Thickness:** Average 200 feet (61 m)
- **Net Sandstone Thickness:** Average 150 feet (46 m)
- **Net/Gross:** 0.75

Hydrocarbon Source, Migration
- **Stratigraphic Unit:** Monterey Shale
- **Time of Maturation:** Early Pliocene
- **Time of Trap Formation:** Middle Pleistocene
- **Time of Migration:** Pleistocene

Hydrocarbons
- **Type:** Oil
- **GOR:** 700:1
- **API Gravity:** 34°
- **FVF:** 1.40
- **Viscosity:** 0.5 cP (5.0 × 10² Pa·s) at reservoir pressure and temperature

Volumetrics
- **In-Place:** 280 MMBO (4.4 × 10⁷ m³)
- **Cumulative Production:** 42 MMBO (6.7 × 10⁶ m³)
- **Ultimate Recovery:** 78 MMBO (1.2 × 10⁷ m³)
- **Recovery Efficiency:** 28%

Wells
- **Spacing:** 1320 feet (402 m); 40 acres (16.2 ha.)
- **Total:** 93
- **Dry Holes:** 8

Typical Well Production
- **Average Daily:** 500 BO (80 m³)
- **Cumulative:** 900 MBO (1.4 × 10⁵ m³)

Other
- **Waterflood:** Initiated in 1978
- **Formation Water Salinity:** 15,000 to 20,000 ppm NaCl

References

Amyx, J.W., Bass, D.M., Jr., and Whiting, R.L., 1960, Petroleum Reservoir Engineering, Physical Properties: New York, McGraw-Hill, 610 p.

Bandy, O.L., and Arnal, R.E., 1969, Middle Tertiary basin development, San Joaquin Valley, California: Geological Society of America Bulletin, v. 80, p. 783–820.

Berg, R.R., 1979, Reservoir sandstones of the Delaware Mountain Group, southeast New Mexico, *in* Guadalu-pian Delaware Mountain Group Symposium: Society of Economic Paleontologists and Mineralogists Permian Basin Section, Publication 79-18, p. 75–95.

Berg, R.R., 1986, Reservoir Sandstones: Englewood Cliffs, New Jersey, Prentice Hall, 481 p.

Bouma, A.H., 1962, Sedimentology of some flysch deposits: A Graphic Approach to Facies Interpretation: Amsterdam, Elsevier Publishing Company, 168 p.

Burzlaff, A.A., 1983, Unitizing and waterflooding the California Yowlumne oil field: Society of Petroleum Engineers 53rd Annual California Regional Meeting,

March 23–25, 1983, Ventura, California, Proceedings, p. 187–193.

Callaway, D.C., 1971, Petroleum potential of San Joaquin basin, California: American Association of Petroleum Geologists Memoir 15, v. 1, p. 239–253.

Clark, J.W., 1987, Development of a description method for the Yowlumne Unit A reservoir [M.S. thesis]: College Station, Texas, Texas A&M University, 106 p.

Dorman, J.H., 1980, Oil and gas in submarine channels in the Great Valley, California: San Joaquin Geological Society Selected Papers, v. 5, p. 38–54.

Ebanks, W.J., 1987, Flow unit concept-integrated approach to reservoir description for engineering projects [abst.]: American Association of Petroleum Geologists Bulletin, v. 71, p. 551–552.

Flood, R.D., and Damuth, J.E., 1987, Quantitative characteristics of sinuous distributary channels on the Amazon Deep-Sea Fan: Geological Society of America Bulletin, v. 98, p. 728–738.

Fortner, D.W., 1988, The effects of composition and bedding on log response, Yowlumne sandstone, Kern County, California [M.S. thesis]: College Station, Texas, Texas A&M University, 195 p.

Graham, S.A., and Williams, L.A., 1985, Tectonic, depositional and diagenetic history of Monterey Formation (Miocene), central San Joaquin basin, California: American Association of Petroleum Geologists Bulletin, v. 69, p. 385–411.

Hackel, O., 1966, Summary of the geology of the Great Valley, California, in Bailey, E.H., ed., Geology of Northern California: California Division of Mines and Geology Bulletin 190, p. 217–238.

Isaacs, C.M., and Petersen, N.F., 1987, Petroleum in the Miocene Monterey Formation, California, in J.R. Hein, ed., Siliceous Sedimentary Rock-Hosted Ores and Petroleum: New York, Van Nostrand Reinhold Co., p. 83–116.

MacPherson, B.A., 1978, Sedimentation and trapping mechanism in upper Miocene Stevens and older turbidite fans of southeastern San Joaquin Valley: American Association of Petroleum Geologists Bulletin, v. 62, p. 2243–2274.

Metz, R.T., and Whitworth, J.L., 1983, Yowlumne oil field: San Joaquin Geological Society Selected Papers, v. 6, p. 3–11.

Petersen, N.F., and Hickey, P.J., 1987, California Plio-Miocene oils: Evidence of early generation: American Association of Petroleum Geologists Studies in Geology 25, p. 351–359.

Phillips, S., 1987, Dipmeter interpretation of turbidite-channel reservoir sandstones, Indian Draw Field, New Mexico: Society of Economic Paleontologists and Mineralogists Special Publication 40, p. 113–128.

Royo, G.R., 1986, Environment of deposition of the Yowlumne sandstone: Internal morphology and rock properties, Kern County, California [M.S. thesis]: College Station, Texas, Texas A&M University, 167 p.

Selley, R.C., 1979, Dipmeter and log motifs in North Sea submarine-fan sands: American Association of Petroleum Geologists Bulletin, v. 63, p. 905–917.

Sullwold, H.H., Jr., 1961, Turbidites in oil exploration, in Peterson, J.A. and Osmond, J.C., eds., Geometry of Sandstone Bodies: Tulsa, Oklahoma, American Association of Petroleum Geologists, p. 63–81.

Taylor, D.S., 1978, California's Yowlumne Field – From basics to barrels: Oil and Gas Journal, March 20, p. 192–200.

Tieh, T.T., Berg, R.R., Popp, R.K., Brasher, J. E., and Pike, J.D., 1986, Deposition and diagenesis of Upper Miocene arkoses, Yowlumne and Rio Viejo fields, Kern County, California: American Association of Petroleum Geologists Bulletin, v. 70, p. 953–969.

Webb, G.W., 1981, Stevens and earlier Miocene turbidite sandstones, southern San Joaquin Valley, California: American Association of Petroleum Geologists Bulletin, v. 65, p. 438–465.

Zieglar, D.L., and Spotts, J.H., 1978, Reservoir and source-bed history of Great Valley, California: American Association of Petroleum Geologists Bulletin, v. 62, p. 813–826.

Key Words

Yowlumne Field, California, San Joaquin basin, Stevens Sandstone, Monterey Shale, late Miocene, submarine fan, turbidite sandstone, hydrodynamics, tilted oil-water contact, dipmeters, flow anisotropy, differential compaction, waterflood.

20

Geological Modeling of a Turbidite Reservoir, Forties Field, North Sea

Alexander A. Kulpecz and Lucia C. van Geuns

Shell U.K. Exploration and Production, London, England; Shell Internationale Petroleum Maatschappij, The Hague, Netherlands

Introduction

Turbidite reservoirs are an important source of production in the U.K. sector of the North Sea, comprising approximately 25% of total recoverable oil and gas reserves (excluding the Southern Gas Basin). Geological modeling of these complex reservoirs plays a crucial role in the efficient recovery of hydrocarbons initially in place. This chapter demonstrates that a depositional model based on chronostratigraphic interpretation helps in the understanding of fluid-flow parameters such as sandstone distribution, internal architecture, vertical transmissibility barriers, and lateral discontinuities. Such a model was used to develop a new reservoir zonation scheme for Forties Field, a scheme which provided the basis for reservoir simulation and economic evaluation.

Forties Field is located in the U.K. sector of the Central North Sea within Blocks 21/10 (British Petroleum - BP) and 22/6a (Shell/Esso)(Figs. 20-1, 20-2). It is one of the largest producing oil fields in the North Sea, with recoverable reserves of approximately 2.5 billion barrels (3.9×10^8 m³). The field was discovered by BP in 1970 with well 21/10-1 drilled to a depth of 7,000 feet (2,135 m) during the early exploration phase of the North Sea. Since 1975, the field has produced 1.9 billion barrels of oil (3.0×10^8 m³) and is in a mature stage of development. More than 125 wells have been drilled in the field, 83 of which are producers.

Water injection was started early in the field life to provide pressure support. The operator of the field, BP, brought a fifth platform on stream in early 1987 (FE platform, Block 22/6a; also known as the Southeast Forties area).

The 1987 final equity redetermination was based on economically recoverable oil and triggered a geological review of the field. A full-field reservoir simulation was required to quantify the additional oil for various development options (development wells, workovers, artificial lift), which were then screened for profitability. The geological model is an integral part of the reservoir simulation and ultimately affects the field development scheme.

The Forties Formation was deposited in a submarine-fan setting. Stratigraphic correlation in the field has always been difficult due to complex lithofacies changes, which isopach and core data suggest as representing intercalated lobe and channel deposits. A chronostratigraphic scheme based on palynological zonations provided a framework for correlation of the reservoir sequence. This scheme was developed by Shell in the early 1970s and is used in exploration and development in the North Sea to establish lateral continuity and sandbody geometry. The chronostratigraphic subdivision of the Forties Formation highlighted the interfingering nature of channel and lobe sequences and formed the basis of a depositional model for reservoir simulation.

Casebooks in Earth Science
Sandstone Petroleum Reservoirs
Eds.: Barwis/McPherson/Studlick
© 1990 by Springer-Verlag New York, Inc.

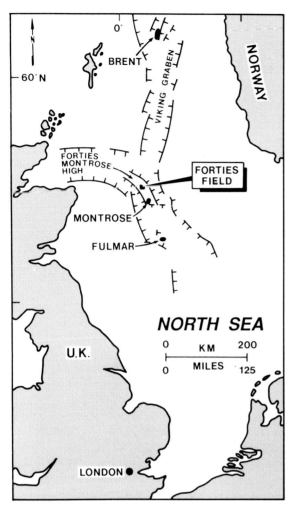

Fig. 20-1. The Forties Field, located in the United Kingdom sector of the North Sea.

Regional Data

The Forties Formation is the uppermost unit of the Paleocene Montrose Group. It directly overlies the Andrew Formation, which is generally a poorer sand-quality submarine-fan deposit (Fig. 20-3). The sandstone reservoir of the Forties Field forms part of a major submarine-fan body which covered the Central Graben area of the North Sea basin during the Paleocene (Fig. 20-4).

Paleocene stratigraphy in the Central Graben represents a renewal of tectonic activity which poured clastics onto the shelf and the relatively stable Late Cretaceous to Danian Chalk basin (Parker, 1975; Hillier et al., 1978). The large composite fan

thins toward the southeast and is a distinctive seismic feature throughout the region. Figure 20-4 is a reconstruction of the late Paleocene paleogeography based on regional seismic stratigraphy showing the sandstone-rich submarine fan in the greater Forties area. A rise of relative sea level may have terminated the coarse clastic influx of the Forties Formation with the deposition of basinal mudstones and tuffaceous mudstones of the Sele and Balder formations. These fine-grained deposits act as the top seal to the Forties reservoir.

The Jurassic Kimmeridge Formation, which is the major source rock in the North Sea, is absent in the Forties geographic area. However, it is believed that the oil is sourced from the Kimmeridge and migrated from the southeast along a path which intersected the Montrose Field.

Upper Tertiary sediments dip gently eastward along the axis of the Central Graben. The Forties Field structure is a broad, low-relief domal anticline with four-way dip closure produced by the draping of deepwater clastics over the Forties-Montrose ridge. Dips at the field closure generally range from 2 to 6°. A field-wide original oil-water contact is recognized at 7,274 feet (2,217 m) subsea. No major faults are recognized within the reservoir sequence. Minor faults and slumping may have occurred but are difficult to detect from well and seismic data.

Reservoir Characterization

Seismic Attributes and Geometry

Forties Field 2-D seismic data were acquired in 1980 and 1981 and, although reprocessed in 1985, provided poor resolution of the Sele-Forties Formation interface, making it difficult to interpret the top Forties marker across the entire field (Fig. 20-5). Because the top Sele Formation reflector can be reasonably picked and mapped throughout the field, the top reservoir structure map was made by seismically mapping the top Sele Formation and subtracting a Sele isochore map derived from well data. The top reservoir depth map shows an oil column approximately 600 feet (180 m) thick at the crest of the field (Fig. 20-6).

Seismic reflectors within the Forties reservoir itself are extremely difficult to correlate, and correlation becomes tenuous beyond reasonable well control. Because these interpretations failed to provide

N
○ = CORED WELLS WITH WELL NO.
● = OIL PRODUCERS
= WATER INJECTORS
OWC
FB
FA
FC
FD
FE
55
63
22
12
52
5
2
08
02
1
8
09
7
0 KM 2
0 MILES 1
21/9 BP
21/10 BP
22/6 SHELL/ESSO

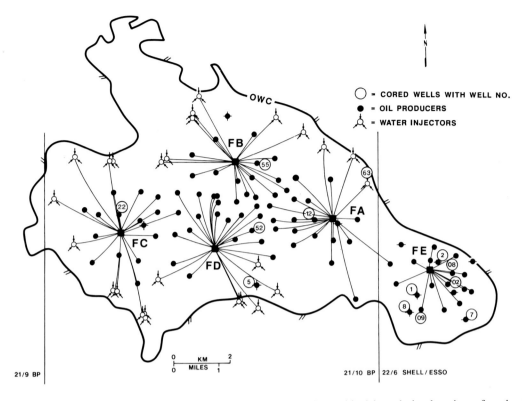

Fig. 20-2. Well location map of the Forties Field showing the offshore license block boundaries, locations of production platforms, wells, and cored wells (circled) used in this chapter.

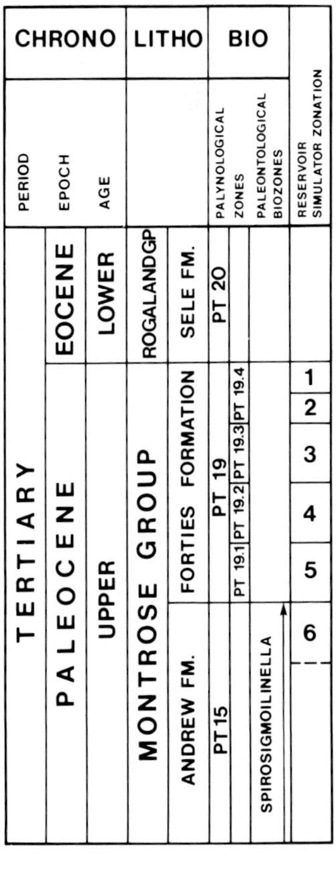

CHRONO
LITHO
BIO
PERIOD
EPOCH
AGE
PALYNOLOGICAL ZONES
PALEONTOLOGICAL BIOZONES
RESERVOIR SIMULATOR ZONATION
TERTIARY
EOCENE
PALEOCENE
LOWER
UPPER
ROGALANDGP
MONTROSE GROUP
SELE FM.
FORTIES FORMATION
ANDREW FM.
PT 20
PT 19
PT 19.1
PT 19.2
PT 19.3
PT 19.4
PT15
SPIROSIGMOILINELLA
1
2
3
4
5
6

Fig. 20-3. Summary of the chronostratigraphic, biostratigraphic, and lithostratigraphic subdivisions for the Central Graben area of the U.K. North Sea.

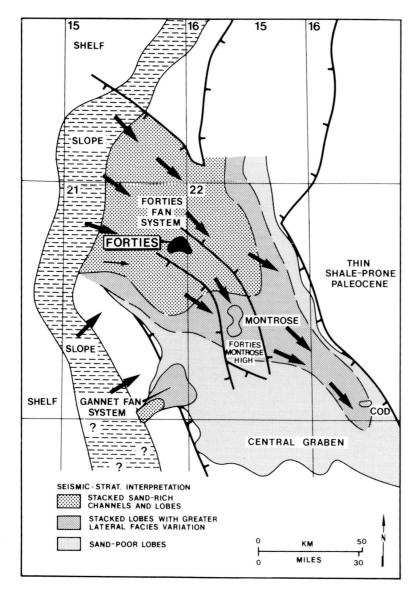

Fig. 20-4. Reconstruction of the upper Paleocene paleogeography of the Central Graben of the U.K. North Sea based on regional seismic interpretation. Sediment source for the Forties submarine fan system was in the northwest.

Fig. 20-6. Structural contour map of the top of the reservoir showing nearly 600 feet (180 m) of relief in the center of the field and an OWC (7,274 ft; 2,217 m) across the structure.

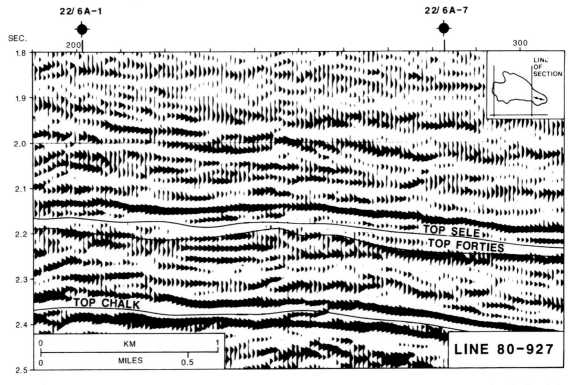

Fig. 20-5. Convential 2-D seismic line (zero-phased migration) across the Southeast Forties area showing the Top Sele Formation, Forties Formation, and Top Chalk picks. Inset is field outline shown in Figure 20-6.

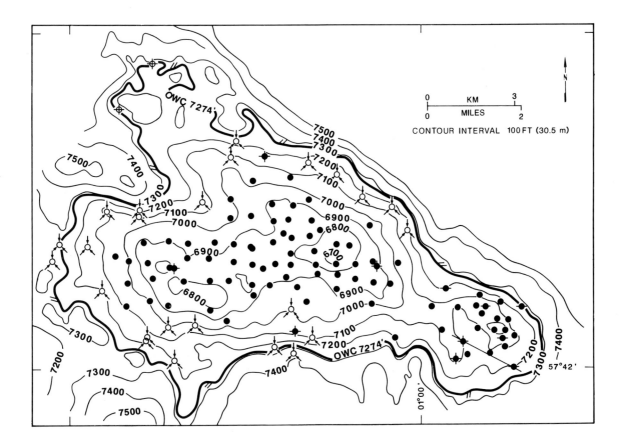

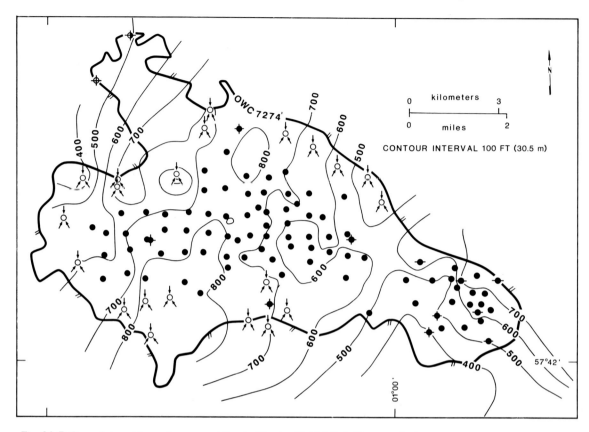

Fig. 20-7. Gross-interval isopach map used for the Forties Field full-field reservoir simulation. The interval mapped is reservoir simulation zone 1 (top reservoir) down to the bottom of zone 6 (within the Andrew Formation).

a reliable basis for correlating impedance and/or lithological units, intrareservoir mapping using the 1985 reprocessed data was not employed for use in this reservoir modeling study.

The gross-interval isopach map represents the cumulative thickness of all sediments within the six reservoir zones of the field (Fig. 20-7). The deepest of these zones, layer 6, was defined for the purpose of aquifer delineation. It is bounded below by an arbitrary but consistent wireline-log pick in the Andrew Formation and above by the Forties-Andrew interface (note the reservoir zonation scheme on Figure 20-3).

Reservoir-quality sandstone is thickest along a northwest-southeast trend (Fig. 20-8). Wells along the northeast rim of the field have extremely high proportions of reservoir-quality sandstone, approaching 100% in the interval above the oil-water contact. The placement of the initial four production platforms was based on this sandstone thickness pattern.

Reservoir Geology

Sedimentological core studies of the Forties reservoir interval have described several lithofacies ranging from structureless, pebbly sandstones to thin-bedded, fine-grained sandstone and laminated mudstone. Broad vertical facies associations could be defined based on these lithofacies descriptions and gamma-ray log patterns (Figs. 20-9, 20-10) (Carman and Young, 1981). Figure 20-9 illustrates the major Forties Field facies associations in conjunction with their corresponding gamma-ray log patterns.

The facies associations in the field fit the classification system of submarine-fan sequences proposed by Mutti and Ricci Lucchi (1978) and by Mutti (1985). The reservoir interval is dominated by multistory channel and lobe sequences. Fan-channel deposits are usually characterized by a "blocky-shaped" log character. From core analyses, these channel-fill sequences consist of composite beds

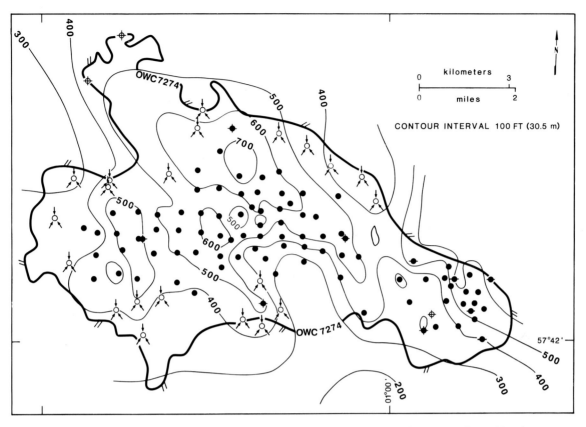

Fig. 20-8. Net-sandstone isopach map of the Forties Field. The interval mapped is the reservoir simulation zone 1 (top reservoir) down to the bottom of zone 6 (within the Andrew Formation). Net sandstone is determined as a normalized gamma-ray response of less than 62.5 API units, greater than 12% porosity, and 1 md or greater air permeability (unstressed).

(thickness of 3–25 ft; 1–8 m) of thick-bedded, medium- to coarse-grained sandstone. The massive channels are commonly the sandstone-dominated lower parts of an upward-thinning and upward-fining sequence interpreted to reflect lateral shifting or progressive abandonment of a channel or channeled lobe. Upward-thickening and upward-coarsening sequences are representative of fan-lobe progradation, often of an unchanneled nature. These deposits are characterized by a "funnel-shaped" log pattern reflecting upward thickening of beds and upward coarsening of grain size. Interbedded mudstone, siltstone, and sandstone (classical turbidites) are found above and below lobe sequences. They are recognized as interchannel or interlobe and basin-plain deposits. These thin-bedded sediments (thickness of beds 0.2–1 ft; 6–30 cm) show a "serrated" gamma-ray log pattern.

The intercalation of lobe and channel sequences is evidenced by abrupt lateral and vertical facies

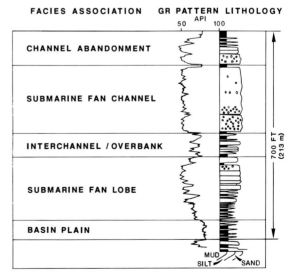

Fig. 20-9. Typical associations of the Forties Formation fan components and their gamma-ray log responses.

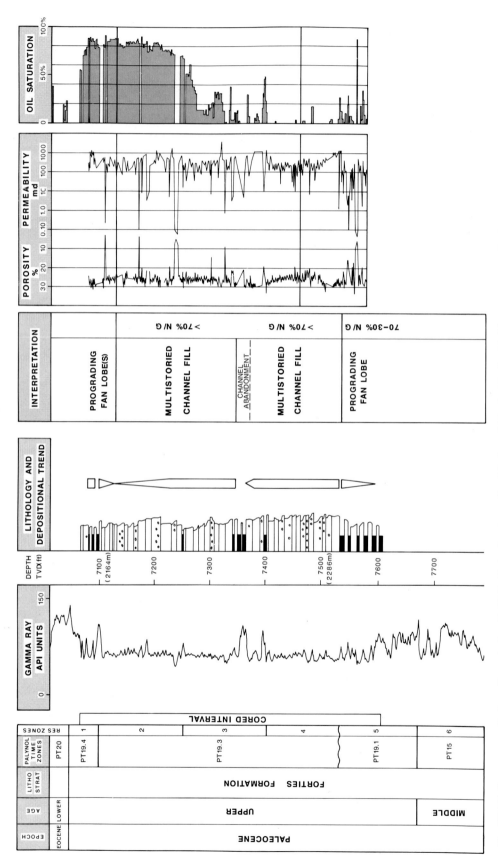

Fig. 20-10. Core composite log of a typical well, 22/6a-2, in the channelized facies within the Forties Field. Average porosity is about 26% and most permeabilities range from 200 to 600 md. N/G = net-to-gross sandstone ratio.

changes. The resultant facies architecture was too complex to attempt a field-wide reservoir subdivision based solely on well-log correlations of facies associations. In a complex of nested channel-fill sandstones, for example, individual units were indistinguishable, because indications of lateral movement of channels are virtually unrecognizable from log character only. However, log responses were helpful in identifying the upward-thickening and upward-thinning sequences associated with prograding lobes or channel abandonment. Because these sequence were seismically unresolvable, a chronostratigraphic framework was required in order to unravel the field-wide distribution of depositional trends.

Chronostratigraphy

The upper Paleocene biostratigraphic subdivision used by Shell is based on qualitative and quantitative analyses of pollen and microplankton (Fig. 20-11). Although the names of individual species have not yet been released for publication, Figure 20-11 demonstrates the chronostratigraphic relationships of these species. Microplankton biofacies are affected by a multitude of environmental factors such as temperature, salinity, oxygen content, and local ocean currents. Combining both pollen and microplankton analyses enabled a more detailed subdivision of the Forties Formation compared with that resulting from microplankton alone.

This biostratigraphic scheme is an exceptionally good basis for chronostratigraphic interpretations in the lower Eocene and upper Paleocene both in the Forties Field vicinity and in larger areas of the Central Graben. Shales from 14 cores, representative of a major area of the Forties Field, were sampled for palynological analysis (the circled wells in Figure 20-2). Using these analyses, a five-layer chronostratigraphic model of the Forties Field reservoir was constructed to develop a depositional model for the reservoir and to provide a chronostratigraphic framework for the detailed field-wide zonation required in a reservoir simulation model. The chronostratigraphic model is based predominantly on correlations between the 14 wells which were cored. Interpretation of areas between sampled .wells was guided by lithostratigraphy based on gamma-ray and sonic log characteristics, established by calibration from the cored wells. The model is summarized by the block diagram in Figure 20-12, which shows the main thickness changes within each chronozone across the field.

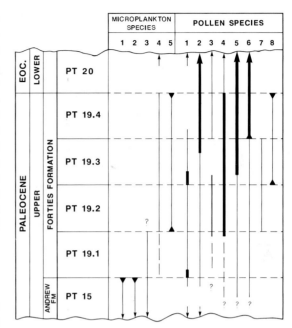

KEY : ▼▲ TOP AND BASE OF SUBZONAL MARKERS
        ━━━ QUANTITATIVE CRITERIA
        ──── QUALITATIVE CRITERIA

Fig. 20-11. Palynological and planktonic subdivision of the Forties Formation (PT19) and the Top Andrew Formation (PT 15). This subdivision forms the basis for the chronostratigraphic interpretation of the Forties Field area.

The broad chronostratigraphic model helped to define more clearly the temporal changes in channel and lobe position and the resultant stacking patterns. The thin-bedded turbidite sequences formed as a result of fine-grained suspension load being deposited on paleohighs. This is evidenced by the close vertical spacing of chronostratigraphic markers. Adjacent to, and in contrast to deposition on these paleohighs, the channel and lobe sequences reveal a significantly expanded section of chronostratigraphic markers (Fig. 20-13). These sand-rich sediments were deposited by a combination of traction and suspension load along preexisting paleotopographic low trends. The morphology of the resulting sediment bodies was thus dependent on paleotopographic control and on contemporaneous differential compaction.

A cross section through Block 22/6a (S.E. Forties area) shows that in a distance of only 0.9 miles (1.4 km), a complete transition occurs from thin-bedded, alternating shale and sandstone of an interchannel or overbank deposit (about 60 ft (18 m) thick) to a

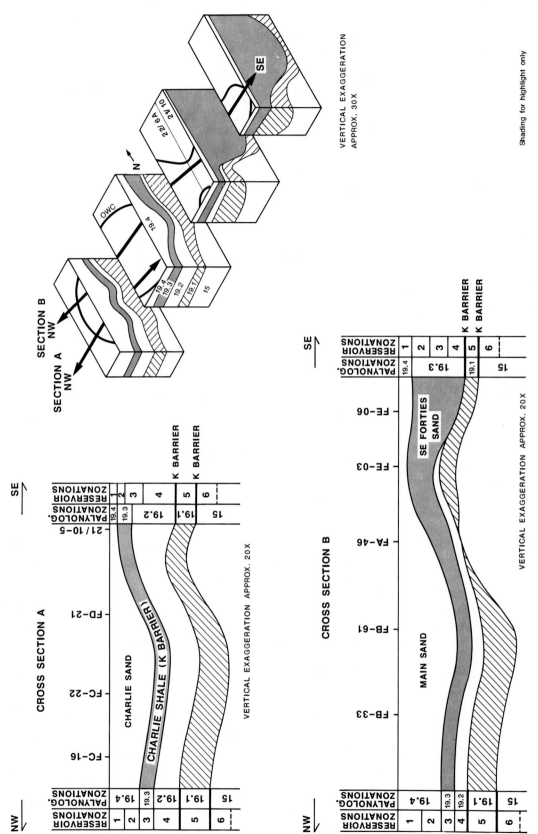

Fig. 20-12. Stratigraphic model based on the palynological/planktonic subdivision of the Forties Formation. Cross sections A and B demonstrate the relative thickness variations across the field for each PT zone.

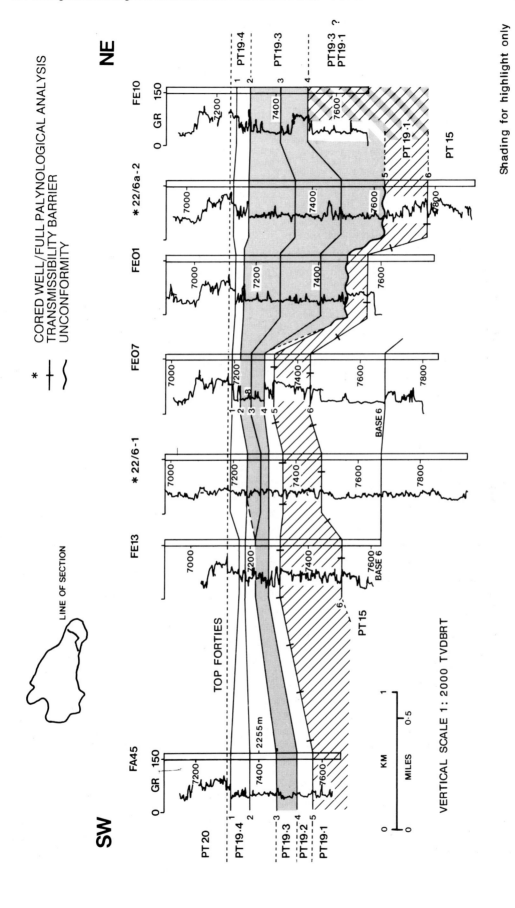

Fig. 20-13. Chronostratigraphic correlation panel through the S.E. Forties area hung on the Top Forties Formation. Note the expansion of the PT 19.3 chronozone from 60 feet (18 m) thick in Wells 22/6-1 and FE 07 to a thick channel deposit (180 ft; 55 m) in Well 22/6a-2.

thick, sandstone-rich sequence of stacked channel-fill deposits (about 180 ft (55 m) thick) (Fig. 20-13).

Depositional History

Three distinct fan-lobe sequences prograded south and southeastward during palynological time (PT) 19. From west to east these are known as the Charlie, Main, and S.E. Forties Sands (Fig. 20-12). These channeled systems acted as sites of major sediment deposition and shifted continuously throughout the depositional history of the Forties Formation.

Fan construction comprised four depositional episodes, as defined by palynological zonation. Facies patterns within these intervals establish a depositional model, which provides the basis for full-field well correlation and for the zonal maps (sandstone thickness, porosity, vertical and horizontal permeability) required for a 3-D reservoir simulation. The depositional model is subdivided for each of the chronozones of the reservoir, as discussed below, in order from oldest to youngest. Figure 20-14 highlights the shifting of depositional units during specified time periods.

PT 19.1. Fan lobes prograding over the predominantly thin-bedded turbidites of the Andrew Formation (PT 15) characterize the major part of this earliest deposition of the Forties Formation. The low sandstone percentage of the uppermost Andrew Formation (PT 5) reduces vertical transmissibility to the overlying PT 19.1 chronozone. This is substantiated by repeat formation tester (RFT) measurements which deviate by 50 psi (345 kPa) or more from the expected pressure-decline gradient during the field production history.

PT 19.2. Production and RFT data confirm the existence of another widespread vertical permeability reduction at the interface of PT 19.2 and the underlying PT 19.1 sediments. Even in areas where both chronozones are sandstone-prone, localized but numerous clay barriers (drapes) intersect with one another and adjacent interchannel deposits, causing minor discontinuities in pressure distribution (< 50 psi, < 345 kPa) and thereby reducing vertical transmissibility.

PT 19.3. A series of sandstone-rich channel fills were deposited in S.E. Forties Field during PT 19.3 time. Sandstone thickness patterns indicate that the Forties-area feeder channel shifted to a more northerly source. In the western portion of the field, an

extensive pressure barrier known as the Charlie Shale was being deposited.

PT 19.4. The youngest sediments of the Forties reservoir belong to the PT 19.4 subdivision. The dominant direction of sand transport reverted to the southeast and two thick sequences were deposited. The first, in the Main area, is a very thick (100–200 ft, 30–60 m) sandstone-rich channel fill which is in pressure communication with the rest of the underlying reservoir. The second is called the Charlie Sand, which is a separate channeled lobe in the west of the field that terminates abruptly southward. Early production history showed it in pressure isolation from the rest of the field.

Prior to this study, it had been thought that the Charlie Sand lobe and underlying Charlie Shale were younger and superimposed on the Main Sand. This chronostratigraphic analysis indicates that the Charlie and Main sands were deposited contemporaneously. The depositional pattern displayed on Figure 20-14, PT 19.4, probably resulted from bifurcation of the feeder system in the northwest and associated differential compaction.

Petrography

Petrographic studies indicate that the reservoir rock is of uniform composition throughout Forties Field. Reservoir sandstones are mainly sublitharenites, with quartz the main detrital component (Fig. 20-15). Feldspar, lithic fragments, and detrital clays occur only in lesser quantities (< 25%). The sandstones are fine- to medium-grained and generally moderately to well-sorted. Diagenetic alterations detrimental to reservoir quality are limited to localized small carbonate concretions and, in some instances, entire sandstone beds with extensive carbonate cement. However, the latter show no evidence of acting as flow barriers in the reservoir and are therefore areally restricted. Carbonate cement was analyzed by cathode luminescence to determine if carbonate growth were related to unique time periods within the reservoir, thus providing a check of the chronostratigraphic timelines. The results showed no unique mineral-growth characteristics that could aid correlation.

Petrophysical Properties

A composite wireline-log section, including core porosity and permeability, of the Forties Formation in Well 22/6a-2 is shown in Figure 20-10. Porosities

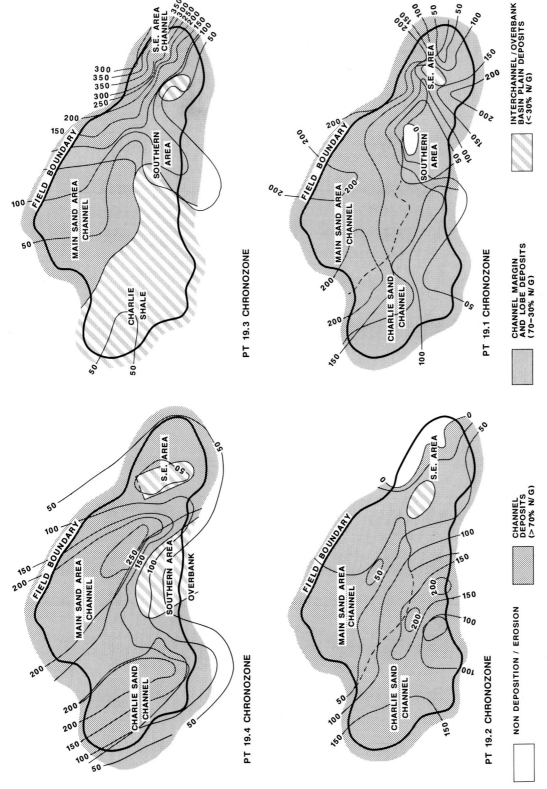

Fig. 20-14. Upper Paleocene chronostratigraphic depositional trend and gross thickness maps of the Forties Field. This series of maps from oldest (lower right corner, PT 19.1) to youngest chronozone (PT 19.4) portrays the switching of channels, lobes, and interchannel depositional sequences over time. Contour interval is 50 feet (15.2 m).

Fig. 20-15. A typical Forties reservoir sandstone in plane polarized light showing loosely packed, moderately to well-sorted detrital quartz grains with minor feldspar and lithic fragments. The blue areas are porosity (26%). Air

permeability is 650 md, and this sample would have a capillary pressure curve similar to curve number 2 in Figure 20-16. Scale bar is 0.5 mm.

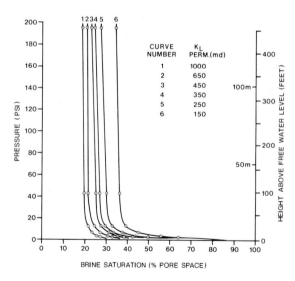

Fig. 20-16. Plot of connate water saturation versus capillary pressure and permeability for samples from the Forties Field well 22/6a-2.

are typically in the range of 24 to 27%, although interchannel/overbank facies generally have porosities lower by several percent. Horizontal air permeabilities are consistently high, in the range of 100s to 1,000s of millidarcies. Lower-permeability sandstones are present only in the thin beds associated with overbank deposition.

The excellent reservoir quality of the Forties Formation is also reflected in the capillary pressure characteristics of the sandstones. Air/brine capillary pressure curves from Well 22/6a-2 are shown in Figure 20-16. All the curves in a permeability range of 150 to 1,000 md show low entry pressures which result in a very sharp transition zone as seen on wireline logs. Connate water saturations typically range from 20 to 35%, with 10% common in the channel-fill sequences which composed nearly two-thirds of the reservoir.

Determination of net sandstone (those sandstones considered to produce in the reservoir) from wireline-log data is problematic due to the influence of clays and calcite cement on the reservoir's log response. The best method to determine net sandstone uses a combination of the gamma-ray log and calculated porosity. To account for the variation in gamma-ray response due to different logging contractors, tool configurations, and mud types, all gamma-ray logs were normalized to a field standard of 30 API units in the cleanest sandstone and 95 API units in the shale. Net sandstone was then defined as having a normalized gamma-ray reading of less than 62.5 API units and a porosity greater than 12%. These cut-offs excluded sandstone with an air permeability of less than 1 md, which were considered as nonreservoir. Three characteristic normalized gamma-ray trends are plotted on Figure 20-17 to show the effect of gamma-ray variation (e.g., shale content) on the predicted permeability.

Although the Forties aquifer is highly saline (about 80,000 ppm NaCl), conductive clay material has a small but significant influence on the resistivity-log response. Consequently, water saturations from wireline logs were calculated according to the Waxman-Smits shaly sand model. These range from 22 to 32%. Although the average net/gross ratio for the field is approximately 75%, the concentration of porous and permeable sandstone in the channels and lobes (approaching 100% reservoir-quality sandstone) helps to create an excellent reservoir sweep, good pressure distribution, and a high recovery factor of approximately 57% fieldwide.

Production Characteristics

Field Performance

The field was first produced in 1975 and reached a plateau of approximately 500 MBOPD (8.0×10^4 m³/D) in 1978. Figure 20-18 is a plot of the daily production of oil and water and of water injection. With nearly 75% of the field reserves of 2.5 billion barrels (3.9×10^8 m³) now depleted, reservoir management of injection rates is significant to ensure maximum sweep efficiency and recovery of remaining reserves. Full field 3-D reservoir simulations play a key role in identifying areas of the reservoir which may contain unswept reserves.

The oil pressure, volume, and temperature (PVT) properties demonstrate a gradation of the gas/oil

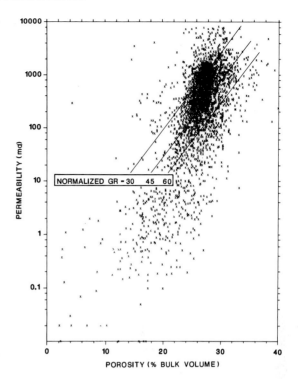

Fig. 20-17. Porosity versus permeability plot. Superimposed on the data are three normalized gamma-ray trends that indicate the effect of shale content (60 API units having the highest content) on the predicted permeability.

ratio from west to east across the field. In the west and southwest, wells have a GOR range of 265 to 340 SCF/bbl (46.6–59.8 m³/m³), while wells to the southeast range up to 410 SCF/bbl (72.1 m³/m³). With no easily recognizable flow restrictions in the field, it is thought that hydrocarbon charging of the field may have been complex. A lighter crude may have been introduced from the southeast after initial charging of the reservoir. This would have resulted in the loss of the slightly heavier crude from the spillpoint in the west of the field. Insufficient time for complete equilibration of the oil properties across this very large field has allowed the areal variation in the reservoir to occur.

Formation volume factors (FVF) and viscosities are consistent with the areal variation of GOR. The western wells ranged near 1.250 RB/STB for FVF with a viscosity of 0.81 cP (8.1×10^2 Pa·s) at 3,230 psi (2.2×10^4 kPa) and 195°F (90°C), with southeastern wells at about 1.285 RB/STB and 0.69 cP (6.9×10^2 Pa·s) and 195°F (90°C).

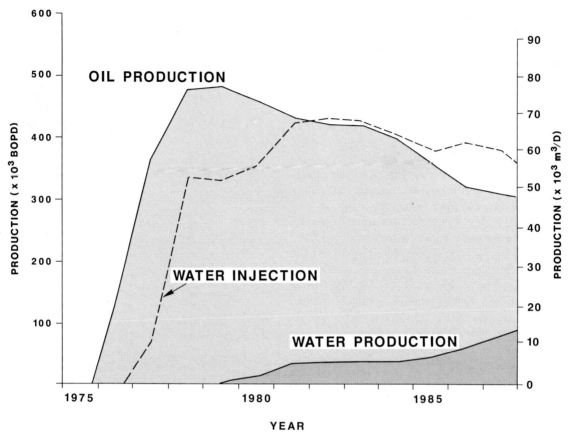

Fig. 20-18. Oil and water production and water-injection profiles of the Forties Field.

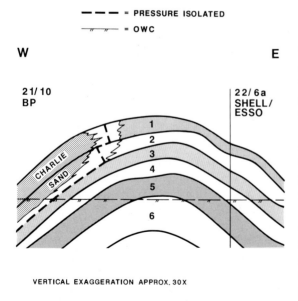

Fig. 20-19. Schematic cross section of the reservoir simulation zonation. Note the pressure isolation of the Charlie Sand from the rest of the field.

Zonal Reservoir Simulation

To develop a practical 3-D reservoir model, the five chronozones (PT 15–PT 19.4) were used as a basis to form a six-zone reservoir simulation model. In areas where one chronozone was very thick (up to 250 ft (75 m)) as in the S.E. Forties channel fill (PT 19.3) or the Main and Charlie sands (PT 19.4), the chronozone was subdivided into two reservoir zones to facilitate the simulation of OWC movement. The bottom of zone 6 was picked to provide further definition for the aquifer (Fig. 20-19).

The chronostratigraphic framework was used to delineate the main geologic features of the reservoir; i.e., channel and lobe trends, channel to paleohigh (overbank) transitions, vertical permeability barriers, and fluid flow (including aquifer size and predicted influx). The lithofacies correlations between cored wells helped to complete the field-wide framework. A schematic cross section of the simulation zonation (1-6) is shown in Figure 20-19.

The overall reservoir-pressure communication in the field is good. However, some permeability "baffles" have been suggested by production and RFT data in the lowermost portions of the oil column (top of zones 5 and 6), and local, intermittent baffles higher in the reservoir. Baffling within large sandy units due to clay draping may cause minor problems in recovery as the field is further depleted. No major diagenetic problems or faults which significantly affect fluid flow have been recognized.

A combination of edge-water and bottom-water drives influences reservoir performance. Extensions of the fan system from the northwest to the southeast provide the best opportunity for aquifer support. Thin-bedded deposition, large shale-outs, and thinner Forties Formation sandstones in the south of Block 21/10 reduce aquifer influx there. Pressure maintenance by water injection was required early in the development of the central and western portions of the field.

The geological input for the reservoir simulation consisted of mapping six layers. For each of these layers, isopach, net/gross, porosity, vertical-to-horizontal permeability ratios (k_v/k_h) and horizontal permeability maps were constructed. The k_v/k_h maps were derived by relating the depositional environment of a particular geological unit (i.e., channels and lobes, channel margins, and thin-bedded interchannel deposits) to a range of net/gross values (Fig. 20-14). The channel-fill and lobe sequences (net/gross values greater than 70%) were assigned k_v/k_h values of 0.1. Lobe/channel edges and thin-bedded deposits were given k_v/k_h values from 0.01 to 0.001. Minor modification of these values occurred during the history-matching phase of the simulation. However, the use of net/gross relationships as a predictor for k_v/k_h ranges was very successful. This method virtually eliminated the common requirement of hand editing of vertical transmissibility reductions into the simulator.

Reservoir engineering data requirements were considerable and included the following important parameters: horizontal and vertical permeability data by layer, full well-by-well historical pressure and production histories, natural and artificial lift well-performance curves, variable PVT modeling, directional relative permeability, and saturation history dependence (to help in modeling water influx). A proprietary, fully implicit, isothermal black/volatile simulator was used for the studies. It consisted of 12,390 active grid blocks each 985 feet by 985 feet (300 m × 300 m). A larger spacing was used on the outermost portions of the grid.

Previous simulations required the introduction of theoretical horizontal and vertical transmissibility barriers to aid the quality of history matching of the production and pressure data. In this simulation, the use of chronostratigraphy in developing the geological input (defining the reservoir geometry, flow units, and likely areas of transmissibility reduction) resulted in the elimination of such theoretical barriers.

Exploration and Production Strategy

In a setting such as the North Sea where seismic identification of lithofacies in the Forties Formation may be difficult, the use of depositional models plays an important role. Such models may also prove to be useful in further Forties Formation plays and those in other submarine fan provinces. The use of chronostratigraphy to develop depositional models in turbidite sequences can be helpful in finding the optimal submarine fan exploration and development target: sand-rich channels and lobes. Thin-bedded sequences with closely spaced chronostratigraphic markers in known sand-rich submarine fans can be representative of paleohighs with channels occurring nearby in lows adjacent to leveed channel sequences. The example of the S.E. Forties area in with the sandstone-rich channels occurring within 0.9 miles (1.4 km) of the thin-bedded, low net/gross ratio wildcat wells (22/6-1) confirms this approach (Fig. 20-13).

The application of reservoir modeling during production from a submarine fan sequence can be extremely beneficial. Accurate history matching or prediction in reservoir simulations due to reliable geological modeling ensures the most efficient exploitation of reserves with optimum well placement, recompletion strategies, water injection planning, and financial management of the project. Poor placement of production and injection wells based on unsubstantiated lithostratigraphic correlations in reservoir simulations could affect ultimate recoveries in fields similar to Forties Field by 5% or more, severely affecting project economics.

Conclusions

The creation of an accurate depositional model is the key to obtaining reliable results in reservoir simulation. The submarine fan environment of the Forties Field is complex, with abrupt transitions from channels and lobes to interchannel facies in terms of bed thickness, net/gross ratios, and shale distribution. The inability to correlate confidently using litho-stratigraphy and seismic data led to the application of palynology and chronostratigraphy.

The use of palynology to develop timelines across the field enabled the definition of individual flow units, the mapping of sandstone distribution in channel-fill and lobe sequences, and the prediction of areas of poorer vertical transmissibility. The ability to map these patterns and model transmissibility reductions with more certainty provided a better foundation for the reservoir simulation and resulted in a reasonable production history match without the introduction of theoretical barriers.

Acknowledgments. The authors thank Shell Internationale Petroleum Maatschappij, Shell U.K. Exploration and Production, Esso U.K., and BP Petroleum Development Limited* for permission to publish this paper. Special thanks to G. Louwaars, T. Schroeder, P. Betts, and W.D. Spearman (Shell Expro), B. Keeling, M. Smith, and G. Irish (Esso), and D.J. Webster and A.M. Carter (BP) for the lively discussion and debate that helped formulate new ideas for the Forties Field. We extend our appreciation to D.J. Stewart for providing the interpretation of upper Paleocene paleogeography (Fig. 20-4). We wish also to thank W.D. Lancaster, J.H. Barwis, and B.K. Levell for their critical review and suggestions for improvement of this chapter.

Reservoir Summary

Field: Forties
Location: Central North Sea, UK
Operators: British Petroleum Development, Limited (Shell U.K. and Esso U.K. are unit partners)
Discovery: 1970
Basin: Central Graben
Tectonic/Regional Paleosetting: Rift basin
Geologic Structure: Domal anticline
Trap Type: Four-way closure
Reservoir Drive Mechanism: Bottom-edge water drive; water injection
 • **Original Reservoir Pressure:** 3,230 psia (2.2×10^4 kPa)
Reservoir Rocks
 • **Age:** Upper Paleocene
 • **Stratigraphic Unit:** Forties Formation
 • **Lithology:** Litharenites
 • **Depositional Environment:** Submarine fan
 • **Productive Facies:** Channel-fill, lobe, and interchannel sandstones
 • **Petrophysics**
  • **Porosity Type:** Primary, intergranular
  • **ϕ:** Average 26%, range 24 to 27%, cutoff 12% (stressed cores)
  • **k:** Average 1,000 md, range 500 to 2,000 md, cutoff 1.0 md (air, unstressed cores)
  • **S_w:** Average 23%, range 22 to 32% (logs)
Reservoir Geometry
 • **Depth:** 7,000 feet (2,135 m)
 • **Areal Dimensions:** 7.5 by 5 miles (12×8.0 km)
 • **Productive Area:** 21,700 acres (8,800 ha.)
 • **Number of Reservoirs:** 2 (Main Reservoir, Charlie Sand Reservoir)
 • **Hydrocarbon Column Height:** NA
 • **Fluid Contacts:** NA
 • **Number of Pay Zones:** 2 (Main Reservoir, Charlie Sand Reservoir)
 • **Gross Sandstone Thickness:** 560 feet (170 m)
 • **Net Sandstone Thickness:** 395 feet (120 m)
 • **Net/Gross:** Average 0.70, range 0.25 to 1.0

---

*BP Petroleum Development Limited may have alternate interpretations to those presented here, and may choose to publish them with an account of production of Southeast Forties (FE platform).

Hydrocarbon Source, Migration
  • **Lithology & Stratigraphic Unit:** Jurassic Kimmeridge shale
  • **Time of Hydrocarbon Maturation:** Early Eocene to Pliocene
  • **Time of Trap Formation:** Eocene or later
  • **Time of Migration:** Neogene
Hydrocarbons
  • **Type:** Oil
  • **GOR:** 340 SCF/STB (59.8 m³/m³)
  • **API Gravity:** 37°
  • **FVF:** 1.250 to 1.285
  • **Viscosity:** 0.69 to 0.81 cP (6.9–8.1 × 10² Pa·s) at 195°F (90°C)
Volumetrics
  • **In-Place:** 4.3 billion BO (6.8 × 10⁸ m³)
  • **Cumulative Production:** 1.9 billion BO (3.0 × 10⁸ m³)
  • **Ultimate Recovery:** 2.5 billion BO (3.9 × 10⁸ m³)
  • **Recovery Efficiency:** 59%
Wells
  • **Spacing:** 3,365 feet (1,025 m); 260 acres (105 ha.)
  • **Total:** 125 total 83 producers
  • **Dry Holes:** 2
Typical Well Production:
  • **Average Daily:** 5,000 BO (795 m³)
  • **Cumulative:** 25 MMBO (4.0 × 10⁶ m³)
Stimulation: Water injection

References

Carman, G.J., and Young, R., 1981, Reservoir geology of the Forties Oilfield, *in* Illing, L.V. and Hobson, G.D., eds., Petroleum Geology of the Continental Shelf of North-West Europe: London, Heyden, p. 371–379.

Hillier, G.R.K., Cobb, R.M., and Dimmock, P.A., 1978, Reservoir development planning for the Forties Field: Proceedings of the European Offshore Conference, Society of Petroleum Engineers, Dallas, Vol. II, p. 325–335.

Mutti, E., 1985, Turbidite systems and their relations to depositional sequences, *in* Zuffa, G.G., ed., Provenance of Arenites: Dordrecht, Reidel Publishing Company, p. 65–93.

Mutti, E., and Ricci Lucchi, F., 1978, Turbidites of the Northern Appennines: Introduction to facies analysis: (English translation by T.H. Nilsen), International Geology Review, v. 20, p. 125–166.

Parker, J.R., 1975, Lower Tertiary sand development in the Central North Sea, *in* Woodland, A.W., ed., Petroleum and the Continental Shelf of Northwest Europe: London, Applied Science Publishers, p. 447–453.

Key Words

Forties Field, North Sea, U.K., Central graben, Forties Formation, Upper Paleocene, submarine fan, turbidite, reservoir simulation, palynological zonations, chronostratigraphy.

21

Reservoir Description of a Miocene Turbidite Sandstone, Midway-Sunset Field, California

Blaine R. Hall and Martin H. Link

Mobil Exploration & Producing U.S., Inc., Denver CO 80217 and Mobil New Exploration Ventures Company, Dallas, TX 75265; and Mobil Research & Development Corporation, Dallas, TX 75244

Introduction

During the past 25 years, numerous models for deepwater sandstone reservoirs have been proposed, many of which are based on comparisons with modern submarine fans. A number of questions have been raised as to the validity of these comparisons (Mutti and Normark, 1987). Few studies have described the reservoir characteristics of individual turbidite sequences as Hsü (1977) did for the Ventura Field, California. With the implementation of enhanced oil recovery (EOR) projects and the need for detailed engineering and geological studies in oil field production, a more thorough understanding of the continuity, distribution, dimensions, and inherent reservoir properties of turbidite sandstones is necessary.

The study presented here for the Webster Zone at Midway-Sunset Field provides a detailed examination of a turbidite sandstone reservoir in a producing oil field (Fig. 21-1). The Webster Zone is composed of unconsolidated, feldspathic (arkosic) sandstone of late Miocene age ("Stevens Equivalent," Monterey Formation) interbedded with diatomaceous mudstone (Fig. 21-2). The reservoirs described here were examined in anticipation of future steam-drive development of the Webster Zone by Mobil Oil Corporation for its MOCO FEE property (Fig. 21-1B).

The Webster reservoir has been on production sporadically for 76 years and is currently under cyclic steam development. Like many old fields, Mobil's MOCO FEE area at Midway-Sunset lacks seismic coverage and its wireline logs and reservoir test data are mainly pre-1960 vintage. Early petrophysical and reservoir information, such as R_t, R_w, S_w, and S_{or} values, capillary pressure curves, dip measurements, and accurate well surveys, are not available for the Webster Zone. In order to use the older wireline logs and to collect pertinent reservoir data for Mobil's MOCO FEE property, two wells (239-D and 327) were drilled, cored, and documented by modern wireline logs (Fig. 21-3). Reservoir petrophysics and volumetrics are still under evaluation and will not be known until full testing and steam-drive development take place.

The purposes of this study are to (1) describe the Webster turbidite reservoirs through the use of logs, cores, and outcrop observations, and (2) map accurately and characterize the properties of the Webster reservoir sand bodies. A **sand or sandstone body** is defined here as a mappable subsurface and/or outcrop sequence of sandstone with minor conglomerate and mudstone interbeds that can be identified and correlated between wells on logs in the field area and/or recognized in outcrop. These sequences are separated by mudstone and shale intervals or may be in depositional or erosional contact with adjacent sand bodies.

Geological Setting

Midway-Sunset Field is on the western side of the southern San Joaquin Valley in Kern County, California (Fig. 21-1A). The field was discovered in 1894 and has produced 1.76 billion barrels of oil

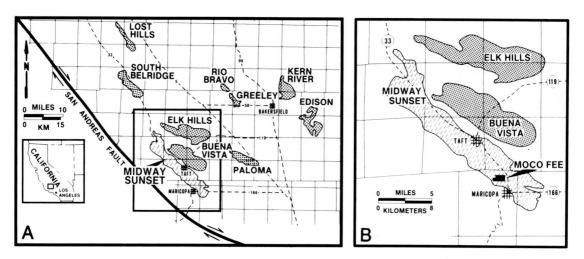

Fig. 21-1. Index maps showing (A) the Midway-Sunset Field and other major oil fields in the southern San Joaquin Valley, and (B) Mobil's MOCO FEE property where the Webster Zone occurs.

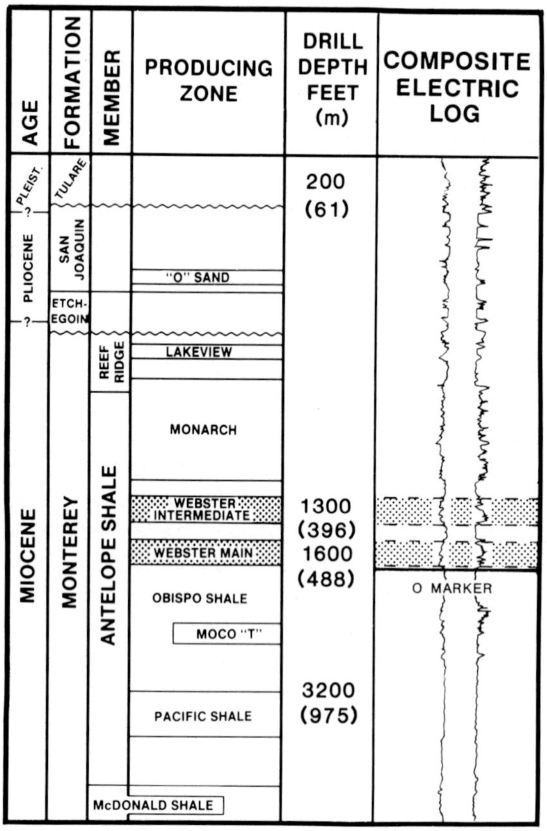

AGE	FORMATION	MEMBER	PRODUCING ZONE	DRILL DEPTH FEET (m)	COMPOSITE ELECTRIC LOG
PLEIST.	TULARE			200 (61)	
PLIOCENE	SAN JOAQUIN		"O" SAND		
	ETCH-EGOIN				
MIOCENE	MONTEREY	REEF RIDGE	LAKEVIEW		
		ANTELOPE SHALE	MONARCH		
			WEBSTER INTERMEDIATE	1300 (396)	
			WEBSTER MAIN	1600 (488)	
			OBISPO SHALE		O MARKER
			MOCO "T"		
			PACIFIC SHALE	3200 (975)	
			McDONALD SHALE		

Fig. 21-2. Generalized stratigraphic column for the MOCO FEE showing the position of the Webster Intermediate and Webster Main reservoir units relative to other producing zones at Midway-Sunset Field, their average drill depths, and composite electric log character. The "O" marker is shown below the Webster Main Zone.

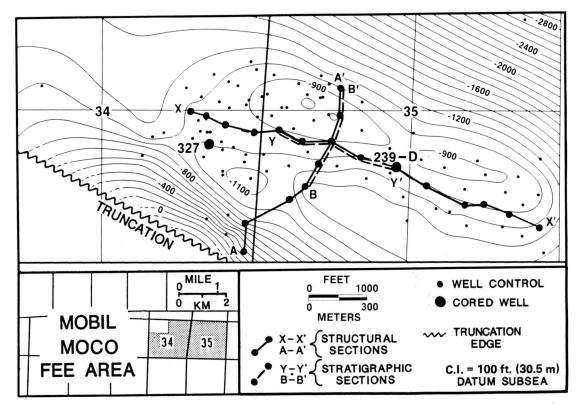

Fig. 21-3. Structure map contoured on the "O" marker, which is a stratigraphic marker just below the Webster Main Zone. The positions of structural sections X-X' and A-A', stratigraphic sections B-B' and Y-Y', and the location of the cored wells (239-D and 327) are shown. Mobil's MOCO FEE property in sections 34 and 35 is shown on the inset map. See Figures 21-5 to 21-8 for the various cross sections.

$(2.8 \times 10^8 \text{ m}^3)$(California Division Mines and Geology, 1987), making it the fourth largest producing field in the United States (International Petroleum Encyclopedia, 1987). The field is approximately 25 miles (40 km) long and 4 miles (6.4 km) wide and consists of a series of en echelon anticlines. It produces from multiple reservoirs ranging in age from Oligocene through Pleistocene, with the main production coming from upper Miocene reservoirs.

The Webster Zone occurs near the southern end of Midway-Sunset Field (Fig. 21-1B). The study area encompasses Mobil's MOCO FEE property, which comprises all of Section 35 and three quarters of Section 34 (Fig. 21-3), a total of approximately 1,120 acres (450 ha.). The Webster Zone, first penetrated in 1910, recorded its first production of 35 BOPD (5.6 m³/D) in 1913 (Ayers, 1942). Subsequent primary development and enhanced recovery of the Webster reservoir have been sporadic, principally because of varying demand for its heavy oil (14° API), resulting in total cumulative production

at the MOCO FEE property of approximately 13 MMBO (2.1×10^6 m³). The source of the oil is the organic-rich shale and mudstone of the Monterey Formation (Isaacs, 1987), the major oil-generating source-bed interval in the San Joaquin basin. The oil has been expelled from the Miocene (7 Ma) to the present (Fischer et al., 1988). Because the interbedded diatomites in the field area are immature (opal A), the oil must have migrated into the MOCO FEE area from a deeper, buried Monterey section to the east and south. Current production by cyclic steam stimulation of 35 wells is 220 BOPD (35 m³/D).

The Webster reservoirs as well as the other Miocene sandstones of the Midway Sunset Field are the initial turbidites that filled an intraslope basin (Fig. 21-4) developed along the western side of the San Joaquin Valley (MacPherson, 1978; Webb, 1981). This basin is subparallel to and within 10 miles (16 km) of the San Andreas fault (Figs. 21-1A, 21-4). The basin formed as a result of regional compression and downwarp directly related to strike-slip move-

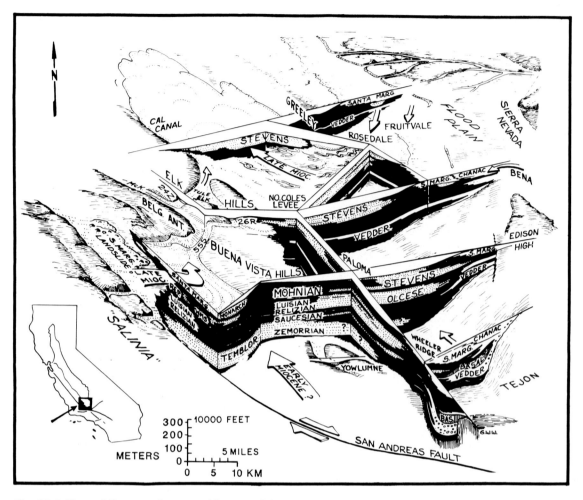

Fig. 21-4. Upper Miocene paleogeographic map of the southern San Joaquin basin showing multiple sediment sources along the margins of the basin. The area of study is in a northwest-trending subbasin between the Belgium and the Buena Vista Hills anticlines and the "Salinia" highland on the western side of the San Joaquin basin.

Paleobathymetric anticlinal highs largely controlled distribution of the upper Miocene sandstone in the western subbasin and central portion of the Stevens depocenter in the central portion of the basin (from Webb, 1981, and reprinted by permission of American Association of Petroleum Geologists).

ment along the fault (Harding, 1976). The Buena Vista Hills and Belgium anticlines formed on the northeastern margin of the intraslope basin, while concurrently to the southwest, a narrow, tectonically active shelf developed parallel to the San Andreas fault (Fig. 21-4). The Webster turbidite sandstones were derived from the southwest from highlands adjacent to the San Andreas fault in the Salinian block. As movement along the fault continued, the basin was deformed into a series of en echelon folds. Subsequent folding of the northeast-trending Webster Zone sandstones across a northwest-trending anticline developed various combinations of stratigraphic and structural petroleum traps.

Stratigraphy and Structure

A generalized stratigraphic column for the MOCO FEE area is shown in Figure 21-2. Productive Miocene sandstones of the Monterey Formation include the Moco "T," Webster Main, Webster Intermediate, Monarch, and Lakeview zones. The Miocene Obispo and Pacific Shales and the Pliocene "O" Sandstone are also locally productive.

The main structure of the MOCO FEE Webster Zone is a central anticline bounded to the southwest by a syncline (Figs. 21-3, 21-5, 21-6). The steeply dipping, upturned southern flank of the syncline has been truncated and is unconformably overlain by the

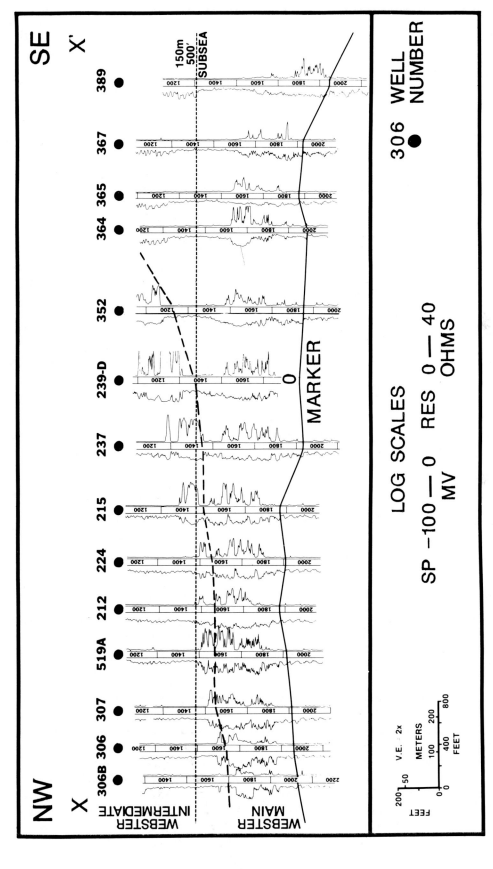

Fig. 21-5. Structural cross section X-X' constructed along the northwest trend of the anticline. The section is hung on the 500-foot (152-m) subsea datum. The position of the "O" marker and a dashed line separating the Webster Main and Intermediate zones are indicated. See Figure 21-3 for location of the section.

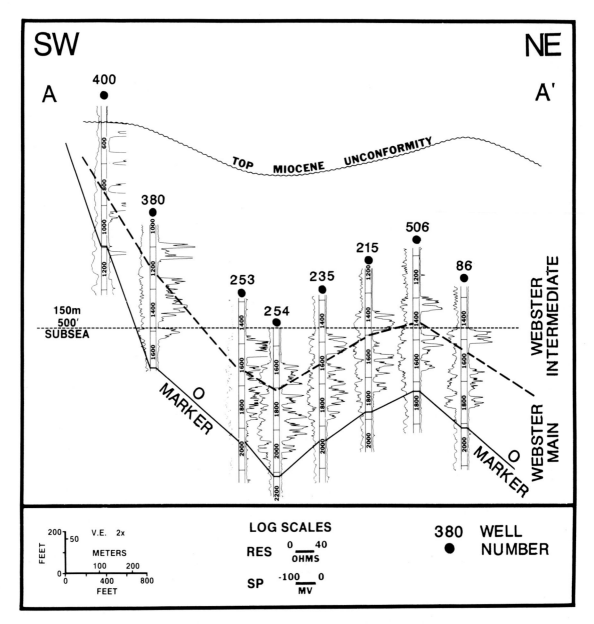

Fig. 21-6. Structural cross section A-A′ constructed parallel to the northeast-trending structural dip. The dashed line separates the Webster Intermediate and Main zones, and the position of the "O" marker and top Miocene unconformity are indicated. See Figure 21-3 for location of the section.

upper Miocene to lower Pliocene (?) Etchegoin Shale. Cross sections X-X ′ (Fig. 21-5) and A-A′ (Fig. 21-6) show the structural position of the Miocene "O" marker. Cross section X-X′ illustrates plunging of the anticline to the northwest and southeast (Fig. 21-5). To the northwest, the Webster Intermediate directly overlies the Webster Main, whereas to the southeast the two units are separated by up to 300 feet (91 m) of shale. The discontinuous nature of the

Webster sandstones as well as their varying log character is also evident. Cross section A-A′ extends from near the truncated edge of the Miocene strata across the syncline and anticline (Fig. 21-6). The angular relationship between the Webster Zone and the Etchegoin Shale is displayed at the southwest end of the section. The Miocene and post-Miocene folding of the "O" marker and top Miocene unconformity is also apparent. The log character of the

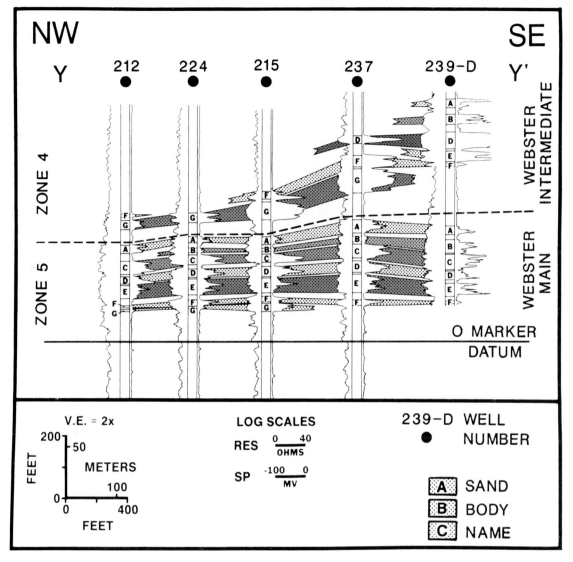

Fig. 21-7. Stratigraphic section Y-Y′ is flattened on the "O" marker datum and constructed across the northwest strike of the inferred slope of the basin. The section shows the Webster Intermediate (zone 4), the Webster Main (zone 5), and the correlation of individual sand bodies (A-G) within each zone. See Figure 21-3 for location of the section.

Webster Zone is more continuous and uniform along the section from southwest to northeast than on cross section X-X′

Sand-Body Geometry

Correlation of more than 140 wireline logs allowed detailed zonation that helped establish sand-body geometry and distribution of the Webster Intermediate (zone 4) and Webster Main (zone 5) at Mobil's MOCO FEE property (Figs. 21-7 to 21-10). The recognition of these sand bodies is based on several factors: mappable geometries determined from wireline log correlations, log character, lithofacies observed in cores, reservoir characteristics, and comparison to nearby outcrops with age-equivalent facies. The Webster Zone has been divided into 15 individual sand bodies: eight within the Webster Intermediate (zones 4A-4H) and seven within the Webster Main (zones 5A-5G) (Figs. 21-7 to 21-10).

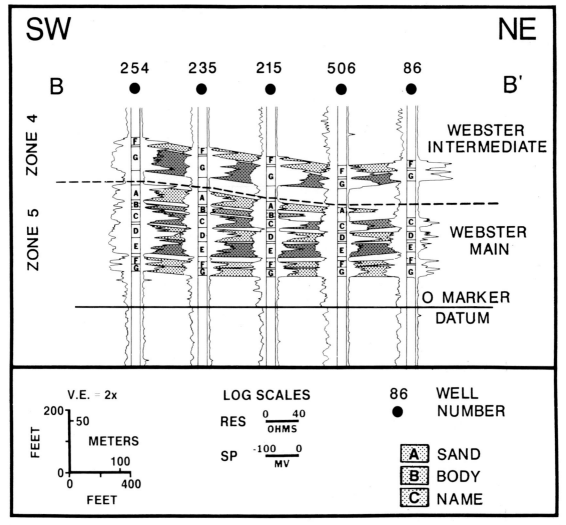

Fig. 21-8. Stratigraphic section B-B' is flattened on the "O" marker datum and constructed parallel to the northeast dip of the inferred slope of the basin. The Webster Intermediate (zone 4), the Webster Main (zone 5), and the correlation of their individual sand bodies (A-G) are shown. See Figure 21-3 for location of the section.

Modern wireline logs are available from only 14 of the wells penetrating the Webster Zone. Therefore, correlation and zonation were strongly dependent on the quality of the older wireline logs. The resistivity signature of the older logs was particularly important because a suppressed SP response, caused by the presence of diatomaceous mudstone, obscures sandstone recognition. No gamma-ray logs were available for the older wells. Correlation and zonation were accomplished by a linked, three-well triangulation grid that was carried across the MOCO FEE property.

Examples of the log character and zonation of the individual sand bodies are shown in detailed stratigraphic cross sections Y-Y' and B-B' (Figs. 21-7, 21-8). Section Y-Y' illustrates the pronounced differences between the individual sand bodies of the Webster Intermediate (zone 4) and the Webster Main (zone 5) (Fig. 21-7). Sand bodies in the Webster Intermediate are less continuous and offlap from northwest to southeast. The log character of individual sand bodies is quite variable. The Webster Main sand bodies have greater lateral continuity across the section and, although the log character of

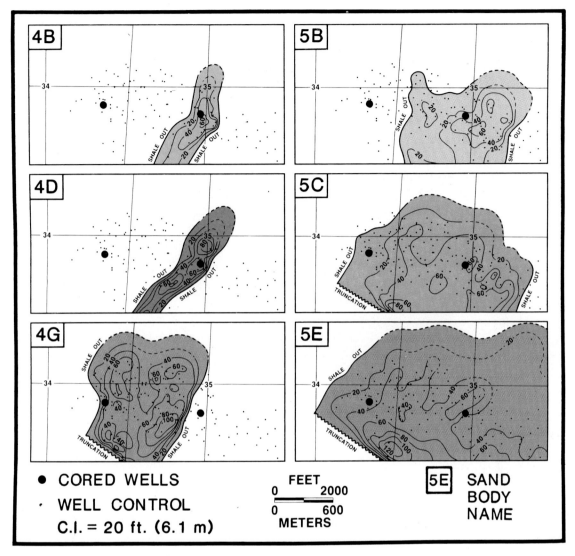

● CORED WELLS

· WELL CONTROL

C.I. = 20 ft. (6.1 m)

FEET
0   2000
0   600
METERS

5E   SAND BODY NAME

Fig. 21-9. Isochore maps of the sand bodies in the Webster Zone, comparing the sand-body geometries and mapped distributions. The variations in shape and thickness suggest that more sheet-like lobe sand bodies occur in the lower Webster Main (5C and 5E) and narrow linear channel-like sand bodies occur in the Upper Webster Intermediate (4B and 4D). Sand bodies 4G and 5B are intermediate between these two types of sand bodies and are interpreted to be channel/lobe transition deposits. See Figures 21-7 and 21-8 for the stratigraphic position of the individual sand bodies and note the position of the 239D and 327 cored wells.

individual sand bodies is variable, variation is sub.stantially less than in the Webster Intermediate sand bodies. Section B-B′ is along the depositional trend and is markedly different from Section Y-Y′ (Fig. 21-8). Individual sand bodies in both the Webster Intermediate and the Webster Main intervals are continuous along Section B-B′, and their log character is much less variable from well to well than those

in Section Y-Y′. The Webster Main zones 5A and 5B show the northeast downdip pinchout characteristic of the sand bodies.

One of the more critical aspects in understanding Webster reservoirs is the mapping of individual sand bodies of the Webster Zone. This mapping utilizes the well-correlation grid and reservoir zonation to construct 15 individual isochore maps (zone 4A-4H,

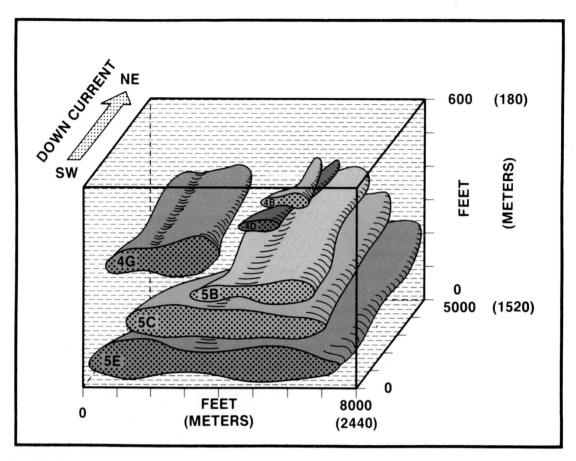

Fig. 21-10. Block diagram depicting the spatial relations and the shapes of the six sand bodies shown on the isochore maps in Figure 21-9.

zone 5A-5G), as illustrated by the six isochore maps shown in Figure 21-9. The isochores are gross composite sandstone interval maps and not net sandstone isopachs. Isopachs could not be constructed here confidently because of the lack of dip and well-survey information to correct for complex and steep structural dips. Dramatic differences are evident in the geometry and distribution of the individual sand bodies (Fig. 21-9). Zone 4B and 4D sand bodies are elongate and lenticular, with their depositional axes offset from one another. Zone 5C and 5E sand bodies are more equidimensional, have sheet-like reservoir geometries, and are stacked one on top of the other. Local, digitate thick and thin areas within zone 5C and 5E sand bodies may reflect the small-scale compensation cycles described by Mutti (1985). Zone 4G and 5B sand bodies are geometrically transitional between the previously described zones. They are offset from one another and appear

to represent the switching of depositional sites. The six individual Webster sand bodies shown in a block diagram (Fig. 21-10) demonstrate the difference in thickness, size, and shape of these individual sand bodies in three dimensions.

Williams Sandstone outcrops (Dibblee, 1973; Gilbert, 1980, 1988) occur within two miles (3.2 km) of MOCO FEE (Fig. 21-11) and were mapped to help clarify and validate subsurface mapping of Webster Zone sand bodies. The two units are of similar age, depositional setting, scale, geometry, and trend. Because the Williams Sandstone dips steeply to the northeast, the outcrops provide cross-sectional views of possible analogues for the Webster sand bodies. The Williams Sandstone is coarser grained, and individual sandstone bodies are thicker than those seen in the Webster reservoirs. These outcrops consist of (1) a lower zone of medium-grained, sheet-like sandstone units about 2,600 to 4,000 feet

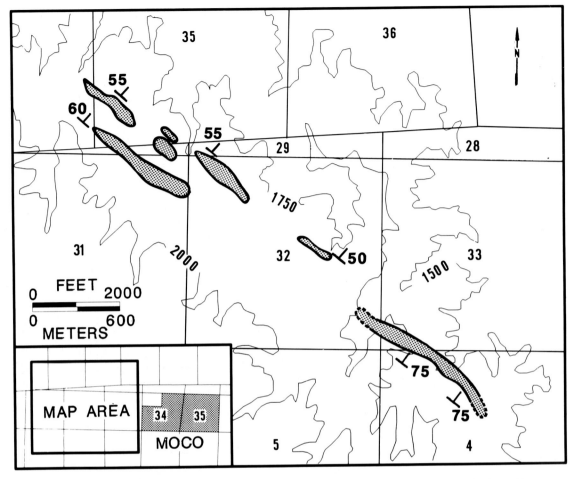

Fig. 21-11. Outcrop map for the Williams Sandstone, which is exposed in the foothills of the Temblor Range less than two miles (3.2 km) from Mobil's MOCO FEE property. Shaded areas are essentially cross sections through northeast-trending sand bodies. These outcrops serve as possible analogs for the Webster Zone reservoirs.

(800–1,220 m) wide and 50 to 100 feet (15–31 m) thick, and (2) an upper interval of lenticular sandstone and conglomerate beds that are 500 to 1,500 feet (152–460 m) wide and 50 to 200 feet (15–61 m) thick (Fig. 21-11). Paleocurrent measurements within the Williams outcrops indicate N81°E sediment transport (Gilbert, 1980), which is similar to the geometric trends displayed by Webster sand bodies.

Reservoir Characterization

Core and Facies Description

Two wells cored in the MOCO FEE Webster Zone recovered more than 800 feet (244 m) of high quality, undisturbed sandstones in 33 cores (Fig. 21-3). This was accomplished in spite of the poorly consolidated nature of the sandstones, which for the most part are held together only by hydrocarbons. Three important results from the study of these cores are (1) good core-to-log correlation, demonstrating the validity of the Webster zonation established from the wireline logs (Figs. 21-12, 21-13), (2) definition of the internal architecture of individual sand bodies, which corresponds directly to their mapped geometries, and (3) sedimentological input for developing a depositional model. In addition, the coring provided log calibration, petrographic evaluation, and direct measurements of porosity, permeability, and saturation.

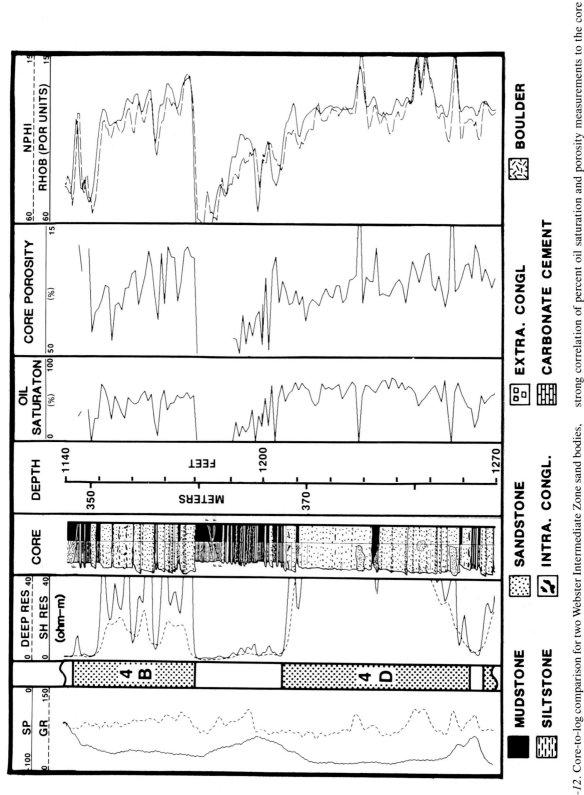

Fig. 21-12. Core-to-log comparison for two Webster Intermediate Zone sand bodies, 4B and 4D in the 239D well, which are interpreted to be channel-fill deposits. The close relation of the zonation established from logs and cores is evident. Note the strong correlation of percent oil saturation and porosity measurements to the core character and the good relationship of porosity measured from the core and from logs. See Figures 21-3, 21-9, and 21-10 for locations of cored wells and sand bodies.

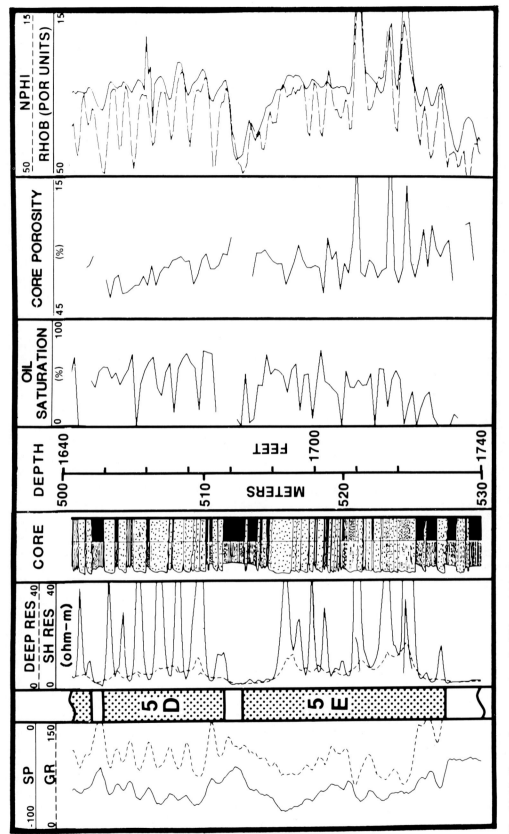

Fig. 21-13. Core-to-log comparison for Webster Main Zone sand bodies, 5D and 5E in the 239D well, which are interpreted to be depositional-lobe deposits inferred largely from the isochore pattern. The coarsening- and thickening-upward log character is poorly developed. Note the strong correlation of oil saturation and porosity to the core character and the good match of porosity measured from the core and from logs. See Figures 21-3, 21-9, and 21-10 for locations of cored wells and sand bodies and Figure 21-12 for lithology symbols.

The two cored wells, well 239D near the center of the Webster Zone and well 327 on the flank of the Webster Zone on the MOCO FEE property, provide a good spatial sampling of the Webster Zone (Figs. 21-3, 21-9). Both the Webster Intermediate and Main intervals were cored completely in each well. In well 327, the Webster Intermediate lies directly on the Webster Main, resulting in a 300-foot (91 m) thick Webster interval. In well 239D, the Webster Intermediate and Main are much thicker, 500 feet (152 m), and are separated by 200 feet (61 m) of mudstone.

Cores from wells 239D and 327 have been calibrated to wireline logs, and Figures 21-12 and 21-13 illustrate the core character and core-to-log calibration for well 239D. Lithology and reservoir properties for selected Webster Intermediate and Main intervals are highlighted. In addition, core photographs show some of the main features of these zones (Figs. 21-14, 21-15).

The Webster Intermediate Zone in both of the cored wells consists of thick-bedded, medium- to very coarse-grained, poorly sorted feldspathic (arkosic) sandstones and conglomerates interbedded with diatomaceous mudstones (Fig. 21-12). Coarse-grained units comprise 50 to 60% of the interval, with individual beds ranging from 0.1 to 22 feet (0.03–6.7 m) thick, averaging 2.5 feet (0.8 m). In places, they contain granitic boulders up to 2 feet (0.6 m) in diameter (Fig. 21-12). The cores also display abundant mudstone intraclasts, load and injection features, slump bedding, and local crudely graded intervals (Bouma, 1962; T_a and T_{ab} subdivisions) (Figs. 21-14, 21-15). Bedding contacts are generally irregular with amalgamation of beds and erosional downcutting of beds up to six inches (15 cm) being common.

Mudstones within the Webster Intermediate Zone, which make up to 50% of the intervals cored

in the two wells, are similar to those in the Webster Main but are more slump folded, faulted, and contorted and contain more resedimented debris flows (Fig. 21-14A). Beds range in thickness from 0.1 to 36 feet (0.03–11.0 m) and are locally interbedded with thin, very fine- to coarse-grained sandstones that are sometimes highly contorted and burrowed (Figs. 21-14A, 21-15C and D). Mudstone interbeds rarely occur within the sandstone sequences.

In contrast to the cores in the Webster Intermediate, the Webster Main cores (Fig. 21-13) consist of alternately thin-bedded, medium- to coarse-grained, moderately to poorly sorted, feldspathic (arkosic) sandstones interbedded with diatomaceous mudstones (Figs. 21-14D, E, F, 21-15F). Webster Main sandstones in the two cored wells make up 62% of the section, and individual beds range from 0.1 to 10.5 feet (0.03–3.2 m) thick, averaging 1.2 feet (0.4 m). The sandstones show internal grading and contain Bouma subdivisions in sequences T_a, T_{ab}, T_{abc}, T_{abcde}, and T_{cde}. Thin-bedded (1 in.; 2.5 cm), hemipelagic and detrital mudstones occur between most of the individual sandstone beds. Sole marks, mudstone intraclasts, load features, dish and pillar structures, and slump bedding are common. Erosional basal contacts occur with minimal downcutting (0.5 in.; 1.3 cm). Some amalgamation is present in a few of the thicker sandstone beds. Both lower and upper bedding contacts are generally sharp, and individual sandstones appear evenly bedded and separated by mudstones in the cores.

Diagenetic, nodular to layered dolomite-cemented zones occur sporadically within these sandstones (Fig. 21-14F). Cemented zones average 1.5 feet (0.5 m) in thickness and either completely or partially cement the host sandstone. These dolomite zones are not continuous between wells. The cemented zones form distinctive spikes on logs, particularly the porosity logs, which allow good log-to-

---

→

Fig. 21-14. Core photographs of the Webster Intermediate and Main zones in the 239D well (MOCO 35 WT 239D). Webster Intermediate cores are at (A) 1,141-, (B) 1,195-, and (C) 1,251-foot depths, and the Webster Main cores are at (D) 1,579-, (E) 1,583-, and (F) 1,605-foot depths. (A) The interchannel deposits are mainly mudstones and slump-folded (arrow) diatomaceous mudstones. (B) and (C) The channel-fill deposits are very coarse-grained and are characterized by numerous mudstone "rip-up"

intraclasts as in (B) and structureless conglomerate clasts as in (C). (D) The Webster Main consists of thin-bedded, medium-grained sandstones that are interpreted to be depositional lobes and channel/lobe transition deposits. They are graded with lower erosional (arrow) contacts; (E) locally have amalgamated (dashed lines) bedding; and (F) contain rare carbonate-cemented (c) zones that are locally faulted (arrows). Vertical scale (for all cores) on lower right is 0.1 feet (3 cm).

core calibration (see Fig. 21-13: 1,710- and 1,721-ft depths).

Mudstones within the Webster Main comprise 38% of the intervals cored in wells 239D and 327, and individual beds range in thickness from 0.1 to more than 28 feet (0.03–8.5 m) (Figs. 21-13, 21-14). They include gray, laminated diatomaceous mudstone, brown massive to laminated detrital mudstone, and resedimented and slumped diatomaceous mudstone and siltstone locally mixed with sandstone and detrital mudstone (Fig. 21-15F). Burrows occur in some of the mudstones and in places at the tops of the sandstone units. Thin-bedded, very fine- to medium-grained sandstone and siltstone are interbedded with the mudstones and are locally contorted and injected into overlying and underlying mudstones. Sandstone and siltstone also occur as burrow fills or are completely mixed with the mudstone in extensively burrowed zones.

Depositional Model

The depositional model is based on interpretation of cores, mapping of individual sand bodies from logs, examination of nearby equivalent outcrops, and the relation of the regional geology to the producing reservoir sandstones in the field. The depositional setting is inferred to have been an intraslope basin in 600 to 4,000 feet (183–1,220 m) of water (MacPherson, 1978; Webb, 1981; Gilbert, 1988). The Webster reservoirs are interpreted as turbidites deposited in a relatively narrow basin that was close to a tectonically active source (Fig. 21-16). These turbidites are interbedded with organic-rich shale and mudstone of the Monterey Formation, which served as both source and seal for oil in the turbidite reservoirs. The Webster Zone lithofacies variations are interpreted to represent a change from depositional lobes in the lower Webster Main to predominantly channel-fill deposits in the upper Webster Intermedi-

ate. A channel/lobe transition zone occurs between these two facies. The overall stacking of turbidite facies in the Webster Zone suggests that the system prograded basinward toward the northeast.

The lower Webster Main Zone contains mapped sand bodies that consist of alternating sandstone and mudstone sequences (Fig. 21-13), some of which contain amalgamated beds at their tops. The sandstone is medium grained and contains well-developed turbidite sequences (Bouma, 1962). Individual sandstone beds are often separated by thin mudstone drapes. These sequences are delineated by isochores (Figs. 21-9, 21-10) as sheet-like bodies that average 5,600 feet (1.7 km) in length, 6,000 feet (1.8 km) in width, and about 28 feet (8.5 m) in thickness. They are interpreted as composite depositional lobes because of their geometries, internal sedimentary features, and vertical sequences seen in cores. These lobes have poorly developed upward-coarsening and thickening cycles, and the wireline logs generally show blocky to spiky profiles. In outcrop, the sand bodies are medium-grained, tabular sheet-like units.

The upper Webster Intermediate Zone sand bodies contain thick sandstone and conglomerate sequences that display upward-thinning and fining sequences with erosionally sharp basal contacts. They are very coarse-grained to conglomeratic, very thick-bedded and amalgamated, and contain few mudstone interbeds. Their associated mudstones are locally slump folded and faulted and generally occur at the tops and between the sandstone units. These sand bodies are mapped as narrow, lenticular, northeast-trending features that average 4,000 feet (1.2 km) in length, 1,100 feet (0.3 km) in width, and about 37 feet (11.3 m) in thickness (Fig. 21-9). They are interpreted as composite channel-fill deposits based on sedimentary structures and vertical sequences seen in cores, the sand-body geometries inferred from isochore maps, blocky wireline log

Fig. 21-15. Core photographs of the Webster Intermediate and Main zones in well 327 (MOCO 34 T-327). Webster Intermediate is at (A) 1,586-, (B) 1,588-, (C) 1,647-, (D) 1,656-, and (E) 1,692-foot depths, and the Webster Main is at (F) 1819-foot depth. (A) and (B) Sandstone dike (1,586-ft depth) and sill (1,588-ft depth) (arrows), with unique oil-stained fractures, that are locally injected into diatomaceous mudstones. (C) and (D) Interbedded oil-impregnated sandstones and burrowed (arrows) diatomites. (E) Gravelly, coarse-grained sandstone cut by fractures (arrows). (F) Resedimented diatomite (d) turbidite or debris-flow deposit, containing black wood fragments and dark gray diatomite intraclasts, is interbedded with dark gray, laminated diatomites. Vertical scale (for all cores) on lower right is 0.1 feet (3 cm).

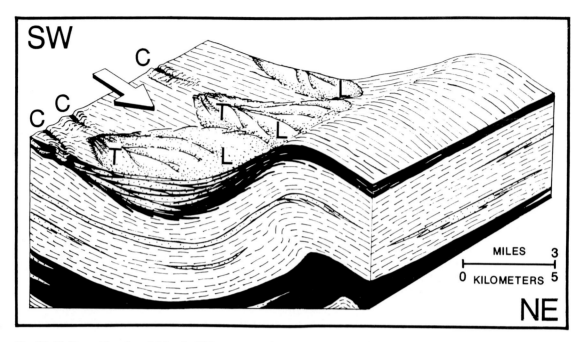

Fig. 21-16. Depositional model for the Webster reservoirs in Mobil's MOCO FEE area of Midway-Sunset Field (adapted from Scott and Tillman, 1981, and reprinted by permission of the Society of Economic Paleontologists and Mineralogists). The Webster was deposited as a series of sandstone bodies on the side of a narrow, slope-dominated basin bounded by a subsea anticline. Sediment transport was downslope, approximately at right angles to the paleoslope. Channel-fill (C), channel/lobe transition (T), and depositional lobe (L) deposits are recognized in this intraslope basin. Structural growth of the bathymetrically high anticline may have been syndepositional. Original figure drawn by M. T. Roberts.

profiles, and thick-bedded amalgamated sandstone character seen in outcrops. These channel-fill sandstones lack well-developed finer-grained levees (Fig. 21-12) and are interpreted to have been deposited in slope gullies.

Between the lower Webster Main and upper Webster Intermediate intervals, several sand bodies occur that are transitional between the depositional lobes and the channel-fill deposits. Two sand bodies (4G and 5B) display similar mapped geometries (Figs. 21-9, 21-10) and core and log signatures that are intermediate between interpreted depositional lobes and channel-fill deposits. They are medium grained, locally conglomeratic and amalgamated, and display scour features and Bouma sequences. These sandstones are sheet-like bodies with lenticular zones and have lengths of 4,200 feet (1.3 km), widths of 3,200 feet (1.0 km), and thicknesses of about 33 feet (10.1 m) (Fig. 21-9). They are interpreted to be channel/lobe transition deposits formed at a break in slope where the channels avulsed and became transitional downfan into depositional lobes. These deposits may be similar to the chan-

nel/lobe transition units discussed by Mutti and Normark (1987).

The lateral continuity and spatial relations of these facies could not be determined with certainty in this study. The turbidite facies appear to be vertically stacked, with depositional lobes overlain by channel/lobe transition deposits, followed by channel-fill deposits, suggesting progradation of facies and switching of depocenters within a narrow two-mile (3.2 km) belt (Figs. 21-9, 21-10). It could not be determined whether the sand bodies are laterally attached in a depositionally downdip direction. The channel-fill deposits appear to thicken abruptly in an updip direction and pinch out in a downdip direction (Fig. 21-10, sand bodies 4B and 4D). The channel-fill sequences do not appear to be connected to depositional lobes or to channel/lobe transition deposits in the field area (Fig. 21-16). The channel/lobe transition and depositional lobe facies are more sheet-like and may be spatially related to one another. They generally thicken updip and broaden and thin in a downdip direction (Figs. 21-9, 21-10; sand bodies 4G, 5B, 5C, and 5E).

The hierarchy of these turbidite facies can be interpreted in two possible ways: (1) tectonically influenced deepwater deposition, where the progradation of the slope and the slope channels occurred over deeper water fan deposits, and (2) eustatically controlled stacking of a lowstand basin-floor fan overlain by the slope-fan or wedge deposits (Vail, 1987). In this model, the lobe and the channel/lobe transition deposits would reflect the lowstand fan, and the channel-fill deposits the channel/levee part of the slope fan or lowstand wedge. The channel fills are interpreted here to be gully deposits and not channel/levee sequences. These systems would correspond to Mutti's (1985) type II and III systems, respectively. The data to answer these questions are not available, and the issues are beyond the scope of this study. However, we believe that tectonics played an important role in focusing deepwater sedimentation in this intraslope basin.

Webster sand bodies are remarkably similar to the slope gully deposits described by Surlyk (1987) for the Upper Jurassic Hareelv Formation in Jameson Land, East Greenland. The Hareelv Formation consists of slope shale with relatively small channel and lobate sand bodies deposited on a paleoslope as gully fills along a tectonically active basin margin. The geometry, scale, and depositional and tectonic setting of these sand bodies are similar to those in the Webster Zone. Surlyk draws analogies to similar Upper Jurassic turbidite reservoirs in the North Sea, with special reference to the giant Magnus Field.

Petrography

The Webster Zone is composed of poorly consolidated feldspathic (arkosic) sandstone, with grain-supported fabric. In intervals of high reservoir quality, the principal "cement" is the oil, which acts as a binder due to its high viscosity. In some of the poorer quality reservoir intervals, the volume of clay matrix is high and acts as a weak cement. Minor compaction textures such as deformation of micas and fracturing of feldspars are evident. Grains are angular to subangular and sorting is poor. In the better reservoir intervals macropores are interconnected, whereas in poorer zones, as the proportion of clay matrix increases, porosity and permeability decrease.

Petrographic and X-ray diffraction studies show that the Webster Zone sandstone contains from 30 to 50% feldspar, 20 to 40% quartz, and less than 5% other minor detrital components, including micas, rock fragments, detrital clays, and heavy minerals. This mineralogy is indicative of a plutonic sediment source. Detrital opaline (silica) material in the form of diatom fragments in the opal A grade is a minor constituent. Authigenic minerals include dolomite, microquartz, heulandite, calcite, pyrite, hematite, and smectite. Typically, these minerals comprise less than 5% of the sample with the exception of the occurrence of dolomite as spheroidal nodules in a few intervals. Illite/mica is the dominant detrital clay in the less than 5 micron size fraction. The main authigenic clay is smectite, and the total clay content is in the range of 2 to 5%. The interbedded mudstones and siltstones are compositionally similar to the sandstones except for higher proportions of clay and opaline silica (opal A).

Petrophysical Properties

The Webster Zone has excellent reservoir properties with an average porosity of about 33%, permeabilities of 800 to 4,000 md, and average oil saturations of 40 to 65% (Table 21-1). The Webster produces heavy oil (14° API) with a viscosity of 600 cP (6.0 × 10⁵ Pa·s) at 113°F (45°C). However, reservoir properties vary between and within individual sand bodies, particularly in a vertical sense. Examples of this variation are shown in the core-to-log comparisons for the Webster Intermediate in Figure 21-12 and for the Webster Main in Figure 21-13.

The Webster Intermediate Zone is sandstone-rich and very thick-bedded, and its reservoir properties are vertically less variable than those of the Webster Main. The distribution of oil saturation and porosity is uniform, resulting in blocky or squared-off wireline-log shapes. The Webster Main's individual sandstones are more layered, have graded beds, and contain more interbedded mudstone and siltstone. Its reservoir properties are more variable (Fig. 21-13). Oil saturation and porosity changes occur within and between individual sandstones in the sand bodies, resulting in more spiky curves on the logs.

The main difference between the Webster Intermediate and Main zones is texture, with the Intermediate being at least one grain-size class larger (coarse vs. medium) than the Main. The potentially most productive reservoirs are those that are coarsest grained and closest to sediment transport pathway(s) in the field area. Porosity is generally uniform in the various turbidite sandstones, whereas permeability appears to control oil saturation and reservoir qual-

Table 21-1. Summary of the reservoir characteristics and properties of the Webster Intermediate and Main zones and the Webster turbidite reservoir facies.

Reservoir character and properties	Webster zones		Turbidite facies		
	Intermediate	Main	Channel fill	Channel/lobe transition	Depositional lobe
Grain Size[1]	Coarse	Medium	V. Coarse	Coarse–Medium	Medium
Sorting[1]	Poor	Moderate	Poor	Poor	Moderate-Poor
Bed Thickness (ft/m)[1]	2.5/0.8	1.2/0.4	3.5/1.1	1.8/0.5	1.6/0.5
Sand Body Thickness (T)[2]	25/7.6	28/8.5	37/11.3	33/10.1	28/8.5
Length (L)[2]	4100/1250	5000/1524	4000/1219	4200/1280	6000/1829
Width (W)[2]	1300/396	5000/1524	1100/335	3200/975	5600/1707
L/W Ratio	3.2:1	1:1	3.6:1	1.3:1	1.1:1
W/T Ratio	52:1	179:1	30:1	97:1	200:1
Volume (ft³/m³) (V = T S L S W)	$1.3\times10^8/3.8\times10^6$	$7.0\times10^8/2.0\times10^7$	$1.6\times10^8/4.6\times10^6$	$4.4\times10^8/1.3\times10^7$	$9.4\times10^8/2.7\times10^7$
Log Shapes	– – – / – – –	– – – / – – –	Blocky Bell-Shaped	Spiky Blocky	Spiky Blocky
Porosity (%)[3]	32	33	31	34	33
Permeability (md)[3]	1100	800	1200	900	700
Oil Saturation (%)[3,4]	65	41	65	47	40
Inferred Geometries	Lenticular	Sheet-Like	Lenticular	Lenticular to Sheet-Like	Sheet-Like
Depositional Setting	Slope Channels	Slope to Basin Lobes	Channel Fills	Channel Upper Lobe Transition	Basin Floor Depositional Lobes

[1]Determined from core measurements of wells 239D and 327.
[2]Analyses are based on isochore mapping and computer-generated contouring and averaging for all the sand bodies.
[3]Based on 930 analyses of core plug samples taken every foot and includes all lithologies within the sand bodies; analyzed by Petroleum Testing Services, Bakersfield, California.
[4]Oil saturation ratio = oil saturation/total fluids.

ity directly. The mineralogical composition of the Webster Intermediate and Main zones is essentially the same.

Reservoir quality of the three turbidite facies (representing channel, lobe, and channel/lobe transition) is excellent. As a group, the three turbidite reservoir facies average 33% porosity, 1,000 md permeability, and 50% oil saturation in the field area. Channel deposits have the best inferred reservoir potential for effective sweep by steam drive due to its high permeability (1,200 md), coarse grain size, and narrowly confined mapped geometry (Table 21-1; Figs. 21-9, 21-10, 21-17). It has the highest oil saturation (65%) of the three facies and can be swept effectively with the fewest wells if they are properly placed to take advantage of sand-body geometry. Depositional-lobe sandstones have a lower reservoir potential for steam-drive sweep

because of lower permeabilities (800 md), lower oil saturations (40%) and finer textures (medium grained) than in the channel sandstones. However, because depositional lobes cover a much larger geographic area, well placement for steam drive is not as critical as for the channel sequences. The reservoir properties of the channel/lobe transition facies are intermediate between the other facies.

Reservoir quality is much better developed in the 239D than in the 327 well. Well 239D appears to be in the center of the sediment transport pathways for many of the sand-body intervals determined from the isochore maps (Fig. 21-9). Well 327 either misses or penetrates many of the sand bodies at their depositional edge. Reservoir quality in well 327 is much lower overall because of finer grain size, poorer sorting, thinner beds and sand bodies, and lower permeability (Table 21-1). Porosity and

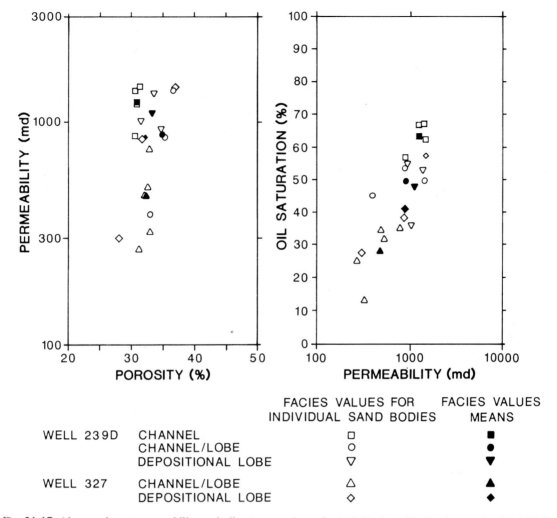

FACIES VALUES FOR   FACIES VALUES
INDIVIDUAL SAND BODIES   MEANS

		INDIVIDUAL SAND BODIES	MEANS
WELL 239D	CHANNEL	□	■
	CHANNEL/LOBE	○	●
	DEPOSITIONAL LOBE	▽	▼
WELL 327	CHANNEL/LOBE	△	▲
	DEPOSITIONAL LOBE	◇	◆

Fig. 21-17. Air porosity vs. permeability and oil saturation vs. permeability cross-plots for the three turbidite facies from the two cored wells in the Webster zones. Mean values of porosity, permeability, and saturation are shown for individual sand bodies (zones 4 and 5, A-G) for a total of 930 data points from the 239D and 327 well cores used in the analysis. Oil saturation = oil saturation/total fluids ratio.

permeability decrease by approximately 40 to 60% away from the inferred sediment pathways.

Production Characteristics

The most important consideration for steam-drive development of the Webster Zone is continuity of flow paths between injectors and producers. Other considerations are the optimum placement of the development pattern over the reservoir, both laterally and vertically, and the effects of steam on the reservoir sandstones.

Individual sand-body thickness maps, such as those shown in Figure 21-9, are critical for predicting flow paths and selecting optimum well locations. When combined with structure maps, the isochore maps are, in effect, flow-path maps for individual sand bodies. These maps provide a sense of the interconnection between sand bodies and offer control for development patterns to penetrate the greatest number of sand bodies in the best positions in the reservoir. Analysis of the core data and individual wireline-log character provide the detailed reservoir information to be used with the maps in placement and interval selection for both the injector and the producer wells. Development in the

MOCO FEE area will begin initially with twelve 5-acre (2.0-ha.) inverted five-spot patterns placed to best exploit the northeast-trending Webster Intermediate and Main reservoirs.

The reaction of Webster Zone reservoir sandstones to steam is a major concern. Because the Webster Intermediate and Main sandstones are essentially unconsolidated and weakly held together by hydrocarbons and some clays, the reservoirs could be subject to formation collapse and reduction of porosity and permeability. In the poorer quality reservoir zones, the migration of "fines" (clay and silt) may reduce reservoir response to steam flooding. The Webster sandstones are composed of relatively reactive minerals (feldspars, rock fragments, clays, and siliceous fossil remains) and are subject to steam-induced diagenesis. An understanding of facies architecture allows these steam-related mineral reactions to be anticipated and to be evaluated in advance by thermal modeling and/or empirical testing.

Exploration and Production Strategy

Webster turbidite reservoirs represent one of the initial deposits of an intraslope basin, with relatively small sand bodies that are dip-oriented and parallel to the northeast-dipping paleoslope. The numerous sand bodies in the Webster reservoir are closely spaced and have a predictable orientation. These channel-fill and depositional-lobe deposits onlap growing subsea structures to the east. The model presented here can be applied to similar exploration and production infill drilling projects along the western side of San Joaquin basin (Fig. 21-16). Care must be taken to recognize that not all turbidite sand bodies in the area are similar. For example, the Webster sand bodies are much different in their size, timing, and trend than the larger Monarch and Spellacy reservoirs, which represent the final phase of the intraslope basin fill. Infill production wells along structural and depositional trend on existing leases or as step-out wells on open acreage can test the concepts presented here.

Many other productive upper Miocene turbidite sandstones, similar in character to the Webster, occur on the western side of San Joaquin basin, including the Potter, Monarch, Spellacy, Republic, Williams, Leutholz, and Moco "T" sandstones, among others (MacPherson, 1978; Webb, 1981; California Oil and Gas Fields, 1985; Gilbert, 1988).

These turbidite sandstones filled a northwest-trending intraslope basin, which was at least 20 miles (32 km) long and 6 miles (9.7 km) wide and about 600 to 4,000 feet (183–1,200 m) deep. Large subsea anticlines formed the northern and eastern margins of the basin and locally blocked turbidite deposition to the east. Major submarine channels such as the 24Z, 26R, and 555 went between and around the growing subsea anticlines and locally connected with the main Stevens depocenter containing sediment derived mainly from the east and southeast (Fig. 21-4).

Turbidite reservoirs like the Webster form excellent stratigraphic traps and can be expected where a narrow, tectonically controlled shelf or shelf-edge delta occurs adjacent to a deep basin that is in close proximity to a major sediment source.

Summary

Old wireline logs from approximately 140 wells were used to subdivide the Webster reservoir of Midway-Sunset Field into two zones and 15 individual sand bodies. Two cored wells in the Webster Intermediate and Webster Main reservoirs were calibrated with modern logs and demonstrated the validity of these zonations and subdivisions. In addition, the cores provided samples for reservoir analyses, including petrography, porosity, permeability, and saturation measurements. These cores allowed calibration of well logs and, in combination with mapping, provided the sedimentological observations for a depositional model of the Webster Zone. In this model, the Webster Zone is interpreted as a turbidite sequence deposited in an intraslope basin. Depositional lobes occur in the lower Webster Main and channel-fill deposits are recognized in the upper Webster Intermediate. Channel/lobe transition deposits occur between the upper Webster Main and the lowermost Webster Intermediate zones.

Based on the mapping of individual Webster sand bodies and analyses of cores, the depositional model provides geological control for steam-flood development of the Webster Zone in Mobil's MOCO FEE property. When combined with engineering studies, this model can be used to predict lateral and vertical sweep efficiency for individual sand bodies, volumetrics and oil in place, and localities for steam drive and well placement. As a direct result of the Webster reservoir description, initial development will utilize twelve 5-acre (2.0-ha.), inverted five-spot patterns placed for coincident and optimum exploitation of the best of the Webster Intermediate and Main reservoirs.

Acknowledgments. We thank J.H. Barwis, P. Braithwaite, I.A. Fischer, W.T. Long, N.L. McIver, J.G. McPherson, B.T. Pearson, J.R.J. Studlick, A. Thomson, R.W. Tillman, and P. Weimer for reviewing this manuscript and J.D. Cocker of Mobil's Denver Petrography Laboratory for providing XRD and SEM analyses. J.W. Ford helped in the preparation of the cross-plot figures, and the South Midway-Sunset Reservoir Management Team contributed valuable insight and background information into this project. Drafting support was provided by Mobil Denver Production Division and Mobil Research & Development, Dallas. This reservoir study was conducted by Mobil Oil Corporation, whose permission to publish is gratefully acknowledged.

Reservoir Summary

Field: Midway-Sunset (Webster Zone)
Location: Kern County, California
Operator: Mobil, Chevron, Texaco, Shell, and others
Discovery: 1894
Basin: San Joaquin basin
Tectonic/Regional Paleosetting: Remnant forearc/transform-margin basin
Geologic Structure: Anticline
Trap Type: Four-way closure and stratigraphic pinch-out
Reservoir Drive Mechanism: Water (steam injection)
• **Original Reservoir Pressure:** 300 psi (2.1×10^3 kPa) at 0 feet (0 m) subsea
Reservoir Rocks
• **Age:** Late Miocene
• **Stratigraphic Unit:** Webster Zone (Stevens equivalent of the Monterey Formation)
• **Lithology:** Medium- to coarse-grained feldsarenite
• **Depositional Environment:** Turbidite, intraslope basin
• **Productive Facies:** Slope channel, channel/lobe transition, and depositional lobes
• **Petrophysics**
  • **Porosity Type:** Total = 33% with primary 33% intergranular & 0% secondary porosity
  • **ϕ:** Average 33%, range 28 to 35% (unstressed cores, cutoff under evaluation)
  • **k:** Average 1,000 md, range 800 to 4,000 md (unstressed cores, cutoff under evaluation)
  • **S_w:** Average 35%, range 26 to 46% (unstressed cores, cutoff under evaluation)
  • **S_{or}:** Under evaluation for steam-drive development
Reservoir Geometry
• **Depth:** 700 to 1,200 feet (210–365 m)
• **Areal Dimensions:** 0.75 by 1.5 miles (1.2×2.4 km)
• **Productive Area:** 1,120 acres (455 ha.)
• **Number of Reservoirs:** 2 reservoir zones with 15 total sand bodies
• **Hydrocarbon Column Height:** 300 to 600 feet (91–183 m)
• **Fluid Contacts:** NA
• **Number of Pay Zones:** 15
• **Gross Sandstone Thickness:** 250 to 500 feet (76–152 m)
• **Net Sandstone Thickness:** 50 to 250 feet (15–76 m)
• **Net/Gross:** 0.6 to 0.8 (average range) at 1,000 feet (305 m) depth
Hydrocarbon Source, Migration
• **Lithology & Stratigraphic Unit:** Basinal shale, Monterey Formation, middle Miocene
• **Time of Hydrocarbon Maturation:** Early Pliocene (?)
• **Time of Trap Formation:** Early Pliocene (?)
• **Time of Migration:** Early to middle Pliocene to present
Hydrocarbons
• **Type:** Oil
• **GOR:** 0

- **API Gravity:** 14°
- **Viscosity:** 600 cP (6.0 × 10⁵ Pa·s) at 113°F (45°C)

Volumetrics
- **In-Place:** Under evaluation for steam-drive development
- **Cumulative Production:** 13 MMBO (2.1 × 10⁶ m³) on MOCO FEE acreage
- **Ultimate Recovery:** Under evaluation for steam-drive development
- **Recovery Efficiency:** Under evaluation for steam-drive development

Wells
- **Spacing:** Irregular at present; plan is for five-acre (2.0-ha.), inverted five-spot pattern
- **Total:** 35 wells currently under cyclic steam stimulation
- **Dry Holes:** 0

Typical Well Production
- **Average Daily:** 11 BO (1.7 m³)
- **Cumulative:** Under evaluation for steam-drive stimulation

Other: The Webster Zone is scheduled for steam-drive development.

References

Ayers, R.N., 1942, Webster area of Midway-Sunset Oil Field: Summary of Operations, California Department of Oil and Gas, v. 26, p. 19–24.

Bandy, O.L., and Arnal, R.E., 1969, Middle Tertiary basin development, San Joaquin Valley, California: Geological Society of America Bulletin, v. 80, p. 783–820.

Bouma, A.H., 1962, Sedimentology of some flysch deposits: Amsterdam, Elsevier, 168 p.

California Oil and Gas Fields, 1985, Central California: California Department of Conservation, Division of Oil and Gas, Sacramento, Third Edition, Vol. 1.

California Division Mines and Geology, 1987, Seventy-Second Report of the State Oil and Gas Supervisor 1986: California Department of Conservation, Division of Oil and Gas, Publication PR-06, p. 70.

Dibblee, T.W., Jr., 1973, Regional geologic map of San Andreas and related faults in Carrizo Plain, Temblor, Caliente, and La Panza Ranges and vicinity, California: U.S. Geological Survey Miscellaneous Geological Investigations Map I-757, scale 1:24,000.

Fischer, K.J., Heasler, H.P., and Surdam, R.C., 1988, Hydrocarbon maturation modeling of the southern San Joaquin basin, California, *in* Graham, S.A. and Olson, H.C., eds., Studies of the Geology of the San Joaquin Basin: Society of Economic Paleontologists and Mineralogists, Pacific Section, p. 53–64.

Gilbert, J.R., Jr., 1980, Stratigraphy and sedimentology of upper Miocene Williams Sand of the San Joaquin Valley, California, with a note on the neighboring Temblor sands [M.S. thesis]: Amherst, University of Massachusetts, 181 p.

Gilbert, J.R., Jr., 1988, Upper Miocene Williams turbidite sandstone west-side southern San Joaquin basin, California, *in* Graham, S. A. and Olson, H.., eds., Studies of the Geology of the San Joaquin Basin: Society of Economic Paleontologists and Mineralogists, Pacific Section, p. 249–260.

Harding, T.P., 1976, Tectonic significance and hydrocarbon trapping consequences of sequential folding synchronous with San Andreas faulting, California: American Association of Petroleum Geologists Bulletin, v. 60, p. 356–378.

Hsü, K.J., 1977, Studies of Ventura Field, California, I: Facies geometry and genesis of lower Pliocene turbidites: American Association of Petroleum Geologists Bulletin, v. 61, p. 137–168.

International Petroleum Encyclopedia, 1987, U.S. fields with reserves exceeding 100 million bbl: New York, Pennwell, v. 20, p. 238.

Isaacs, C.M., 1987, Sources and deposition of organic matter in the Monterey Formation, south-central coastal basins of California, *in* Meyer, R.F., ed., Exploration for heavy crude and natural bitumen: American Association of Petroleum Geologists Studies in Geology 25, p. 193–205.

MacPherson, B.A., 1978, Sedimentation and trapping mechanism in upper Miocene Stevens and older turbidite fans of southeastern San Joaquin Valley: American Association of Petroleum Geologists Bulletin, v. 62, p. 2243–2274.

Mutti, E., 1985, Turbidite systems and their relations to depositional sequences, *in* Zuffa, G.G., ed., Provenance of Arenites: Dordrecht, Netherlands, D. Reidel Publication Company, p. 65–93.

Mutti, E., and Normark, W.R., 1987, Comparing examples of modern and ancient turbidite systems: Problems and concepts, *in* Legget, J.K. and Zuffa, G.G., eds., Marine Clastic Sedimentology: Concepts and Case Histories: Lancaster, England, Kluwer Academic Publications, p. 1–38.

Scott, R.M., and Tillman, R.W., 1981, Stevens Sandstone (Miocene), San Joaquin basin, California, *in* Siemers, C.T., Tillman, R.W., and Williamson, C.R., eds., Deepwater clastic sediments—a core workshop: S.E.P.M. Core Workshop No. 2: Society of Economic Paleontologists and Mineralogists, p. 116–248.

Surlyk, F., 1987, Slope and deep shelf gully sandstones, Upper Jurassic, East Greenland: American Association of Petroleum Geologists Bulletin, v. 71, p. 464–475.

Vail, P.R., 1987, Seismic stratigraphy interpretation procedure, *in* Bally, A.W., ed., Atlas of Seismic Stratigraphy: American Association of Petroleum Geologists Studies in Geology 27, v. 1, p. 1–10.

Webb, G.W., 1981, Stevens and earlier Miocene turbidite sandstones, southern San Joaquin Valley, California: American Association of Petroleum Geologists Bulletin, v. 65, p. 438–465.

Key Words

Midway-Sunset Field, California, San Joaquin basin, Webster Zone, Monterey Formation, late Miocene, turbidite, submarine fan channel, submarine fan lobe, submarine fan channel/lobe transition, slope-gully deposits, steam-drive development, EOR.

Authors' Postscript

In the period 1989 to 1990 following the writing of this chapter, Mobil Oil Corporation drilled an additional 200 wells (72 Webster wells) in a 60-acre (24.3 ha.) area at the crest of the 35 Anticline in section 35, for enhanced heavy oil recovery (Fig. 21-3). Extensive new wireline-log and core data, including selected temperature surveys, were collected to further characterize the reservoirs. Twenty-nine new wells were drilled in the Webster Intermediate and 43 wells in the Webster Main zones with all the new wells testing the lateral (northwest-southeast) variability of the reservoirs. Three wells were cored in the Webster Zone, recovering over 1,300 feet (43.6 m) of core. An additional 200 wells are planned for Mobil's MOCO FEE property over the next five years and these wells will test the updip and downdip continuity of individual sand bodies.

These new data strongly support the interpretations and predictions made in this chapter. In particular, the sand-body geometries and trends have been validated. In addition, the new data allowed for further subdivisions of the sand bodies and the recognition of a greater percentage of boulder conglomerate in the Webster Intermediate (Zone 4 sand bodies) than was previously recognized. These conglomerate zones have porosity values of about 10-15%, pointing to a greater variability in the reservoir's quality, than was originally thought. Wet zones and oil-water contacts can now be defined using the new suite of wireline logs to improve the net pay and oil-in-place calculations.

22

Deep-Sea-Fan Channel-Levee Complexes, Arbuckle Field, Sacramento Basin, California

Douglas P. Imperato and Tor H. Nilsen

Applied Earth Technology, Inc., Redwood City, California 94063

Introduction

Mud-rich deep-sea-fan systems throughout the world yield significant accumulations of hydrocarbons. However, these distinctive turbidite deposits remain as yet poorly understood. Although several modern mud-rich deep-sea fans such as Mississippi fan (Kastens and Shor, 1985; Stelting et al., 1985; Bouma et al., 1986), Amazon fan (Damuth et al., 1983; Flood and Damuth, 1987; Damuth et al., 1988; Manley and Flood, 1988), Bengal fan (Emmel and Curray, 1985), and Rhone fan (Droz and Bellaiche, 1985) have been extensively studied, relatively few ancient mud-rich deep-sea fans have been described in detail. McCabe (1978) and Melvin (1986) provided useful descriptions of two Carboniferous delta-fed mud-rich fan sequences, although no ancient fans of this type have been directly related to hydrocarbon accumulations. In this chapter, we present a detailed stratigraphic study of facies associations, sandstone distribution, and sandstone geometries and the relation of these to hydrocarbon production in the Forbes Formation, an ancient mud-rich turbidite system of the California Great Valley forearc basin.

The mud-rich turbidite sequence of the Upper Cretaceous (Santonian to Campanian) Forbes Formation in the Sacramento basin is one of the most prolific producers of nonassociated gas in California and is one of the best-studied ancient mud-rich turbidite sequences in the world. Drilling activity for Forbes targets has remained steady despite recent nationwide industry decreases (Weagant, 1986). Analyses of a large number of wells (more than 5,000), many miles of seismic-reflection data, cores, and nearby outcrops of gas-productive strata have contributed to our knowledge of this depositional system, which may serve as an analogue for similar, less-understood systems in other parts of the world.

Cumulative gas production from Campanian reservoirs in the Sacramento basin through 1984 exceeded 1.3 trillion cubic feet (3.7×10^{10} m³), with gas reserves being more than 250 billion cubic feet (7.1×10^9 m³) (California Division of Oil and Gas, 1984). The Arbuckle Gas Field (Fig. 22-1) in Colusa County in the central Sacramento basin (T.13-14N.,R.2W.), approximately 50 miles (80 km) northwest of Sacramento, has produced more than 70 BCFG (2.0×10^9 m³) from reservoir sandstones deposited as channel-levee complexes of the mud-rich deep-sea-fan system of the middle member of the Forbes Formation. An anticlinal feature near the town of Arbuckle was originally identified seismically and penetrated by several shallow wells beginning in 1950. The Arbuckle Gas Field was discovered in 1957 when the Western Gulf Oil Company successfully completed the Arbuckle Unit C-1 well at 5,581 to 5,608 feet (1,701–1,709 m) and 5,873 to 5,910 feet (1,790–1,801 m), with a flow pressure of 1,245 psi (8.6×10^3 kPa) through a 1/2-inch (1.3-cm) choke and an IP of 7.9

Casebooks in Earth Science
Sandstone Petroleum Reservoirs
Eds.: Barwis/McPherson/Studlick
© 1990 by Springer-Verlag New York, Inc.

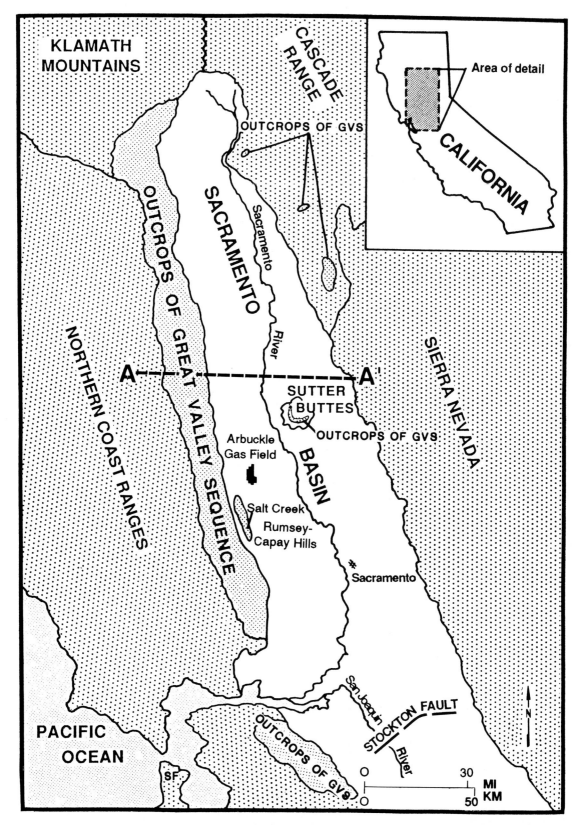

Fig. 22-1. Sacramento basin and surrounding areas showing location of the Arbuckle Gas Field and outcrops of Great Valley sequence. Abbreviations: GVS, Great Valley sequence; SF, San Francisco. Cross section A-A' is shown in Figure 22-2.

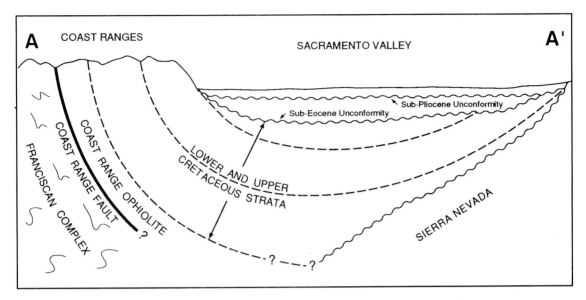

Fig. 22-2. Generalized structure and stratigraphy of the Sacramento basin. Stratigraphic and structural relations are not to scale. See Figure 22-1 for location of section.

MMCFGPD (2.2 × 10⁵ m³/D) from the middle member of the Forbes Formation.

Western Gulf Oil Company successfully completed 12 more gas wells through 1958, using an established 160-acre (64.8-ha.) spacing. Occidental Petroleum Corporation began exploration in the Arbuckle area in 1959 and successfully completed a dual-zone producer in the Forbes Formation. These two companies dominated development of the Arbuckle Gas Field until the mid-1960s, at which time several other companies became active. Although nine wells have been completed in the field since 1983, only three wells were completed during the prior 20 years. Due to a lack of state law governing well offsets and spacing and the discontinuous nature of Forbes Formation reservoirs, recent drilling activity has proceeded without constraints on well spacing. Presently, the major operators in the field are Chevron USA, Inc. and Occidental, USA.

The California Division of Oil and Gas (1988) reports a proven acreage of 3,280 (1,328 ha.) with 78 wells in the field: 29 gas wells (six shut-in), 21 abandoned gas wells, and 28 "dry holes"; four wells were being drilled. The number of producing wells has remained relatively constant at 25 to 30 throughout the history of the field. Peak production was in 1961, reaching more than 8.0 BCFG (2.2 × 10⁸ m³) that year. Remaining gas reserves are estimated at 4.0 BCFG (1.1 × 10⁸ m³) (California Division of Oil and Gas, 1988).

Regional Framework

Tectonic and Structural Setting

The Late Cretaceous tectonic framework of California was dominated by convergence resulting from subduction of the Farallon plate (Atwater, 1970; Dickinson, 1976; Ingersoll, 1979; Nilsen, 1986). The Sacramento basin formed part of the larger, narrow, north-trending Great Valley forearc basin in the arc-trench gap region, which developed between the oceanic trench to the west and the Sierran magmatic arc to the east. Huge volumes of sediment were supplied to the basin from granitic and metamorphic highland source areas in the Klamath Mountains to the north and Sierra Nevada to the east (Ingersoll, 1979, 1983). The Franciscan assemblage accumulated by tectonic and sedimentary processes in the trench and associated accretionary wedge to the west (Bailey et al., 1964; Dickinson, 1976). The western margin of the forearc basin during the Cretaceous was probably formed by uplifted Franciscan rocks that formed a submarine topographic barrier.

In response to continued subduction of the Farallon plate and possibly accretion of tectonostratigraphic terranes to the west, uplift of both the subduction complex and the western margin of the forearc basin was initiated in the Paleogene. The present synclinal structural axis of the basin trends north-south; steeply east-dipping deep-marine strata of the Great Valley sequence crop out along the western erosional margin of the basin, and gently west-dipping shallow-marine and nonmarine strata outcrop along the eastern margin of the basin (Figs. 22-1, 22-2).

The northward migration of the Mendocino triple junction and associated faulting and volcanism in the late Neogene have resulted in the formation of many large structures along the length of the Sacramento basin. These structures are dominated by a series of steeply dipping, generally northwest-trending en-echelon faults, with variable strike-slip and dip-slip components. These faults and related features (such as flower structures and folds) are important in trapping gas throughout the basin. Development of the anticline in the Arbuckle Gas Field probably resulted from compression related to transform tectonics of the northerly migrating Mendocino triple junction.

A local north-trending anticline along the generally eastward-dipping western margin of the basin provides as much as 200 feet (61 m) of closure at a level near the top of the productive interval in the southern part of the Arbuckle Gas Field (Fig. 22-3). Structure-contour maps of various horizons indicate flexure along an axis which trends 160 degrees. Minimal closure decreases with increasing depth, indicating that the anticline has the form of a similar fold. The sub-Eocene unconformity above the Cretaceous strata is folded, yet the sub-Plio-Pleistocene unconformity is undeformed. We have found no evidence for Cretaceous deformation in the Arbuckle area as suggested by Vaughan (1962, 1968).

A nearly conjugate system of predominantly high-angle normal faults with relatively minor offset (less than 500 feet (152 m) was mapped in the Arbuckle Field by Vaughan (1962, 1968). He deduced that the Arbuckle anticline consisted of several half grabens. Apparent offset gas-water contacts between some wells suggest permeability barriers that may be due to faulting, although these offsets can also be explained by stratigraphic "pinch-outs" of sandstone bodies.

We have observed several east-dipping thrust faults in the vicinity of the Arbuckle Field both on seismic-reflection data and in outcrop. A steep structural gradient in the northwest corner of the field (Fig. 22-3) may be interpreted as a thrust fault, although this has not been defined seismically or in well-log correlations. It is possible that thrust faults may be present in the field at depths below well penetrations.

Regional Stratigraphy

The Sacramento basin was the northern depocenter of the Cretaceous Great Valley forearc basin and contains a thick sequence of Cretaceous strata deposited primarily in slope, deep-sea-fan, and basin-plain environments. These turbidite sequences thin abruptly to the east, where they onlap igneous and metamorphic basement rocks of the Sierra Nevada and generally grade into shallow-marine facies (Fig. 22-2). During the Tertiary, when the Sacramento basin underwent uplift and erosion (Dickinson et al., 1979), deposition was dominated by shallow-marine and nonmarine facies. Alluvial deposits of the Plio-Pleistocene Tehama Formation cap the Tertiary sequence throughout most of the basin.

The Occidental Petroleum Corporation No. 1 Arbuckle Section Four Unit (Sec. 4, T.13N., R.2W.) is the deepest well in the Arbuckle Gas Field. It reached a total depth of 12,007 feet (3,660 m) after penetrating more than 5,500 feet (1,680 m) of Forbes Formation strata (Fig. 22-4). The Forbes Formation was deposited as a mud-rich, basin-plain, deep-sea-fan, and slope turbidite system which prograded southward down the plunging axis of the Cretaceous forearc basin. We have divided it informally into three members (Fig. 22-4): (1) the Dobbins Shale Member (Kirby, 1943), basin-plain deposits; (2) the unnamed middle member, deep-sea-fan deposits; and (3) the unnamed upper member, slope mudstone and sandstone-filled gully deposits.

Fig. 22-3. Structure-contour map on middle member of the Forbes Formation wireline-log marker in the Arbuckle Gas Field (see Fig. 22-4). This marker generally corresponds to the top of the productive interval, which is approximately 1,000 to 1,500 feet (305–460 m) thick. Contour interval is 100 feet (30.5 m).

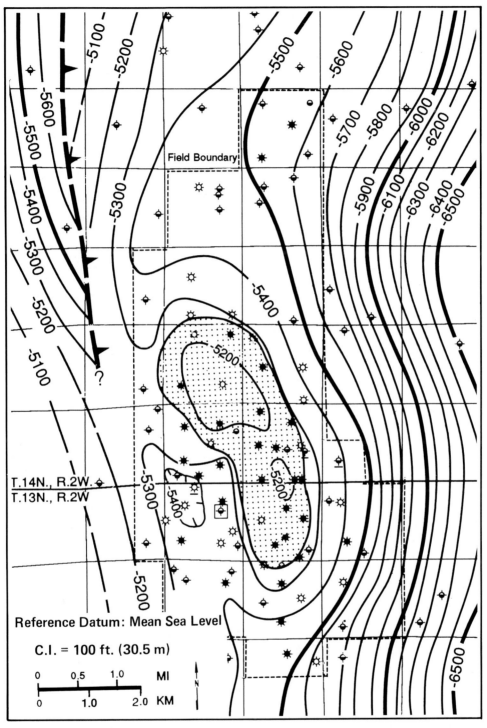

Reference Datum: Mean Sea Level

C.I. = 100 ft. (30.5 m)

Field Boundary

T.14N., R.2W.
T.13N., R.2W

▲ — ▲ Inferred Thrust Fault
(Barbs on Hanging Wall)

Area of Structural Closure

□   Type Log

⊙   Drilling in Progress
☼   Abandoned Gas Well
⚲   Directionally Drilled Well
✦   Dry and Abandoned
✳   Gas Producing Well

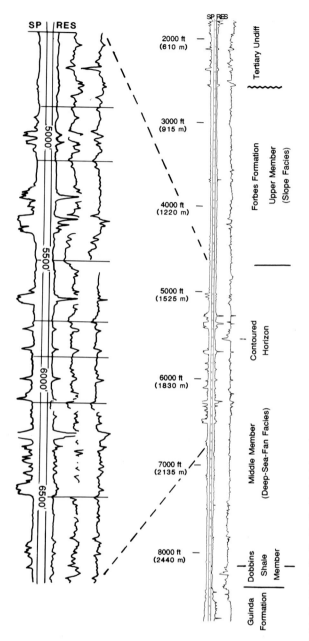

Fig. 22-4. Type log of the Arbuckle Gas Field showing formation correlations, member subdivisions, and the contoured horizon (Fig. 22-3). The productive interval (shown in detail) is between approximately 5,000 and 6,500 feet (1,525–1,980 m). The underlying Guinda Formation and older units are not shown; the total depth of the well is 12,007 feet (3,660 m). See Figure 22-3 for well location.

Southerly prograding basin-plain, deep-sea-fan, and slope sequences of the Forbes Formation are expressed as distinct seismic facies. Basin-plain deposits are relatively thin and form one of the most continuous, prominent set of reflectors in the section. Deep-sea-fan deposits are generally represented as relatively flat, subparallel, and continuous reflectors. The slope sequence can be identified as a relatively transparent unit, with short, discontinuous reflectors and local low-angle clinoforms.

Deltaic deposits of the Kione Formation conformably overlie and interfinger with the upper member of the Forbes Formation throughout most of the basin. An angular unconformity at the base of the Tehama Formation truncates Forbes Formation strata in the western part of the Arbuckle Field (Fig. 22-4). In the eastern part of the field, a thin interval of Kione Formation is truncated by the Eocene Capay Shale, which is truncated by the Tertiary sequence.

Reservoir Characterization

Depositional Facies and Facies Associations

Mud-rich deep-sea-fan turbidite systems such as the Forbes Formation differ markedly from mixed-sediment or sand-rich fans in many ways (Nelson and Nilsen, 1984). Inner-fan, middle-fan, and outer-fan sequences, as defined by Normark (1970), Mutti and Ricci Lucchi (1972), and Walker (1978) for other types of fans, are not easily recognized. Numerous relatively small gullies fed sediment to the fan system from the Kione delta system to the north. These gullies shifted abruptly in location through time, preventing the development of stable, permanent, large canyons and inner-fan channels. The fan consists of a system of numerous channels and associated levees which gradually become thinner and narrower distally. Most of the sediment transported by the channels is deposited as overbank material, and channels are flanked by extensive levee systems. Lobes are absent or not well developed. Sandstone reservoirs in such systems are thus formed by ribbon-like sandstone beds that are extremely limited in lateral extent, especially along depositional strike.

Outcrops of the Forbes Formation in the Rumsey-Capay Hills, approximately 5 miles (8 km) west of

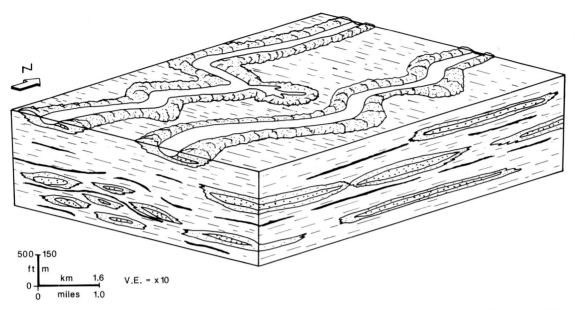

Fig. 22-5. Block diagram showing the geometries of typical channel-levee sequences of the deep-sea-fan facies of the middle member of the Forbes Formation. Heavy lines in the cross sections indicate shale resistivity markers.

the Arbuckle Gas Field, and selected cores within the basin provide an excellent opportunity to gain a more detailed understanding of the Forbes Formation deep-sea-fan system. We have measured and described in detail several nearly continuous well-exposed sections as thick as 2,500 feet (760 m). Bed-by-bed descriptions and interpretations of Bouma (1962) subdivisions and of Mutti and Ricci Lucchi (1972) facies and facies associations were made for each measured outcrop and core section. From these, we have identified the lateral and vertical distribution of facies deposited by a migrating channel-levee complex (Figs. 22-5 to 22-9).

The middle member of the Forbes Formation comprises a series of sequences deposited as channel-levee complexes (Fig. 22-5). The lower part of each sequence consists of several vertically "stacked" levee deposits separated by interchannel deposits. This relatively shale-rich interval generally has an average thickness of approximately 50 feet (15.2 m) and is overlain by increasingly sandstone-rich channel deposits which range in thickness from 25 to 100 feet (7.6–30.5 m) and may be stacked to form composite sequences as thick as 300 to 400 feet (91–122 m). The sequence deposited by a migrating channel-levee complex thus ranges in thickness from approximately 100 to 500 feet (30.5–152 m) and generally forms a coarsening-, then upward-fining "symmetrical" cycle (Fig. 22-6).

Levee sequences are dominated by thin-bedded siltstones and fine-grained sandstones (Fig. 22-7) that generally represent progradation (into interchannel areas) and commonly form thin upward-coarsening log signatures (Fig. 22-6). Levee deposits generally average 5 to 25 feet (1.5–7.6 m) in thickness and typically become thinner and "pinch out" away from channels into associated interchannel mudstones. At Salt Creek (Fig. 22-6), sandstone-to-shale ratios in levee sequences range from 1:5 to 1:1, and average bed thickness is approximately 1 to 2 inches (2.5–5.1 cm; Fig. 22-6). Laterally continuous, planar-to ripple-laminated beds (Facies D beds of Mutti and Ricci Lucchi, 1972), which range in thickness from less than 1 inch to 1 foot (2.5–30.5 cm), are common (Fig. 22-8). Laterally discontinuous, poorly sorted coarser-grained beds, which exhibit "pinch-and-swell" lenticular bedding, are also common in levee sequences (Facies E beds of Mutti and Ricci Lucchi, 1972), primarily in the proximal, most sandstone-rich part of the levee sequence (Fig. 22-8). Sandstone bed thicknesses of distal-levee deposits are commonly less than can be resolved by well logs, so they may be difficult to distinguish from more shale-rich interchannel deposits (Fig. 22-6).

Channel sequences of the middle Forbes Formation (Fig. 22-9) are dominated by fine- to coarse-grained sandstone beds with massive bedding and

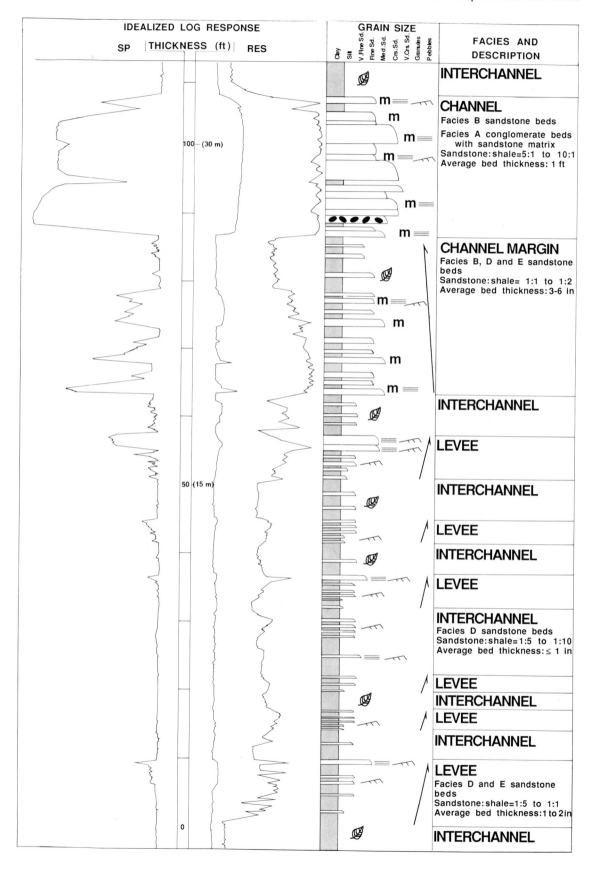

Fig. 22-7. A levee sequence in the middle member of the Forbes Formation cropping out in Salt Creek, Rumsey-Capay Hills. See Figure 22-6 for description.

Fig. 22-8. Typical facies D and E beds (classification of Mutti and Ricci Lucchi, 1972) in levee sequence of the middle member of the Forbes Formation in outcrops in the Rumsey-Capay Hills.

Fig. 22-6. Part of measured section of the middle member of the Forbes Formation cropping out along Salt Creek in the Rumsey-Capay Hills, approximately 5 miles (8 km) west of the Arbuckle Gas Field, showing idealized wireline-log signatures, generalized bed descriptions, and facies interpretations. The lower part of this sequence is dominated by thickening- and coarsening-upward levee deposits, which are overlain by channel-margin and channel deposits. Thicknesses of thin-bedded sandstones in levee sequences are shown schematically.

Fig. 22-9. A channel and channel-margin sequences in the middle member of the Forbes Formation cropping out in Salt Creek, Rumsey-Capay Hills, and described in measured section (Fig. 22-6). Sandstone-rich sequence is approximately 25 to 30 feet (7.6–9.1 m) thick.

Fig. 22-10. Typical facies B beds (classification of Mutti and Ricci Lucchi, 1972) in channel sequence of the middle member of the Forbes Formation in outcrops in the Rumsey-Capay Hills. Staff is approximately 5 feet (1.5 m) long.

subparallel laminations (divisions T_a and T_b of Bouma, 1962) which can be classified as Facies B beds of Mutti and Ricci Lucchi (1972; Fig. 22-10). Measurements of flute casts and sole marks within channel sequences indicate consistent transport directions from north to south, down the axis of the basin. Dish structures and other evidence of post-depositional fluid expulsion are common. Conglomerate beds composed of shale rip-up and ferruginous ironstone-concretion clasts are present locally (Facies A beds of Mutti and Ricci Lucchi, 1972). These laterally discontinuous beds display abrupt lateral thickness changes and erosional bases.

Sandstone-to-shale ratios in channel sequences generally range from 5:1 to 10:1 and are highest in the axes of channels where sandstone beds are commonly amalgamated. The average sandstone-bed thickness in the channel sequence at Salt Creek (Fig. 22-1) is approximately 1 foot (0.3 m; Fig. 22-6). In channel-margin areas, where sandstone beds are separated by shale intervals, sandstone-to-shale ratios are lower, generally ranging from 1:1 to 1:2. The average thickness of channel-margin sandstone beds at Salt Creek is 3 to 6 inches (7.6–15.2 cm; Fig. 22-6). Thin-bedded Facies E beds are typically associated with shale-rich intervals within channel-margin deposits. Log signatures in channel sequences are thus "blocky" or "bell-shaped"; channel-margin areas typically have "serate" log signatures (Fig. 22-6). Shale rip-up clast conglomerate at the base of some channel sequences alters log signatures such that they may appear to coarsen upward.

Interchannel deposits consist of very thinly interbedded sandstone and shale. Sandstone-to-shale ratios generally range from 1:5 to 1:10, and sandstone beds have an average thickness of approximately 1 inch (2.5 cm). Sandstone-to-shale ratios are highest in interchannel areas closest to channels and may decrease away from channels to near zero. Interchannel deposits typically consist of interbedded Facies D sandstone and mudstone. Bioturbation is abundant, probably as a result of low sediment accumulation rate. On well logs, thin-bedded interchannel deposits generally appear as sequences composed entirely of shale because sandstone beds are commonly below tool resolution, even adjacent to channels where interchannel deposits are relatively sandstone-rich (Fig. 22-6).

Petrography and Diagenesis

The following petrographic descriptions of sandstones from the Forbes Formation are based on unpublished work by Karl A. Mertz for Applied Earth Technology, Inc. and RPI, Inc. and two published abstracts (Mertz, 1987, 1988). Forbes Formation reservoir sandstones are best classified as compositionally and texturally submature. They have undergone significant compaction and burial diagenesis. Grains are generally subangular to subrounded and moderately to poorly sorted; average grain size ranges from fine to medium sand (0.12–0.50 mm).

Reservoir sandstones are composed primarily of quartzofeldspathic and volcanic-lithic detritus, reflecting their derivation from a volcanic-plutonic magmatic-arc system to the north. Composition closely parallels that of the Upper Cretaceous "Rumsey" petrofacies of Dickinson and Rich (1972) and Ingersoll (1983), and sandstones are "lithic feldsarenites" according to the classification of Folk and others (1970). They are dominated by abundant microcrystalline quartz, feldspar, and mica (mainly biotite); plagioclase and K-feldspar are generally preserved in roughly equal proportions. Lithic fragments (predominantly volcanic lithics) constitute approximately 10 to 20% and chert averages 3 to 5% of total framework grains. Polycrystalline quartz and metamorphic lithics are present in small amounts.

Diagenetic studies indicate that reservoir porosity is chiefly secondary in origin. Although log-derived porosity values generally range from 20 to 25%,

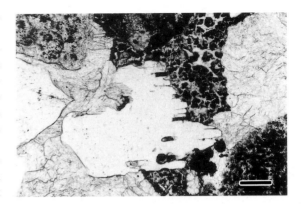

Fig. 22-11. Photomicrograph of sandstone from the middle member of the Forbes Formation. Sample is from a core in the central Sacramento basin (T.12N.,R.1W.). Note the highly etched plagioclase-feldspar grain engulfed in the second generation of carbonate cements: ferroan calcite (dark cement at right) and a low magnesium calcite occluding most porosity (clear cement). Scale bar is 1 mm.

tightly cemented "bone beds" with porosity less than 5% are present locally and represent sandstones lacking secondary porosity. The diagenetic assemblage and preservation of most framework constituents suggest that Forbes Formation strata have not been exposed to temperatures higher than those typical of intermediate burial diagenesis, approximately 230 to 284°F (110–140°C).

Early compaction, dissolution, replacement, and cementation resulted in significant reduction of primary porosity. Extensive mechanical compaction features are evident, including grain breakage, fractures, and deformation. Grain-to-grain contacts are abundant. Dissolution of unstable framework grains is apparent, especially plagioclase, micas, and volcanic and metamorphic lithic fragments. Reduction of primary porosity results largely from continued compaction and early cementation by Fe-calcite, K-feldspar and quartz overgrowths, and authigenic illite/smectite (Figs. 22-11 to 22-13). Dissolution and development of secondary porosity in reservoir sandstones probably resulted from extensive mesodiagenetic leaching associated with early thermal maturation of organic matter and generation of organic acids, a process described by Surdam and others (1989). Secondary porosity is partially occluded by authigenic kaolinite, siderite, chlorite, zeolites, and continued generation of illite/smectite.

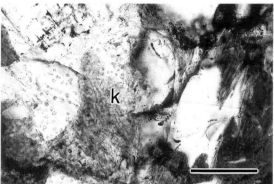

Fig. 22-12. Photomicrograph of sandstone from the middle member of the Forbes Formation. Sample is from a core in the central Sacramento basin (T.10N.,R.1E.) and shows partially dissolved plagioclase-feldspar grain (center) with intragranular secondary porosity partially occluded by diagenetic illite/smectite (darkish clay material along twin planes). Intergranular secondary porosity is visible at upper left of plagioclase grain. Scale bar is 0.1 mm.

Fig. 22-13. Photomicrograph of sandstone from outcrop of the middle member of the Forbes Formation on the western margin of the central Sacramento basin, showing typical dissolution and porosity textures (dark areas). Note the highly etched plagioclase grain at right, the partially etched potassium-feldspar grains at left and upper left, and diagenetic kaolinite (k) partially occluding secondary porosity. Most secondary porosity in Forbes sandstones results from dissolution of unstable aluminosilicate grains and early carbonate cements. Scale bar is 0.1 mm.

Genetic Sequences and Net-Sandstone Geometry

Subdivision of the productive zone in the Arbuckle Gas Field into genetic sequences provides an understanding of the relationships between sandstone-body geometries, facies distribution, and gas production. Gas is produced from many separate, discontinuous reservoir sandstone bodies which were deposited by a series of channel-levee complexes. Net-sandstone maps of relatively thin, genetic intervals, deposited by individual channel-levee complexes, reveal the geometries of reservoir sandstones which cannot be interpreted from gross-interval net-sandstone maps. Net-sandstone values were determined by summing thicknesses of sandstone with SP-responses greater than 10 millivolts from the shale baseline.

Seven shale-resistivity markers have been correlated to subdivide the productive interval of the middle member of the Forbes Formation into six intervals, each generally representing the sequence deposited by a single channel-levee complex (Fig. 22-14). Markers generally bound individual "symmetrical" channel-levee complex cycles. Interval thickness variations or "pinch-outs," facies changes, and erosion by subsequent channels make correlations of markers extremely difficult and laborious. The markers correlated in this study approximate

time lines and may locally represent "condensed" sections of pelagic or hemipelagic shales deposited between turbidite flows or, in some cases, bentonitic layers. There are no indications of significant unconformities within the Forbes Formation in the vicinity of the Arbuckle Gas Field.

Correlation of well logs with seismic-reflection data reveals that the more prominent wireline-log markers correspond with seismic reflectors. Downlapping reflectors are evident primarily on east-west (stratigraphic strike) seismic lines. These likely represent lateral stratigraphic "pinch-out" of channel-levee sequences and help to explain the lateral discontinuity of the shale-resistivity markers. Downlapping reflectors are especially well developed where deep-sea-fan strata of the middle member of the Forbes Formation onlap the eastern slope of the Sacramento basin.

Six genetically related channel-levee complexes comprise the productive interval of the Forbes Formation (Fig. 22-14). Channel-levee complex intervals range in thickness from less than 200 feet (61 m) to more than 500 feet (152 m). The lowermost interval is extremely sandstone-rich and probably represents deposition by more than one channel system. Significant variations in thickness are evident within some of the intervals bounded by shale-resistivity markers. For example, the interval directly

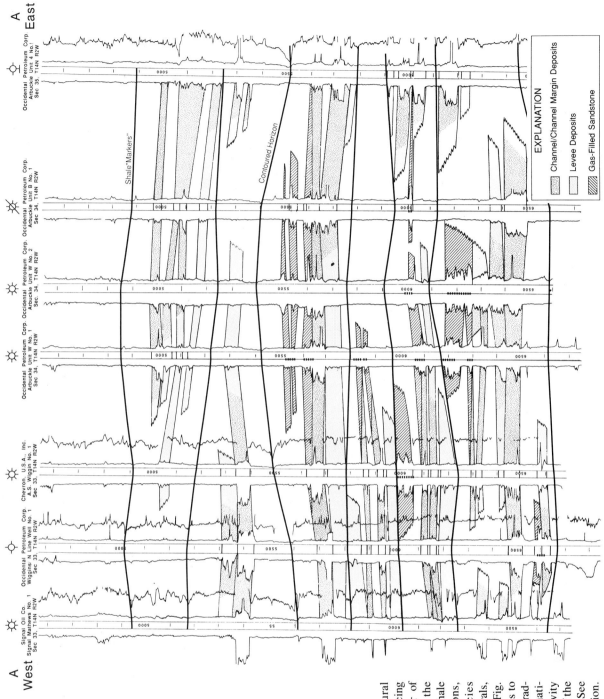

Fig. 22-14. East-west structural cross section of the main producing interval in the middle member of the Forbes Formation across the Arbuckle anticline showing shale resistivity marker correlations, sandstone correlations, facies interpretations, producing intervals, and the contoured horizon (see Fig. 22-4). The transition from levees to interchannel areas is generally gradual, although it is shown schematically as abrupt. SP and resistivity traces are on the left and right of the depth column, respectively. See Figure 22-15 for location of section.

above the contoured horizon in Figure 22-14, which averages about 300 feet (91 m), thins eastward by more than 240 feet (73 m) in less than 2 miles (3.2 km). Such thickness variations likely result from differential compaction of sandstones and shales, because interval thickness commonly corresponds to net-sandstone thickness, although differences in sediment-accumulation rate (depending on proximity to channels) may influence interval thickness as well. Levee deposits are not evident at the base of all sequences because they may consist of sandstone beds too thin to be resolved on wireline logs.

Net-sandstone distribution in the interval directly below the contoured horizon (see Figs. 22-3, 22-4, 22-14) represents deposition by a single channel-levee complex (Fig. 22-15). The interval ranges in thickness from 200 to 300 feet (61–91 m) and is overlain by a relatively thick interchannel sequence. Basal levee deposits are generally well developed in this interval and have a total thickness of about 50 feet (15 m). A westerly shift in the location of the channel axis is evident from the presence of two "stacked" channel sequences separated by a shale interval in the easternmost part of the cross section A-A' (Fig. 22-14). The main channel sequence is as thick as 130 feet (40 m), averaging approximately 50 feet (15 m). Channel-margin and levee deposits cap the sequence in the vicinity of the axis of the Arbuckle anticline (Fig. 22-14).

The distribution of sandstone in the interval directly below the contoured horizon has a generally north-south orientation (Fig. 22-15). It is relatively thick (100–130 ft; 30–40 m) and narrow (about 1 mi (1.6 km) wide) in the northern part of the field and progressively widens to approximately 4 miles (6.4 km) and thins to 50 to 100 feet (15–30 m) in the south. The more sandstone-rich nature of the interval in the eastern part of the field can be attributed to "stacking" of channel sequences, resulting from the westerly shift in location of the channel axis (Figs. 22-14, 22-15).

Comparison of net-sandstone maps of several channel-levee complexes within the middle member of the Forbes Formation indicates that areas of thickest sandstone distribution are commonly laterally offset from from one another, probably a result of updip channel avulsion. Turbidite flows are focused into the topographically low interchannel areas. Differential compaction of sandstone and shale may accentuate the differences in relative topography between the channel and the interchannel areas.

Although it is difficult to demonstrate conclusively the interconnectedness of all sandstones within individual channel-levee complexes, it is likely that sandstones within individual migrating channel-levee complexes are generally interconnected and separated from sandstones of other channel-levee sequences by impermeable shale-rich interchannel sequences. Up-channel avulsion may result in abrupt shifts in channel locations. In such instances, channel sandstone bodies may be separated by impermeable shale intervals. An understanding of trapping mechanisms and production characteristics helps to demonstrate the degree of interconnectedness of sandstone bodies within and between the channel-levee complexes.

Trapping Mechanisms

Gas production from the middle member of the Forbes Formation in the Arbuckle Gas Field is from a sandstone-rich interval that is more than 1,500 feet (460 m) thick at a depth of approximately 5,000 to 6,500 feet (1,520–1,980 m; Figs. 22-4, 22-14). Although the source of gas in the Forbes Formation is unclear (Jenden and Kaplan, 1989), the timing of gas migration can be constrained by our understanding of the timing of structural deformation, since structural features are important in trapping gas in the Forbes Formation throughout the Sacramento basin. Some structures have been interpreted as forming in the Late Cretaceous, but surface and subsurface mapping indicates that the majority of structures that are important in trapping gas, including the Arbuckle anticline, were formed in the late Tertiary, probably in the late Miocene or early Pliocene.

Although most of the gas production in the Arbuckle Gas Field is trapped within the structural closure of the Arbuckle anticline, stratigraphic traps and combination structural-stratigraphic traps have been recognized as well. Approximately half of the 50 completed gas wells in the field are within the area of structural closure (at the level of the contoured horizon which is approximately the top of the productive interval; Fig. 22-3). Therefore, about half of the gas wells in the Arbuckle Field produce from at least partial stratigraphic traps.

Within the closure of the Arbuckle anticline, it appears that gas accumulates primarily in the uppermost (youngest) channel or channel-margin sandstone of a channel-levee sequence (Fig. 22-14). This is evident in the interval directly below the con-

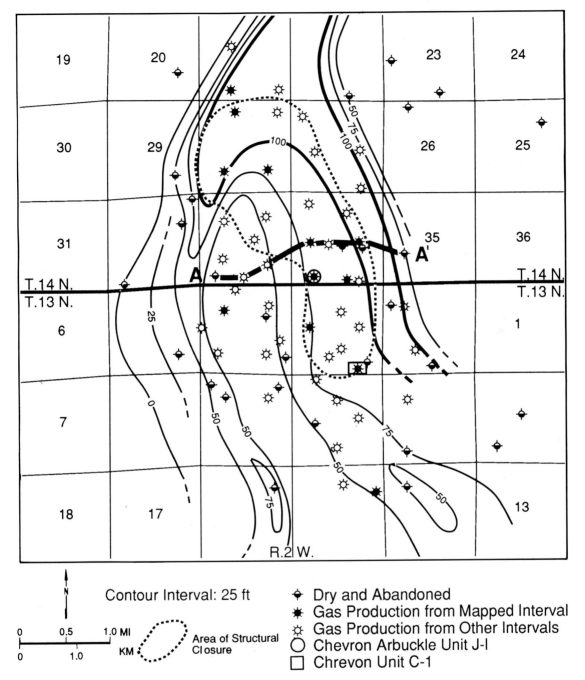

Fig. 22-15. Net-sandstone isopach map of productive interval directly below the çontoured horizon (Fig. 22-10) in the middle member of the Forbes Formation. The interval ranges in thickness from 200 to 300 feet (61–91 m) and generally represents the distribution of sandstone deposited by a single channel-levee complex. Decline curves for the two indicated wells are shown in Figure 22-16. Cross section is shown in Figure 22-14.

toured horizon (Figs. 22-14, 22-15). Since this sandstone is present throughout the area of structural closure, it is probably capable of producing commercial volumes of gas in other wells which have been perforated in other reservoirs. Consistent accumulations of gas in the uppermost sandstone beds of channel-levee sequences within structural closure suggest that sandstones within the genetic sequence are interconnected and are sealed and separated from overlying sequences by shale-rich interchannel deposits.

Outside of the structural closure of the Arbuckle anticline, gas traps are provided primarily by updip "pinch-outs" of levee and channel-margin sandstones into interchannel shales. Such traps are formed within structural closure as well but account for less production than pure structural traps. Downdip amalgamated channel sandstones are commonly water saturated and are interconnected updip with gas-saturated channel-margin and levee sandstones. The updip hydrocarbon seal is provided by shale-rich interchannel deposits of thinly interbedded sandstone and shale. Three wells are productive from the mapped interval (Fig. 22-15) in stratigraphically trapped reservoirs.

The transition from distal-levee to interchannel deposits is gradual, and it is commonly difficult to distinguish thinly interbedded sandstone-rich distal-levee sequences from the thinly interbedded shale-rich interchannel sequences using wireline logs. As the distal part of levees may be the most structurally updip part of the reservoir, significant volumes of gas are capable of being produced from distal-levee sequences which cannot be distinguished on wireline logs from interchannel shale deposits.

Production Characteristics

Per-well reserves in the Forbes Formation reservoirs are extremely variable. In the Arbuckle Gas Field, commercial reserves range from less than 1 BCFG (2.8×10^7 m³) to several BCFG. Initial production rates are similarly very erratic, ranging from several hundred MCFGPD (2.8×10^3 m³) to more than 20 MMCFGPD (5.6×10^5 m³/D). The decision to complete or to abandon wells with substantial gas shows is commonly difficult due to the largely inconclusive nature of drill-stem tests. Reserves are also difficult to evaluate or assign.

Although cumulative gas-production records for individual wells are not readily available, we have compiled these data as production-decline curves for selected wells (Fig. 22-16). The Chevron-Unit J No. 1 (Sec. 34, T.14N.,R.2W.) is a dual-zone completion and has produced more than 1 BCF (2.8×10^7 m³) since it was completed in 1958. The field discovery well, the Chevron-Unit C No. 1 (Sec. 3, T.13N.,R.2W.), has produced nearly 5.5 BCFG (1.6×10^8 m³) since it was completed in 1957. Both wells are productive from structurally trapped reservoirs and generally represent the range of cumulative gas production from wells within structural closure. Well-production histories from wells throughout the Sacramento basin indicate that middle Forbes Formation reservoirs are purely pressure-driven.

The average annual per-well production for the field to date, calculated from annual field-production statistics (California Division of Oil and Gas, 1988), is approximately 80 MMCFG (2.3×10^6 m³). After initially high rates in the late 1950s and early 1960s, the average annual per-well production declined steadily until the early 1980s, when increased drilling activity increased production (Fig. 22-16). Producing wells in the Arbuckle Gas Field have an average time-on-production of approximately 20 years; excluding the nine wells completed in the past five years, the average is nearly 28 years.

Sixteen of the 29 producing wells, but only three of the 21 abandoned gas wells, are within the closure of the Arbuckle anticline, suggesting that wells completed in stratigraphically trapped reservoirs produce for shorter periods of time and may therefore be less productive. However, several wells outside of structural closure have been producing for more than 25 years, indicating that there is significant production from stratigraphically trapped reservoirs. Outside of structural closure, annual well production averages about 20 MMCFG (5.7×10^5 m³), although some of these wells produce as much as 50 MMCFG (1.4×10^6 m³) annually.

Standard drilling and completion techniques are successful in the Arbuckle Gas Field. Most completed wells are perforated within a single channel-levee complex. Wells which are perforated in more than one channel-levee complex generally produce from structurally trapped reservoirs. Bottom-hole formation pressures at depths greater than 7,000 feet (2,130 m) can exceed 6,000 psi (4.1×10^4 kPa); thus, pressure gradients may be as high as 0.85 psi/ft (19.2 kPa/m). Vaughan (1962, 1968) attributes such high geostatic pressure gradients to compaction

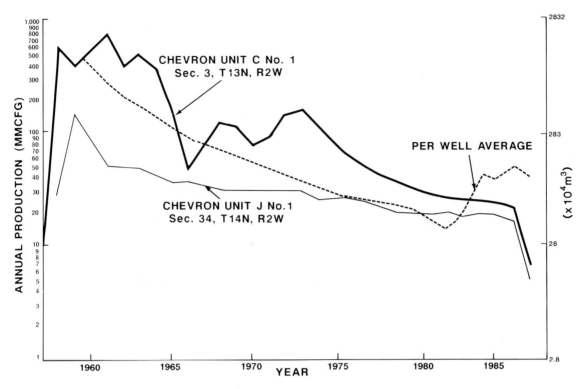

Fig. 22-16. Production-decline curves for selected wells in the Arbuckle Gas Field, and the average annual per-well production in the Arbuckle Gas Field. Locations of selected wells are shown in Figure 22-15.

caused by prior greater burial depths, but abnormally high pressures may also result from tectonic effects.

Tightly cemented layers, or "bone beds," present both in outcrop and in the subsurface (where they appear as high-resistivity "spikes") may locally reduce the porosity and permeability of reservoir sandstones to near zero. In outcrop, such layers have been observed throughout thin and thick sandstone sequences and laterally within highly porous and permeable beds. The drilling rate is significantly decreased when such beds are encountered in the subsurface, and they may significantly affect production. The lack of secondary porosity development in these layers may result from variations in flow or content of pore fluids.

Exploration and Production Strategy

Successful exploration strategies for Forbes Formation targets utilize detailed well-log correlations in conjunction with geophysical techniques. As gas production in the Forbes Formation is primarily from deep-sea-fan deposits of the middle member, determination of the regional distribution of basinal deep-sea-fan and slope facies associations, using seismic-reflection data in conjunction with available well logs, provides a critical stratigraphic framework. The resolution of regional and local sandstone distribution within the fan facies then forms the basis for the development of exploration and production strategies. As channel-levee sandstone bodies are lenticular and extremely discontinuous, they are not necessarily apparent on seismic-reflection data and are best resolved using well-log correlations and projections of regional trends.

Delineation of potential locations for wells requires the subdivision of the mud-rich deep-sea-fan sequence into numerous, relatively thin, genetically related depositional units through the correlation of wireline-log shale "markers" and the construction of net-sandstone maps of these units. These techniques are especially effective in field development, as a high density of wells enables detailed mapping of channel-levee sequences. Markers generally represent time lines, probably formed by condensed sections of hemipelagic or pelagic shale deposited between turbidite events. They are relatively laterally continuous and generally correspond with abrupt acoustic velocity changes between shale-rich intervals. Because the resultant seismic reflections from these interfaces may aid in the critical process of subdividing the deep-sea-fan interval for detailed sandstone mapping, accurate correlation between well and seismic-reflection data is extremely important. Delineation of genetically related intervals is critical in both exploration and development, as it appears that sandstones within individual channel-levee complexes are generally interconnected and are separated from sandstones of other channel-levee sequences by impermeable shale-rich interchannel sequences.

Detailed correlations and construction of several net-sandstone maps within the productive interval in the Arbuckle Field indicate that several wells, primarily those within structural closure, are capable of production from intervals other than those which are perforated. Opportunities thus exist for successful recompletions of producing or abandoned wells within the structural closure of the Arbuckle anticline. In Figure 22-14, two sandstones in the Arbuckle Unit W No. 2 directly below the contoured horizon are probably capable of producing commercial volumes of gas although they have not been perforated.

Amplitude-anomaly (or "bright spot") and amplitude-with-offset seismic techniques are extremely effective tools in prospect generation in the Forbes Formation, especially when used in conjunction with detailed well-log correlations. Amplitude-with-offset analyses are necessary to corroborate amplitude anomalies as caused by gas-saturated sandstones, as anomalies may result from organic-rich layers, bentonitic layers, or "bone beds" (Weagant, 1986; Weagant and Sterling, 1988). Seismic-amplitude techniques are somewhat less useful in the Arbuckle area than in other parts of the Sacramento basin, however, primarily because reservoirs are deeper and do not necessarily yield strong amplitude anomalies. Additionally, as in other parts of the basin, zones which yield good anomalies may prove to be uneconomic because of low gas saturations and high water saturations. Nevertheless, the combination of detailed geological and geophysical analyses is the most effective means of reducing drilling risks.

Although the majority of favorable structural features in the Sacramento basin has already been recognized, tremendous possibilities exist for discovery of untested stratigraphically trapped reservoirs

throughout the basin. Documentation of stratigraphic traps within the Arbuckle Gas Field implies great potential for extension of the field. Future drilling in the largely untested area south of the field may substantially increase reserves.

Conclusions

Deep-sea-fan deposits of the middle member of the Forbes Formation consist of sequences of genetically related channel-levee-complexes. The lower part of the sequence consists of about 50 feet (15 m) of vertically "stacked" levee deposits intercalated with interchannel deposits. This relatively shale-rich interval is overlain by sandstone-rich channel-margin and channel facies associations, which range in thickness from 25 to 100 feet (8–30 m) and may form "stacked" vertical sequences as thick as 300 to 400 feet (91–122 m). The sequence deposited by a migrating channel-levee complex thus ranges in thickness from approximately 100 to 500 feet (30–152 m) and generally consists of a shale-rich interval which gradually coarsens then fines upward, forming a generally "symmetrical" cycle (Fig. 22-6).

Subdivision of deep-sea-fan deposits into individual, genetically related channel-levee complexes by correlating shale resistivity markers is critical to understanding the distribution of reservoir sandstone bodies. As reservoirs are consistently formed by the most updip sandstone body in the genetic sequence, it is likely that sandstones within individual channel-levee complexes are generally interconnected but are separated from sandstones of other channel-levee sequences by impermeable shale-rich interchannel sequences. Within structural closure, reservoirs are primarily formed by the channel and channel-margin deposits which cap the channel-levee-complex sequence. Outside of structural closure, traps are formed by the updip stratigraphic "pinch-outs" of levee and channel-margin deposits into interchannel shales.

Successful strategies for exploration and development in the Forbes Formation integrate geophysical with well-log correlation techniques. Detailed sandstone mapping used in conjunction with amplitude-anomaly ("bright spot") and amplitude-with-offset techniques greatly reduces drilling risks. In the Arbuckle Gas Field, our preliminary analyses indicate that within the area of structural closure, significant potential exists for recompletions or infill drilling to develop previously unrecognized reserves. The recognition of significant gas accumulations in reservoir-quality sandstones outside of structural closure implies great potential for additional, stratigraphically trapped reserves in the extension of the field to the north and south. A thorough understanding of the geometries, orientations, trapping mechanisms, and production histories of known reservoir sandstones can thus be used to predict reservoir potential in more poorly understood areas.

Acknowledgments. The authors thank John Barwis, Fred Keller, and Charles Winker for their helpful comments on the manuscript. Karl A. Mertz, Jr. and Vincent R. Ramirez provided valuable information regarding petrographic and structural aspects of the study, respectively. Donald W. Moore helped with the block diagram reconstruction, and Mary Imperato drafted some of the figures. We thank RPI International, Inc. for permission to publish this chapter and various companies for supplying samples.

Reservoir Summary

Field: Arbuckle Gas Field
Location: Colusa County, California
Operators: Western Gulf, Occidental, Chevron, and others
Discovery: 1957
Basin: Sacramento basin
Tectonic/Regional Paleosetting: Forearc basin
Geologic Structure: Anticline on regional dip
Trap Types: Structural closure, stratigraphic pinch- outs
Reservoir Drive Mechanism: Depletion
 • Original Reservoir Pressure: 2,200 to 4,800 psi (1.5–3.3×10^4 kPa) at approximately 6,000 feet (1,830 m) subsea
Reservoir Rocks
 • **Age:** Late Cretaceous (Campanian)
 • **Stratigraphic Unit:** Forbes Formation
 • **Lithology:** Fine- to coarse-grained lithic feldsarenite
 • **Depositional Environment:** Mud-rich deep-sea fan
 • **Productive Facies:** Channel-levee complexes
 • **Petrophysics**
  • **Porosity Type:** Secondary
  • ϕ: Average 23%, range 20 to 25% (logs)
  • **k:** NA
  • S_w: Average 55% (logs)
Reservoir Geometry
 • **Depth:** 5,000 to 6,500 feet (1,525–1,980 m)
 • **Areal Dimensions:** 6 by 3 miles (9.7×4.8 km)
 • **Productive Area:** 3,280 acres (1,330 ha.)
 • **Number of Reservoirs:** 5 to 10
 • **Hydrocarbon Column Height:** NA
 • **Fluid Contacts:** NA
 • **Number of Pay Zones:** Many
 • **Gross Sandstone Thickness:** 10 to 200 feet (3.0–61 m)
 • **Net Sandstone Thickness:** 10 to 150 feet (3.0–46 m)
 • **Net/Gross:** 0.5 to 1.0
Hydrocarbon Source, Migration
 • **Lithology and Stratigraphic Unit:** Basinal shale (?), Dobbins Shale (?)
 • **Time of Gas Maturation:** ?
 • **Time of Trap Formation:** Late Miocene to early Pliocene (?)
 • **Time of Migration:** Late Miocene to early Pliocene (?)
Hydrocarbons
 • **Type:** Gas
Volumetrics
 • **In-Place:** NA

• **Cumulative Production:** 70.9 BCFG (2.0×10^9 m³)
• **Ultimate Recovery:** 75 BCFG (2.1×10^9 m³)
• **Recovery Efficiency:** NA
Wells
• **Spacing:** Initially 160 acres (64.8 ha.); none presently
• **Total:** 78
• **Dry Holes:** 28
Typical Well Production
• **Average Daily:** 75 to 300 MCFG ($2.1–8.5 \times 10^3$ m³)
• **Cumulative:** 1 to 2 BCFG ($2.8–5.7 \times 10^7$ m³)

References

Atwater, T., 1970, Implications of plate tectonics for the Cenozoic tectonic evolution of western North America: Geological Society of America Bulletin, v. 81, p. 3513–3536.

Bailey, E.H., Irwin, W.P., and Jones, D.L., 1964, Franciscan and related rocks, and their significance in the geology of western California: California Division of Mines and Geology Bulletin 183, 177 p.

Bouma, A.H., 1962, Sedimentology of some flysch deposits: Amsterdam, Elsevier, 168 p.

Bouma, A.H., Coleman, J.M., and Meyer, A.W., 1986, Introduction, objectives and principal results of Deep Sea Drilling Project Leg 96, in Bouma, A.H., Coleman, J.M., and others, Initial Reports of the Deep Sea Drilling Project: Washington, D.C., U.S. Government Printing Office, v. 96, p. 15–36.

California Division of Oil and Gas, 1984, 69th Annual Report of the State Oil and Gas Supervisor: California Department of Conservation, Division of Oil and Gas Publication PR06, 147 p.

California Division of Oil and Gas, 1988, 72nd Annual Report of the State Oil and Gas Supervisor: California Department of Conservation, Division of Oil and Gas Publication PR06, 167 p.

Damuth, J.E., Kowsmann, R.O., Flood, R.D., Belderson, R.H., and Gorini, M.A., 1983, Age relationships of distributary channels on Amazon deep-sea fan: Implications for fan growth pattern: Geology, v. 11, p. 470–473.

Damuth, J.E., Flood, R.D., Kowsmann, R.O., Belderson, R.H., and Gorini, M.A., 1988, Anatomy and growth pattern of Amazon deep-sea fan as revealed by long-range side-scan sonar (GLORIA) and high-resolution seismic studies: American Association of Petroleum Geologists Bulletin, v. 72, p. 885–911.

Dickinson, W.R., 1976, Sedimentary basins developed during evolution of Mesozoic-Cenozoic arc-trench system, in Western North America: Canadian Journal of Earth Sciences, v. 13, p. 1268–1370.

Dickinson, W.R., and Rich, E.I., 1972, Petrologic inter-

vals and petrofacies in the Great Valley sequence, Sacramento Valley, California: Geological Society of America Bulletin, v. 83, p. 3004–3024.

Dickinson, W.R., Helmhold, K.P., and Stein, J.A., 1979, Mesozoic lithic sandstones in central Oregon: Journal of Sedimentary Petrology, v. 49, p. 501–516.

Droz, L., and Bellaiche, G., 1985, Rhone deep-sea fan: morphostructure and growth pattern: American Association of Petroleum Geologists Bulletin, v. 69, p. 460–479.

Emmel, F.J., and Curray, J.R., 1985, Bengal fan, Indian Ocean, in Bouma, A.H., Normark, W.R., and Barnes, N.E., eds., Submarine Fans and Related Turbidite Systems: New York, Springer-Verlag, p. 107–112.

Flood, R.D., and Damuth, J.E., 1987, Quantitative characteristics of sinuous distributary channels on the Amazon deep-sea fan: Geological Society of America Bulletin, v. 98, p. 728–738.

Folk, R.L., Andrews, P.B., and Lewis, D.W., 1970, Detrital sedimentary rock classification and nomenclature for use in New Zealand: New Zealand Journal of Geology and Geophysics, v. 13, p. 937–968.

Ingersoll, R.V., 1979, Evolution of the Late Cretaceous forearc basin, northern and central California: Geological Society of America Bulletin, v. 90, p. 813–826.

Ingersoll, R.V., 1983, Petrofacies and provenance of late Mesozoic forearc basin, northern and central California: American Association of Petroleum Geologists Bulletin, v. 67, p. 1125–1142.

Jenden, P.D., and Kaplan, I.R., 1989, Origin of natural gas in the Sacramento basin, California: American Association of Petroleum Geologists Bulletin, v. 73, p. 431–453.

Kastens, K.A., and Shor, A.N., 1985, Depositional processes of a meandering channel on Mississippi fan: American Association of Petroleum Geologists Bulletin, v. 27, p. 190–202.

Kirby, J.M., 1943, Upper Cretaceous stratigraphy of west side of Sacramento Valley south of Willows, Glenn County, California: American Association of Petroleum Geologists Bulletin, v. 27, p. 279–305.

Manley, P.L., and Flood, R.D., 1988, Cyclic sediment deposition within Amazon deep-sea fan: American

Association of Petroleum Geologists Bulletin, v. 72, p. 912–925.

McCabe, P.J., 1978, The Kinderscoutian delta (Carboniferous) of northern England: A slope influenced by density currents, *in* Stanley, D.J. and Kelling, G., eds., Sedimentation in Submarine Canyons, Fans, and Trenches: Stroudsbourg, Pennsylvania, Dowden, Hutchinson and Ross, p. 116–126.

Melvin, J., 1986, Upper Carboniferous fine-grained turbiditic sandstones from southwest England: A model for growth in an ancient, delta-fed subsea fan: Journal of Sedimentary Petrology, v. 56, p. 19–34.

Mertz, K.A., Jr., 1987, Provenance and diagenetic trends in feldspar and lithic-rich sandstones, Forbes Formation (Cretaceous), Sacramento basin, California: Society of Economic Paleontologists and Mineralogists Midyear Meeting Abstracts, v. 4, p. 56–57.

Mertz, K.A., Jr., 1988, Diagenetic events and development and occlusion of secondary porosity in Upper Cretaceous feldspathic sandstones: northern Sacramento basin, California: Society of Economic Paleontologists and Mineralogists Midyear Meeting Abstracts, v. 5, p. 35–36.

Mutti, E., and Ricci Lucchi, F., 1972, Le torbiditi dell' Appennine settentrionale: Introduzione all'analisi di facies: Memorie Societa Geologica Italiana, v. 11, p. 161–199. (English translation by Nilsen, T.H., 1978, International Geology Review, v. 20, p. 125–166.)

Nelson, C.H., and Nilsen, T.H., 1984, Modern and ancient deep-sea fan deposits: Society of Economic Paleontologists and Mineralogists Short Course No. 14, 404 p.

Nilsen, T.H., 1986, Cretaceous paleogeography of western North America, *in* Abbott, P.L., ed., Cretaceous Stratigraphy, Western North America: Pacific

Section, Society of Economic Paleontologists and Mineralogists, p. 1–40.

Normark, W.R., 1970, Growth patterns of deep-sea fans: American Association of Petroleum Geologists Bulletin, v. 54, p. 2170–2195.

Stelting, C.E. and DSDP Shipboard Scientists, 1985, Migratory characteristics of a mid-fan meander belt, Mississippi fan, *in* Bouma, A.H., Normark, W.R., and Barnes, N.E., eds., Submarine Fans and Related Turbidite Systems: New York, Springer-Verlag, p. 283–290.

Surdam, R.C., Crossey, L.J., Hagen, E.S., and Heasler, H.P., 1989, Organic-inorganic interactions and sandstone diagenesis: American Association of Petroleum Geologists Bulletin, v. 73, p. 1–23.

Vaughan, R.H., 1962, The Arbuckle gas field, California, *in* Bowen, O.E. Jr., ed., Geologic Guide to the Gas and Oil Fields of Northern California: California Division of Mines and Geology Bulletin, v. 181, p. 112–119.

Vaughan, R.H., 1968, Arbuckle gas field, Colusa County, California, *in* Natural Gases of North America—Part 2, Natural Gases in Rocks of Mesozoic Age: American Association of Petroleum Geologists Memoir 9, p. 646–652.

Walker, R.G., 1978, Deep-water sandstone facies and ancient submarine fans: Models for exploration for stratigraphic traps: American Association of Petroleum Geologists Bulletin, v. 62, p. 932–966.

Weagant, F.E., 1986, Exploring for Forbes Sands not easy, but profitable: World Oil, December, p. 53–58.

Weagant, F.E., and Sterling, R.H., 1988, South Tisdale Gas Field, Sacramento basin, California: Discovery by use of seismic amplitude variation with offset analysis [abst.]: American Association of Petroleum Geologists Bulletin, v. 72, p. 397.

Key Words

Arbuckle Gas Field, California, Sacramento basin, Forbes Formation, Late Cretaceous, turbidite, mud-rich deep-sea fan deposits, channel-levee complexes, differential compaction, secondary porosity.

Appendix

GEOLOGIC TIME SCALE

PRECAMBRIAN

PALEOZOIC

MESOZOIC

CENOZOIC

Grain Size Scales

The grade scale most commonly used for sediments is the Wentworth scale, which is a logarithmic scale in that each grade limit is twice as large as the next smaller grade limit. For more detailed work, sieves have been constructed at intervals $\sqrt[2]{2}$ and $\sqrt[4]{2}$.

The ϕ (phi) scale, devised by Krumbein, is a more convenient way of presenting data than if the values are expressed in millimeters, and is used almost entirely in recent work. $\phi = -\log_2 d$, where $d =$ diameter (mm).

U.S. Standard Sieve Mesh #		Millimeters (mm)	Microns (μm)	Phi (ϕ)	Wentworth size class
		4096		-12	
		1024		-10	Boulder (-8 to -12ϕ)
Use		256		-8	
wire					Cobble (-6 to -8ϕ)
		64		-6	
squares		16		-4	Pebble (-2 to -6ϕ)
5		4		-2	
6		3.36		-1.75	
7		2.83		-1.5	Granule
8		2.38		-1.25	
10		2.00		-1.0	
12		1.68		-0.75	
14		1.41		-0.5	Very coarse sand
16		1.19		-0.25	
18		1.00		0.0	
20		0.84		0.25	
25		0.71		0.5	Coarse sand
30		0.59		0.75	
35	1/2	0.50	500	1.0	
40		0.42	420	1.25	
45		0.35	350	1.5	Medium sand
50		0.30	300	1.75	
60	1/4	0.25	250	2.0	
70		0.210	210	2.25	
80		0.177	177	2.5	Fine sand
100		0.149	149	2.75	
120	1/8	0.125	125	3.0	
140		0.105	105	3.25	
170		0.088	88	3.5	Very fine sand
200		0.074	74	3.75	
230	1/16	0.0625	62.5	4.0	
270		0.053	53	4.25	
325		0.044	44	4.5	Coarse silt
		0.037	37	4.75	
	1/32	0.031	31	5.0	
Analyzed	1/64	0.0156	15.6	6.0	Medium silt
by	1/128	0.0078	7.8	7.0	Fine silt
	1/256	0.0039	3.9	8.0	Very fine silt
Pipette		0.0020	2.0	9.0	
or		0.00098	0.98	10.0	Clay
		0.00049	0.49	11.0	
Hydrometer		0.00024	0.24	12.0	
		0.00012	0.12	13.0	
		0.00006	0.06	14.0	

(Published with permission from Folk, R.L., 1980, Petrology of Sedimentary Rocks: Hemphill Publishing, Austin, TX, 182 p.)

(A)

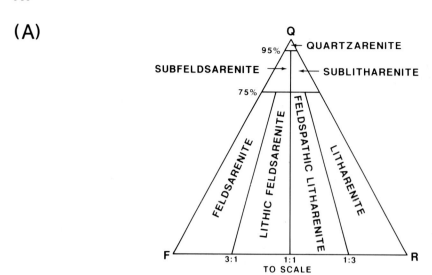

Q = monocrystalline and polycrystalline quartz (excluding chert)
F = monocrystalline feldspar
R = rock fragments (igneous, metamorphic, and sedimentary,
      including chert)

(B)

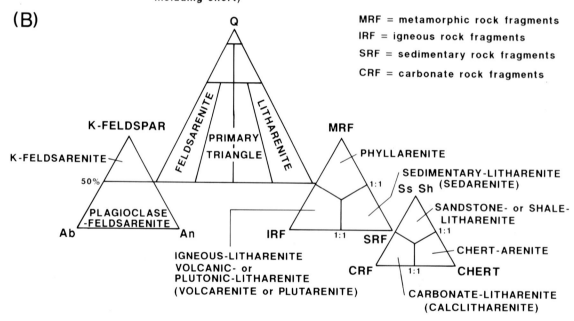

The sandstone compositional classification of Folk and others (1970) which is used throughout this book. (A) The primary arenite triangle. (B) The second- and third-order arenite triangles that may be devised to refine the nomenclature given in the primary triangle.

Reservoir Summary Terminology (Examples Shown in Parentheses)

Reservoir Drive Mechanism – Dominant pressure (i.e., recovery) mechanism for the subject reservoir (water, depletion, partial water, etc.).
  • **Original Reservoir Pressure** – Preproduction pressure at a specified depth for the subject reservoir, in psi and feet and in Pascals and meters (4,500 psi at 9,050 ft subsea).

Petrophysics
  • **Porosity Type(s)** – Mode(s) of void space, with percentages (ϕ_{total} = 10%, with primary = 2% and secondary intragranular = 8%).
  • **ϕ** – Porosity, expressed as a percentage of bulk volume (28%) from conventional core, sidewall sample, or wireline-log data. Typically expressed as average, range, and cutoff of the reservoir rocks. Average is the arithmetic mean, range is the lowest to the highest, and cutoff is the economic minimum for productivity or pay.
  • **k** – Permeability, in millidarcies (46 md), is normally unstressed air permeability from conventional core plugs and is the property of a rock to transmit fluids. Typically expressed as average, range, and cutoff of the reservoir rocks. Average is the arithmetic mean, range is the lowest to the highest, cutoff is the economic minimum for productivity or pay.
  • **S_w** – The rock's water saturation, normally from wireline data, as a volume percentage (42%) of total fluids. Typically expressed as average, range, and cutoff of the reservoir rocks. Average is the arithmetic mean, range is the lowest to the highest, cutoff is the economic minimum for productivity or pay.
  • **S_{or}** – The residual oil saturation, as a volume percentage (21%) of total fluids; the oil remaining after production. Usually greater than the irreducible oil saturation.

Reservoir Geometry
  • **Areal Dimensions** – Measured horizontal distances, in miles and kilometers.
  • **Number of Reservoirs** – Number of isolated reservoirs for the sandstone or sand of interest.
  • **Hydrocarbon Column Height** – The vertical distance from the base of the top seal to the observed or calculated water contact.
  • **Number of Pay Zones** – The number of economically producible layers within the reservoir of interest.
  • **Gross Sandstone or Sand Thickness** – The measured thickness from the top of an individual sandstone or sand to the base of that sandstone or sand, including intervening nonreservoir lithologies, in feet and meters.
  • **Net Sandstone or Sand Thickness** – The measured thickness of sandstone or sand only (gross thickness minus thickness of shale and other intervening lithologies), in feet and meters. May also indicate total thickness of porous sandstone or sand only, in which case a cutoff porosity must be given.
  • **Net/Gross** – Net sandstone or sand thickness divided by the gross sandstone or sand thickness, expressed as a decimal.

Hydrocarbon(s):
  • **Type** – Gas, oil or condensate.
  • **GOR** – The gas/oil ratio; standard cubic feet of gas per barrel of oil (1,600:1) or cubic meters of gas per cubic meter of oil.
  • **API Gravity** – A standard measure of oil density (40°), compared to that of water; see selected conversion factors.
  • **FVF (Formation-Volume Factor)** – The factor applied to convert a barrel of gas-free oil in a stock tank at the surface into an equivalent amount of oil in the reservoir. Abbreviated as B_o or called the "shrinkage factor" (1.25). Usually ranges between 1.14 and 1.60. The gas FVF (B_g) is reservoir cubic feet divided by standard cubic feet or reservoir cubic meters divided by standard cubic meters.
  • **Viscosity** – A numerical index of the internal resistance of the fluid to flow, in centipoise or Pascal·seconds at a specified temperature (28 cP at 90°F).

Volumetrics (field and/or reservoir):
  • **In-Place** – Total "tank" volume (900 MMBO).
  • **Cumulative Production** – Total production to date (300 MMBO).
  • **Ultimate Recovery** – Expected total production when the field/reservoir is abandoned (500 MMBO). May include oil from supplemental recovery methods (secondary and tertiary production).
  • **Recovery Efficiency** – The Ultimate Recovery divided by the In-Place volume, as a percentage (56%).

Wells: Excludes those wells drilled as injectors, sidetracks, or replacements, unless specified.
- **Spacing** – The nominal distance between wells in feet and meters and either the regulatory unit or approximate drainage area per well, in acres and hectares.
- **Total** – The total number of wells drilled into the reservoir.
- **Dry Holes** – The number of wells in the reservoir which were expected to produce but did not, i.e., Total minus the successful wells.

Selected Conversion Factors*

To convert:	To:	Multiply by:
Linear Units		
inches (in.)	centimeters (cm)	2.54
feet (ft)	meters (m)	0.3048
miles (mi)	kilometers (km)	1.609
Area Units		
square miles (mi²)	square kilometers (km²)	2.590
acres (ac)	hectares (ha.)	0.405
Volume Units		
cubic feet (ft³)	cubic meters (m³)	0.028
barrels (bbl)	cubic meters (m³)	0.159
thousand cubic feet (MCF)	cubic meters (m³)	28.317
million cubic feet (MMCF)	thousand cubic meters ($\times 10^3$ m³)	28.317
billion cubic feet (BCF)	million cubic meters ($\times 10^6$ m³)	28.317
trillion cubic feet (TCF)	billion cubic meters ($\times 10^9$ m³)	28.317
gallons (gal)	liters (L)	3.785
Mass Units		
pounds (lb)	kilograms (kg)	0.454
Viscosity Units		
centipoise (cP)	Pascal-seconds (Pa•s)	1000.0
Pressure Units		
pound-force per square inch (psi)	kilopascals (kPa)	6.895
pound-force per square inch (psi)	bars (bar)	0.069
Gradient Units		
pound-force per square inch per foot (psi/ft)	kilopascals per meter (kPa/m)	22.62
Richness Units		
barrels of oil per acre (BO/ac)	cubic meters per hectare (m³/ha.)	0.393
Formation–Volume Factor Units		
standard cubic feet of gas per barrel of oil (SCF/bbl)	standard cubic meters of gas per cubic meter of oil (m³/m³)	0.176
Temperature		
degrees Fahrenheit (°F)	degrees Celsius (°C)	(°F − 32) × 5/9

Miscellaneous

1°F/100 ft = 1.8°C/100 m
2.9°F/mi = 1.0°C/km
API Gravity = (141.5/specific gravity) − 131.5

(Modified from and reprinted by permission of the American Association of Petroleum Geologists, 1989)

*A useful discussion of units and conversion is *The SI Metric System of Units and SPE Metric Standard* (1982), published by the Society of Petroleum Engineers, 39 p.

Abbreviations

B	billion	m.y.	million-year interval
bbl	barrels	NA	not available and/or not known
BCPD	barrels condensate per day	NFG	net feet of gas
B_g	formation-volume factor for gas	NFO	net feet of oil
B_o	formation-volume factor for oil	NGL	natural gas liquids
BGC	barrels of gas condensate	NP	not penetrated
BH	bottom hole	OIP	oil in place
BLG	barrels of liquid gas	OOIP	original oil in place
BO	barrels of oil	OWC	oil-water contact
BOPD	barrels of oil per day	ppg	pounds per gallon
BTU	British thermal unit	ppm	parts per million
BW	barrels of water	psi	pounds per square inch
BWPD	barrels of water per day	PVT	pressure-volume-temperature
CFG	cubic feet of gas	RFT	repeat formation tester
CFGPD	cubic feet of gas per day	R_t	formation resistivity
cP	centipoise	R_w	water resistivity
CUM	cumulative volume produced	s	second
D	day	SCFG	standard cubic feet of gas
EOR	enhanced oil recovery	SITP	shut-in tubing pressure
EUR	estimated ultimate recovery	S_g	gas saturation
FVF	formation-volume factor	S_o	oil saturation
FTP	flowing tubing pressure	S_{oi}	initial oil saturation
GOC	gas-oil contact	S_{or}	residual oil saturation
GOR	gas-oil ratio	S_w	water saturation
GR	gamma ray	SP	spontaneous potential
GWC	gas-water contact	ss	sandstone
IP	initial production	STB	stock-tank barrel
Ka	thousand years before present	STB/D	stock-tank barrels per day
M	thousand	TD	total depth
MD	measured depth	TDS	total dissolved solids
MM	million	TOC	total organic carbon
Ma	million years before present	TVD	true vertical depth
md	millidarcies	URE	ultimate recovery efficiency
msec	milliseconds		

Depositional Environment Plate Captions

Fluvial Environments

P. 6 top: Infrared photograph of point bars (looking upstream) on the Brazos River, near Richmond, Texas. These are the point bars that Bernard used for his 1950s classic work on modern meander-belt systems. Accretionary ridges (scroll bars) are delineated by stands of oak trees, and two oxbow lakes flank the main channel. The reservoir-quality sand in this meander belt forms a ribbon about 60 feet (18 m) thick and 1.5 miles (2.5 km) wide. *(Photo by John H. Barwis)*; **bottom:** Stacked meander-belt sequences from the Pennsylvanian Breathitt Formation, Louisa, Kentucky. Point-bar lateral accretion bedding is visible in several sequences. Note the wide range of vertical connectivity. *(Photo by John H. Barwis)*

Desert Environments

P. 132 top: Satellite image of the Namib Sand Sea of southern Africa. Three main dune types are shown, and their distribution is primarily a function of the wind regime. The coastal belt (light colored sand), up to 12 miles (20 km) wide, contains transverse dunes oriented perpendicular to the prevailing S–SW winds. The central Sand Sea contains large, simple and complex linear dunes resulting from a bidirectional wind pattern and oriented in a N–S to NW direction. These linear dunes have heights of up to 500 feet (150 m) and interdune spacings averaging 1 to 2 miles (1–3 km). To the east the linear dunes are low, partly vegetated, and are reddened. In the area of the Tsauchab River incursion into the Sand Sea (right center of photo) are the high star dunes which reach heights of about 1,000 feet (300 m) above the dry lake (playa) floor. Image taken March 4, 1985. *(Reproduced by permission of Earth Observation Satellite Company, Lanham, Maryland, U.S.A.)*; **bottom:** Eolian sandstone interpreted to have been deposited by dune bedform climb. Note the first-order bounding surfaces separating the individual dune sets. Several scales of dune sets are represented, and many are very large (see the person at bottom center for scale). Navajo Sandstone (Jurassic), Checkerboard Mesa, Zion National Park, Utah. *(Photo by John G. McPherson)*

Deltaic Environments

P. 226 top: The modern birdfoot delta of the Mississippi River, showing the main distributary channels and the active crevasse subdeltas or bay-fills. The sediment plumes extend far beyond the river mouth and carry the fine suspended load many miles out into the Gulf of Mexico. This modern lobe of the Mississippi delta was built in the last 600 years and is much more elongate and confined than were the older lobes of the delta. As a result, the lobe is five times as thick (300 ft; 100 m). North is to the upper right, and the image width spans 40 miles (60 km). Image taken March 25, 1984. *(Reproduced by permission of Earth Observation Satellite Company, Lanham, Maryland, U.S.A.)*; **bottom:** A typical upward-coarsening deltaic sequence, from the Ferron Sandstone (Late Cretaceous), near Emery, east-central Utah. The sequence shows prodelta shale and thin-bedded distal delta-front (distal bar) sandstones at the base, overlain by thicker bedded and coarser grained sandstones of the proximal delta front (distributary mouth-bar), which are overlain by thick, in places massive, distributary channel sandstones (light colored) that continue to the top of the outcrop. *(Photo by John G. McPherson)*

Estuarine/Barrier Environments

P. 318 top: The Chandeleur Islands flanking the Mississippi delta in the Gulf of Mexico. These barrier islands represent the reworking of an old lobe of the Mississippi delta, the St. Bernard lobe, which was active 1,800 years ago. Prominent are the washover fans of sand that overtop the organic-rich back barrier bay. The islands are continually moving in a landward direction (to the left), as evidenced by the emergence of former back-barrier deposits on the beach foreshore. *(Photo by John G. McPherson)*; **bottom:** A stacked foreshore and shoreface sequence from the Blackhawk Formation (Cretaceous) of the Book Cliffs, Utah. The lower shoreface is thin-bedded and fine-grained. It passes upward into thick, cross-bedded upper-shoreface sandstones, which are capped by parallel-laminated sandstone (gray) of the foreshore. *(Photo by John G. McPherson)*

Shelf Environments

P. 414 top: A heat-sensitive image, enhanced with color, to show the Gulf Stream current on the Atlantic Shelf in the area of Cape Hatteras, North Carolina. The image is from an Advanced Very High Resolution Radiometer (AVHRR) carried onboard a low orbiting weather satellite. Cool water appears blue and purple, whereas the warm water is yellow and orange. Note the eddies generated at the boundary between the four miles (6 km) per hour north-bound (to the upper right), warm Gulf Stream and the colder coastal waters. The image spans 30 miles (50 km) and was taken at 1:21 PM EST, March 12, 1989. *(Image courtesy of R. L. Borger);* **bottom:** A shelf sequence in the Twowells Tongue of the Dakota Sandstone (Upper Cretaceous), near San Ysidro, northwest New Mexico. The sequence comprises a cross-bedded sand-ridge facies at the top (upper blocky sandstone), overlying a ripple cross-laminated sand-plume facies (middle backweathered unit), which overlies a marine shale with thin beds of storm-generated sandstone (partially scree-covered lower unit). *(Photo by Dag Nummedal)*

Turbidite Environments

P. 466 top: GLORIA side-scan sonar image of meandering distributary channels (red) on the middle Amazon deep-sea fan [water depth is approximately 10,500 ft (3,200 m)]. The image spans approximately 20 miles by 60 miles (30 × 100 km). Downslope is toward the left. The largest channel (red) is about 0.5 miles (1 km) wide and 230 feet (70 m) deep and is perched atop a broad (15 mi; 25 km), natural levee system (purple). *(Sonograph courtesy of John E. Damuth);* **bottom:** Thin-bedded turbidite sandstone, with interbedded hemipelagic mudstone (facies D of Mutti and Ricci Lucchi classification), showing non-cyclic trends and remarkable lateral continuity. This sequence (lower Eocene, with stratigraphic top to the left) is the basin-plain facies of a deep-sea fan complex. Note people for scale at lower right. Zumaya beach, San Sebastian, Spain. *(Photo by G. Shanmugam, and reproduced with permission of the Geological Society of America)*

Author Index

Subject Index

Page numbers in *italics* refer to figures.